T0189247

Lecture Notes in Computer Science 13845

Founding Editors

Gerhard Goos
Juris Hartmanis

Editorial Board Members

The series Lecture Notes in Computer Science (LNCS), including its subseries Lecture Notes in Artificial Intelligence (LNAI) and Lecture Notes in Bioinformatics (LNBI), has established itself as a medium for the publication of new developments in computer science and information technology research, teaching, and education.

LNCS enjoys close cooperation with the computer science R & D community, the series counts many renowned academics among its volume editors and paper authors, and collaborates with prestigious societies. Its mission is to serve this international community by providing an invaluable service, mainly focused on the publication of conference and workshop proceedings and postproceedings. LNCS commenced publication in 1973.

Lei Wang · Juergen Gall · Tat-Jun Chin ·
Imari Sato · Rama Chellappa
Editors

Computer Vision – ACCV 2022

16th Asian Conference on Computer Vision
Macao, China, December 4–8, 2022
Proceedings, Part V

 Springer

Editors
Lei Wang 🆔
University of Wollongong
Wollongong, NSW, Australia

Juergen Gall 🆔
University of Bonn
Bonn, Germany

Tat-Jun Chin 🆔
University of Adelaide
Adelaide, SA, Australia

Imari Sato
National Institute of Informatics
Tokyo, Japan

Rama Chellappa 🆔
Johns Hopkins University
Baltimore, MD, USA

ISSN 0302-9743   ISSN 1611-3349 (electronic)
Lecture Notes in Computer Science
ISBN 978-3-031-26347-7   ISBN 978-3-031-26348-4 (eBook)
https://doi.org/10.1007/978-3-031-26348-4

This Springer imprint is published by the registered company Springer Nature Switzerland AG
The registered company address is: Gewerbestrasse 11, 6330 Cham, Switzerland

Preface

The 16th Asian Conference on Computer Vision (ACCV) 2022 was held in a hybrid mode in Macau SAR, China during December 4–8, 2022. The conference featured novel research contributions from almost all sub-areas of computer vision.

For the main conference, 836 valid submissions entered the review stage after desk rejection. Sixty-three area chairs and 959 reviewers made great efforts to ensure that every submission received thorough and high-quality reviews. As in previous editions of ACCV, this conference adopted a double-blind review process. The identities of authors were not visible to the reviewers or area chairs; nor were the identities of the assigned reviewers and area chairs known to the authors. The program chairs did not submit papers to the conference.

After receiving the reviews, the authors had the option of submitting a rebuttal. Following that, the area chairs led the discussions and final recommendations were then made by the reviewers. Taking conflicts of interest into account, the area chairs formed 21 AC triplets to finalize the paper recommendations. With the confirmation of three area chairs for each paper, 277 papers were accepted. ACCV 2022 also included eight workshops, eight tutorials, and one grand challenge, covering various cutting-edge research topics related to computer vision. The proceedings of ACCV 2022 are open access at the Computer Vision Foundation website, by courtesy of Springer. The quality of the papers presented at ACCV 2022 demonstrates the research excellence of the international computer vision communities.

This conference is fortunate to receive support from many organizations and individuals. We would like to express our gratitude for the continued support of the Asian Federation of Computer Vision and our sponsors, the University of Macau, Springer, the Artificial Intelligence Journal, and OPPO. ACCV 2022 used the Conference Management Toolkit sponsored by Microsoft Research and received much help from its support team.

All the organizers, area chairs, reviewers, and authors made great contributions to ensure a successful ACCV 2022. For this, we owe them deep gratitude. Last but not least, we would like to thank the online and in-person attendees of ACCV 2022. Their presence showed strong commitment and appreciation towards this conference.

December 2022

Lei Wang
Juergen Gall
Tat-Jun Chin
Imari Sato
Rama Chellappa

Organization

General Chairs

Gérard Medioni   University of Southern California, USA
Shiguang Shan   Chinese Academy of Sciences, China
Bohyung Han   Seoul National University, South Korea
Hongdong Li   Australian National University, Australia

Program Chairs

Rama Chellappa   Johns Hopkins University, USA
Juergen Gall   University of Bonn, Germany
Imari Sato   National Institute of Informatics, Japan
Tat-Jun Chin   University of Adelaide, Australia
Lei Wang   University of Wollongong, Australia

Publication Chairs

Wenbin Li   Nanjing University, China
Wanqi Yang   Nanjing Normal University, China

Local Arrangements Chairs

Liming Zhang   University of Macau, China
Jianjia Zhang   Sun Yat-sen University, China

Web Chairs

Zongyuan Ge   Monash University, Australia
Deval Mehta   Monash University, Australia
Zhongyan Zhang   University of Wollongong, Australia

AC Meeting Chair

Chee Seng Chan   University of Malaya, Malaysia

Area Chairs

Aljosa Osep   Technical University of Munich, Germany
Angela Yao   National University of Singapore, Singapore
Anh T. Tran   VinAI Research, Vietnam
Anurag Mittal   Indian Institute of Technology Madras, India
Binh-Son Hua   VinAI Research, Vietnam
C. V. Jawahar   International Institute of Information Technology,
                                          Hyderabad, India
Dan Xu   The Hong Kong University of Science and
                                          Technology, China
Du Tran   Meta AI, USA
Frederic Jurie   University of Caen and Safran, France
Guangcan Liu   Southeast University, China
Guorong Li   University of Chinese Academy of Sciences,
                                          China
Guosheng Lin   Nanyang Technological University, Singapore
Gustavo Carneiro   University of Surrey, UK
Hyun Soo Park   University of Minnesota, USA
Hyunjung Shim   Korea Advanced Institute of Science and
                                          Technology, South Korea
Jiaying Liu   Peking University, China
Jun Zhou   Griffith University, Australia
Junseok Kwon   Chung-Ang University, South Korea
Kota Yamaguchi   CyberAgent, Japan
Li Liu   National University of Defense Technology,
                                          China
Liang Zheng   Australian National University, Australia
Mathieu Aubry   Ecole des Ponts ParisTech, France
Mehrtash Harandi   Monash University, Australia
Miaomiao Liu   Australian National University, Australia
Ming-Hsuan Yang   University of California at Merced, USA
Palaiahnakote Shivakumara   University of Malaya, Malaysia
Pau-Choo Chung   National Cheng Kung University, Taiwan

Yung-Yu Chuang		National Taiwan University, Taiwan
Zhaoxiang Zhang		Chinese Academy of Sciences, China
Ziad Al-Halah		University of Texas at Austin, USA
Zuzana Kukelova		Czech Technical University, Czech Republic

Additional Reviewers

Abanob E. N. Soliman	Atsushi Shimada	Chao Liu
Abdelbadie Belmouhcine	Attila Szabo	Chao Shi
Adrian Barbu	Aurelie Bugeau	Chaowei Tan
Agnibh Dasgupta	Avatharam Ganivada	Chaoyi Li
Akihiro Sugimoto	Ayan Kumar Bhunia	Chaoyu Dong
Akkarit Sangpetch	Azade Farshad	Chaoyu Zhao
Akrem Sellami	B. V. K. Vijaya Kumar	Chen He
Aleksandr Kim	Bach Tran	Chen Liu
Alexander Andreopoulos	Bailin Yang	Chen Yang
Alexander Fix	Baojiang Zhong	Chen Zhang
Alexander Kugele	Baoquan Zhang	Cheng Deng
Alexandre Morgand	Baoyao Yang	Cheng Guo
Alexis Lechervy	Basit O. Alawode	Cheng Yu
Alina E. Marcu	Beibei Lin	Cheng-Kun Yang
Alper Yilmaz	Benoit Guillard	Chenglong Li
Alvaro Parra	Beomgu Kang	Chengmei Yang
Amogh Subbakrishna	Bin He	Chengxin Liu
Adishesha	Bin Li	Chengyao Qian
Andrea Giachetti	Bin Liu	Chen-Kuo Chiang
Andrea Lagorio	Bin Ren	Chenxu Luo
Andreu Girbau Xalabarder	Bin Yang	Che-Rung Lee
Andrey Kuehlkamp	Bin-Cheng Yang	Che-Tsung Lin
Anh Nguyen	BingLiang Jiao	Chi Xu
Anh T. Tran	Bo Liu	Chi Nhan Duong
Ankush Gupta	Bohan Li	Chia-Ching Lin
Anoop Cherian	Boyao Zhou	Chien-Cheng Lee
Anton Mitrokhin	Boyu Wang	Chien-Yi Wang
Antonio Agudo	Caoyun Fan	Chih-Chung Hsu
Antonio Robles-Kelly	Carlo Tomasi	Chih-Wei Lin
Ara Abigail Ambita	Carlos Torres	Ching-Chun Huang
Ardhendu Behera	Carvalho Micael	Chiou-Ting Hsu
Arjan Kuijper	Cees Snoek	Chippy M. Manu
Arren Matthew C.	Chang Kong	Chong Wang
Antioquia	Changick Kim	Chongyang Wang
Arjun Ashok	Changkun Ye	Christian Siagian
Atsushi Hashimoto	Changsheng Lu	Christine Allen-Blanchette

Christoph Schorn
Christos Matsoukas
Chuan Guo
Chuang Yang
Chuanyi Zhang
Chunfeng Song
Chunhui Zhang
Chun-Rong Huang
Ci Lin
Ci-Siang Lin
Cong Fang
Cui Wang
Cui Yuan
Cyrill Stachniss
Dahai Yu
Daiki Ikami
Daisuke Miyazaki
Dandan Zhu
Daniel Barath
Daniel Lichy
Daniel Reich
Danyang Tu
David Picard
Davide Silvestri
Defang Chen
Dehuan Zhang
Deunsol Jung
Difei Gao
Dim P. Papadopoulos
Ding-Jie Chen
Dong Gong
Dong Hao
Dong Wook Shu
Dongdong Chen
Donghun Lee
Donghyeon Kwon
Donghyun Yoo
Dongkeun Kim
Dongliang Luo
Dongseob Kim
Dongsuk Kim
Dongwan Kim
Dongwon Kim
DongWook Yang
Dongze Lian

Dubing Chen
Edoardo Remelli
Emanuele Trucco
Erhan Gundogdu
Erh-Chung Chen
Rickson R. Nascimento
Erkang Chen
Eunbyung Park
Eunpil Park
Eun-Sol Kim
Fabio Cuzzolin
Fan Yang
Fan Zhang
Fangyu Zhou
Fani Deligianni
Fatemeh Karimi Nejadasl
Fei Liu
Feiyue Ni
Feng Su
Feng Xue
Fengchao Xiong
Fengji Ma
Fernando Díaz-del-Rio
Florian Bernard
Florian Kleber
Florin-Alexandru
    Vasluianu
Fok Hing Chi Tivive
Frank Neumann
Fu-En Yang
Fumio Okura
Gang Chen
Gang Liu
Gao Haoyuan
Gaoshuai Wang
Gaoyun An
Gen Li
Georgy Ponimatkin
Gianfranco Doretto
Gil Levi
Guang Yang
Guangfa Wang
Guangfeng Lin
Guillaume Jeanneret
Guisik Kim

Gunhee Kim
Guodong Wang
Ha Young Kim
Hadi Mohaghegh
    Dolatabadi
Haibo Ye
Haili Ye
Haithem Boussaid
Haixia Wang
Han Chen
Han Zou
Hang Cheng
Hang Du
Hang Guo
Hanlin Gu
Hannah H. Kim
Hao He
Hao Huang
Hao Quan
Hao Ren
Hao Tang
Hao Zeng
Hao Zhao
Haoji Hu
Haopeng Li
Haoqing Wang
Haoran Wen
Haoshuo Huang
Haotian Liu
Haozhao Ma
Hari Chandana K.
Haripriya Harikumar
Hehe Fan
Helder Araujo
Henok Ghebrechristos
Heunseung Lim
Hezhi Cao
Hideo Saito
Hieu Le
Hiroaki Santo
Hirokatsu Kataoka
Hiroshi Omori
Hitika Tiwari
Hojung Lee
Hong Cheng

Hong Liu
Hu Zhang
Huadong Tang
Huajie Jiang
Huang Ziqi
Huangying Zhan
Hui Kong
Hui Nie
Huiyu Duan
Huyen Thi Thanh Tran
Hyung-Jeong Yang
Hyunjin Park
Hyunsoo Kim
HyunWook Park
I-Chao Shen
Idil Esen Zulfikar
Ikuhisa Mitsugami
Inseop Chung
Ioannis Pavlidis
Isinsu Katircioglu
Jaeil Kim
Jaeyoon Park
Jae-Young Sim
James Clark
James Elder
James Pritts
Jan Zdenek
Janghoon Choi
Jeany Son
Jenny Seidenschwarz
Jesse Scott
Jia Wan
Jiadai Sun
JiaHuan Ji
Jiajiong Cao
Jian Zhang
Jianbo Jiao
Jianhui Wu
Jianjia Wang
Jianjia Zhang
Jianqiao Wangni
JiaQi Wang
Jiaqin Lin
Jiarui Liu
Jiawei Wang

Jiaxin Gu
Jiaxin Wei
Jiaxin Zhang
Jiaying Zhang
Jiayu Yang
Jidong Tian
Jie Hong
Jie Lin
Jie Liu
Jie Song
Jie Yang
Jiebo Luo
Jiejie Xu
Jin Fang
Jin Gao
Jin Tian
Jinbin Bai
Jing Bai
Jing Huo
Jing Tian
Jing Wu
Jing Zhang
Jingchen Xu
Jingchun Cheng
Jingjing Fu
Jingshuai Liu
JingWei Huang
Jingzhou Chen
JinHan Cui
Jinjie Song
Jinqiao Wang
Jinsun Park
Jinwoo Kim
Jinyu Chen
Jipeng Qiang
Jiri Sedlar
Jiseob Kim
Jiuxiang Gu
Jiwei Xiao
Jiyang Zheng
Jiyoung Lee
John Paisley
Joonki Paik
Joonseok Lee
Julien Mille

Julio C. Zamora
Jun Sato
Jun Tan
Jun Tang
Jun Xiao
Jun Xu
Junbao Zhuo
Jun-Cheng Chen
Junfen Chen
Jungeun Kim
Junhwa Hur
Junli Tao
Junlin Han
Junsik Kim
Junting Dong
Junwei Zhou
Junyu Gao
Kai Han
Kai Huang
Kai Katsumata
Kai Zhao
Kailun Yang
Kai-Po Chang
Kaixiang Wang
Kamal Nasrollahi
Kamil Kowol
Kan Chang
Kang-Jun Liu
Kanchana Vaishnavi
    Gandikota
Kanoksak Wattanachote
Karan Sikka
Kaushik Roy
Ke Xian
Keiji Yanai
Kha Gia Quach
Kibok Lee
Kira Maag
Kirill Gavrilyuk
Kohei Suenaga
Koichi Ito
Komei Sugiura
Kong Dehui
Konstantinos Batsos
Kotaro Kikuchi

Kouzou Ohara
Kuan-Wen Chen
Kun He
Kun Hu
Kun Zhan
Kunhee Kim
Kwan-Yee K. Wong
Kyong Hwan Jin
Kyuhong Shim
Kyung Ho Park
Kyungmin Kim
Kyungsu Lee
Lam Phan
Lanlan Liu
Le Hui
Lei Ke
Lei Qi
Lei Yang
Lei Yu
Lei Zhu
Leila Mahmoodi
Li Jiao
Li Su
Lianyu Hu
Licheng Jiao
Lichi Zhang
Lihong Zheng
Lijun Zhao
Like Xin
Lin Gu
Lin Xuhong
Lincheng Li
Linghua Tang
Lingzhi Kong
Linlin Yang
Linsen Li
Litao Yu
Liu Liu
Liujie Hua
Li-Yun Wang
Loren Schwiebert
Lujia Jin
Lujun Li
Luping Zhou
Luting Wang

Mansi Sharma
Mantini Pranav
Mahmoud Zidan
    Khairallah
Manuel Günther
Marcella Astrid
Marco Piccirilli
Martin Kampel
Marwan Torki
Masaaki Iiyama
Masanori Suganuma
Masayuki Tanaka
Matan Jacoby
Md Alimoor Reza
Md. Zasim Uddin
Meghshyam Prasad
Mei-Chen Yeh
Meng Tang
Mengde Xu
Mengyang Pu
Mevan B. Ekanayake
Michael Bi Mi
Michael Wray
Michaël Clément
Michel Antunes
Michele Sasdelli
Mikhail Sizintsev
Min Peng
Min Zhang
Minchul Shin
Minesh Mathew
Ming Li
Ming Meng
Ming Yin
Ming-Ching Chang
Mingfei Cheng
Minghui Wang
Mingjun Hu
MingKun Yang
Mingxing Tan
Mingzhi Yuan
Min-Hung Chen
Minhyun Lee
Minjung Kim
Min-Kook Suh

Minkyo Seo
Minyi Zhao
Mo Zhou
Mohammad Amin A.
    Shabani
Moein Sorkhei
Mohit Agarwal
Monish K. Keswani
Muhammad Sarmad
Muhammad Kashif Ali
Myung-Woo Woo
Naeemullah Khan
Naman Solanki
Namyup Kim
Nan Gao
Nan Xue
Naoki Chiba
Naoto Inoue
Naresh P. Cuntoor
Nati Daniel
Neelanjan Bhowmik
Niaz Ahmad
Nicholas I. Kuo
Nicholas E. Rosa
Nicola Fioraio
Nicolas Dufour
Nicolas Papadakis
Ning Liu
Nishan Khatri
Ole Johannsen
P. Real Jurado
Parikshit V. Sakurikar
Patrick Peursum
Pavan Turaga
Peijie Chen
Peizhi Yan
Peng Wang
Pengfei Fang
Penghui Du
Pengpeng Liu
Phi Le Nguyen
Philippe Chiberre
Pierre Gleize
Pinaki Nath Chowdhury
Ping Hu

Ping Li
Ping Zhao
Pingping Zhang
Pradyumna Narayana
Pritish Sahu
Qi Li
Qi Wang
Qi Zhang
Qian Li
Qian Wang
Qiang Fu
Qiang Wu
Qiangxi Zhu
Qianying Liu
Qiaosi Yi
Qier Meng
Qin Liu
Qing Liu
Qing Wang
Qingheng Zhang
Qingjie Liu
Qinglin Liu
Qingsen Yan
Qingwei Tang
Qingyao Wu
Qingzheng Wang
Qizao Wang
Quang Hieu Pham
Rabab Abdelfattah
Rabab Ward
Radu Tudor Ionescu
Rahul Mitra
Raül Pérez i Gonzalo
Raymond A. Yeh
Ren Li
Renán Rojas-Gómez
Renjie Wan
Renuka Sharma
Reyer Zwiggelaar
Robin Chan
Robin Courant
Rohit Saluja
Rongkai Ma
Ronny Hänsch
Rui Liu

Rui Wang
Rui Zhu
Ruibing Hou
Ruikui Wang
Ruiqi Zhao
Ruixing Wang
Ryo Furukawa
Ryusuke Sagawa
Saimunur Rahman
Samet Akcay
Samitha Herath
Sanath Narayan
Sandesh Kamath
Sanghoon Jeon
Sanghyun Son
Satoshi Suzuki
Saumik Bhattacharya
Sauradip Nag
Scott Wehrwein
Sebastien Lefevre
Sehyun Hwang
Seiya Ito
Selen Pehlivan
Sena Kiciroglu
Seok Bong Yoo
Seokjun Park
Seongwoong Cho
Seoungyoon Kang
Seth Nixon
Seunghwan Lee
Seung-Ik Lee
Seungyong Lee
Shaifali Parashar
Shan Cao
Shan Zhang
Shangfei Wang
Shaojian Qiu
Shaoru Wang
Shao-Yuan Lo
Shengjin Wang
Shengqi Huang
Shenjian Gong
Shi Qiu
Shiguang Liu
Shih-Yao Lin

Shin-Jye Lee
Shishi Qiao
Shivam Chandhok
Shohei Nobuhara
Shreya Ghosh
Shuai Yuan
Shuang Yang
Shuangping Huang
Shuigeng Zhou
Shuiwang Li
Shunli Zhang
Shuo Gu
Shuoxin Lin
Shuzhi Yu
Sida Peng
Siddhartha Chandra
Simon S. Woo
Siwei Wang
Sixiang Chen
Siyu Xia
Sohyun Lee
Song Guo
Soochahn Lee
Soumava Kumar Roy
Srinjay Soumitra Sarkar
Stanislav Pidhorskyi
Stefan Gumhold
Stefan Matcovici
Stefano Berretti
Stylianos Moschoglou
Sudhir Yarram
Sudong Cai
Suho Yang
Sumitra S. Malagi
Sungeun Hong
Sunggu Lee
Sunghyun Cho
Sunghyun Myung
Sungmin Cho
Sungyeon Kim
Suzhen Wang
Sven Sickert
Syed Zulqarnain Gilani
Tackgeun You
Taehun Kim

Takao Yamanaka
Takashi Shibata
Takayoshi Yamashita
Takeshi Endo
Takeshi Ikenaga
Tanvir Alam
Tao Hong
Tarun Kalluri
Tat-Jen Cham
Tatsuya Yatagawa
Teck Yian Lim
Tejas Indulal Dhamecha
Tengfei Shi
Thanh-Dat Truong
Thomas Probst
Thuan Hoang Nguyen
Tian Ye
Tianlei Jin
Tianwei Cao
Tianyi Shi
Tianyu Song
Tianyu Wang
Tien-Ju Yang
Tingting Fang
Tobias Baumgartner
Toby P. Breckon
Torsten Sattler
Trung Tuan Dao
Trung Le
Tsung-Hsuan Wu
Tuan-Anh Vu
Utkarsh Ojha
Utku Ozbulak
Vaasudev Narayanan
Venkata Siva Kumar
    Margapuri
Vandit J. Gajjar
Vi Thi Tuong Vo
Victor Fragoso
Vikas Desai
Vincent Lepetit
Vinh Tran
Viresh Ranjan
Wai-Kin Adams Kong
Wallace Michel Pinto Lira

Walter Liao
Wang Yan
Wang Yong
Wataru Shimoda
Wei Feng
Wei Mao
Wei Xu
Weibo Liu
Weichen Xu
Weide Liu
Weidong Chen
Weihong Deng
Wei-Jong Yang
Weikai Chen
Weishi Zhang
Weiwei Fang
Weixin Lu
Weixin Luo
Weiyao Wang
Wenbin Wang
Wenguan Wang
Wenhan Luo
Wenju Wang
Wenlei Liu
Wenqing Chen
Wenwen Yu
Wenxing Bao
Wenyu Liu
Wenzhao Zheng
Whie Jung
Williem Williem
Won Hwa Kim
Woohwan Jung
Wu Yirui
Wu Yufeng
Wu Yunjie
Wugen Zhou
Wujie Sun
Wuman Luo
Xi Wang
Xianfang Sun
Xiang Chen
Xiang Li
Xiangbo Shu
Xiangcheng Liu

Xiangyu Wang
Xiao Wang
Xiao Yan
Xiaobing Wang
Xiaodong Wang
Xiaofeng Wang
Xiaofeng Yang
Xiaogang Xu
Xiaogen Zhou
Xiaohan Yu
Xiaoheng Jiang
Xiaohua Huang
Xiaoke Shen
Xiaolong Liu
Xiaoqin Zhang
Xiaoqing Liu
Xiaosong Wang
Xiaowen Ma
Xiaoyi Zhang
Xiaoyu Wu
Xieyuanli Chen
Xin Chen
Xin Jin
Xin Wang
Xin Zhao
Xindong Zhang
Xingjian He
Xingqun Qi
Xinjie Li
Xinqi Fan
Xinwei He
Xinyan Liu
Xinyu He
Xinyue Zhang
Xiyuan Hu
Xu Cao
Xu Jia
Xu Yang
Xuan Luo
Xubo Yang
Xudong Lin
Xudong Xie
Xuefeng Liang
Xuehui Wang
Xuequan Lu

Xuesong Yang
Xueyan Zou
XuHu Lin
Xun Zhou
Xupeng Wang
Yali Zhang
Ya-Li Li
Yalin Zheng
Yan Di
Yan Luo
Yan Xu
Yang Cao
Yang Hu
Yang Song
Yang Zhang
Yang Zhao
Yangyang Shu
Yani A. Ioannou
Yaniv Nemcovsky
Yanjun Zhu
Yanling Hao
Yanling Tian
Yao Guo
Yao Lu
Yao Zhou
Yaping Zhao
Yasser Benigmim
Yasunori Ishii
Yasushi Yagi
Yawei Li
Ye Ding
Ye Zhu
Yeongnam Chae
Yeying Jin
Yi Cao
Yi Liu
Yi Rong
Yi Tang
Yi Wei
Yi Xu
Yichun Shi
Yifan Zhang
Yikai Wang
Yikang Ding
Yiming Liu

Yiming Qian
Yin Li
Yinghuan Shi
Yingjian Li
Yingkun Xu
Yingshu Chen
Yingwei Pan
Yiping Tang
Yiqing Shen
Yisheng Zhu
Yitian Li
Yizhou Yu
Yoichi Sato
Yong A.
Yongcai Wang
Yongheng Ren
Yonghuai Liu
Yongjun Zhang
Yongkang Luo
Yongkang Wong
Yongpei Zhu
Yongqiang Zhang
Yongrui Ma
Yoshimitsu Aoki
Yoshinori Konishi
Young Jun Heo
Young Min Shin
Youngmoon Lee
Youpeng Zhao
Yu Ding
Yu Feng
Yu Zhang
Yuanbin Wang
Yuang Wang
Yuanhong Chen
Yuanyuan Qiao
Yucong Shen
Yuda Song
Yue Huang
Yufan Liu
Yuguang Yan
Yuhan Xie
Yu-Hsuan Chen
Yu-Hui Wen
Yujiao Shi

Yujin Ren
Yuki Tatsunami
Yukuan Jia
Yukun Su
Yu-Lun Liu
Yun Liu
Yunan Liu
Yunce Zhao
Yun-Chun Chen
Yunhao Li
Yunlong Liu
Yunlong Meng
Yunlu Chen
Yunqian He
Yunzhong Hou
Yuqiu Kong
Yusuke Hosoya
Yusuke Matsui
Yusuke Morishita
Yusuke Sugano
Yuta Kudo
Yu-Ting Wu
Yutong Dai
Yuxi Hu
Yuxi Yang
Yuxuan Li
Yuxuan Zhang
Yuzhen Lin
Yuzhi Zhao
Yvain Queau
Zanwei Zhou
Zebin Guo
Ze-Feng Gao
Zejia Fan
Zekun Yang
Zelin Peng
Zelong Zeng
Zenglin Xu
Zewei Wu
Zhan Li
Zhan Shi
Zhe Li
Zhe Liu
Zhe Zhang
Zhedong Zheng

Zhenbo Xu
Zheng Gu
Zhenhua Tang
Zhenkun Wang
Zhenyu Weng
Zhi Zeng
Zhiguo Cao
Zhijie Rao
Zhijie Wang
Zhijun Zhang
Zhimin Gao
Zhipeng Yu
Zhiqiang Hu
Zhisong Liu
Zhiwei Hong
Zhiwei Xu

Zhiwu Lu
Zhixiang Wang
Zhixin Li
Zhiyong Dai
Zhiyong Huang
Zhiyuan Zhang
Zhonghua Wu
Zhongyan Zhang
Zhongzheng Yuan
Zhu Hu
Zhu Meng
Zhujun Li
Zhulun Yang
Zhuojun Zou
Ziang Cheng
Zichuan Liu

Zihan Ding
Zihao Zhang
Zijiang Song
Zijin Yin
Ziqiang Zheng
Zitian Wang
Ziwei Yao
Zixun Zhang
Ziyang Luo
Ziyi Bai
Ziyi Wang
Zongheng Tang
Zongsheng Cao
Zongwei Wu
Zoran Duric

Contents – Part V

Datasets and Performance Analysis

Recognition: Feature Detection, Indexing, Matching, and Shape Representation

Improving Few-shot Learning by Spatially-aware Matching and CrossTransformer

Hongguang Zhang[1]([✉])(iD), Philip H. S. Torr[4], and Piotr Koniusz[2,3](iD)

[1] Systems Engineering Institute, AMS, Shanghai, China
`zhang.hongguang@outlook.com`
[2] Data61/CSIRO, Sydney, Australia
`piotr.koniusz@data61.csiro.au`
[3] Australian National University, Canberra, Australia
[4] Oxford University, Oxford, UK

Abstract. Current few-shot learning models capture visual object relations in the so-called meta-learning setting under a fixed-resolution input. However, such models have a limited generalization ability under the scale and location mismatch between objects, as only few samples from target classes are provided. Therefore, the lack of a mechanism to match the scale and location between pairs of compared images leads to the performance degradation. The importance of image contents varies across coarse-to-fine scales depending on the object and its class label, *e.g.*, generic objects and scenes rely on their global appearance while fine-grained objects rely more on their localized visual patterns. In this paper, we study the impact of scale and location mismatch in the few-shot learning scenario, and propose a novel Spatially-aware Matching (SM) scheme to effectively perform matching across multiple scales and locations, and learn image relations by giving the highest weights to the best matching pairs. The SM is trained to activate the most related locations and scales between support and query data. We apply and evaluate SM on various few-shot learning models and backbones for comprehensive evaluations. Furthermore, we leverage an auxiliary self-supervisory discriminator to train/predict the spatial- and scale-level index of feature vectors we use. Finally, we develop a novel transformer-based pipeline to exploit self- and cross-attention in a spatially-aware matching process. Our proposed design is orthogonal to the choice of backbone and/or comparator.

Keywords: Few-shot · Multi-scale · Transformer · Self-supervision

1 Introduction

CNNs are the backbone of object categorization, scene classification and fine-grained image recognition models but they require large amounts of labeled data. In contrast, humans enjoy the ability to learn and recognize novel objects and complex visual concepts from very few samples, which highlights the superiority of biological vision over CNNs. Inspired by the brain ability to learn in the few-samples regime, researchers study the so-called problem of few-shot learning for

© The Author(s), under exclusive license to Springer Nature Switzerland AG 2023
L. Wang et al. (Eds.): ACCV 2022, LNCS 13845, pp. 3–20, 2023.
https://doi.org/10.1007/978-3-031-26348-4_1

Fig. 1. Scale mismatch in *mini*-ImageNet. Top row: support samples randomly selected from episodes. Remaining rows: failure queries are marked by red boxes. We estimated that ∼30% mismatches are due to the object scale mismatch, which motivates the importance of scale and region matching in few-shot learning.

which networks are adapted by the use of only few training samples. Several proposed relation-learning deep networks [1–4] can be viewed as performing a variant of metric learning, which they are fail to address the scale- and location-mismatch between support and query samples as shown in Fig. 1. We follow such models and focus on studying how to capture the most discriminative object scales and locations to perform accurate matching between the so-called query and support representations.

A typical relational few-shot learning pipeline consists of (i) feature encoder (backbone), (ii) pooling operator which aggregates feature vectors of query and support images of an episode followed by forming a relation descriptor, and (iii) comparator (base learner). In this paper, we investigate how to efficiently apply spatially-aware matching between query and support images across different locations and scales. To this end, we propose a Spatially-aware Matching (SM) scheme which scores the compared locations and scales. The scores can be regularized to induce sparsity and used to re-weight similarity learning loss operating on $\{0, 1\}$ labels (different/same class label). Note that our SM is orthogonal to the choice of baseline, therefore it is applicable to many existing few-shot learning pipelines.

As the spatial size (height and width) of convolutional representations vary, pooling is required before feeding the representations into the comparator. We compare several pooling strategies, *i.e.*, average, max and second-order pooling (used in object, texture and action recognition, fine-grained recognition, and few-shot learning [4–10]) which captures covariance of features per region/scale. Second-order Similarity Network (SoSN) [4] is the first work which validates the usefulness of autocorrelation representations in few-shot learning. In this paper, we employ second-order pooling as it is permutation-invariant w.r.t. the spatial location of aggregated vectors while capturing second-order statistics which are more informative than typical average-pooled first-order features. As second-order pooling can aggregate any number of feature vectors into a fixed-size representation, it is useful in describing regions of varying size for spatially-aware matching.

Though multi-scale modeling has been used in low-level vision problems, and matching features between regions is one of the oldest recognition tools [11,12], relation-based few-shot learning (similarity learning between pairs of images) has not used such a mechanism despite clear benefits.

In addition to our matching mechanism, we embed the self-supervisory discriminators into our pipeline whose auxiliary task has to predict scale and spatial annotations, thus promoting a more discriminative training of encoder, attention and comparator. This is achieved by the use of Spatial-aware Discriminator (SD) to learn/predict the location and scale indexes of given features. Such strategies have not been investigated in matching, but they are similar to pretext tasks in self-supervised learning.

Beyond using SM on classic few-shot learning pipelines, we also propose a novel transformer-based pipeline, Spatially-aware Matching CrossTransformer (SmCT), which learns the object correlations over locations and scales via cross-attention. Such a pipeline is effective when being pre-trained on large-scale datasets.

Below we summarize our contributions:

  i. We propose a novel spatially-aware matching few-shot learning strategy, compatible with many existing few-shot learning pipelines. We form possible region- and scale-wise pairs, and we pass them through comparator whose scores are re-weighted according to the matching score of region- and scale-wise pairs obtained from the Spatial Matching unit.
 ii. We propose self-supervisory scale-level pretext tasks using second-order representations and auxiliary label terms for locations/scales, *e.g.*, scale index.
iii. We investigate various matching strategies, *i.e.*, different formulations of the objective, the use of sparsity-inducing regularization on attention scores, and the use of a balancing term on weighted average scores.
 iv. We propose a novel and effective transformer-based cross-attention matching strategy for few-shot learning, which learns object matching in pairs of images according to their respective locations and scales.

2   Related Work

One- and few-shot learning has been studied in shallow [13–18] and deep learning setting [1,2,2–4,10,19–32]. Early works [17,18] employ generative models. Siamese Network [19] is a two-stream CNN which can compare two streams.

Matching Network [1] introduces the concept of support set and L-way Z-shot learning protocols to capture the similarity between a query and several support images in the episodic setting which we adopt. Prototypical Net [2] computes distances between a query and prototypes of each class. Model-Agnostic Meta-Learning (MAML) [20] introduces a meta-learning model which can be considered a form of transfer learning. Such a model was extended to Gradient Modulating MAML (ModGrad) [33] to speed up the convergence. Relation Net [3] learns the relationship between query and support images by a deep comparator that produces relation scores. SoSN [4] extends Relation Net [3] by second-order pooling. SalNet [24] is a saliency-guided end-to-end sample halluci-nating model. Graph Neural Networks (GNN) [34–38] can also be combined with few-shot learning [21,25,26,39–41]. In CAN [27], PARN [28] and RENet [42], self-correlation and cross-attention are employed to boost the performance. In contrast, our work studies explicitly matching over multiple scales and locations of input patches instead of features. Moreover, our SM is the first work study-ing how to combine self-supervision with spatial-matching to boost the perfor-mance. SAML [29] relies on a relation matrix to improve metric measurements between local region pairs. DN4 [30] proposes the deep nearest neighbor neural network to improve the image-to-class measure via deep local descriptors. Few-shot learning can also be performed in the transductive setting [43] and applied to non-standard problems, *e.g.*, keypoint recognition [44].

Second-order pooling has been used in texture recognition [45] by Region Covariance Descriptors (RCD), in tracking [5] and object category recogni-tion [8,9]. Higher-order statistics have been used for action classification [46,47], domain adaptation [48,49], few-shot learning [4,10,24,50,51], few-shot object detection [52–55] and even manifold-based incremental learning [56].

We employ second-order pooling due to its (i) permutation invariance (the ability to factor out spatial locations of feature vectors) and (ii) ability to capture second-order statistics.

Notations. Let $\mathbf{x} \in \mathbb{R}^d$ be a d-dimensional feature vector. $\mathcal{I}_N$ stands for the index set $\{1, 2, ..., N\}$. Capitalized boldface symbols such as $\boldsymbol{\Phi}$ denote matrices. Lowercase boldface symbols such as $\boldsymbol{\phi}$ denote vectors. Regular fonts such as Φ_{ij}, ϕ_i, n or Z denote scalars, *e.g.*, Φ_{ij} is the $(i, j)^{\text{th}}$ coefficient of $\boldsymbol{\Phi}$. Finally, $\delta(\mathbf{x} - \mathbf{y}) = 1$ if $\mathbf{x} = \mathbf{y}$ and 0 otherwise.

3   Approach

Although spatially-aware representations have been studied in low-level vision, *e.g.*, deblurring, they have not been studied in relation-based learning (few-shot learning). Thus, it is not obvious how to match feature sets formed from pairs of images at different locations/resolutions.

In conventional image classification, high-resolution images are known to be more informative than their low-resolution counterparts. However, extracting the discriminative information depends on the most expressive scale which varies between images. When learning to compare pairs of images (the main mechanism of relation-based few-shot learning), one has to match correctly same/related objects represented at two different locations and/or scales.

Inspired by such issues, we show the importance of spatially-aware matching across locations and scales. To this end, we investigate our strategy on classic few-shot learning pipelines such as Prototypical Net, Relation Net and Second-order Similarity Network which we refer to as (*PN+SM*), (*RN+SM*) and (*SoSN+SM*) when combined with our Spatially-aware Matching (SM) mechanism.

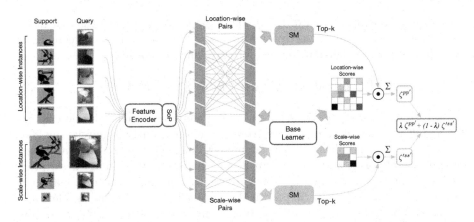

Fig. 2. The pipeline of *SoSN+SM*. We downsample input images twice (3 scales) and extract 5 sub-regions from the original image. Thus, 8 sub-images are passed through our encoder, and intermediate feature vectors are obtained and aggregated with SoP into matrices (red and blue blocks). We obtain 5+5 location-wise support/query matrices $\boldsymbol{\Psi}_k^i$ and $\boldsymbol{\Psi}_q^i$ per support/query images k and q, where $i \in \{1, ..., 5\}$. We also obtain 3+3 scale-wise matrices $\boldsymbol{\Psi}_k'^i$ and $\boldsymbol{\Psi}_q'^i$, where $i \in \{1, ..., 3\}$. We pair them via relation operator ϑ (*e.g.*, concatenation) into 25 and 9 relation descriptors passed to the attention mechanism and relation network, which produces weight scores $w_{pp'}$ (5 × 5) and $w'_{ss'}$ (3 × 3), and relation scores $\zeta_{pp'}$ (5 × 5) and $\zeta'_{ss'}$ (3 × 3), respectively. Finally, relation scores are re-weighted by attention scores and aggregated into the final score.

3.1 Spatially-aware Few-shot Learning

Below, we take the *SoSN+SM* pipeline as an example to illustrate (Fig. 2) how we apply SM on the SoSN few-shot learning pipeline. We firstly generate spatially-aware image sequences from each original support/query sample, and feed them into the pipeline. Each image sequence includes 8 images, *i.e.*, 5 location-wise crops and 3 scale-wise instances. Matching over such support-query sequences requires computing correlations between 8 × 8 = 64 pairs, which leads to significant training overhead when the model is trained with a large batch size. Thus, we decouple spatially-aware matching into location-wise and scale-wise matching steps to reduce the computational cost.

Specifically, let $\mathbf{I}^1, ..., \mathbf{I}^4$ be four corners cropped from $\mathbf{I}$ of 84 × 84 size without overlap and $\mathbf{I}^5$ be a center crop of $\mathbf{I}$. We refer to such an image sequence by $\{\mathbf{I}^p\}_{p \in \mathcal{I}_5}$, and they are of 42 × 42 resolution. Let $\mathbf{I}'^1$ be equal to the input image $\mathbf{I}$ of 84 × 84 size, and $\mathbf{I}'^2$ and $\mathbf{I}'^3$ be formed by downsampling $\mathbf{I}$ to resolutions 42 × 42 and 21 × 21. We refer to these images by $\{\mathbf{I}'^s\}_{s \in \mathcal{I}_3}$. We pass these images via the encoding network $f(\cdot)$:

$$\boldsymbol{\Phi}^p = f(\mathbf{I}^p; \mathcal{F}) \quad \text{and} \quad \boldsymbol{\Phi}'^s = f(\mathbf{I}'^s; \mathcal{F}). \tag{1}$$

where $\mathcal{F}$ denotes parameters of encoder network, $\boldsymbol{\Phi}^p$ and $\boldsymbol{\Phi}'^s$ are feature maps at the location p and scale s, respectively. As feature maps vary in size, we apply second-order pooling from Eq. (2) to these maps. We treat the channel mode as D-dimensional vectors and spatial modes $H \times W$ as HW such vectors. As $N = HW$ varies, we define it to be N for crops and N'^s for scales. Then we form

$$\boldsymbol{\Psi}^p = \eta\left(\frac{1}{N}\boldsymbol{\Phi}^p\boldsymbol{\Phi}^{pT}\right) \quad \text{and} \quad \boldsymbol{\Psi}'^s = \eta\left(\frac{1}{N'^s}\boldsymbol{\Phi}'^s\boldsymbol{\Phi}'^{sT}\right). \tag{2}$$

Subsequently, we pass the location- and scale-wise second-order descriptors $\boldsymbol{\Psi}^p$ and $\boldsymbol{\Psi}'^s$ to the relation network (comparator) to model image relations. For the L-way 1-shot problem, we have a support image (k^{th} index) with its image descriptors $(\boldsymbol{\Phi}_k^p, \boldsymbol{\Phi}'_k^s)$ and a query image (q^{th} index) with its image descriptors $(\boldsymbol{\Phi}_q^p, \boldsymbol{\Phi}'_q^s)$. Moreover, each of the above descriptors belong to one of L classes in the subset $\{c_1, ..., c_L\} \subset \mathcal{I}_C$ that forms the so-called L-way learning problem and the class subset $\{c_1, ..., c_L\}$ is chosen at random from $\mathcal{I}_C \equiv \{1, ..., C\}$. Then, the L-way 1-shot relation requires relation scores:

$$\zeta_{kq}^{pp'} = r\left(\vartheta\left(\boldsymbol{\Psi}_k^p, \boldsymbol{\Psi}_q^p\right); \mathcal{R}\right) \quad \text{and} \quad \zeta_{kq}'^{ss'} = r\left(\vartheta\left(\boldsymbol{\Psi}_k'^s, \boldsymbol{\Psi}_q'^s\right); \mathcal{R}\right), \tag{3}$$

where ζ and ζ' are relation scores for a (k, q) image pair at locations (p, p') and scales (s, s'). Moreover, $r(\cdot)$ is the relation network (comparator), $\mathcal{R}$ are its trainable parameters, $\vartheta(\cdot, \cdot)$ is the relation operator (we use concatenation along the channel mode). For Z-shot learning, this operator averages over Z second-order matrices representing the support image before concatenating with the query matrix.

A naive loss for the location- & scale-wise model is given as:

$$L = \sum_{k,q,p} w_p \left(\zeta_{kq}^{pp} - \delta\left(l_k - l_q\right)\right)^2 + \lambda \sum_{k,q,s} w_s' \left(\zeta_{kq}'^{ss} - \delta\left(l_k - l_q\right)\right)^2, \tag{4}$$

where l_k and l_q refer to labels for support and query samples, w_p and w_s' are some priors (weight preferences) w.r.t. locations p and scales s. For instance, $w_p = 1$ if $p = 5$ (center crop), $w_p = 0.5$ otherwise, and $w_s = 1/2^{s-1}$.

A less naive formulation assumes a modified loss which performs matching between various regions and scales, defined as:

$$L = \sum_{k,q} \sum_{p,p'} \left(w_{pp'}^{kq}\right)^\gamma \left(\zeta_{kq}^{pp'} - \delta\left(l_k - l_q\right)\right)^2 + \lambda \sum_{k,q} \sum_{s,s'} \left(w_{ss'}'^{kq}\right)^\gamma \left(\zeta_{kq}'^{ss'} - \delta\left(l_k - l_q\right)\right)^2, \tag{5}$$

where $w_{pp'}$ and $w_{ss'}'$ are some pair-wise priors (weight preferences) w.r.t. locations (p, p') and scales (s, s'). We favor this formulation and we strive to learn $w_{pp'}^{kq}$ and $w_{ss'}'^{kq}$ rather than just specify rigid priors for all (k, q) support-query pairs. Finally, coefficient $0 \leq \gamma \leq \infty$ balances the impact of re-weighting. If $\gamma = 0$, all weights are equal one. If $\gamma = 0.5$, lower weights contribute in a balanced way. If $\gamma = 1$, we obtain regular re-weighting. If $\gamma = \infty$, the largest weight wins.

Spatially-Aware Matching. As our feature encoder processes images at different scales and locations, the model should have the ability to select the best matching locations and scales for each support-query pair. Thus, we propose a pair-wise attention mechanism to re-weight (activate/deactivate) different

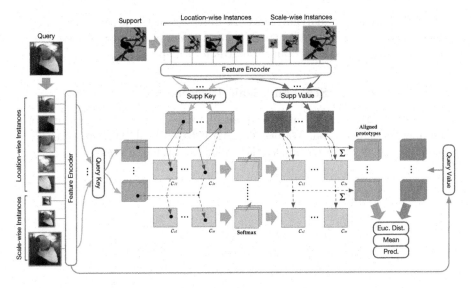

Fig. 3. Our Spatially-aware Matching CrossTransformer (SMCT) is built upon the cross-transformer [57]. We introduce the spatially-aware (location- and scale-wise) image sequences as inputs to exploit cross-attention during feature matching.

matches withing support-query pairs when aggregating the final scores of comparator. Figure 2 shows this principle.

Specifically, as different visual concepts may be expressed by their constituent parts (mixture of objects, mixture of object parts, *etc.*.), each appearing at a different location or scale, we perform a soft-attention which selects $w_{pp'} \geq 0$ and $w'_{ss'} \geq 0$. Moreover, as co-occurrence representations are used as inputs to the attention network, the network selects a mixture of dominant scales and locations for co-occurring features (which may correspond to pairs of object parts).

The Spatially-aware Matching (SM) network (two convolutional blocks and an FC layer) performs the location- and scale-wise matching respectively using shared model parameters. We opt for a decoupled matching in order to reduce the training overhead. We perform $5 \times 5 + 3 \times 3 = 34$ matches per support-query pair rather than $5 \times 5 \times 3 \times 3 = 225$ matches but a full matching variant is plausible (and could perform better). We have:

$$w_{pp'}^{kq} = m\left(\vartheta(\cdot \boldsymbol{\Psi}_k^p, \boldsymbol{\Psi}_q^{p'}); \boldsymbol{\mathcal{M}}\right) \quad \text{and} \quad w_{ss'}^{\prime kq} = m\left(\vartheta(\cdot \boldsymbol{\Psi}_k^{\prime s}, \boldsymbol{\Psi}_q^{\prime s'}); \boldsymbol{\mathcal{M}}\right), \quad (6)$$

where $m(\cdot)$ is the Spatially-aware Matching network, $\boldsymbol{\mathcal{M}}$ denotes its parameters, (k, q) are query-support sample indexes. We impose a penalty to control the sparsity of matching:

$$\Omega = \sum_{k,q} \sum_{p,p'} \left| w_{pp'}^{kq} \right| + \sum_{k,q} \sum_{s,s'} \left| w_{ss'}^{\prime kq} \right|. \quad (7)$$

Our spatially-aware matching network differs from the feature-based attention mechanism as we score the match between pairs of cropped/resized regions (not individual regions) to produce the attention map.

3.2 Self-supervised Scale and Scale Discrepancy

Scale Discriminator. To improve discriminative multi-scale representations, we employ self-supervision. We design a MLP-based Scale Discriminator (SD) as shown in Fig. 4 which recognizes the scales of training images.

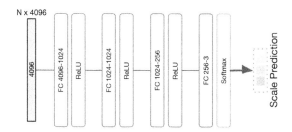

Fig. 4. Scale Discriminator with 3 fully-connected layers.

Specifically, we feed second-order representations to the SD module and assign labels 1, 2 or 3 for 256×256, 128×128 or 64×64 images, respectively. We apply cross-entropy loss to train the SD module and classify the scale corresponding to given second-order feature matrix.

Given $\boldsymbol{\Psi}_i^s$ which is the second-order representation of $\mathbf{X}_i^s$, we vectorize them via (:) and forward to the SD module to predict the scale index s. We have:

$$\mathbf{p}_i^s = \mathrm{sd}(\boldsymbol{\Psi}_{i(:)}^s; \boldsymbol{\mathcal{C}}), \tag{8}$$

where $\mathrm{sd}(\cdot)$ refers to the scale discriminator, $\boldsymbol{\mathcal{C}}$ denotes parameters of $\mathrm{sd}(\cdot)$, and $\mathbf{p}$ are the scale prediction scores for $\boldsymbol{\Psi}$. We go over all i corresponding to support and query images in the mini-batch and we use cross-entropy to learn the parameters of Scale Discriminator:

$$L_{sd} = -\sum_{i,s} \log\left(\frac{\exp(\mathbf{p}_i^s[s])}{\sum_{s'} \exp(\mathbf{p}_i^s[s'])}\right), \tag{9}$$

where $s, s' \in \mathcal{I}_S$ enumerate over scale indexes.

Discrepancy Discriminator. As relation learning requires comparing pairs of images, we propose to model scale discrepancy between each support-query pair by assign a discrepancy label to each pair. Specifically, we assign label $\Delta_{s,s^*} = s - s^* + 1$ where s and s^* denote the scales of a given support-query pair. Then we train so-called Discrepancy Discriminator (DD) to recognize the discrepancy between scales. DD uses the same architecture as SD while the input dimension is doubled due to concatenated support-query pairs on input. Thus:

$$\mathbf{p}_{ij}^{s,s*} = \mathrm{dd}(\vartheta(\boldsymbol{\Psi}_i^s, \boldsymbol{\Psi}_j^{s*}); \mathcal{D}), \tag{10}$$

where $\mathrm{dd}(\cdot)$ refers to scale discrepancy discriminator, $\mathcal{D}$ are the parameters of dd, $\mathbf{p}_{ij}^{s,s*}$ are scale discrepancy prediction scores, ϑ is concat. in mode 3. We go over all i, j support+query image indexes in the mini-batch and we apply the cross-entropy loss to learn the discrepancy labels:

$$L_{dd} = -\sum_{i,s}\sum_{j,s^*} \log\left(\frac{\exp(\mathbf{p}_{ij}^{s,s^*}[\Delta_{s,s^*}])}{\sum_{s'}\sum_{s'^*}\exp(\mathbf{p}_{ij}^{s,s^*}[\Delta_{s',s'^*}])}\right), \tag{11}$$

where where $s, s', s^*, s'^* \in \mathcal{I}_S$ enumerate over scale indexes.

Final Loss. The total loss combines the proposed Scale Selector, Scale Discriminator and Discrepancy Discriminator:

$$\underset{\mathcal{F},\mathcal{R},\mathcal{M},\mathcal{C},\mathcal{D}}{\arg\min} \quad \alpha\Omega + L + \beta L_{sd} + \gamma L_{dd}, \tag{12}$$

where α, β, γ are the hyper-parameters that control the impact of the regularization and each individual loss component.

3.3   Transformer-Based Spatially-Aware Pipeline

Our SM network can be viewed as an instance of attention, whose role is to re-weight numbers of spatial pairs to improve the discriminative relation learning between support and query samples. Recently, transformers have proven very effective in learning the discriminative representations in both natural language processing and computer vision tasks. Inspired by the self-attention [57], which can naturally be used to address the feature matching problem, we further develop a novel Spatially-aware Matching CrossTransformer (SmCT) to match location- and scale-wise support-query patches in few-shot learning. Figure 3 shows the architecture of SmCT, which consists of 4 heads, namely the support key and value heads, and the query key and value heads. In contrast to using SM in classic pipelines, where we decoupled the matching into location-wise and scale-wise steps, all possible matching combinations are considered in the SmCT pipeline. Thus, we do not use symbols $\mathbf{I}^p$ and $\mathbf{I}'^s$ in this section. The image sequence is simply given by $\{\mathbf{I}^s\}_{s\in\mathcal{I}_8}$.

We pass the spatial sequences of support and query samples $\mathbf{I}_k^s$ and $\mathbf{I}_q^s$ into the backbone and obtain $\boldsymbol{\Phi}_k^s, \boldsymbol{\Phi}_q^{\tilde{s}} \in \mathbb{R}^{D\times N}$. Due to different scales of feature maps for $s = 6, 7, 8$, we downsample larger feature maps and obtain $N' = 10 \times 10$ feature vectors. We feed them into the key/value heads to obtain keys $\mathbf{K}_k^s, \mathbf{K}_q^{\tilde{s}} \in \mathbb{R}^{d_k\times N'}$ and values $\mathbf{V}_k^s, \mathbf{V}_q^{\tilde{s}} \in \mathbb{R}^{d_v\times N'}$. Then we multiply support-query spatial key pairs followed by SoftMax in order to obtain the normalized cross-attention scores $\mathbf{C}_{s\tilde{s}} = \mathrm{SoftMax}(\mathbf{K}_k^{s\,T}\mathbf{K}_q^{\tilde{s}}) \in \mathbb{R}^{N'\times N'}$, which are used as correlations in

Table 1. Evaluations on *mini*-ImageNet and *tiered*-ImageNet (5-way acc. given).

Model		Backbone	mini-ImageNet		tiered-ImageNet	
			1-shot	5-shot	1-shot	5-shot
MN	[1]	-	43.56 ± 0.84	55.31 ± 0.73	-	-
PN	[2]	Conv-4-64	49.42 ± 0.78	68.20 ± 0.66	53.31 ± 0.89	72.69 ± 0.74
MAML	[20]	Conv-4-64	48.70 ± 1.84	63.11 ± 0.92	51.67 ± 1.81	70.30 ± 1.75
RN	[3]	Conv-4-64	50.44 ± 0.82	65.32 ± 0.70	54.48 ± 0.93	71.32 ± 0.78
GNN	[21]	Conv-4-64	50.30	66.40	-	-
MAML++	[58]	Conv-4-64	52.15 ± 0.26	68.32 ± 0.44	-	-
SalNet	[24]	Conv-4-64	57.45 ± 0.86	72.01 ± 0.75	-	-
SoSN	[4]	Conv-4-64	52.96 ± 0.83	68.63 ± 0.68	-	-
TADAM	[59]	ResNet-12	58.50 ± 0.30	76.70 ± 0.30	-	-
MetaOpt	[60]	ResNet-12	61.41 ± 0.61	77.88 ± 0.46	65.99 ± 0.72	81.56 ± 0.53
DeepEMD	[61]	ResNet-12	65.91 ± 0.82	82.41 ± 0.56	-	-
PN+SM		Conv-4-64	52.01 ± 0.80	69.92 ± 0.67	54.37 ± 0.82	75.13 ± 0.77
RN+SM		Conv-4-64	54.99 ± 0.87	68.57 ± 0.63	57.01 ± 0.91	75.04 ± 0.78
SoSN+SM		Conv-4-64	57.11 ± 0.84	71.98 ± 0.63	61.58 ± 0.90	78.64 ± 0.75
SoSN+SM		ResNet-12	62.36 ± 0.85	78.86 ± 0.63	66.35 ± 0.91	82.21 ± 0.71
DeepEMD+SM		ResNet-12	$\mathbf{66.93 \pm 0.84}$	$\mathbf{84.34 \pm 0.61}$	$\mathbf{70.19 \pm 0.89}$	$\mathbf{86.98 \pm 0.74}$

Table 2. Evaluations on the Open MIC dataset (Protocol I) (1-shot learning accuracy). (http://users.cecs.anu.edu.au/~koniusz/openmic-dataset).

Model	way	p1→p2	p1→p3	p1→p4	p2→p1	p2→p3	p2→p4	p3→p1	p3→p2	p3→p4	p4→p1	p4→p2	p4→p3
RN [3]		71.1	53.6	63.5	47.2	50.6	68.5	48.5	49.7	68.4	45.5	70.3	50.8
SoSN$_{(84)}$ [4]	5	81.4	65.2	75.1	60.3	62.1	77.7	61.5	82.0	78.0	59.0	80.8	62.5
SoSN$_{(256)}$		84.1	69.3	82.5	64.9	66.9	82.8	65.8	85.1	81.1	65.1	83.9	66.6
SoSN+SM		**85.6**	**73.6**	**85.0**	**67.7**	**69.6**	**83.1**	**68.2**	**86.9**	**82.9**	**67.4**	**84.7**	**68.4**
RN [3]		40.1	30.4	41.4	23.5	26.4	38.6	26.2	25.8	46.3	23.1	43.3	27.7
SoSN$_{(84)}$ [4]	20	61.5	42.5	61.0	36.1	38.3	56.3	38.7	59.9	59.4	37.4	59.0	38.6
SoSN$_{(256)}$		63.9	49.2	65.9	43.1	44.6	62.6	44.2	63.9	64.1	43.8	63.1	44.3
SoSN+SM		65.5	51.1	67.6	45.2	44.6	64.5	46.3	66.2	67.0	45.3	65.8	47.1
RN [3]		37.8	27.3	39.8	22.1	24.3	36.7	24.5	23.7	44.2	21.4	41.5	25.5
SoSN$_{(84)}$ [4]	30	60.6	40.1	58.3	34.5	35.1	54.2	36.8	58.6	56.6	35.9	57.1	37.1
SoSN$_{(256)}$		61.7	46.6	64.1	41.4	40.9	60.3	41.6	61.0	60.0	42.4	61.2	41.4
SoSN+SM		62.6	47.3	65.2	41.9	41.7	61.5	43.1	61.8	61.0	43.1	62.1	42.3

p1: shn+hon+clv, p2: clk+gls+scl, p3: sci+nat, p4: shx+rlc.
Notation $x{\rightarrow}y$ means training on exhibition x and testing on y

aggregation of support values w.r.t. each location and scale. We obtain the aligned spatially-aware prototypes $\tilde{\mathbf{V}}_k^s \in \mathbb{R}^{d_v \times N'}$:

$$\tilde{\mathbf{V}}_k^s = \sum_{\tilde{s}} \mathbf{V}_k^s \mathbf{C}_{s\tilde{s}}. \tag{13}$$

We measure Euclidean distances between the aligned prototypes and corresponding query values, which act as the final similarity between sample $\mathbf{I}_k$ and $\mathbf{I}_q$:

$$L = \sum_{k,q} \left(\zeta_{kq} - \delta(l_k - l_q) \right)^2 \quad \text{where} \quad \zeta_{kq} = \sum_s \| \tilde{\mathbf{V}}_k^s - \tilde{\mathbf{V}}_q^s \|_F^2 . \tag{14}$$

4   Experiments

Below we demonstrate usefulness of our proposed Spatial- and Scale-matching Network by evaluations (one- and few-shot protocols) on *mini*-ImageNet [1], *tiered*-ImageNet [63], Meta-Dataset [62] and fine-grained datasets.

Table 3. The experiments on selected subsets of Meta-Dataset (train-on-ILSVRC setting). We compare our Spatially-aware Matching pipelines with recent baseline models. Following the training steps in [57], we also apply the SimCLR episodes to train our SmCT (which uses the ResNet-34 backbone).

		ImageNet	Aircraft	Bird	DTD	Flower	Avg
k-NN	[62]	41.03	46.81	50.13	66.36	83.10	57.49
MN	[1]	45.00	48.79	62.21	64.15	80.13	60.06
PN	[2]	50.50	53.10	68.79	66.56	85.27	64.84
RN	[3]	34.69	40.73	49.51	52.97	68.76	49.33
SoSN	[4]	50.67	54.13	69.02	66.49	87.21	65.50
CTX	[57]	62.76	79.49	80.63	75.57	95.34	78.76
PN+SM		53.12	57.06	72.01	70.23	88.96	68.28
RN+SM		41.07	46.03	53.24	58.01	72.98	54.27
SoSN+SM		52.03	56.68	70.89	69.03	90.36	67.80
SMCT		**64.12**	**81.03**	**82.98**	**76.95**	**96.45**	**80.31**

Table 4. Evaluations on fine-grained recognition datasets, Flower-102, CUB-200-2011 and Food-101 (5-way acc. given). See [3,4] for details of baselines listed in this table.

	Flower-102		CUB-200-2011		Food-101	
Model	1-shot	5-shot	1-shot	5-shot	1-shot	5-shot
PN	62.81	82.11	37.42	51.57	36.71	53.43
RN	68.52	81.11	40.36	54.21	36.89	49.07
SoSN	76.27	88.55	47.45	63.53	43.12	58.13
RN+SM	71.69	84.45	45.79	58.67	45.31	55.67
SoSN+SM	**81.69**	**91.21**	**54.24**	**70.85**	**48.86**	**63.67**

Setting. We use the standard 84×84 image resolution for *mini*-ImageNet and fine-grained datasets, and 224×224 resolution for Meta-Dataset for fair comparisons. Hyper-parameter α is set to 0.001 while β is set to 0.1 via cross-validation on *mini*-ImageNet. Note that SmCT uses ResNet-34 backbone on Meta-Dataset, and ResNet-12 on *mini*-ImageNet and *tiered*-ImageNet. The total number of training episode is 200000, and the number of testing episode is 1000.

4.1 Datasets

Below, we describe our experimental setup and datasets.

***mini*-ImageNet** [1] consists of 60000 RGB images from 100 classes. We follow the standard protocol (64/16/20 classes for training/validation/testing).

***tiered*-ImageNet** [63] consists of 608 classes from ImageNet. We follow the protocol that uses 351 base classes, 96 validation and 160 test classes.

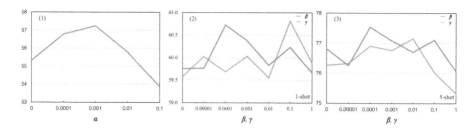

Fig. 5. Accuracy w.r.t.α (subfigure 1), β (the blue curve in subfigure 2&3) and γ (the red curve in subfigure 2&3), which control the impact of SD and DD. (Color figure online)

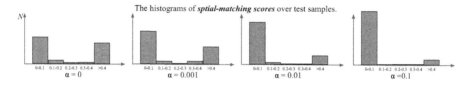

Fig. 6. The histograms of spatial-matching scores with different α values, which demonstrate how Ω induces sparsity and improves the performance.

Open MIC is the Open Museum Identification Challenge (Open MIC) [64], a dataset with photos of various museum exhibits, *e.g.*, paintings, timepieces, sculptures, glassware, relics, science exhibits, natural history pieces, ceramics, pottery, tools and indigenous crafts, captured from 10 museum spaces according to which this dataset is divided into 10 subproblems. In total, it has 866 diverse classes and 1–20 images per class. We combine ($shn+hon+clv$), ($clk+gls+scl$), ($sci+nat$) and ($shx+rlc$) into subproblems $p1$, ..., $p4$. We form 12 possible pairs in which subproblem x is used for training and y for testing (x→y).

Meta-Dataset [62] is a recently proposed benchmark consisting of 10 publicly available datasets to measure the generalized performance of each model. The Train-on-ILSVRC setting means that the model is merely trained via ImageNet training data, and then evaluated on the test data of remaining 10 datasets. In this paper, we follow the Train-on-ILSVRC setting. We choose 5 datasets to measure the overall performance.

Flower-102 [65] contains 102 fine-grained classes of flowers. Each class has 40–258 images. We randomly select 80/22 classes for training/testing.

Caltech-UCSLD-Birds 200-2011 (CUB-200-2011) [66] has 11788 images of 200 fine-grained bird species, 150 classes a for training and the rest for testing.

Food-101 [67] has 101000 images of 101 fine-grained classes, and 1000 images per category. We choose 80/21 classes for training/testing.

4.2 Performance Analysis

Table 1 shows our evaluations on *mini*-ImageNet [1] and *tiered*-ImageNet. Our approach achieves the state-of-the-art performance among all methods based on both 'Conv-4-64' and 'ResNet-12' backbones at 84×84 input resolution. By adding SM (Eq. (3)) and self-supervisory loss (Eq. (9)), we achieve 57.11 and 71.98 scores for 1- and 5-shot protocols with the SoSN baseline (Conv-4-64 backbone), and 66.93 and 85.34 with the DeepEMD baseline (ResNet-12 backbone) on *mini*-ImageNet. The average improvements gained from SM are 4% and 3.5% with the Conv-4-64 backbone, 1% and 1.9% with the ResNet-12 backbone for 1- and 5-shot protocols. These results outperform results in previous works, which strongly supports the benefit of our spatially-aware matching. On *tiered*-ImageNet, our proposed model obtains 3% and 4% improvement for 1-shot and 5-shot protocols with both the Conv-4-64 and the ResNet-12 backbones. Our performance on *tiered*-ImageNet is also better than in previous works. However, we observed that SmCT with the ResNet-12 backbone does not perform as strongly on the above two datasets. A possible reason is that the complicated transformer architecture overfits when the scale of training dataset is small (in contrast to its performance on Meta-Dataset).

We also evaluate our proposed network on the Open MIC dataset [64]. Table 2 shows that our proposed method performs better than baseline models. By adding SM and self-supervised discriminators into baseline models, the accuracies are improved by $1.5\% - 3.0\%$.

Table 3 presents results on the Train-on-ISLVRC setting on 5 of 10 datasets of Meta-Dataset. Applying SM on classic simple baseline few-shot leaning methods brings impressive improvements on all datasets. Furthermore, our SmCT achieves the state-of-the-art results compared to previous methods. This obser-

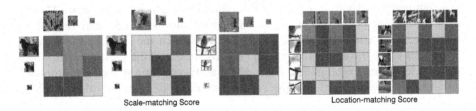

Scale-matching Score   Location-matching Score

Fig. 7. The histograms of spatial-matching scores with different α values, which demonstrate how Ω induces sparsity and improves the performance.

Table 5. Ablation studies w.r.t. the location- and scale-wise inputs (no matching used). Region crops and scale selection is done either in the image space (*img*) or on the feature maps (*feat.*), (5-way 1-shot accuracy, 'SoSN+SM' with ResNet-12 backbone).

Scale-wise (*img*)	Scale-wise (*feat.*)	Loc.-wise (*img*)	Loc.-wise (*feat.*)
59.51	58.54	60.14	58.16

vation is consistent with our analysis that transformer-based networks are likely to be powerful when being trained on a large-scale dataset.

Table 4 shows that applying SM on classic baseline models significantly outperforms others on fine-grained Flower-102, CUB-200-2011 and Food-101.

Spatially-Aware Matching. The role of SM is to learn matching between regions and scales of support-query pairs. As shown in Table 5, using the Spatially-aware Matching Network can further improve results for both 1-shot and 5-shot learning on *mini*-ImageNet by 1.5% and 1.0%, respectively. The results on Flower-102, Food-101 yield $\sim 1.5\%$ gain for SmN. Figure 5 shows the impact of α, β and γ on the performance. Figure 6 verifies the usefulness of Ω regularization term. To this end, we show histograms of spatial-matching scores to demonstrate how α affects the results *vs.* sparsity of matching scores. For instance, one can see the results are the best for moderate $\alpha = 0.001$, and the bin containing null counts also appears larger compared to $\alpha = 0$ (desired behavior). Figure 7 visualizes the matching scores. We randomly sample support-query pairs to show that visually related patches have higher match scores (red) than unrelated pairs (blue). Do note the location- and scale- discrimination is driven by matching both similar and dissimilar support-query pairs driven by the relation loss.

Self-supervised Discriminators. Pretext tasks are known for their ability to boost the performance of image classification due to additional regularization they provide. Applying Scale Discriminator (SD) and Discrepancy Discriminator (DD) is an easy and cheap way to boost the representational power of our network. Pretext tasks do not affect the network complexity or training times. According to our evaluations, the SD improves the 1-shot accuracy by 1.1% and 5-shot accuracy by 1.0%, while the DD improves the accuracy by 1.2% and 1.1% 0.9% for 1- and 5-shot respectively on the *mini*-ImageNet dataset.

In summary, without any pre-training, combining our Spatially-aware Matching strategy brings consistent improvements on the Conv-4-64 and ResNet-12 backbones and various few-shot learning methods, with the overall accuracy outperforming state-of-the-art methods on all few-shot learning datasets. Our novel transformer-based SmCT model also performs strongly on the recently proposed Meta-Dataset, which further supports the usefulness of spatial modeling.

5   Conclusions

We have proposed the Spatially-aware Matching strategy for few-shot learning, which is shown to be orthogonal to the choice of baseline models and/or backbones. Our novel feature matching mechanism helps models learn a more accurate similarity due to matching multiple locations and coarse-to-fine scales of support-query pairs. We show how to leverage a self-supervisory pretext task based on spatial labels. We have also proposed a novel Spatially-aware Matching CrossTransformer to perform matching via the recent popular self- and cross-attention strategies. Our experiments demonstrate the usefulness of the proposed

SM strategy and SmCT in capturing accurate image relations. Combing our SM with various baselines outperforms previous works in the same testbed. SmCT achieves SOTA results on large-scale training data.

Acknowledgements. This work is supported by National Natural Science Fundation of China (No. 62106282), and Young Elite Scientists Sponsorship Program by CAST (No. 2021-JCJQ-QT-038). Code: https://github.com/HongguangZhang/smfsl-master.

References

1. Vinyals, O., Blundell, C., Lillicrap, T., Wierstra, D., et al.: Matching networks for one shot learning. In: NIPS, pp. 3630–3638 (2016)
2. Snell, J., Swersky, K., Zemel, R.: Prototypical networks for few-shot learning. In: NeurIPS, pp. 4077–4087 (2017)
3. Sung, F., Yang, Y., Zhang, L., Xiang, T., Torr, P.H., Hospedales, T.M.: Learning to compare: Relation network for few-shot learning. arXiv preprint arXiv:1711.06025 (2017)
4. Zhang, H., Koniusz, P.: Power normalizing second-order similarity network for few-shot learning. In: WACV, pp. 1185–1193. IEEE (2019)
5. Porikli, F., Tuzel, O.: Covariance tracker. In: CVPR (2006)
6. Guo, K., Ishwar, P., Konrad, J.: Action recognition from video using feature covariance matrices. Trans. Imgage Process. **22**, 2479–2494 (2013)
7. Carreira, J., Caseiro, R., Batista, J., Sminchisescu, C.: Semantic segmentation with second-order pooling. In: Fitzgibbon, A., Lazebnik, S., Perona, P., Sato, Y., Schmid, C. (eds.) ECCV 2012. LNCS, vol. 7578, pp. 430–443. Springer, Heidelberg (2012). https://doi.org/10.1007/978-3-642-33786-4_32
8. Koniusz, P., Yan, F., Gosselin, P.H., Mikolajczyk, K.: Higher-order occurrence pooling for bags-of-words: visual concept detection. IEEE Trans. Pattern Anal. Mach. Intell. **39**, 313–326 (2017)
9. Koniusz, P., Zhang, H., Porikli, F.: A deeper look at power normalizations. In: CVPR, pp. 5774–5783 (2018)
10. Wertheimer, D., Hariharan, B.: Few-shot learning with localization in realistic settings. In: CVPR, pp. 6558–6567 (2019)
11. Lazebnik, S., Schmid, C., Ponce, J.: Beyond bags of features: Spatial pyramid matching for recognizing natural scene categories. CVPR **2**, 2169–2178 (2006)
12. Yang, J., Yu, K., Gong, Y., Huang, T.: Linear spatial pyramid matching using sparse coding for image classification. In: CVPR, pp. 1794–1801 (2009)
13. Miller, E.G., Matsakis, N.E., Viola, P.A.: Learning from one example through shared densities on transforms. In: CVPR, vol. 1, pp. 464–471 (2000)
14. Li, F.F., VanRullen, R., Koch, C., Perona, P.: Rapid natural scene categorization in the near absence of attention. Proc. Natl. Acad. Sci. **99**, 9596–9601 (2002)
15. Fink, M.: Object classification from a single example utilizing class relevance metrics. In: NIPS, pp. 449–456 (2005)
16. Bart, E., Ullman, S.: Cross-generalization: Learning novel classes from a single example by feature replacement. In: CVPR, pp. 672–679 (2005)
17. Fei-Fei, L., Fergus, R., Perona, P.: One-shot learning of object categories. IEEE Trans. Pattern Anal. Mach. Intell. **28**, 594–611 (2006)
18. Lake, B.M., Salakhutdinov, R., Gross, J., Tenenbaum, J.B.: One shot learning of simple visual concepts. CogSci (2011)

19. Koch, G., Zemel, R., Salakhutdinov, R.: Siamese neural networks for one-shot image recognition. In: ICML Deep Learning Workshop, vol. 2 (2015)

20. Finn, C., Abbeel, P., Levine, S.: Model-agnostic meta-learning for fast adaptation of deep networks. In: ICML, pp. 1126–1135 (2017)

21. Garcia, V., Bruna, J.: Few-shot learning with graph neural networks. arXiv preprint arXiv:1711.04043 (2017)

22. Rusu, A.A., et al.: Meta-learning with latent embedding optimization. arXiv preprint arXiv:1807.05960 (2018)

23. Gidaris, S., Komodakis, N.: Dynamic few-shot visual learning without forgetting. In: CVPR, pp. 4367–4375 (2018)

24. Zhang, H., Zhang, J., Koniusz, P.: Few-shot learning via saliency-guided hallucination of samples. In: CVPR, pp. 2770–2779 (2019)

25. Kim, J., Kim, T., Kim, S., Yoo, C.D.: Edge-labeling graph neural network for few-shot learning. In: CVPR (2019)

26. Gidaris, S., Komodakis, N.: Generating classification weights with GNN denoising autoencoders for few-shot learning. In: CVPR (2019)

27. Hou, R., Chang, H., Ma, B., Shan, S., Chen, X.: Cross attention network for few-shot classification. In: NeurIPS, vol. 32 (2019)

28. Wu, Z., Li, Y., Guo, L., Jia, K.: Parn: Position-aware relation networks for few-shot learning. In: ICCV, pp. 6659–6667 (2019)

29. Hao, F., He, F., Cheng, J., Wang, L., Cao, J., Tao, D.: Collect and select: semantic alignment metric learning for few-shot learning. In: CVPR, pp. 8460–8469 (2019)

30. Li, W., Wang, L., Xu, J., Huo, J., Gao, Y., Luo, J.: Revisiting local descriptor based image-to-class measure for few-shot learning. In: CVPR, pp. 7260–7268 (2019)

31. Zhang, H., Li, H., Koniusz, P.: Multi-level second-order few-shot learning. IEEE Trans. Multimed. **99**, 1–16 (2022)

32. Ni, G., Zhang, H., Zhao, J., Xu, L., Yang, W., Lan, L.: ANF: attention-based noise filtering strategy for unsupervised few-shot classification. In: Pham, D.N., Theeramunkong, T., Governatori, G., Liu, F. (eds.) PRICAI 2021. LNCS (LNAI), vol. 13033, pp. 109–123. Springer, Cham (2021). https://doi.org/10.1007/978-3-030-89370-5_9

33. Simon, C., Koniusz, P., Nock, R., Harandi, M.: On modulating the gradient for meta-learning. In: Vedaldi, A., Bischof, H., Brox, T., Frahm, J.-M. (eds.) ECCV 2020. LNCS, vol. 12353, pp. 556–572. Springer, Cham (2020). https://doi.org/10.1007/978-3-030-58598-3_33

34. Sun, K., Koniusz, P., Wang, Z.: Fisher-Bures adversary graph convolutional networks. UAI **115**, 465–475 (2019)

35. Zhu, H., Koniusz, P.: Simple spectral graph convolution. In: ICLR (2021)

36. Zhu, H., Sun, K., Koniusz, P.: Contrastive Laplacian eigenmaps. In: NeurIPS, pp. 5682–5695 (2021)

37. Zhang, Y., Zhu, H., Meng, Z., Koniusz, P., King, I.: Graph-adaptive rectified linear unit for graph neural networks. In: The Web Conference (WWW), pp. 1331–1339. ACM (2022)

38. Zhang, Y., Zhu, H., Song, Z., Koniusz, P., King, I.: COSTA: covariance-preserving feature augmentation for graph contrastive learning. In: KDD, pp. 2524–2534. ACM (2022)

39. Wang, L., Liu, J., Koniusz, P.: 3d skeleton-based few-shot action recognition with JEANIE is not so naïve. arXiv preprint arXiv: 2112.12668 (2021)

40. Wang, L., Koniusz, P.: Temporal-viewpoint transportation plan for skeletal few-shot action recognition. In: ACCV (2022)

41. Wang, L., Koniusz, P.: Uncertainty-DTW for time series and sequences. In: Avidan, S., Brostow, G., Cissé, M., Farinella, G.M., Hassner, T. (eds.) Computer Vision–ECCV 2022. ECCV 2022. LNCS, vol. 13681, pp. 176–195. Springer, Cham (2022). https://doi.org/10.1007/978-3-031-19803-8_11

42. Kang, D., Kwon, H., Min, J., Cho, M.: Relational embedding for few-shot classification. In: ICCV, pp. 8822–8833 (2021)

43. Zhu, H., Koniusz, P.: EASE: unsupervised discriminant subspace learning for transductive few-shot learning. In: CVPR (2022)

44. Lu, C., Koniusz, P.: Few-shot keypoint detection with uncertainty learning for unseen species. In: CVPR (2022)

45. Tuzel, O., Porikli, F., Meer, P.: Region covariance: a fast descriptor for detection and classification. In: Leonardis, A., Bischof, H., Pinz, A. (eds.) ECCV 2006. LNCS, vol. 3952, pp. 589–600. Springer, Heidelberg (2006). https://doi.org/10.1007/11744047_45

46. Koniusz, P., Cherian, A., Porikli, F.: Tensor representations via kernel linearization for action recognition from 3d skeletons. In: Leibe, B., Matas, J., Sebe, N., Welling, M. (eds.) ECCV 2016. LNCS, vol. 9908, pp. 37–53. Springer, Cham (2016). https://doi.org/10.1007/978-3-319-46493-0_3

47. Koniusz, P., Wang, L., Cherian, A.: Tensor representations for action recognition. IEEE Trans. Pattern Anal. Mach. Intell. **44**, 648–665 (2022)

48. Koniusz, P., Tas, Y., Porikli, F.: Domain adaptation by mixture of alignments of second-or higher-order scatter tensors. In: CVPR, vol. 2 (2017)

49. Tas, Y., Koniusz, P.: CNN-based action recognition and supervised domain adaptation on 3d body skeletons via kernel feature maps. In: BMVC, p. 158. BMVA Press (2018)

50. Zhang, H., Koniusz, P., Jian, S., Li, H., Torr, P.H.S.: Rethinking class relations: absolute-relative supervised and unsupervised few-shot learning. In: CVPR, pp. 9432–9441 (2021)

51. Koniusz, P., Zhang, H.: Power normalizations in fine-grained image, few-shot image and graph classification. IEEE Trans. Pattern Anal. Mach. Intell. **44**, 591–609 (2022)

52. Zhang, S., Luo, D., Wang, L., Koniusz, P.: Few-shot object detection by second-order pooling. In: Ishikawa, H., Liu, C.-L., Pajdla, T., Shi, J. (eds.) ACCV 2020. LNCS, vol. 12625, pp. 369–387. Springer, Cham (2021). https://doi.org/10.1007/978-3-030-69538-5_23

53. Yu, X., Zhuang, Z., Koniusz, P., Li, H.: 6DoF object pose estimation via differentiable proxy voting regularizer. In: BMVC, BMVA Press (2020)

54. Zhang, S., Wang, L., Murray, N., Koniusz, P.: Kernelized few-shot object detection with efficient integral aggregation. In: CVPR, pp. 19207–19216 (2022)

55. Zhang, S., Murray, N., Wang, L., Koniusz, P.: Time-rEversed DiffusioN tEnsor Transformer: a new TENET of few-shot object detection. In: Avidan, S., Brostow, G., Cissé, M., Farinella, G.M., Hassner, T. (eds.) Computer Vision–ECCV 2022. ECCV 2022. LNCS, vol. 13680, pp. 310–328. Springer, Cham (2022). https://doi.org/10.1007/978-3-031-20044-1_18

56. Simon, C., Koniusz, P., Harandi, M.: On learning the geodesic path for incremental learning. In: CVPR, pp. 1591–1600 (2021)

57. Doersch, C., Gupta, A., Zisserman, A.: Crosstransformers: spatially-aware few-shot transfer. arXiv preprint arXiv:2007.11498 (2020)

58. Antoniou, A., Edwards, H., Storkey, A.: How to train your maml. arXiv preprint arXiv:1810.09502 (2018)

59. Oreshkin, B., Lopez, P.R., Lacoste, A.: TADAM: task dependent adaptive metric for improved few-shot learning. In: Advances in Neural Information Processing Systems, pp. 721–731 (2018)

60. Lee, K., Maji, S., Ravichandran, A., Soatto, S.: Meta-learning with differentiable convex optimization. In: CVPR, pp. 10657–10665 (2019)

61. Zhang, C., Cai, Y., Lin, G., Shen, C.: DeepEMD: few-shot image classification with differentiable earth mover's distance and structured classifiers. In: CVPR, pp. 12203–12213 (2020)

62. Triantafillou, E., et al.: Meta-dataset: A dataset of datasets for learning to learn from few examples. arXiv preprint arXiv:1903.03096 (2019)

63. Ren, M., et al.: Meta-learning for semi-supervised few-shot classification. In: ICLR (2018)

64. Koniusz, P., Tas, Y., Zhang, H., Harandi, M., Porikli, F., Zhang, R.: Museum exhibit identification challenge for the supervised domain adaptation and beyond. In: Ferrari, V., Hebert, M., Sminchisescu, C., Weiss, Y. (eds.) ECCV 2018. LNCS, vol. 11220, pp. 815–833. Springer, Cham (2018). https://doi.org/10.1007/978-3-030-01270-0_48

65. Nilsback, M.E., Zisserman, A.: Automated flower classification over a large number of classes. In: Proceedings of the Indian Conference on Computer Vision, Graphics and Image Processing (2008)

66. Wah, C., Branson, S., Welinder, P., Perona, P., Belongie, S.: The Caltech-UCSD Birds-200-2011 Dataset. Technical report CNS-TR-2011-001, California Institute of Technology (2011)

67. Bossard, L., Guillaumin, M., Van Gool, L.: Food-101 – mining discriminative components with random forests. In: Fleet, D., Pajdla, T., Schiele, B., Tuytelaars, T. (eds.) ECCV 2014. LNCS, vol. 8694, pp. 446–461. Springer, Cham (2014). https://doi.org/10.1007/978-3-319-10599-4_29

AONet: Attentional Occlusion-Aware Network for Occluded Person Re-identification

Guangyu Gao$^{(\boxtimes)}$ [ID], Qianxiang Wang [ID], Jing Ge [ID], and Yan Zhang [ID]

Beijing Institute of Technology, Beijing, China
guangyugao@bit.edu.cn

Abstract. Occluded person Re-identification (Occluded ReID) aims to verify the identity of a pedestrian with occlusion across non-overlapping cameras. Previous works for this task often rely on external tasks, *e.g.*, pose estimation, or semantic segmentation, to extract local features over fixed given regions. However, these external models may perform poorly on Occluded ReID, since they are still open problems with no reliable performance guarantee and are not oriented towards ReID tasks to provide discriminative local features. In this paper, we propose an Attentional Occlusion-aware Network (AONet) for Occluded ReID that does not rely on any external tasks. AONet adaptively learns discriminative local features over latent landmark regions by the trainable pattern vectors, and softly weights the summation of landmark-wise similarities based on the occlusion awareness. Also, as there are no ground truth occlusion annotations, we measure the occlusion of landmarks by the awareness scores, when referring to a memorized dictionary storing average landmark features. These awareness scores are then used as a soft weight for training and inferring. Meanwhile, the memorized dictionary is momenta updated according to the landmark features and the awareness scores of each input image. The AONet achieves 53.1% mAP and 66.5% Rank1 on the Occluded-DukeMTMC, significantly outperforming state-of-the-arts without any bells and whistles, and also shows obvious improvements on the holistic datasets Market-1501 and DukeMTMC-reID, as well as the partial datasets Partial-REID and Partial-iLIDS. The code and pre-trained models will be released online soon.

Keywords: Occluded ReID · Occlusion-aware · Landmark · Orthogonal

1 Introduction

Most Person Re-identification (ReID) [2,12,30] approaches focus more on holistic pedestrian images, and tend to fail in real-world scenarios where a pedestrian is partially visible, *e.g.*, occluded by other objects. The Occluded person Re-identification (Occluded ReID) is then investigated which aims to handle the occlusion distractions. Some previous Occluded ReID methods perform part-to-part matching based on fine-grained external local features [13,24], *e.g.*, with body parts assigned larger weights and occlusion parts smaller weights.

L. Wang et al. (Eds.): ACCV 2022, LNCS 13845, pp. 21–36, 2023.
https://doi.org/10.1007/978-3-031-26348-4_2

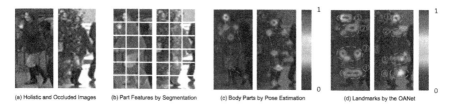

(a) Holistic and Occluded Images   (b) Part Features by Segmentation   (c) Body Parts by Pose Estimation   (d) Landmarks by the OANet

Fig. 1. Illustration of local feature responses. (a) Holistic and occluded images. (b) Local features by partitioning over segmentation mask. (c) Local features based on pose estimation. (d) Landmark features by our AONet.

The key to solving Occluded ReID is to locate landmark regions and then extract well-aligned features from non-occluded landmark regions, while reasonably reducing or prohibiting the use of features from occluded landmark regions. Some Occluded ReID works use body parts attained from pose estimation for local feature extraction [3,4,24], and suppress or exclude the local features of some occluded body parts with low pose confidence. However, the reliability of pose estimation is not guaranteed (*e.g.*, failure on the knees and waist in Fig. 1(c)). Moreover, pose features are often not necessarily adapted to ReID tasks due to cross-task variance. Another group of methods [13,21,22,32] extracts local features directly on uniformly partitioned grids on pedestrian images, and measures the occlusion of each grid guided by the semantic segmentation task, as shown in Fig. 1(b). However, due to the different poses and non-rigid deformation of the human body, these models cannot accurately perform part alignment and thus often fail. In addition, there are some methods that achieve Occluded ReID by locating occluded parts or measuring the occlusion degree using the pose estimation [4,27] or semantic segmentation tasks [3,32].

However, the aforementioned methods rely on external tasks, such as pose estimation or semantic segmentation, to extract local features on fixed given regions of the human body. On one hand, the results of these external tasks may be imprecise; on the other hand, the obtained local features are usually not discriminative enough for Occluded ReID. [10] presented the Matching on Sets (MoS), positioning Occluded ReID as a set matching task without using external models. Compared to this work, we go further to adaptively extract more discriminative local features as well as more accurately sense and measure the occlusions. We then propose an Attentional Occlusion-aware Network (AONet) with the Landmark Activation Layer and the Occlusion Awareness (OA) component. The latent landmark features refer to features of ReID oriented local parts (*i.e.*, latent landmarks), and are resistant to landmark occlusion. The occlusion awareness score measures the visibility of each landmark according to the average landmark features in the memorized dictionary. Besides, to prevent the model collapse problem that multiple landmarks focus on the same region, we involve the orthogonality constraints among landmarks features.

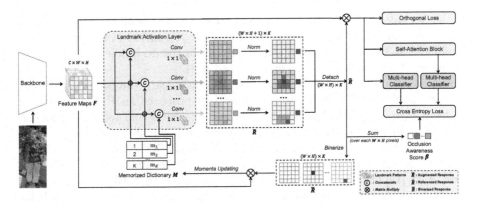

Fig. 2. The framework of AONet, including a Landmark Activation (LA) layer to extract the landmark features, and the Occlusion Awareness (OA) score to measure the occlusion. The responses of occluded pixels will be lower than the corresponding average response passing through the LA layer. Then, the normalization over all pixels and the average responses will further scale down these occluded responses (the green branch). Finally, the normalized pixel responses are summed up as the occlusion awareness score, and used to update the memorized dictionary. (Color figure online)

Our main contributions can be summarized as follows:

- Instead of relying on any external tasks, we only use a learnable parameter matrix (*i.e.*, the landmark patterns) and a memorized dictionary storing the average landmark features, to guide the extraction of landmark features that are more discriminative and resistant to occlusion.
- Furthermore, we define the occlusion awareness score to sense and measure the occlusion of each landmark explicitly, especially by referring to the average landmark features in the memorized dictionary.
- Our AONet achieves excellent performance on not only the occluded dataset Occluded-DukeMTMC, but also the holistic and partial datasets, *i.e.*, Duke-MTMC-reID, Market-1501, Partial-REID, and Partial iLIDS, significantly outperforming state-of-the-art.

2 Related Works

Person ReID has been studied in terms of both feature representation learning [22,29,33] and distance metric learning [1,8,23]. However, most ReID methods focus on matching the holistic pedestrian images, and do not perform well on occlusion images [13,24], which limits their applicability in real-world scenarios.

Occluded ReID [10,24] is aimed at matching occluded person images to holistic ones across dis-joint cameras, which is challenging due to distracting factors like cluttered scenes or dense crowd. To solve it, [7] proposed an occlusion-robust alignment-free model, using an occlusion-sensitive foreground probability generator with guidance from a semantic segmentation model. [13] refined the setup of the Occluded ReID problem to be more realistic, *i.e.*, both probe and gallery images containing occlusion. They introduced a PGFA method that exploits pose landmarks to disentangle useful information from the occlusion noise. Here, we tackle an Occluded ReID problem as defined in [13].

Later, [4] proposed a PVPM method to jointly learn the pose-guided local features and the self-mined part visibility. [24] proposed an HOReID method to learn high-order relations for discriminative features and topology information for robust alignment, by an external human key-points prediction model. In [32], a Semantic-aware Occlusion-Robust Network (SORN) was proposed that exploits the intrinsic relationship between person ReID and semantic segmentation. Also, [14] proposed a Semantic Guided Shared Feature Alignment (SGSFA) method to extract features focusing on the non-occluded parts, using guidance from external human parsing and pose estimation models. The above works require guidance information from external tasks (*e.g.*, semantic segmentation, pose estimation) either for local feature extraction or occlusion measurement. Recently, [10] presented the Matching on Sets (MoS) method, viewing Occluded ReID as a set matching task without requiring spatial alignment.

3 Attentional Occlusion-Aware Network

The Attentional Occlusion-aware Network (AONet) mainly includes the extraction of the attentional landmark features, and the calculation of the Occlusion Awareness (OA) score, as shown in Fig. 2. Meanwhile, a learnable matrix is used to explicitly represent the *landmark patterns* for the more discriminative features. A *memorized dictionary* is defined as a strong reference, which stores the average landmark features and is dynamically updated in a momentum way. The discriminative local features, *i.e.*, the landmark features, are extracted adaptively according to both the memorized dictionary and the landmark patterns.

3.1 Landmark Patterns and Memorized Dictionary

Landmark Patterns. We define the *landmark patterns* $I \in \mathbb{R}^{C \times K}$ as trainable parameters to attend to specific discriminative landmarks, *i.e.*, the attentional latent landmarks. We expect the learned *landmark patterns* to encode local patterns, which help explain the inputs (feature maps F).

Memorized Dictionary. We also define the *memorized dictionary* $M \in \mathbb{R}^{C \times K}$ to store the average features of the K latent landmarks. M is zero-initialized but momentum updated under the guidance of landmark patterns batch by batch. Moreover, the updating considers the occlusion of each landmark, *i.e.*, using the referenced response maps in the calculation of the occlusion awareness scores (see details in Sect. 3.4). Namely, given the *referenced response map* $\widehat{R}_k \in \mathbb{R}^{W \times H}$ for the k_{th} landmark, we binarize $\widehat{R}_k$ as $\tilde{R}_k \in \mathbb{R}^{W \times H}$ by setting all pixels corresponding to the maximum value to 1 and the rest to 0. Then, given a

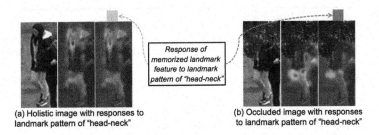

(a) Holistic image with responses to   (b) Occluded image with responses
landmark pattern of "head-neck"   to landmark pattern of "head-neck"

Fig. 3. Visualization of the effect *with* or *without* the reference of *average response* on the example pattern 'head-neck'. The use of *average response* has no particular impact on the holistic image in (a), but is effective in suppressing false alarms on the occluded image in (b). In (b), facing the occlusion of 'head-neck', the response map has false alarms on 'knees' without referring to average response (the 2nd image), but this gets obviously alleviated with the reference (the 3rd image).

batch of B images, we use momentum updating to get the updated memorized dictionary M_k^{t+1} with a balance weight (α) as:

$$M_k^{t+1} = \alpha M_k^t + (1 - \alpha)\frac{1}{B}\sum_{b=1}^{B} F_b \tilde{R}_k. \tag{1}$$

3.2 Attentional Latent Landmarks

The learnable landmark patterns $I = \{i_k\}_{k=1}^K, i_k \in \mathbb{R}^C$ should be trained together with other parameters of the network. The 1×1 convolution can be seen as an operation where a $1 \times 1 \times C$ sized filter is applied over the input and then weighted to generate several activation maps. That is, the 1×1 filter can be thought of as some type of pattern matching to create a linear projection of a stack of feature maps. Therefore, we realize the landmark patterns by the 1×1 filters, as shown in the Landmark Activation layer of Fig. 2.

In details, the 1×1 convolution layer appended after the CNN backbone network takes $F \in \mathbb{R}^{C \times H \times W}$ (feature maps of an input image) as input, and outputs K landmark-specific response maps $R = \{R_k\} \in \mathbb{R}^{K \times W \times H}$. We normalize these response maps among all pixels to form the basic normalized response maps $\check{R} \in \mathbb{R}^{K \times W \times H}$, then the value of pixel (w, h) in the k_{th} map is calculated as,

$$\check{R}_k(w, h) = \frac{\phi(i_k, F(w, h))}{\sum_{(i,j)=(1,1)}^{(W,H)} \phi(i_k, F(i,j))}, \tag{2}$$

and $\phi(i_k, F(i,j)) = \exp(i_k^T F(i,j))$ is the similarity based response.

After that, without considering occlusion awareness, we easily obtain the Standard Landmark-specific (SL) features of $\bar{f}_k \in \mathbb{R}^C$ for the k_{th} landmark by $\bar{f}_k = F\check{R}_k$. However, the SL features cannot accurately reflect the response of landmarks in the occluded image. As shown in Fig. 3(b), when the example landmark (seems to be the 'head-neck' parts) is occluded, this landmark still has

---

Algorithm 1. The Main Flowchart of the AONet.

Input: Batch of feature maps $\mathcal{F} = \{F^b \in \mathbb{R}^{W \times H \times C}\}_{b=1}^{B}$; landmark patterns $I = \{i_k\} \in \mathbb{R}^{C \times K}$; where B, C and K are the size of batch, channel and the number of landmarks.

Output: The awareness scores $\beta = \{\beta_k\}$ and updated memorized dictionary M;

1: Initialize the memorized dictionary $M_t = \{m_k\} \in \mathbb{R}^{C \times K}$;
   (*noting: superscript b is omitted until step 11 for convenience.*)
2: **for** $t = 1$ to T **do**
3:     Response maps $R = \phi(I, F) \in \mathbb{R}^{K \times W \times H}$, where $R_k \in \mathbb{R}^{W \times H}$ is the k_{th} landmark's response on F;
4:     Average responses $a = \{a_k\}$, where $a_k = \phi(i_k, m_k)$.
5:     Each augmented response map $\overline{R}_k = \{R_k; a_k\} \in \mathbb{R}^{W \times H+1}$.
6:     Normalizing $\overline{R}_k$ with Eq. 3.
7:     Referenced response map $\widehat{R}_k = \{\widehat{R}_k(w, h)\} \in \mathbb{R}^{W \times H}$, i.e., detaching the value corresponding to s_k.
8:     Calculating the awareness scores (*e.g.*, β_k^b) based on Eq. 4;
9:     The k_{th} OA feature $f_k = F\widehat{R}_k$.
10:    Binarization over $\widehat{R}_k$ to get binarized response map $\tilde{R}_k$.
11:    Updating the K memorized landmark features:
       $M_k^{t+1} = \alpha M_k^t + (1 - \alpha)\frac{1}{B}\sum_{b=1}^{B} F_b\tilde{R}$;
12: **end for**
13: **return** β and M respectively;

---

large activated regions (*i.e.*, the false alarm on the parts of 'knees'). Thus, we adopt the landmark features that characterize the occlusion awareness, *i.e.*, the OA features (see Sect. 3.4) instead of the SL features finally.

3.3   Referenced Response Map

Meanwhile, a special feature map, *i.e.*, the *referenced response map*, is defined to measure the occlusion awareness and represent the discriminative feature. We first calculate the similarity-based response between each landmark pattern (*e.g.*, $i_k \in I$) and its corresponding memorized average feature (*e.g.*, $m_k \in M$), which is named as the *average response* (*e.g.*, $a_k = \phi(x_k, m_k)$). While the memorized average features are the statistical representation of each landmark, the *average responses* can be used as some real and strong reference values to suppress false alarms, *e.g.*, scaling down responses of false alarms through uniform normalization, as shown in Fig. 3. More details can be seen in Algorithm 1.

Then, for an input image, given a landmark-specific response map $R_k = \phi(i_k, F)$, if all responses in R_k are significantly lower than the average response a_k, it means that this landmark is not present in this image, and the area corresponding to this landmark is occluded. Thus, we concatenate each a_k to the corresponding R_k to form the augmented response map $\overline{R}_k$, which is then normalized in a similar way as Eq. (2) (but on $W \times H + 1$ elements). That is, the normalized response of the k_{th} landmark pattern on the n_{th} pixel by the similarity function ϕ is calculated by

$$\widehat{\boldsymbol{R}}_k(w,h) = \frac{\phi(i_k, \boldsymbol{F}(w,h))}{a_k + \sum_{(i,j)=(1,1)}^{(W,H)} \phi(i_k, \boldsymbol{F}(i,j))}. \tag{3}$$

Given a landmark pattern i_k, pixel responses $\phi(i_k, x_n)$ that are far below the average response a_k are normalized to small values or even 0. Consequently, the responses of the occluded pixels, and pixels unrelated to the landmark patterns i_k are greatly suppressed, as shown in Fig. 3. Finally, we detach out the normalized response value of the average features, and rename the remained part as the *referenced response map*, i.e., $\widehat{\boldsymbol{R}}_k = \{\widehat{\boldsymbol{R}}_k(w,h)\} \in \mathbb{R}^{W \times H}$.

3.4 Occlusion Awareness

Although we do not have true annotations about the occlusion of each landmark, we can utilize the memorized dictionary $\boldsymbol{M}$ that stores the average feature of each landmark, as a special strong reference to measure the occlusion, as shown in Algorithm 1. Namely, the pixels that refer to a particular landmark should have a large response to this landmark's pattern. Namely, the feature of pixels referring to a particular landmark should be similar to the memorized average feature of that landmark, and also, both of them have comparable similarity based responses to the corresponding landmark pattern.

Occlusion Awareness Score. Ideally, if the regions referring to a landmark are occluded, there should be no responses to this landmark, and all pixels should not be used for learning any learnable landmark pattern. However, the network itself is not aware of the occlusion, and the responses of this landmark (*e.g.,* 'head-neck') will transfer to other unoccluded but wrong regions (*e.g.,* 'knees') to extract features (see Fig. 3). An intuitive idea for addressing this problem is to accurately measure the degree of occlusion by some metric (*i.e.,* awareness score), and then use it to suppress the impact of occluded regions in training and inference. Therefore, we further explicitly define an occlusion awareness score to measure the degree of occlusion based on the *referenced response map* $\widehat{\boldsymbol{R}}$. Specifically, we define the awareness score of the k_{th} landmark as,

$$\beta_k = \sum_{(1,1)}^{(W,H)} \widehat{\boldsymbol{R}}_k(w,h). \tag{4}$$

Then, β_k is used to reduce the weight of the occluded landmarks not only in training but also in inference.

Occlusion Awareness Feature. We need not only to sense the awareness score of each landmark but also involve such occlusion awareness in feature representation. Thus, the *referenced response maps* are also used as very crucial guidance to generate the more discriminative landmark features, *i.e.,* Occlusion Awareness (OA) features. Specifically, we replace $\check{\boldsymbol{R}}_k$ to the referenced response map $\widehat{\boldsymbol{R}}_k$, and get the k_{th} OA feature as $f_k = \boldsymbol{F}\widehat{\boldsymbol{R}}_k$.

3.5   Training and Inference

Training Losses. We use the cross-entropy loss weighted by the occlusion awareness to constrain each landmark. Specifically, we perform the classification loss on both the OA features before and after the self-attention block, *i.e.*,

$$L_{cls} = -\frac{1}{K}\left(\sum_{k=1}^{K} \beta_k \log p_k^1 + \sum_{k=1}^{K} \beta_k \log p_k^2\right), \tag{5}$$

where β_k is the k_{th} landmark's awareness score, and p_k^1 and p_k^2 are the predicted probability of the k_{th} landmark features before and after the self-attention block.

Without any other constraints but only the classification loss, different landmarks are easy to collapse to focus on the same part. Thus, we propose the *orthogonal loss* to ensure spatial diversity between landmark features. In detail, when the cosine similarity is calculated between two landmark features (f_i and f_j), the *orthogonal loss* is defined as

$$L_{ot} = -\frac{1}{K^2}\sum_{i=1}^{k}\sum_{j=i+1}^{K} \log(1 - |cosine(f_i, f_j)|_+). \tag{6}$$

where $|\cdot|_+$ means the ramp function, *i.e.*, $max(0, \cdot)$.

Finally, the overall objective function is formulated by

$$L_{AONet} = L_{cls} + \lambda_{ot} L_{ot}, \tag{7}$$

where L_{cls} is the cross-entropy based classification loss, and L_{ot} refers to the *orthogonal loss* among landmark features before the self-attention block, and λ_{ot} is the balance weight.

Inference. For inference, given a pair of images (im^1 and im^2) and their feature maps (F^1 and F^2), as well as their landmark features (*e.g.*, f_k^1 and f_k^2), their similarity is calculated based on the cosine similarity $cosine(\cdot)$ by

$$\text{sim}(im^1, im^2) = \frac{1}{K}\sum_{k=1}^{K} \beta_k^1 \beta_k^2 cosine(f_k^1, f_k^2), \tag{8}$$

where β_k^1 and β_k^2 are the occlusion awareness scores of the k_{th} landmark.

4   Experiments

4.1   Datasets and Implementations

We mainly evaluate AONet on the most popular occluded ReID dataset, *i.e.*, Occluded-DukeMTMC [13], where both the probe and gallery images have occlusion. In addition, we also experiment on holistic person ReID datasets: Market-1501 [34] and DukeMTMC-reID [15], as well as the partial ReID datasets: Partial-REID [35] and Partial-iLIDS [6]. All experiments are performed based on a single query image and without re-ranking [36].

Table 1. Comparison of performance on metrics of Ranks and mAP on the Occluded-DukeMTMC dataset.

Methods	Rank1	Rank5	Rank10	mAP
PGFA[ICCV19]	51.4	68.6	74.9	37.3
HOReID[cvpr20]	55.1	–	–	43.8
SORN[TCSVT20]	57.6	73.7	79.0	46.3
SGSFA[ACML20]	62.3	77.3	82.7	47.4
DIM[arXiv17]	21.5	36.1	42.8	14.4
PartAligned[ICCV17]	28.8	44.6	51.0	20.2
RandErasing[AAAI20]	40.5	59.6	66.8	30.0
HACNN[CVPR18]	34.4	51.9	59.4	26.0
AOS[CVPR18]	44.5	–	–	32.2
PCB[ECCV18]	42.6	57.1	62.9	33.7
PartBilinear[ECCV18]	36.9	–	–	–
FD-GAN[NeurIPS18]	40.8	–	–	–
DSR[CVPR18]	40.8	58.2	65.2	30.4
MoS[AAAI21]	61.0	–	–	49.2
AONet	**66.5**	**79.4**	**83.8**	**53.1**
MoS$_{w/ibn}$[AAAI21]	66.6	–	–	55.1
AONet$_{w/ibn}$	**68.8**	**81.4**	**85.8**	**57.3**

We use ResNet50 [5] pre-trained on ImageNet as the backbone network. For a fair comparison, we also incorporate the *instance batch normalization (ibn)* into ResNet50 (*i.e.*, AONet$_{w/ibn}$) as in [10]. To acquire high-resolution feature maps, the stride of conv4_1 is set to 1. We resize original images into 256×128, with a half probability of flipping them horizontally. Then, the images are padded by 10 pixels and randomly cropped back to 256×128, and then randomly erased with a half probability. We use the Adam optimizer [11] with a learning rate of $3.5e - 4$, warm up the training in the first 20 epochs and decay the learning rate with 0.1 in the $50th$ and $90th$ epoch. The batch size is 64, 4 images per person, and a total of 120 epochs are trained end-to-end. The weight of orthogonal loss, *i.e.*, λ_{ot}, is set to 0.01 and the momentum α for memorized dictionary updating is set to 0.9. If not specified, the number of landmarks is set as 6.

4.2 Comparisons to State-of-the-Arts

Results on Occluded ReID Dataset. Table 1 shows the performance of AONet and several competing methods, including *methods without external models*: DIM [31], PartAligned [33], RandErasing [37], HACNN [12], AOS [9], PCB [22], PartBilinear [19], FD-GAN [18], DSR [6], MoS [10], *methods with external models*: PGFA [13], SORN [32], SGSFA [14], HOReID [24], and the most related *set matching based method* [10], on the Occluded-DukeMTMC dataset.

Table 2. Performance comparison on Holistic Person ReID datasets of the Market-1501 and DukeMTMC-reID.

Methods	Market-1501		DukeMTMC-reID	
	Rank1	mAP	Rank1	mAP
PCB+RPP[ECCV18]	93.8	81.6	83.3	69.2
MGN[MM18]	95.7	86.9	88.7	78.4
VPM[CVPR19]	93.0	80.8	83.6	72.6
SORN[TCSVT20]	94.8	84.5	86.9	74.1
PDC[ICCV17]	84.2	63.4	–	–
PSE[CVPR18]	87.7	69.0	27.3	30.2
PGFA[ICCV19]	91.2	76.8	82.6	65.5
HOReID[CVPR20]	94.2	84.9	86.9	75.6
PartAligned[ICCV17]	81.0	63.4	–	-
HACNN[CVPR18]	91.2	75.6	80.5	63.8
CAMA[CVPR19]	94.7	84.5	85.8	72.9
MoS[AAAI21]	94.7	86.8	88.7	77.0
AONet	95.2	86.6	88.7	77.4

Our AONet shows a significant advantage over other methods. Note that our AONet uses no external models as in most of the previous works. Moreover, on Rank1 and mAP, AONet improves 4.2% and 5.7% over the SOTA method *SGSFA* (with external models). AONet also improves 5.5% and 3.9% over *MoS* (without external models). In MoS, a pre-trained backbone network IBN is used to achieve better results, so we also propose the **AONet**$_{w/ibn}$ utilizing IBN, which also achieves better results.

Results on Holistic ReID Datasets. Many related methods achieved good performance on Occluded ReID datasets, but they perform unsatisfactorily on holistic person ReID datasets and cannot be applied widely [13]. The AONet is also evaluated on the holistic person ReID datasets (*i.e.*, Market-1501 and DukeMTMC-reID) and compared with three groups of competing methods: *uniform-partition based* (PCB [22], VPM [21], MGN [25], *pose-guided based* (PDC [17], PSE [16], PGFA [13], HOReID [24]) and *attention-guided based* methods (PartAligned [33], HACNN [12], CAMA [28]). As shown in Table 2, the AONet produces satisfactory results in holistic cases even using an occluded oriented network. Meanwhile, the methods of different groups all perform well on holistic datasets and without large performance gaps. The reason could be, that almost all body parts are visible in holistic datasets, offering a greater possibility to locate all parts and thus obtain discriminative features easily. Meanwhile, AONet not only achieves SOTA performance on the occluded dataset, but also achieves competitive results on holistic datasets.

Results on Patial ReID Datasets. To fully validate the effectiveness of the AONet, we also conduct experiments on the partial person ReID datasets of

Table 3. Performance comparison on Partial ReID datasets of Partial-REID and Partial-iLIDS.

Methods	Partial-REID		Partial-iLIDS	
	Rank1	Rank3	Rank1	Rank3
DSR[CVPR18]	50.7	70.0	58.8	67.2
VPM[CVPR19]	67.7	81.9	65.5	84.7
HOReID[CVPR20]	**85.3**	91.0	72.6	86.4
AONet (*crop*)	85.0	**92.7**	68.1	84.9
PGFA[ICCV19]	68.0	80.0	69.1	80.9
SGSFA[ACML20]	68.2	–	–	–
SORN [TCSVT20]	76.7	84.3	79.8	86.6
AONet (*whole*)	75.3	86.3	**80.7**	**86.6**

Partial-REID and Partial-iLIDS. Since these two datasets are always only used as the test dataset, existing works are trained on other ReID dataset (*e.g.*, Market-1501). Not only that, but existing works also use the partial ReID dataset in two different ways (the two groups in Table 3), the main difference being whether the visible pedestrian area is cropped out separately as a new image. For example, the SOTA methods of HOReID [24] and SORN [32] are evaluated on partial datasets with images of the whole pedestrian or the cropped visible parts, respectively. We use **AONet**(*whole*) and **AONet**(*crop*) to refer to the performance of AONet on partial datasets in these two ways. As shown in Table 3, our **AONet**(*whole*) and **AONet**(*crop*) achieve the best performance on datasets of Patial-REID and Partial-iLIDS respectively, which proved the efficiency of our approach.

4.3   Ablation Studies

We test the effect of components in AONet with the below variants on Occluded-DukeMTMC: i) *SLFea* (Standard Landmark-specific features), extracting landmark features without occlusion awareness. ii) *SAtt* (Self-Attention), enabling information interaction between landmark features. iii) *OAFea* (Occlusion Awareness Features), being similar to *SLFea*, but referring to the memorized features. iv) *OAScore* (Occlusion Awareness Score), measuring the occlusion degree of each landmark. v) *OLoss* (orthogonal loss), constraining over different pairs of landmark features.

With or Without Each Component. We define a basic baseline *Base_GAP*, which includes the backbone of ResNet, a Global Average Pooling (GAP) layer, and a softmax layer. As shown in Table 4, compared to *Base_GAP*, utilizing *SLFea* achieves significantly better performance, reflecting the advantage of set matching over global feature matching. Utilizing *SAtt* gains an extra improvement of 1.3% on Rank1 and 1.4% on mAP. Besides, with the simple combination of *SLFea* and *SAtt*, Rank1 of 58.6% is achieved, better than most previous methods as shown in Table 1. We argue that our attentional landmarks would facilitate reconstructing the information of the occluded landmark using other landmarks.

Table 4. The ablation study of the components in AONet.

Method	SLFea	SAtt	OAFea	OAScore	OLoss	Rank1	mAP
Base_GAP						49.4	40.0
Base_SLFea	✓					57.3	43.9
Base_SAtt	✓	✓				58.6	45.3
AONet[†]		✓	✓			62.6	51.3
AONet[‡]		✓	✓	✓		65.6	52.5
AONet		✓	✓	✓	✓✓	**66.5**	**53.1**

Table 5. Ablation study of the components compared to baselines on the Occluded-DukeMTMC dataset.

Methods	Rank1	Rank5	Rank10	mAP
Base_GAP	49.4	63.7	68.9	40.0
Base_Pose	52.1	66.2	71.1	42.3
Base_OAFea (AONet[†])	62.6	77.1	81.6	51.3
AONet[†]	62.6	77.1	81.6	51.3
+Base_Max	63.4	77.3	82.1	51.7
+OAScore (AONet[‡])	65.6	79.1	83.6	52.5
AONet[‡]	65.6	79.1	83.6	52.5
+Base_RegL	65.9	78.8	83.7	52.7
+OLoss (AONet)	**66.5**	**79.4**	**83.8**	**53.1**

When we replace *SLFea* with *OAFea* that incorporates occlusion awareness, we obtain an improvement on Rank1 (4%) and mAP (6%) (Base_SLFea *vs* AONet[†]). Meanwhile, the involving of *OAScore* achieves an extra improvement on Rank1 (3%) and mAP (1.2%) (AONet[†] *vs* AONet[‡]), which means both *OAScore* and *OAFea* effectively mitigate feature mislocalization caused by occlusion. Furthermore, utilizing *OLoss* does work very effectively with an improvement of 0.9% and 0.6% on Rank1 and mAP respectively.

Comparisons to Various Baselines. Firstly, we construct three comparable baselines: i) *Base_Pose* refers to method using external pose features [20] as the pedestrian representation. ii) *Base_Max* means directly choosing the maximum value in normalized responses without referring to memory. iii) *Base_RegL* is the method of position regularization loss [26].

As shown in Table 5, the method utilizing landmark features of *OAFea* obviously gains better performance. Thus, it is indeed crucial to involve occlusion awareness in landmark representation. Meanwhile, the performance improves obviously by comprehensively weakening the occluded landmark features, no matter the method with *OAScore* or Base_Max. That is, a reasonable awareness of occlusion indeed brings better performance. However, simply taking the maximum value cannot accurately sense occlusion, but the *OAScore* referring

Table 6. The performance of AONet with different numbers of landmarks on Occluded Person ReID.

Methods	Rank1	Rank5	Rank10	mAP
AONet (K = 2)	64.3	79.7	84.5	52.3
AONet (K = 4)	65.2	79.0	84.3	53.4
AONet (K = 6)	66.5	79.4	83.8	53.1
AONet (K = 8)	66.2	79.6	83.9	53.0
AONet (K = 10)	65.7	79.8	84.3	52.7

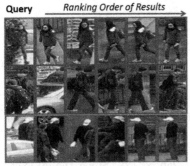

(a) Visualization of retrieval results.   (b) Visualization of landmarks.

Fig. 4. (a) Visualization of retrieval results. The 1st image in each row is the query, and the next five images are returned images with descending ranking. Green and red rectangles indicate correct and error results. (b) Visualization of landmarks in the original pedestrian image. Each column of images refers to the visualization of a specific landmark by its corresponding response map. (Color figure online)

to memorized features can effectively handle this problem with 2.2% and 0.8% improvements on both Rank1 and mAP.

Besides, we evaluate the comparable efficiency of *OLoss* by the orthogonal loss and *RegL* with the position regularization loss [26]. As shown in Table 5, the *OLoss* does show obvious improvements on performance. The reason may be that, the position regularization loss, while enabling different landmarks to indicate different regions, does not guarantee attention to select discriminative landmark features, which is however what our orthogonal constraint is good at.

Influence of Number of Landmarks. As shown in Table 6, the performance improves at first as the number of landmarks increases, possibly because more local features provide more robustness to occlusion. However, too many landmarks lead the network to focus on more fine-grained local features, or even background noise, which lacks sufficient discrimination for identification.

Visualization Analysis. We visualize the image retrieval results of the AONet approach in Fig. 4(a). We get the correct image by AONet for both horizontal

Table 7. Comparison of costs with the state-of-the-arts.

Methods	FLOPs(G)	#Params(M)
PGFA[ICCV19]	29.51	57.51
HOReID[CVPR20]	35.80	109.23
SORN[TCSVT20]	24.73	41.96
SGSFA[ACML20]	16.13	47.71
MoS[AAAI21]	12.57	24.22
AONet	6.22	30.22

and vertical occlusion, as well as to object and pedestrian occlusion. However, when the effective region is too small, the retrieval easily makes mistakes. We also visualize the landmark response map. As shown in Fig. 4(b), each landmark focuses on a different unique semantic pattern.

Cost Evaluation. To more clearly quantify the advantages of our model over other state-of-the-art models. As shown in Table 7, we compare the number of model parameters "#Param" and floating-point operations "FLOPs", where FLOPs are calculated at an input size of 256×128. Since the AONet does not use any additional models, such as models of pose estimation and semantic segmentation, it has a smaller time and space overhead. Besides, our AONet has good parallel computing properties while not relying on any additional model, so it is computed at the fastest speed.

5   Conclusion

Previous works for occluded person ReID often rely on external tasks, *e.g.*, pose estimation or semantic segmentation, to extract local features over fixed given regions. In this paper, we propose an end-to-end Attentional Occlusion-aware Network (AONet), including a Landmark Activation layer to extract the landmark features, and an Occlusion Awareness (OA) score to explicitly measure the occlusion. Without any external information by extra tasks, we adaptively extract discriminate anti-occlusion local features with the landmark patterns. The OA is the focus of this paper, on the one hand, providing occlusion reference information to prevent landmark patterns from focusing on the wrong region, and on the other hand, to generate occlusion awareness scores to reduce the weight of the occluded landmark features in classification loss and image matching.

Acknowledgements. This work was supported by the National Natural Science Foundation of China under Grant No. 61972036.

References

1. Chen, D., Xu, D., Li, H., Sebe, N., Wang, X.: Group consistent similarity learning via deep crf for person re-identification, pp. 8649–8658 (2018)
2. Chen, T., et al.: Abd-net: attentive but diverse person re-identification, pp. 8351–8361 (2019)
3. Gao, L., Zhang, H., Gao, Z., Guan, W., Cheng, Z., Wang, M.: Texture semantically aligned with visibility-aware for partial person re-identification, pp. 3771–3779 (2020)
4. Gao, S., Wang, J., Lu, H., Liu, Z.: Pose-guided visible part matching for occluded person reid, pp. 11744–11752 (2020)
5. He, K., Zhang, X., Ren, S., Sun, J.: Deep residual learning for image recognition, pp. 770–778 (2016)
6. He, L., Liang, J., Li, H., Sun, Z.: Deep spatial feature reconstruction for partial person re-identification: alignment-free approach, pp. 7073–7082 (2018)
7. He, L., Wang, Y., Liu, W., Zhao, H., Sun, Z., Feng, J.: Foreground-aware pyramid reconstruction for alignment-free occluded person re-identification, pp. 8450–8459 (2019)
8. Hermans, A., Beyer, L., Leibe, B.: In defense of the triplet loss for person re-identification. arXiv preprint arXiv:1703.07737 (2017)
9. Huang, H., Li, D., Zhang, Z., Chen, X., Huang, K.: Adversarially occluded samples for person re-identification, pp. 5098–5107 (2018)
10. Jia, M., Cheng, X., Zhai, Y., Lu, S., Ma, S., Zhang, J.: Matching on sets: conquer occluded person re-identification without alignment (2021)
11. Kingma, D.P., Ba, J.L.: Adam: a method for stochastic gradient descent, pp. 1–15 (2015)
12. Li, W., Zhu, X., Gong, S.: Harmonious attention network for person re-identification, pp. 2285–2294 (2018)
13. Miao, J., Wu, Y., Liu, P., Ding, Y., Yang, Y.: Pose-guided feature alignment for occluded person re-identification, pp. 542–551 (2019)
14. Ren, X., Zhang, D., Bao, X.: Semantic-guided shared feature alignment for occluded person re-identification, pp. 17–32 (2020)
15. Ristani, E., Solera, F., Zou, R., Cucchiara, R., Tomasi, C.: Performance measures and a data set for multi-target, multi-camera tracking, pp. 17–35 (2016)
16. Sarfraz, M.S., Schumann, A., Eberle, A., Stiefelhagen, R.: A pose-sensitive embedding for person re-identification with expanded cross neighborhood re-ranking, pp. 420–429 (2018)
17. Su, C., Li, J., Zhang, S., Xing, J., Gao, W., Tian, Q.: Pose-driven deep convolutional model for person re-identification, pp. 3960–3969 (2017)
18. Suh, Y., Wang, J., Tang, S., Mei, T., Lee, K.M.: Fd-gan: pose-guided feature distilling gan for robust person re-identification, pp. 1229–1240 (2018)
19. Suh, Y., Wang, J., Tang, S., Mei, T., Lee, K.M.: Part-aligned bilinear representations for person re-identification, pp. 402–419 (2018)
20. Sun, K., Xiao, B., Liu, D., Wang, J.: Deep high-resolution representation learning for human pose estimation, pp. 5693–5703 (2019)
21. Sun, Y., et al.: Perceive where to focus: learning visibility-aware part-level features for partial person re-identification, pp. 393–402 (2019)
22. Sun, Y., Zheng, L., Yang, Y., Tian, Q., Wang, S.: Beyond part models: person retrieval with refined part pooling (and a strong convolutional baseline), pp. 480–496 (2018)

23. Tian, M., et al.: Eliminating background-bias for robust person re-identification, pp. 5794–5803 (2018)
24. Wang, G., et al.: High-order information matters: learning relation and topology for occluded person re-identification, pp. 6449–6458 (2020)
25. Wang, G., Yuan, Y., Chen, X., Li, J., Zhou, X.: Learning discriminative features with multiple granularities for person re-identification, pp. 274–282 (2018)
26. Xie, W., Shen, L., Zisserman, A.: Comparator networks, pp. 782–797 (2018)
27. Xu, J., Zhao, R., Zhu, F., Wang, H., Ouyang, W.: Attention-aware compositional network for person re-identification, pp. 2119–2128 (2018)
28. Yang, W., Huang, H., Zhang, Z., Chen, X., Huang, K., Zhang, S.: Towards rich feature discovery with class activation maps augmentation for person re-identification, pp. 1389–1398 (2019)
29. Yao, H., Zhang, S., Hong, R., Zhang, Y., Xu, C., Tian, Q.: Deep representation learning with part loss for person re-identification. IEEE Trans. Image Process. **28**(6), 2860–2871 (2019)
30. Ye, M., Shen, J., Lin, G., Xiang, T., Shao, L., Hoi, S.C.: Deep learning for person re-identification: a survey and outlook. IEEE Trans. Pattern Anal. Mach. Intell. **44**, 2872–2893 (2021)
31. Yu, Q., Chang, X., Song, Y.Z., Xiang, T., Hospedales, T.M.: The devil is in the middle: exploiting mid-level representations for cross-domain instance matching (2018)
32. Zhang, X., Yan, Y., Xue, J.H., Hua, Y., Wang, H.: Semantic-aware occlusion-robust network for occluded person re-identification. IEEE Trans. Circ. Syst. Video Technol. **31**(7), 2764–2778 (2020)
33. Zhao, L., Li, X., Zhuang, Y., Wang, J.: Deeply-learned part-aligned representations for person re-identification, pp. 3219–3228 (2017)
34. Zheng, L., Shen, L., Tian, L., Wang, S., Wang, J., Tian, Q.: Scalable person re-identification: a benchmark, pp. 1116–1124 (2015)
35. Zheng, W.S., Li, X., Xiang, T., Liao, S., Lai, J., Gong, S.: Partial person re-identification, pp. 4678–4686 (2015)
36. Zhong, Z., Zheng, L., Cao, D., Li, S.: Re-ranking person re-identification with k-reciprocal encoding, pp. 1318–1327 (2017)
37. Zhong, Z., Zheng, L., Kang, G., Li, S., Yang, Y.: Random erasing data augmentation, vol. 34, pp. 13001–13008 (2020)

FFD Augmentor: Towards Few-Shot Oracle Character Recognition from Scratch

Xinyi Zhao⃝, Siyuan Liu⃝, Yikai Wang⃝, and Yanwei Fu$^{(\boxtimes)}$⃝

School of Data Science, Fudan University, Shanghai, China
{xyzhao19,liusiyuan19,yikaiwang19,yanweifu}@fudan.edu.cn

Abstract. Recognizing oracle characters, the earliest hieroglyph discovered in China, is recently addressed with more and more attention. Due to the difficulty of collecting labeled data, recognizing oracle characters is naturally a Few-Shot Learning (FSL) problem, which aims to tackle the learning problem with only one or a few training data. Most current FSL methods assume a disjoint but related big dataset can be utilized such that one can transfer the related knowledge to the few-shot case. However, unlike common phonetic words like English letters, oracle bone inscriptions are composed of radicals representing graphic symbols. Furthermore, as time goes, the graphic symbols to represent specific objects were significantly changed. Hence we can hardly find plenty of prior knowledge to learn without negative transfer. Another perspective to solve this problem is to use data augmentation algorithms to directly enlarge the size of training data to help the training of deep models. But popular augment strategies, such as dividing the characters into stroke sequences, break the orthographic units of Chinese characters and destroy the semantic information. Thus simply adding noise to strokes perform weakly in enhancing the learning capacity.

To solve these issues, we in this paper propose a new data augmentation algorithm for oracle characters such that (1) it will introduce informative diversity for the training data while alleviating the loss of semantics; (2) with this data augmentation algorithm, we can train the few-shot model from scratch without pre-training and still get a powerful recognition model with superior performance to models pre-trained with a large dataset. Specifically, our data augmentation algorithm includes a B-spline free form deformation method to randomly distort the strokes of characters but maintain the overall structures. We generate 20–40 augmented images for each training data and use this augmented training set to train a deep neural network model in a standard pipeline. Extensive experiments on several benchmark datasets demonstrate the effectiveness of our augmentor. Code and models are released in https://github.com/Hide-A-Pumpkin/FFDAugmentor.

This paper is the final project of Neural Network and Deep Learning (DATA130011.01, Course Instructor: Dr. Yanwei Fu; TA: Yikai Wang), School of Data Science, Fudan University.

L. Wang et al. (Eds.): ACCV 2022, LNCS 13845, pp. 37–53, 2023.
https://doi.org/10.1007/978-3-031-26348-4_3

Keywords: Oracle character recognition · Few-shot learning · Data augmentation · Free form deformation

1 Introduction

Oracle inscriptions are one of the earliest pictographs in the world. Dating back to Shang Dynasty [1], ancient Chinese carved oracle characters on animal bones or tortoise shells to record the present and divine the future. Due to the gradual abandonment of oracle characters and the loss of oracle samples in the long history, we now can only discover very limited oracle bone inscriptions, and most of the characters are incompletely preserved. What is worse, as oracle bone inscriptions were created by people of different ethnic groups in different regions and were written on nail plates of various shapes and curvatures, the oracle characters are hard to recognize and distinguish from each other.

In the early time, archaeologists [1] could identify some widely used and easily identifiable characters. Then, with recognition models, researchers [2–4] can identify some new characters after training on annotated oracle characters. However, the ultimate goal of fully understanding the oracle inscription system is far from attainable since there remain a lot of undecipherable oracle pieces. For those deciphered words, mostly we only collect very limited characters [5] that is far from enough to train a powerful recognition model. To go further, we can formulate the oracle recognition task as a Few-Shot Learning (FSL) [6,7] problem targeted at recognizing oracle characters in the limited data case.

FSL is a popular machine learning topic that aims to train a learning model with only a few data available. Based on the motivation to solve the limited training data issue, we can roughly classify FSL algorithms into the following categories: (1) Learning to learn, or meta-learning algorithms [8–14], aims to train the few-shot model to learn the capacity of learning from a few examples by simulating the few-shot scenario in the training stage; (2) Data-augmentation algorithms [5,15,16] directly expand the training set by generating new data based on the few training samples; (3) Semi-supervised algorithms [17–23] have the access to the additional unlabeled dataset and try to utilize this unlabeled knowledge to help train the few-shot model.

However, most FSL algorithms assume the existence of related large labeled datasets to pre-train the few-shot model. But as we do not have such datasets, these algorithms are not suitable for the oracle character recognition problem. Current algorithms to tackle the oracle character recognition problem in the FSL pipeline mainly focus on data-augmentation, including using hierarchical representation [3], dividing characters into structure and texture components [24], converting characters to stroke vectors [5].Nevertheless, they still fail to solve the problem of limited training data for their under-utilization of structured information and the mining of stroke information is limited and costly.

In this paper, we propose a new data augmentation approach by adopting Free Form Deformation (FFD) [25] which is initially used in the field of non-rigid registration [26]. FFD deforms the image by manipulating an underlying

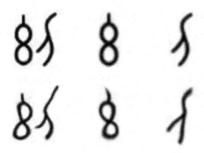

Fig. 1. The oracle character 'Ci' (in the left column) is made up of two parts, an eight-like character (in the middle column) and a knife-like radical (in the right column). After the FFD transformation (from top to bottom), though the strokes are distorted, the overall structure remains unchanged. We can still re-recognize the character 'Ci' based on them.

mesh of control points [25] and calculates the displacement rule for each pixel in the image individually. When it is applied to oracle character images, each pixel that makes up a stroke moves by their each displacement rule, which leads to the distortion of strokes and the corruption of local information. Meanwhile, the two adjacent pixels' displacement rules are similar in general, which maintains the consistency of radicals and stability of the global information (See Fig. 1 as an example). By corrupting local information while maintaining global information, FFD well preserves the radical features of oracle bone inscriptions and randomly distorts stokes, making the augmented image more representative.

With this FFD Augmentor, we can now tackle the few-shot oracle characters recognition by utilizing the online Free Form Deformation algorithm as our augmentor and generating a bunch of augmented data from each annotated training instance for each category. As the generated training data is of high quality and diversity, we now make it possible to train the few-shot model in a standard supervised manner with our generated data *from scratch*. To better show the effectiveness of our FFD Augmentor, we select a powerful few-shot training method called EASY [27], which is composed of widely used training modules without specifically designed algorithms, and train the model with the generated data in a standard pipeline to ensure that our proposed algorithm can be utilized as a general module for the oracle recognition task. Extensive experiments on benchmark datasets, Oracle-FS [5] and HWOBC [28], verify the effectiveness of our proposed algorithm. We further conduct experiments on a sketch dataset [29] to show that the effectiveness of our augmentor is not limited to the oracle characters.

The main contributions of our work are as follows:

(1) To the best of our knowledge, we are the first to apply the non-rigid transformation, namely FFD, to the field of data augmentation.

(2) Our generated training data are diverse and informative such that the deep model can be trained with generated data from scratch without the help of an additional large unlabeled or labeled dataset.

(3) We demonstrate the effectiveness of our approach through comprehensive experiments on Oracle-FS and HWOBC datasets, reaching a significant accuracy improvement over competitors.

2   Related Works

2.1   Oracle Character Recognition

Here we mainly survey machine learning algorithms for oracle character recognition. Conventional researches [30–32] regarded the characters as a non-directional graph and utilized its topological properties as features for graph isomorphism. Guo et al. [3] proposed a hierarchical representation and released a dataset of hand-printed oracle characters from 261 different categories.

In the deep learning era, neural networks gradually became the mainstream recognition method. Yang et al. [33] studied several different networks and suggested using the Gabor method for feature extraction. Zhang et al. [34] utilized the nearest-neighbor classifier to reject unseen categories and configure new categories. Other models like capsule network [4] and generative model [24] are also proposed to solve this task.

However, few researchers focus on few-shot learning of oracle character recognition. Orc-Bert [5] converts character images to stroke vectors and learns the stoke features from large unlabeled source data. The method requires large unlabeled source data and the generated augmented stroke data performs average compared with the benchmark. Instead, we show that a single oracle character is informative enough to train the recognition model via our proposed data augmentation algorithm without pretraining on related large datasets.

2.2   Few-Shot Learning

Few-Shot Learning (FSL) [35,36] aims to train a machine learning model with only a few training data. Gidaris et al. [6] created an attention-based few-shot classification weight generator with a small number of gradient steps. MAML [8] searched for the best initial weights to accelerate the learning process reducing the risk of over-fitting. Chen et al. [37] applied self-supervised learning in a generalized embedding network to provide a robust representation for downstream tasks. LEO [38] reduced the complexity by learning a low-dimension model embedding and used the nearest neighbor criterion to classify. Chen et al. [39] generated multiple features at different scales and selected the most important local representations among the entire task under the complex background.

In this paper, we used the EASY [27] framework, which combined several simple approaches like backbone training, and featured vectors projection in the above literature, and reached the state-of-art performance in training without huge computation cost.

2.3   Data Augmentation Approaches

Traditional data augmentation methods are mainly focused on rigid transformation such as flipping [40], rotation [9], shearing [41] and cropping [40]. Blending or occluding [42] parts of images is also widely used, but they require expert domain knowledge to prevent critical information from being disrupted. In the context of few-shot learning, some efficient data augmentation algorithms like AutoAugment [43] are not applicable for the lack of large-scale labeled data.

Distinguished from the traditional image, the character image is regarded as an intermediate between text and image. As an image, it retains textual features, such as the ability to be reconstructed into a sequence of strokes. Han et al. [5] captured the stroke order information of Chinese characters and utilized the pre-trained Sketch-Bert model to augment few-shot labeled oracle characters. Yue et al. [44] designed a dynamic dataset augmentation method using a Generative Adversarial Network to solve the data imbalance problem. However, none of the state-of-art data augmentation approaches addressed the problem of oracle bone characters recognition from the overall structure.

As the ancestor of Chines characters, the oracle character is at least a logographic language and the upper-level orthographic units, like radicals, should contain richer information. Simply introducing noise to strokes cannot enhance or may even weaken the model's ability to recognize radical components.

2.4   Non-Rigid Transformation

Non-Rigid transformation is a critical technique in the field of image registration, which focuses on finding the optimal transformation from one input image to another [26]. In the medical field, extensive research has been conducted on non-Rigid transformation for its essential role in analyzing medical effects over time. Existing non-rigid transformation techniques include AIR [45], Diffeomorphic Demons [46], and FFD [47].

FFD is a commonly used algorithm for image registration [25]. The registration from moving image to fixed image is modeled by the combined motion of a global affine transformation and a free-form deformation based on B-splines. Compared with rigid transformation, FFD has a higher degree of freedom which can better model the motion between two images. When applied to data augmentation, this can also bring greater flexibility to enlarge the training dataset.

3   Methodology

3.1   Problem Formulation

Here we define the few-shot oracle character learning problem without related large datasets for pre-training, thus degenerating to the standard supervised learning manner. We are provided with labeled dataset, $\mathcal{D}$, which comprises the category set $\mathcal{C}$, $|\mathcal{C}| = n$. For a certain k-shot learning task, our augmentor and classifier would only have access to k annotated training instances for each

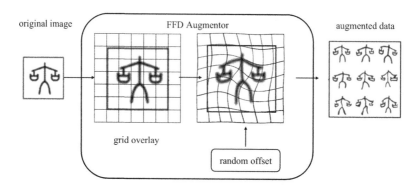

Fig. 2. Illustration of our FFD Augmentor. We generate random offsets for each control point and construct a deformed image based on the recalculation of the world coordinates of each vertex of the image.

category. We randomly sample k and q images for each category $C_i \in \mathcal{C}, i = 1, ...n$ to construct the training set $\mathcal{S}$ and the evaluation set $\mathcal{Q}$ respectively. We aim to train on $\mathcal{S}$ and generalize to $\mathcal{Q}$. We take accuracy on $\mathcal{Q}$ as the evaluation metric.

3.2   Overview of Framework

As shown in Fig. 2, our data augmentation method, FFD Augmentor, consists of several parts. For each image, we first create a local coordinate system by splitting the whole image into local patches using grids. The grid vertexes are used as the control points to define the local position of the neighboring pixels. Then random offsets are generated to shift the control points, thus shifting the neighboring pixels. Due to random offsets being used to shift control points, the whole image is modified with destroyed local information.

Using our proposed FFD Augmentor, we generate several augmented data for each training image and then store them to expand the training set. With this expanded training set, we now can train the few-shot model from scratch.

3.3   FFD Augmentor

Though there are few studies on the composition of oracle characters, as the ancestor of Chinese characters [48], oracle characters intuitively retain similar characteristics. For example, compared with strokes, radicals contain richer information. This motivates us to perform a data augmentation algorithm to generate local diversity while preserving global structures.

In non-rigid registration, researchers [25,49,50] applies FFD to achieve this goal by calculates displacement rule for each pixel in the image individually. When it is applied to oracle character images, each pixel that makes up a stroke moves according to their each displacement rules. Thus it leads to the distortion of strokes and the corruption of local information. On the other hand, thanks to the continuity of these features and physical rules, the displacement rules for two adjacent pixels are similar in general. Globally, the relative positions of radicals remain consistent and the character's structure is well preserved. See Fig. 1 for an illustration. Hence in this paper, we adopt FFD to generate new training data for few-shot oracle character recognition problem.

Free Form Deformation. As the oracle characters can be represented as a grayscale image, we implement the 2D version of Free Form Deformation based on B-splines [25]. Specifically, for the oracle character grayscale image $x \in \mathbb{R}^{h \times w}$, we design a two-dimensional mapping

$$T : (x_1, x_2) \quad \rightarrow \quad (x_1', x_2'), \tag{1}$$

to simulate the non-rigid transformation. We decouple T by a global deformation mapping and a local deformation mapping as:

$$T(x_1, x_2) = T_{\text{global}}(x_1, x_2) + T_{\text{local}}(x_1, x_2). \tag{2}$$

The global deformation mapping is a simple affine transformation defined as:

$$T_{\text{global}}(x_1, x_2) = \begin{pmatrix} \theta_{11} & \theta_{12} \\ \theta_{21} & \theta_{22} \end{pmatrix} \begin{pmatrix} x_1 \\ x_2 \end{pmatrix} + \begin{pmatrix} \theta_{13} \\ \theta_{23} \end{pmatrix}, \tag{3}$$

while the local deformation mapping is the major concern in our algorithm. Specifically, we first distribute a series of grid points over the image at a certain spacing based on the predetermined patch number. Denote the area of the oracle character grayscale image as $\Omega = \{(x_1, x_2) \mid 0 \le x_1 \le X_1, 0 \le x_2 \le X_2\}$, we split it by control points $\Phi = \{\phi_{i,j}\}$ into several patches of size $n_1 \times n_2$, where n_i is the distance between adjacent control points in the i-th dimension. Then we can define the local deformation mapping as the product of B-splines functions:

$$T_{\text{local}}(x_1, x_2) = \sum_{l=0}^{3} \sum_{m=0}^{3} B_t(u) B_m(v) \phi_{i+l, j+m}, \tag{4}$$

where $i = \lfloor x_1/n_1 \rfloor - 1, j = \lfloor x_2/n_2 \rfloor - 1, u = x_1/n_1 - \lfloor x_1/n_1 \rfloor, v = x_2/n_2 - \lfloor x_2/n_2 \rfloor$, and the B-splines functions are defined as:

$$B_0(u) = \frac{(1-u)^3}{6}, B_1(u) = \frac{3u^3 - 6u^2 + 4}{6},$$
$$B_2(u) = \frac{-3u^3 + 3u^2 + 3u + 1}{6}, B_3(u) = \frac{u^3}{6}. \tag{5}$$

Then the augmentation comes when we randomly apply offsets with a pre-defined range $O = [O_{\min}, O_{\max}]$ to shift the control points. For a specific grid

---

Algorithm 1. FFD Augmentor

Require: Training Image x, patch size of $n_1 \times n_2$, offset range O.
Ensure: Augmented Image.
1: Initialize the new image with empty pixel value: $\tilde{x}$;
2: For each control point C, random initialize its degree of shift based on Eq. (6);
3: **for** all pixel $x_i = (x_{i,1}, x_{i,2})$ in x **do**
4:     Generate the new coordinates $\tilde{x}_i = (\tilde{x}_{i1}, \tilde{x}_{i2})$ based on Eq. (4);
5: Use bi-linear interpolation algorithm to complete unvisited pixels in $\tilde{x}$.
6: **return** $\tilde{x}$

---

point $\phi_{i,j} = (x_{\phi_i}, x_{\phi_j})$, we randomly initialize its degree of shift within the offset range.

$$\mathrm{T} : (x_{\phi_i}, x_{\phi_j}) \quad \rightarrow \quad (x_{\phi_i} + \Delta x_{\phi_i}, x_{\phi_j} + \Delta x_{\phi_j}), \quad \Delta x_{\phi_i}, \Delta x_{\phi_j} \in O. \quad (6)$$

Then for each pixel (x_1, x_2) within the image, we calculate the deformed location based on Eq. (4) with the shifted control points deformed in Eq. (6). After the displacement transformation rules for all pixels are determined, we finally re-sample the image by pixels according to their rules to achieve non-rigid deformation. If the transformed pixel coordinates exceed the image size, the grayscale value will be replaced by 255. Finally, for the unvisited pixels in the generated image, we use a bi-linear interpolation algorithm [51] to fill these empty holes.

Augmentor. As mentioned in Sect. 1, non-rigid transformation can destroy local information while maintaining global information. By generating multiple FFD-augmented training samples, the model extracts and learns the structured information of the oracle characters, rather than relying on some particular strokes to classify the character. This is critical in the task of few-shot oracle character recognition since it will alleviate the problem of bias and overfitting caused by the limited training samples.

Algorithm 1 illustrates the detailed pseudo-code for our FFD Augmentor. An ablation study about the number of FFD-augmented training samples and the selection of FFD hyperparameters will be further discussed in Sect. 4.3.

3.4   Training with FFD Augmentor

To show the effectiveness of the FFD Augmentor, we adopt a popular training algorithm, the Ensemble Augmented-Shot Learning method (EASY) [27] to combine with our proposed FFD Augmentor.

Specifically, we test our algorithm on several widely used CNN architectures, including ResNet-12 [52], ResNet-18, ResNet-20, and WideResNet [53], respectively. For the FFD augmented training set, we also apply standard data augmentation strategies, including cropping, flipping, and color jittering. When training, each mini-batch is divided into two parts: the first part is input to the

Table 1. Accuracy (%) of oracle characters recognition on **Oracle-FS** under all three few-shot settings with classifiers ResNet18 including Orc-Bert, EASY with and without FFD augmentor. Because we only share the ResNet18 classifier in common with Orc-Bert, we compare with their best-performance method, i.e. Orc-Bert Augmentor with point-wise displacement on ResNet18.

k-shot	Orc-Bert	EASY	FFD + EASY
1	31.9	55.17	**78.90**
3	57.2	79.45	**93.36**
5	68.2	90.34	**96.47**

Table 2. Accuracy (%) of oracle characters recognition on **Oracle-FS** under all three few-shot settings with different architectures. The Basic model is the pure model without any augmentation method involved.

k-shot	model	Basic	FFD+Basic	EASY	FFD + EASY
1	ResNet12	14.95	33.95	58.46	**76.79**
	ResNet18	21.22	26.47	55.17	**78.90**
	ResNet20	8.61	30.28	54.67	**77.09**
	WideResNet	15.35	33.85	53.77	**77.75**
3	ResNet12	33.06	68.83	84.91	**92.89**
	ResNet18	35.90	57.25	79.45	**93.36**
	ResNet20	32.28	66.08	82.53	**92.57**
	WideResNet	40.23	71.12	84.81	**93.42**
5	ResNet12	54.53	77.92	91.30	**95.38**
	ResNet18	46.70	70.00	90.34	**96.47**
	ResNet20	55.82	77.60	91.55	**95.26**
	WideResNet	60.62	78.38	92.43	**97.59**

standard classifier with the feature augmentation strategy Manifold-MixUp [54]; the second part is with the rotation transformation and input to both heads. For details on training the EASY model, we suggest to read the original paper [27].

4 Experiments

We conduct extensive experiments to validate the effectiveness of our FFD augmentor and provide ablation studies to analysis each part of our algorithm.

4.1 Experimental Settings

Datasets. We demonstrate the effectiveness of our FFD Augmentor on Oracle-FS [5] and HWOBC [28]. Oracle-FS contains 200 oracle character categories. We run experiments on 3 different few-shot settings, including k-shot for $k = 1, 3, 5$ where for each category we only have access to k labeled training data. To evaluate the performance, we randomly select 20 instances to construct the

Table 3. Accuracy (%) of oracle characters recognition on **HWOBC** under 1-shot setting with different architectures.

Model	Basic	FFD+Basic	EASY	FFD + EASY
ResNet12	14.92	55.72	65.24	**98.53**
ResNet18	15.73	38.43	41.85	**98.97**
ResNet20	7.65	54.63	52.32	**98.60**
WideResNet	10.2	63.37	62.16	**99.52**

testing dataset for each category. HWOBC consists of 3881 oracle character categories, each containing 19 to 25 image samples. We randomly selected 200 categories, and each category is divided into 1-shot training sets and 15-sample test sets. Because the accuracy of our model in the 1-shot setting is high enough, we did not test on more k-shot settings.

Competitors. We mainly compare our results with Orc-Bert [5], the SOTA algorithm for the few-shot oracle character recognition task. Orc-Bert masks some strokes, predicts them by a pre-trained model and make additional noise on each stroke to generate multiple augmented images. We also train EASY [27] for both with FFD augmentor and without FFD augmentor to compare the results.

Implementation Details. We implement FFD Augmentor training methods using PyTorch [55]. The number of training epochs is 100 with a batch size of 64. Unless otherwise specified, We follow the hyper-parameters for training EASY in their default settings. We conducted experiments with the FFD Augmentor of 5 patch num, 11 max offset, and 30 augmented samples. For 1-shot and 3-shot, we used a cosine scheduled learning rate beginning with 0.1, while we used a learning rate of 0.01 for 5-shot. The images are resized to 50×50.

4.2   Evaluation of FFD Augmented Training

It can be clearly noticed in Table 1 that our FFD-based data augmentor can defeat the state-of-the-art method by more than 30% under all few-shot settings. Also can be seen in Table 2, our data augmentation method plays a decisive role in improving accuracy.

On Oracle-FS, our 1-shot accuracy reaches 76.5% for all the classifiers, which outperforms EASY without being augmented by 20%. Our 3-shot accuracy achieved 93.42%, exceeding the accuracy of all the 5-shot models without FFD augmentation. For the 5-shot setting, our model's accuracy reaches 97.59% on WideResNet.

On the HWOBC dataset, the effect of our data augmentation tool is more prominent. We compare the result with EASY and Conventional Data Augmentation. As seen from the Table 3, our FFD augmentor improves the accuracy from the original 65.24% to 99.52% for the 1 shot setting.

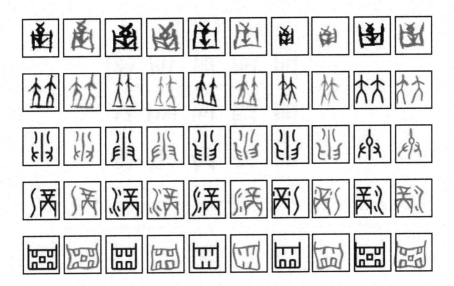

Fig. 3. Examples of oracle character images and the FFD-augmented samples.

4.3 Further Analysis of FFD Augmentor

Visualization. Here we provide more visualization of the oracle characters and FFD augmented images in Fig. 3. Clearly, for all kinds of oracle characters, the FFD Augmentor will consistently generate diverse and informative images. Hence we can provide many realistic augmented images to alleviate the lack of training data in the few-shot oracle character recognition task.

Ablation Study. In this part, we conduct more experiments to evaluate our FFD augmentor, including the min-max offset, the number of patches, and augmented samples. These experiments are running with ResNet18. To make our results more accurate, all experiments on hyperparameters are conducted twice and we average the results of the two experiments as the final accuracy.

(1)Max Offset Value: In our FFD augmentor, the random offset value was generated through a uniform distribution in the interval between the minimal and the maximal offset value. The maximal offset value is set as a hyperparameter while we negate it as the minimal offset. They together limit the movement range of the offset. The closer they are, the smaller the deformation of the image. In our experiment, we tested the maximum value from 0 to 15. Max offset value of 0 indicates that no free form transformation is performed on the original image, i.e., as an ablation experiment for our data augmenter.

(2)Num of Patches: The number of patches influences the number of control points of the grid. The more patches in the FFD transformation, the more transformation control points there are, and the more complex the deformation can be. In our realization, we test the number of patches from 3, 5 to 7. Considering the effectiveness of the transformation and the time overhead, we did not test

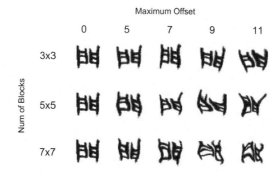

Fig. 4. Illustration of num of patches and maximum offset difference.

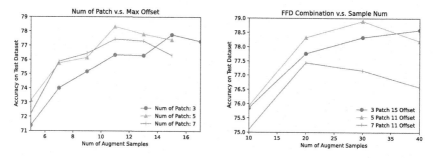

Fig. 5. Left: Different combination of Patch numbers and max offsets varies in accuracy. **Right:** Num of Augmented Sample influences the model accuracy.

on more patch nums. The time of generating the augmented dataset is proportional to the square of patch nums. When the number of patches increases to a large amount, the enhancement effect of the picture is not obvious but will take considerable time, much longer than the training cost.

Take Fig. 4 as an example, when the max offset value and num of patches are both limited, the deformation is closer to rigid transformation and the overall shape of the text remains unchanged. But when the num of patches and maximum offset value becomes too large, the deformation is so complex that the overall structure is severely damaged and the oracle character is hard to identify. As max offset value and num of patches influence the deformation process in different ways, we experimented on the trade-off between max offset value and num of patches. The results are shown in Fig. 5. With the rise of the max offset value, the accuracy of all patch nums increases. However, their accuracy begins to decrease when the offset exceeds a certain threshold. As can be seen, the top three accuracy combinations of patch Num and max offset is $(3,11)$, $(5,15)$ and $(7,11)$.

Table 4. Top-1 Accuracy under different learning rates in the 5-shot task.

Learning rate	0.1	0.05	0.01	0.00	0.001
Accuracy	95.82	96.31	**96.47**	89.08	87.05

Table 5. Top-1 Accuracy under different support samples with no FFD augmentation.(k-shot=1).

Aug-sample	1	5	10	20	40
Accuracy	52.94	53.39	**53.52**	53.40	53.36

Num of Augmented Sample. We then experimented on the num of augmented samples to figure out whether the larger num of augmented samples contributes to the better performance of our model. Here, we mean generate num of augmented images for each training image. For example, under the 3-shot setting, if the num of Augmented Sample equals 30, we generate 90 augmented images.

Intuitively, with more augmented samples, the accuracy will be higher. However, the results of experiments (See Fig. 5) show that due to the limited number of samples, too many augmented images will lead to overfitting, i.e., the test accuracy will become lower with the decrease of training loss. Two FFD combinations show an increasing trend followed by a decreasing trend and the growth trend is also diminishing for the combination of 3 patches and 15 offset value. When the number of augmented samples equals 30, the FFD combination of 5 patch num and 11 max offset reaches the maximum accuracy.

Besides, the computation time of data augmentation is a crucial factor to be considered. FFD is a time-intensive transformation that increases with the size of the image and the number of patches. The flaw of FFD is less fatal in few-shot learning for a small number of images. Our FFD augmentor takes about 0.4 to 0.5s to generate each image of size 50×50 for 5 patch num. Due to the expensive time cost for data augmentation, we trade off both performance and computation time. Combining all the results above, we find that with 5 patch num, 11 max offset and 30 augment samples, our model achieves the best performance of 78.9% in 1-shot.

Learning Rate. Different learning rates affect the convergence speed and accuracy of the model. A low learning rate may cause the model to overfit the training dataset or converge too slowly. A high learning rate may prevent the model from convergence. We experimented on the influence of different learning rates under different k-Shot settings. As shown in Table 4, for $k=5$, learning rate of 0.01 reaches the highest accuracy.

Num of Augmented Samples Using Random Crop. Our experiments also test the number of image samples processed by random crop and flip before backbone training. The results in Table 5 show that the accuracy rate is highest when the size is 10, and there is a risk of overfitting when the size is larger.

4.4   Applicability to Other Problems

Though our paper is intended to tackle oracle character recognition, the innovative augmentor we proposed has much broader applications. To better demonstrate its versatility, we here report more experiments on the sketch recognition task. The task takes a sketch s as input and predicts its category label c. We did the toy experiment on the sketch dataset [29], which contains 20,000 unique sketches evenly distributed over 250 object categories, with the same dataset setting as before. The experiment results in Table 6 demonstrate superior performance after adopting the FFD Augmentor, which is 15% higher than the EASY without FFD augmentation. More applications of the augmentor will be explored in future works.

Table 6. Accuracy (%) of sketch recognition on the sketch dataset. The FFD Augmentor is with 5 patch num, 11 max offset, and 30 augmented samples.

k-shot	Model	EASY	FFD + EASY
1	ResNet12	35.38	**49.75**
	ResNet18	38.51	**54.27**
	ResNet20	30.48	**53.91**
	WideResNet	40.27	**55.17**
3	ResNet12	61.72	**71.38**
	ResNet18	55.03	**73.38**
	ResNet20	62.21	**75.23**
	WideResNet	59.22	**76.04**
5	ResNet12	72.98	**80.76**
	ResNet18	67.00	**81.77**
	ResNet20	69.49	**80.87**
	WideResNet	70.87	**84.38**

5   Conclusion

We address the task for oracle character recognition with a few labeled training samples in this study. We propose a new data augmentation tool for few-shot oracle recognition problems, FFD Augmentor, which is based on the free form deformation method commonly used the registration. FFD Augmentor generates a series of augmented images by random FFD on the original images for the classifier for training. Numerous experiments in three few-shot scenarios support the efficacy of our FFD Augmentor. Our generated training data are so efficient and informative that the deep model can be trained with generated data from scratch, without any additional large unlabeled dataset for pretraining. Our model has broad prospects in the field of written character recognition field. A wider range of applications will be explored in future studies.

References

1. Keightley, D.N.: Graphs, words, and meanings: three reference works for Shang oracle-bone studies, with an excursus on the religious role of the day or sun. J. Am. Oriental Soc. **117**, 507–524 (1997)
2. Xing, J., Liu, G., Xiong, J.: Oracle bone inscription detection: a survey of oracle bone inscription detection based on deep learning algorithm. In: Proceedings of the International Conference on Artificial Intelligence, Information Processing and Cloud Computing. AIIPCC 2019, New York, NY, USA, Association for Computing Machinery (2019)
3. Guo, J., Wang, C., Roman-Rangel, E., Chao, H., Rui, Y.: Building hierarchical representations for oracle character and sketch recognition. IEEE Trans. Image Process. **25**, 104–118 (2015)
4. Lu, X., Cai, H., Lin, L.: Recognition of oracle radical based on the capsule network. CAAI Trans. Intell. Syst. **15**, 243–254 (2020)
5. Han, W., Ren, X., Lin, H., Fu, Y., Xue, X.: Self-supervised learning of Orc-Bert augmentor for recognizing few-shot oracle characters. In: Ishikawa, H., Liu, C.-L., Pajdla, T., Shi, J. (eds.) ACCV 2020. LNCS, vol. 12627, pp. 652–668. Springer, Cham (2021). https://doi.org/10.1007/978-3-030-69544-6_39
6. Vinyals, O., Blundell, C., Lillicrap, T., Wierstra, D., et al.: Matching networks for one shot learning. In: Advances in Neural Information Processing Systems 29 (2016)
7. Snell, J., Swersky, K., Zemel, R.: Prototypical networks for few-shot learning (2017)
8. Finn, C., Abbeel, P., Levine, S.: Model-agnostic meta-learning for fast adaptation of deep networks. In: International conference on machine learning, pp. 1126–1135. PMLR (2017)
9. Santoro, A., Bartunov, S., Botvinick, M., Wierstra, D., Lillicrap, T.: Meta-learning with memory-augmented neural networks. In: Balcan, M.F., Weinberger, K.Q. (eds.) Proceedings of The 33rd International Conference on Machine Learning. Volume 48 of Proceedings of Machine Learning Research., New York, New York, USA, PMLR, pp. 1842–1850 (2016)
10. Oreshkin, B., López, P.R., Lacoste, A.: TADAM: task dependent adaptive metric for improved few-shot learning (2018)
11. Sung, F., Yang, Y., Zhang, L., Xiang, T., Torr, P.H., Hospedales, T.M.: Learning to compare: Relation network for few-shot learning (2018)
12. Nichol, A., Achiam, J., Schulman, J.: On first-order meta-learning algorithms. arXiv preprint arXiv:1803.02999 (2018)
13. Rusu, A.A., et al.: Meta-learning with latent embedding optimization (2019)
14. Mishra, N., Rohaninejad, M., Chen, X., Abbeel, P.: A simple neural attentive meta-learner (2018)
15. Chen, Z., Fu, Y., Chen, K., Jiang, Y.G.: Image block augmentation for one-shot learning. Proceed. AAAI Conf. Artif. Intell. **33**, 3379–3386 (2019)
16. Antoniou, A., Storkey, A., Edwards, H.: Augmenting image classifiers using data augmentation generative adversarial networks. In: Kůrková, V., Manolopoulos, Y., Hammer, B., Iliadis, L., Maglogiannis, I. (eds.) ICANN 2018. LNCS, vol. 11141, pp. 594–603. Springer, Cham (2018). https://doi.org/10.1007/978-3-030-01424-7_58
17. Ren, M., et al.: Meta-learning for semi-supervised few-shot classification (2018)
18. Liu, Y., et al.: Learning to propagate labels: transductive propagation network for few-shot learning (2019)
19. Li, X., et al.: Learning to self-train for semi-supervised few-shot classification (2019)

20. Wang, Y., Xu, C., Liu, C., Zhang, L., Fu, Y.: Instance credibility inference for few-shot learning. In: IEEE/CVF Conference on Computer Vision and Pattern Recognition (CVPR) (2020)
21. Wang, Y., Zhang, L., Yao, Y., Fu, Y.: How to trust unlabeled data? instance credibility inference for few-shot learning. IEEE Transactions on Pattern Analysis and Machine Intelligence (2021)
22. Yalniz, I.Z., Jégou, H., Chen, K., Paluri, M., Mahajan, D.: Billion-scale semi-supervised learning for image classification. arXiv preprint arXiv:1905.00546 (2019)
23. Springenberg, J.T.: Unsupervised and semi-supervised learning with categorical generative adversarial networks. arXiv preprint arXiv:1511.06390 (2015)
24. Wang, M., Deng, W., Liu, C.L.: Unsupervised structure-texture separation network for oracle character recognition. IEEE Trans. Image Process. **31**, 3137–3150 (2022)
25. Rueckert, D., Sonoda, L., Hayes, C., Hill, D., Leach, M., Hawkes, D.: Nonrigid registration using free-form deformations: application to breast MR images. IEEE Trans. Med. Imaging **18**, 712–721 (1999)
26. Keszei, A.P., Berkels, B., Deserno, T.M.: Survey of non-rigid registration tools in medicine. J. Digit. Imaging **30**, 102–116 (2017)
27. Bendou, Y., et al.: Easy: ensemble augmented-shot y-shaped learning: State-of-the-art few-shot classification with simple ingredients. arXiv preprint arXiv:2201.09699 (2022)
28. Li, B., Dai, Q., Gao, F., Zhu, W., Li, Q., Liu, Y.: HWOBC-a handwriting oracle bone character recognition database. J. Phys: Conf. Ser. **1651**, 012050 (2020)
29. Eitz, M., Hays, J., Alexa, M.: How do humans sketch objects? ACM Trans. Graph. **31**, 1–10 (2012)
30. Zhou, X.L., Hua, X.C., Li, F.: A method of Jia Gu wen recognition based on a two-level classification. In: Proceedings of 3rd International Conference on Document Analysis and Recognition, vol. 2, pp. 833–836. IEEE (1995)
31. Li, F., Woo, P.-Y.: The coding principle and method for automatic recognition of Jia Gu wen characters. Int. J. Human-Comput. Stud. **53**(2), 289–299 (2000)
32. Qingsheng, L.: Recognition of inscriptions on bones or tortoise shells based on graph isomorphism. Computer Engineering and Applications (2011)
33. Yang, Z., et al.: Accurate oracle classification based on deep convolutional neural network. In: 2018 IEEE 18th International Conference on Communication Technology (ICCT), pp. 1188–1191 (2018)
34. Zhang, Y.K., Zhang, H., Liu, Y.G., Yang, Q., Liu, C.L.: Oracle character recognition by nearest neighbor classification with deep metric learning. In: 2019 International Conference on Document Analysis and Recognition (ICDAR), pp. 309–314 (2019)
35. Sung, F., Yang, Y., Zhang, L., Xiang, T., Torr, P.H., Hospedales, T.M.: Learning to compare: relation network for few-shot learning. In: Proceedings of the IEEE Conference on Computer Vision and Pattern Recognition, pp. 1199–1208 (2018)
36. Qiao, S., Liu, C., Shen, W., Yuille, A.L.: Few-shot image recognition by predicting parameters from activations. In: Proceedings of the IEEE Conference on Computer Vision and Pattern Recognition, pp. 7229–7238 (2018)
37. Chen, D., Chen, Y., Li, Y., Mao, F., He, Y., Xue, H.: Self-supervised learning for few-shot image classification. In: ICASSP 2021–2021 IEEE International Conference on Acoustics, Speech and Signal Processing (ICASSP), pp. 1745–1749. IEEE (2021)
38. Rusu, A.A., et al.: Meta-learning with latent embedding optimization. arXiv preprint arXiv:1807.05960 (2018)

39. Chen, H., Li, H., Li, Y., Chen, C.: Multi-scale adaptive task attention network for few-shot learning. CoRR arXiv:abs/2011.14479 (2020)

40. Qi, H., Brown, M., Lowe, D.G.: Low-shot learning with imprinted weights. In: Proceedings of the IEEE Conference on Computer Vision and Pattern Recognition (CVPR) (2018)

41. Shyam, P., Gupta, S., Dukkipati, A.: Attentive recurrent comparators. In: Precup, D., Teh, Y.W. (eds.) Proceedings of the 34th International Conference on Machine Learning. Volume 70 of Proceedings of Machine Learning Research, pp. 3173–3181. PMLR (2017)

42. DeVries, T., Taylor, G.W.: Improved regularization of convolutional neural networks with cutout. arXiv preprint arXiv:1708.04552 (2017)

43. Cubuk, E.D., Zoph, B., Mane, D., Vasudevan, V., Le, Q.V.: AutoAugment: learning augmentation strategies from data. In: Proceedings of the IEEE/CVF Conference on Computer Vision and Pattern Recognition, pp. 113–123 (2019)

44. Yue, X., Li, H., Fujikawa, Y., Meng, L.: Dynamic dataset augmentation for deep learning-based oracle bone inscriptions recognition. J. Comput. Cult. Herit. **15**, 3532868 (2022)

45. Woods, R.P., Grafton, S.T., Holmes, C.J., Cherry, S.R., Mazziotta, J.C.: Automated image registration: I. general methods and intrasubject, intramodality validation. J. Comput. Assist. Tomogr. **22**, 139–152 (1998)

46. Vercauteren, T., Pennec, X., Perchant, A., Ayache, N.: Diffeomorphic demons: efficient non-parametric image registration. Neuroimage **45**, S61–S72 (2009)

47. Oliveira, F.P., Tavares, J.M.R.: Medical image registration: a review. Comput. Methods Biomech. Biomed. Engin. **17**, 73–93 (2014). PMID: 22435355

48. Ziyi, G., et al.: An improved neural network model based on inception-v3 for oracle bone inscription character recognition. Scientific Programming (2022)

49. Jiang, J., Luk, W., Rueckert, D.: FPGA-based computation of free-form deformations in medical image registration. In: Proceedings. 2003 IEEE International Conference on Field-Programmable Technology (FPT)(IEEE Cat. No. 03EX798), pp. 234–241. IEEE (2003)

50. Rohlfing, T., Maurer, C.R.: Nonrigid image registration in shared-memory multiprocessor environments with application to brains, breasts, and bees. IEEE Trans. Inf Technol. Biomed. **7**, 16–25 (2003)

51. Gribbon, K., Bailey, D.: A novel approach to real-time bilinear interpolation. In: Proceedings. DELTA 2004. Second IEEE International Workshop on Electronic Design, Test and Applications, pp. 126–131 (2004)

52. He, K., Zhang, X., Ren, S., Sun, J.: Deep residual learning for image recognition. In: Proceedings of the IEEE Conference on Computer Vision and Pattern Recognition, pp. 770–778 (2016)

53. Zagoruyko, S., Komodakis, N.: Wide residual networks. CoRR arXiv:abs/1605.07146 (2016)

54. Mangla, P., Kumari, N., Sinha, A., Singh, M., Krishnamurthy, B., Balasubramanian, V.N.: Charting the right manifold: manifold mixup for few-shot learning. In: Proceedings of the IEEE/CVF Winter Conference on Applications of Computer Vision, pp. 2218–2227 (2020)

55. Paszke, A., et al.: PyTorch: an imperative style, high-performance deep learning library. In: Advances in Neural Information Processing Systems 32, pp. 8024–8035. Curran Associates, Inc. (2019)

Few-shot Metric Learning: Online Adaptation of Embedding for Retrieval

Deunsol Jung, Dahyun Kang, Suha Kwak, and Minsu Cho[✉]

Pohang University of Science and Technology (POSTECH), Pohang, South Korea
{deunsol.jung,dahyun.kang,suha.kwak,mscho}@postech.ac.kr

Abstract. Metric learning aims to build a distance metric typically by learning an effective embedding function that maps similar objects into nearby points in its embedding space. Despite recent advances in deep metric learning, it remains challenging for the learned metric to generalize to unseen classes with a substantial domain gap. To tackle the issue, we explore a new problem of few-shot metric learning that aims to adapt the embedding function to the target domain with only a few annotated data. We introduce three few-shot metric learning baselines and propose the *Channel-Rectifier Meta-Learning* (CRML), which effectively adapts the metric space online by adjusting channels of intermediate layers. Experimental analyses on *mini*ImageNet, CUB-200-2011, MPII, as well as a new dataset, *mini*DeepFashion, demonstrate that our method consistently improves the learned metric by adapting it to target classes and achieves a greater gain in image retrieval when the domain gap from the source classes is larger.

1 Introduction

The ability of measuring a reliable distance between objects is crucial for a variety of problems in the fields of artificial intelligence. Metric learning aims to learn such a distance metric for a type of input data, *e.g.*, images or texts, that conforms to semantic distance measures between the data instances. It is typically achieved by learning an embedding function that maps similar instances to nearby points on a manifold in the embedding space and dissimilar instances apart from each other. Along with the recent advance in deep neural networks, deep metric learning has evolved and applied to a variety of tasks such as image retrieval [31], person re-identification [6] and visual tracking [34]. In contrast to conventional classification approaches, which learn category-specific concepts using explicit instance-level labels for predefined classes, metric learning learns the general concept of distance metrics using relational labels between samples in the form of pairs or triplets. This type of learning is natural for information retrieval, *e.g.*, image search, where the goal is to return instances that are most similar to a query, and is also a powerful tool for open-set problems where we match or classify instances of totally new classes based on the learned metric.

Supplementary Information The online version contains supplementary material available at https://doi.org/10.1007/978-3-031-26348-4_4.

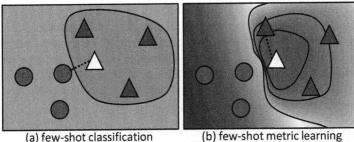

(a) few-shot classification   (b) few-shot metric learning

Fig. 1. Few-shot classification vs. few-shot metric learning. (a) A few-shot classifier learns to construct a decision boundary between support examples of different classes (red and blue). The query (white) is correctly classified as the red class, but fails to be placed away from blue samples. (b) A *few-shot metric learner* learns improved distance relations by adapting the embedding space online using the support examples, where the contour line represents the points with the same distance from the query.

For this reason, metric learning has focused on generalization to unseen classes, that have never been observed during training [15,31,44]. Despite recent progress of deep metric learning, however, it remains exceedingly challenging for the learned embedding to generalize to unseen classes with a substantial domain gap.

To bridge the generalization gap in metric learning, we investigate the problem of few-shot metric learning that aims to *adapt an embedding function on the fly* to target classes with only a few annotated data. While the problem of few-shot learning has been actively studied for classification [21,30], the problem for metric learning has never been directly investigated so far to the best of our knowledge. Few-shot metric learning and classification problems share the goal of adapting to scarce labeled data, but diverge in terms of their training objectives and evaluation protocols, thus requiring different approaches. As illustrated in Fig. 1, few-shot classification focuses on forming a decision boundary between samples of different classes and often fails to measure the relative distance relations between the samples, which are crucial for a retrieval task. Our analysis also shows that the improvement of classification accuracy does not necessarily lead to that of retrieval accuracy (Table 7). While one of the main approaches to few-shot classification is to learn a metric space for nearest-neighbor classification to generalize to unseen classes [7,18,21,24,33,38,39,45–47], it exploits the learned metric in testing without adapting the embedding space to the few-shot instances online. In this sense, metric-based few-shot classification is very different from few-shot metric learning that we investigate in this paper.

In this work we introduce three baselines for few-shot metric learning by adapting existing methods and propose the Channel-Rectifier Meta-Learning (CRML), which effectively adapts the metric space online by adjusting channels of intermediate layers. We compare them to conventional metric learning as well as few-shot classification counterparts on *mini*ImageNet [30], CUB-200-2011 [40], and MPII [23]. We also introduce a new multi-attribute dataset for image retrieval, dubbed *mini*DeepFashion, where two instances may have an opposite similarity relationship over multiple attributes (perspectives), and thus embedding functions are required to adapt to target attributes. Experiments show that CRML as well as the three baselines significantly improve image retrieval quality by adapting the model to target classes. Notably, such improvement is significant when the gap between training and testing domains arises, which conventional metric learning often fails to bridge.

2   Related Work

2.1   Metric Learning

Metric learning has been studied extensively in the past decades [32], and has shown a great success recently using deep embedding networks trained with new losses. One of the widely studied losses is the pair-wise loss [8,31,35,43] which minimizes the distance between two instances that have the same label and separates them otherwise. Such losses include contrastive loss [4,8,12], triplet loss [31,41], lifted structured loss [35], and multi-similarity loss [43]. Unlike pair-based losses, proxy-based losses [2,19,28,29] associate proxy embeddings with each training class as a part of learnable parameters and learn semantic distance between an instance and the proxies. These previous methods, which we refer to as *conventional* metric learning, emphasize the generalization performance on unseen classes that have never been observed during training. However, they often suffer from a significant gap between source and target classes [27]. Although it is very practical to utilize a few labeled data from the target classes on the fly, online adaptation of the metric has never been explored so far to the best of our knowledge. Recently, Milbich *et al.* [27] showcase the effect of few-shot adaptation as a mean of out-of-distribution deep metric learning, their work does not present a problem formulation and a method dedicated to few-shot learning while our work does both of them. In many practical applications of metric learning, a metric can also be learned with continuous labels of similarity , which are more informative but costly to annotate [11,20,23,36]. Online adaptation of metric learning may be particularly useful for such a scenario where we need to adapt the metric to the target classes with their few yet expensive labels.

2.2   Few-shot Classification

Few-shot learning has been actively investigated for classification problems, and recent work related to ours is roughly categorized into three types: metric-based, optimization-based, and transfer-learning methods. The key idea of metric-based methods [7,18,21,24,33,38,39,45–47] is to learn an embedding space via episodic training so that the class membership of a query is determined based on its nearest class representations in the embedding space. Although the metric-based few-shot classification and few-shot metric learning share the terminology "metric", they clearly differ from each other in terms of their learning objectives and practical aspects. While metric-based few-shot classifiers construct a decision boundary using a learned metric without online adaptation of embeddings, few-shot metric learners learn improved metric function by adapting the embeddings online. In this aspect, N-way 1-shot retrieval task, proposed in Triantafillou *et al.* [38], is different from our task, few-shot metric learning. The N-way 1-shot retrieval task in [38] does not perform any online adaptation of embedding in the inference time; thus, it is exactly the same as the conventional deep metric learning. In this work, we focus on instance retrieval problems on an adaptive embedding space of a few examples, while class discrimination is out of our interest.

The optimization-based few-shot classification methods [10,30,37] learn how to learn a base-learner using a few annotated data. While two aforementioned lines of work follow meta-learning frameworks, recent studies suggest that the standard transfer learning is a strong baseline for few-shot classification [5,9, 16,25,42]. Such transfer learning methods pre-train a model using all available training classes and leverage the model for testing.

The contribution of this paper is four-fold:

- We introduce a new problem of *few-shot metric learning* that aims to adapt an embedding function to target classes with only a few annotated data.
- We present three few-shot metric learning baselines and a new method, Channel-Rectifier Meta-Learning (CRML), which tackles the limitations of the baselines.
- We extensively evaluate them on standard benchmarks and demonstrate that the proposed methods outperform the conventional metric learning and few-shot classification approaches by a large margin.
- We introduce *mini*DeepFashion, which is a challenging multi-attribute retrieval dataset for few-shot metric learning.

3   Few-shot Metric Learning

The goal of few-shot metric learning is to learn an embedding function for target classes with a limited number of labeled instances. In this section, we first revisit conventional metric learning, and then introduce the problem formulation and setup of few-shot metric learning.

3.1   Metric Learning Revisited

Let us assume data $\mathcal{X}$ of our interest, *e.g.*, a collection of images. Given an instance $x \in \mathcal{X}$, we can sample its positive example x^+, which is from the same class with x, and its negative example x^-, which belongs to a different class from x. The task of metric learning is to learn a distance function d such that

$$\forall(x, x^+, x^-), \ d(x, x^+; \theta) < d(x, x^-; \theta). \tag{1}$$

Deep metric learning solves the problem by learning a deep embedding function $f(\cdot, \theta)$, parameterized by θ, that projects instances to a space where the

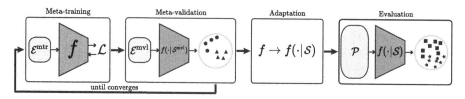

Fig. 2. The problem formulation of few-shot metric learning as an episodic training.

Euclidean distance is typically used as a distance function d:

$$d(x, x'; \theta) = \| f(x; \theta) - f(x'; \theta) \|_2^2. \qquad (2)$$

Note that metric learning focuses on unseen class generalization. The conventional setup for metric learning [35] assumes a set of training classes $\mathcal{C}^{\mathrm{tr}}$ and its dataset $\mathcal{D}^{\mathrm{tr}} = \{(x_t, y_t) | y_t \in \mathcal{C}^{\mathrm{tr}}\}_t$ that contains labeled instances x_t of the training classes. The task of metric learning then is to learn an embedding model using the dataset $\mathcal{D}^{\mathrm{tr}}$ so that it generalizes to a dataset of *unseen classes* $\mathcal{D}^{\mathrm{un}} = \{(x_u, y_u) | y_u \in \mathcal{C}^{\mathrm{un}}\}_u$, which contains instances from the classes not observed in training, *i.e.*, $\mathcal{C}^{\mathrm{tr}} \cap \mathcal{C}^{\mathrm{un}} = \emptyset$.

3.2   Problem Formulation of Few-shot Metric Learning

In contrast to the conventional metric learning, few-shot metric learning uses annotated instances of the target classes (thus, not totally unseen any more) but only a few instances per class, which is reasonable in most real-world scenarios. For simplicity and fair comparison, we applied N-way K-shot setting. Let us assume a target class set $\mathcal{C}$, which contains N classes of our interest, and its *support set* $\mathcal{S} = \{(x_s, y_s) | y_s \in \mathcal{C}\}_{s=1}^{NK}$, which contains K labeled instances for each of the target classes; K is supposed to be small. The task of N-way K-shot metric learning is to learn an embedding model using the support set $\mathcal{S}$ so that it generalizes to a *prediction set* $\mathcal{P} = \{(x_p, y_p) | y_p \in \mathcal{C}\}_{p=1}^{M}$, which contains M unseen instances, *i.e.*, $\mathcal{S} \cap \mathcal{P} = \emptyset$, from the target classes. The generalization performance is evaluated using instance retrieval on $\mathcal{P}$.

Our goal is to learn such a few-shot metric learning model using a set of labeled instances from *non-target* classes, which can be viewed as a meta-learning problem. In training a few-shot metric learning model, we thus adopt the episodic training setup of meta-learning [10,30,33,39] as illustrated in Fig. 2. In this setup, we are given a meta-train set $\mathcal{D}^{\mathrm{mtr}}$ and a meta-validation set $\mathcal{D}^{\mathrm{mvl}}$. They both consist of labeled instances from non-target classes but their class sets, $\mathcal{C}^{\mathrm{mtr}}$ and $\mathcal{C}^{\mathrm{mvl}}$, are disjoint, *i.e.*, $\mathcal{C} \cap \mathcal{C}^{\mathrm{mtr}} = \mathcal{C} \cap \mathcal{C}^{\mathrm{mvl}} = \mathcal{C}^{\mathrm{mtr}} \cap \mathcal{C}^{\mathrm{mvl}} = \emptyset$.

A meta-train episode is constructed from $\mathcal{D}^{\mathrm{mtr}}$ by simulating a support set and its prediction set; (1) a support set $\mathcal{S}^{\mathrm{mtr}}$ is simulated by sampling N classes from $\mathcal{C}^{\mathrm{mtr}}$ then K instances for each of the N classes, and (2) a prediction set $\mathcal{P}^{\mathrm{mtr}}$, which is disjoint from $\mathcal{S}^{\mathrm{mtr}}$, is simulated by sampling other K' instances for each of the N classes. A meta-validation episode is constructed from $\mathcal{D}^{\mathrm{mvl}}$

likewise. The meta-trained few-shot metric learning model is tested on meta-test set $\mathcal{E} = \{(\mathcal{S}_m, \mathcal{P}_m)\}_m$ from the target classes $\mathcal{C}$.

4   Methods

We introduce three baselines for few-shot metric learning (Sect. 4.1–4.3) by adapting existing few-shot learning methods, representative for few-shot classification and appropriate for few-shot metric learning in the sense that it adapts the embedding spaces online. We then discuss the limitations of the baselines and propose our method that overcomes the limitations (Sect. 4.4).

4.1   Simple Fine-Tuning (SFT)

As the first simple baseline for few-shot metric learning, we use the standard procedure of inductive transfer learning. In training, the embedding model f is trained on meta-training set $\mathcal{D}^{\mathrm{mtr}}$ from scratch with a metric learning loss; this is done in the same way with conventional metric learning, not involving any episodic training. In testing with an episode $(\mathcal{S}, \mathcal{P})$, the trained model $f(x; \theta_0)$ is simply fine-tuned using the target support set $\mathcal{S}$ by computing gradients with respect to a metric learning loss on $\mathcal{S}$:

$$\theta' = \theta_0 - \alpha \nabla_\theta \mathcal{L}(S; \theta_0). \tag{3}$$

After fine-tuning, the model is tested on $\mathcal{P}$. We choose the number of updates that shows the highest performance on the (meta-)validation set.

4.2   Model-Agnostic Meta-Learning (MAML)

As the second baseline for few-shot metric learning, we employ MAML [10], which meta-learns a good initialization for few-shot adaptation via episodic training. Given a meta-train episode $(\mathcal{S}_k^{\mathrm{mtr}}, \mathcal{P}_k^{\mathrm{mtr}})$, the meta-training process consists of inner and outer loops. In the inner loop, the parameters of a *base-learner* are updated using a *meta-learner* and the support set $\mathcal{S}_k^{\mathrm{mtr}}$. In MAML, the base-learner corresponds to the embedding model f, and the meta-learner corresponds to the initializer θ_0 for the embedding model. The inner loop updates the base learner using $\mathcal{S}_k^{\mathrm{mtr}}$ by a few gradient descent steps:

$$\theta_1 = \theta_0 - \alpha \nabla_\theta \mathcal{L}(\mathcal{S}_k^{\mathrm{mtr}}; \theta_0), \tag{4}$$

where α is the step size of inner-loop updates. In the outer loop, the meta-learner is updated using the loss of the updated base-learner on the prediction set $\mathcal{P}_k^{\mathrm{mtr}}$:

$$\theta_0' = \theta_0 - \eta \nabla_\theta \mathcal{L}(\mathcal{P}_k^{\mathrm{mtr}}; \theta_1), \tag{5}$$

where η is the step size of outer-loop updates. This meta-optimization with episodic training seeks to learn the initialization of the embedding model that generalizes well to unseen classes. For computational efficiency, we use the first-order approximation of meta-gradient update as in [10].

The meta-testing procedure with an episode $(\mathcal{S}, \mathcal{P})$ is the same as that of the inner loop; the meta-trained model $f(x; \theta_0')$ is evaluated by fine-tuning it on $\mathcal{S}$ with a few gradient steps and testing it on $\mathcal{P}$. We choose the number of steps that shows the highest performance on the $\mathcal{D}^{\mathrm{mvl}}$.

4.3   Meta-Transfer Learning (MTL)

As the third baseline, we adapt MTL [37] for few-shot metric learning. MTL first pre-trains an embedding model f on the meta-training set $\mathcal{D}^{\mathrm{mtr}}$ and then conducts an optimization-based meta-learning process while freezing all the layers except for the last fully-connected one. Here, the base-learner is the last fully-connected layer ψ and the meta-learner consists of two groups: a set of channel scaling/shifting parameters $\Phi = \{(\gamma^l, \beta^l)\}_l$ and the initialization of the last fully-connected layer ψ_0. The channel-scaling/shifting parameters are applied to the frozen convolutional layers of the embedding model f so that the parameters of each conv layer are scaled by γ^l and shifted by β^l:

$$\mathrm{Conv}_{\mathrm{CSS}}^l(X; W^l, b^l, \gamma^l, \beta^l) = (W^l \odot \gamma^l) * X + (b^l + \beta^l), \tag{6}$$

where W^l and b^l are the weight and bias of each convolution layer, $\odot$ is channel-wise multiplication, and $*$ is convolution. For example, if the 3×3 convolution kernel is size of $128 \times 64 \times 3 \times 3$, then channel-scaling parameter is size of $128 \times 64 \times 1 \times 1$.

In the inner loop of the meta-training, the last fully-connected layer is fine-tuned on $\mathcal{S}_k^{\mathrm{mtr}}$ from the initialization ψ_0. In the outer loop, the set of channel-scaling/shifting parameters Φ and the initialization of the last fully-connected layer ψ_0 are meta-updated using the prediction loss.

In meta-testing with an episode $(\mathcal{S}, \mathcal{P})$, following the process of the inner loop, only the last layer ψ is updated via a few gradient steps from the meta-learned initialization ψ_0 using $\mathcal{S}$, and the fine-tuned model is tested on $\mathcal{P}$.

4.4   Channel-Rectifier Meta-Learning (CRML)

In this subsection we discuss limitations of the aforementioned baselines and then propose a simple yet effective method for few-shot metric learning.

The main challenge in few-shot metric learning is how to effectively adapt the vast metric space using only a few examples while avoiding the danger of overfitting; as expected, the issue of overfitting is particularly critical in few-shot learning since only a few annotated examples are given for adaptation online. Updating all learnable parameters during meta-testing, as typically done in simple fine-tuning (Sect. 4.1) and MAML (Sect. 4.2), often causes quick over-fitting. A reasonable alternative is to fine-tune a part of the network only, e.g., the output layer as in MTL (Sect. 4.3), with all the other parts frozen. This partial update strategy is shown to be effective for classification problems where class

---

Algorithm 1. Channel-Rectifier Meta-Learning

Input: Meta training set $\mathcal{D}^{\mathrm{mtr}}$, learning rate α, η, pre-trained embedding model f
Output: The set of initialization of channel-scaling/shifting parameter $\Phi_0 = \{(\gamma_0^l, \beta_0^l)\}_l$

1: Initialize $\gamma^l \leftarrow \mathbf{1}, \beta^l \leftarrow \mathbf{0}$
2: **for** $(\mathcal{S}_k^{\mathrm{mtr}}, \mathcal{P}_k^{\mathrm{mtr}}) \in D^{\mathrm{mtr}}$ **do**
3:     $\Phi_1 \leftarrow \Phi_0 - \alpha \nabla_\Phi \mathcal{L}(\mathcal{S}_k^{\mathrm{mtr}}; f, \Phi_0)$ // Inner loop
4:     $\Phi_0 \leftarrow \Phi_0 - \eta \nabla_\Phi \mathcal{L}(\mathcal{P}_k^{\mathrm{mtr}}; f, \Phi_1)$ // Outer loop
5: **end for**

---

decision boundaries can be easily affected by the specific part. In metric learning, however, fine-tuning the output layer or other specific layer online turns to hardly change the metric space, *i.e.*, distance relations among embeddings (Sect. 5.2).

To tackle the issue, we propose the *Channel-Rectifier Meta-Learning* (CRML) that meta-learns *how to rectify* channels of intermediate feature layers. The main idea is inspired by channel scaling/shifting of MTL (Eq. (6)), which is shown to be effective in adapting pretrained layers but never used in online adaptation. Unlike MTL, we propose to leverage the channel scaling/shifting module, dubbed *channel rectifier*, for online adaptation. In other words, we set the channel rectifier $\Phi = \{(\gamma^l, \beta^l)\}_l$ as a base-learner and its initialization Φ_0 as a meta-learner. In this setup, we pre-train an embedding model f on the meta-training set $\mathcal{D}^{\mathrm{mtr}}$ and all the pre-trained parameters are frozen during the subsequent meta-learning process. Instead, in the meta-training stage, we update the channel rectifier Φ in the inner loop (Eq. (4)) while updating the initialization of the channel rectifier Φ_0 in the outer loop. This meta-learning process of CRML is summarized in Algorithm 1, where we describe a single-step inner loop with meta-batch of size 1 for the sake of simplicity.

In meta-testing with an episode $(\mathcal{S}, \mathcal{P})$, only the channel rectifier Φ is fine-tuned by the support set $\mathcal{S}$ with a few gradient steps from the learned initialization Φ_0, and the fine-tuned model is tested on $\mathcal{P}$. Note that the CRMLallows the channel rectifier to effectively exploit the support set to adapt the embedding function online.

5 Experiments

5.1 Experimental Settings

Datasets and Scenarios. We evaluate CRML and three baselines on two standard few-shot learning datasets, *mini*ImageNet [30] and CUB-200-2011 [40]. The *mini*ImageNet dataset is a subset of the ImageNet [22] and consists of 60,000 images categorized into 100 classes with 600 images each. We use the splits divided into 64/16/20 classes for (meta-)training (meta-)validation, and (meta-)testing, which has been introduced by [30]. The CUB-200-2011 (CUB) is a fine-grained classification dataset of bird species. It consists of 200 classes with 11,788 images in total. Following the evaluation protocol of [14], the split is divided into

100/50/50 species for (meta-)training, (meta-)validation, and (meta-)testing. We also use the MPII dataset, a human pose dataset, to train the metric learning model with continuous labels. Following the split of [23], 22,285 full-body pose images are divided into 12,366 for (meta-)training and 9,919 for (meta-)testing. Following [20], the label for a pose pair is defined as a pose distance, the sum of Euclidean distances between body-joint locations. To verify the influence of the domain gap between the source and the target classes, we conduct cross-domain experiments [5], which is designed to have a significant domain gap between training and evaluation; (meta-)training set consists of all samples in *mini*ImageNet where each (meta-)validation and (meta-)test set consists of 50 classes from CUB. Lastly, we propose a new multi-attribute dataset for few-shot metric learning, dubbed *mini*DeepFashion. The *mini*DeepFashion is built on DeepFashion [26] dataset, which is a multi-label classification dataset with six fashion attributes. It consists of 491 classes and the number of all instances in the dataset amounts to 33,841. The details of *mini*DeepFashion is in Sect. 5.5.

Evaluation Metrics. We use two standard evaluation metrics, mAP (mean value of average precision) [3] and Recall@k [17], to measure image retrieval performances. Recall@k evaluates the retrieval quality beyond k nearest neighbors while mAP evaluates the full ranking in retrieval. Since the MPII dataset for the human pose retrieval is labeled with continuous real values, we employee two metrics defined on continuous labels following [20]: mean pose distance (mPD) and a modified version of normalized discounted cumulative gain (nDCG). The mPD_k evaluates the mean pose distance between a query and k nearest images. The modified $nDCG_k$ evaluates the rank of the k nearest images and their relevance scores. The details about evaluation metrics are specified in the supplementary material.

Implementation Details. We use ResNet-18 [13] for the main backbone from scratch. We append a fully-connected layer with the embedding size of 128 followed by $l2$ normalization on top of the backbone. We use the multi-similarity loss [43] for training all the baselines and ours, CRML. For the human pose retrieval task on the MPII, we use ResNet-34 with the embedding size of 128, which is pre-trained on ImageNet [22] for a fair comparison with [20]. We fine-tune the network with log-ratio loss [20] using from 25 to 300 pairs out of all possible $\binom{12366}{2} \approx 7.6 \times 10^7$ pairs. Complete implementation details are specified in the supplementary material.

5.2 Effectiveness of Few-shot Metric Learning

Few-shot Metric Learning is effective on Discrete-Label Benchmarks. We first compare few-shot metric learning to conventional metric learning (DML) in Tables 1, 2, and 3. All the few-shot metric learning methods consistently outperform DML not only on the 5-way 5-shot setting, which is standard for few-shot learning but also on the full-way 5-shot setting, which is standard for image

Table 1. Performance on *mini*ImageNet.

Method	5-way 5-shot				20-way 5-shot			
	mAP	R@1	R@2	R@4	mAP	R@1	R@2	R@4
DML	46.9	73.5	84.1	91.2	21.0	49.2	62.6	74.2
FSML$_{SFT}$	63.3	78.6	85.7	91.8	29.9	52.6	65.1	75.3
FSML$_{MAML}$	65.9	79.8	87.5	92.2	28.6	52.7	65.6	76.3
FSML$_{MTL}$	56.5	77.2	86.7	92.6	24.0	50.0	64.6	76.2
CRML	**69.2**	**83.2**	**89.9**	**93.8**	**30.7**	**56.3**	**68.6**	**78.5**

Table 2. Performance on CUB-200-2011.

Method	5-way 5-shot				50-way 5-shot			
	mAP	R@1	R@2	R@4	mAP	R@1	R@2	R@4
DML	57.8	81.4	88.6	93.6	26.3	51.8	62.5	72.2
FSML$_{SFT}$	79.9	87.7	91.6	93.8	31.1	55.3	66.2	74.8
FSML$_{MAML}$	82.0	89.5	93.2	95.4	33.2	54.6	66.3	75.7
FSML$_{MTL}$	71.9	86.2	91.4	94.5	30.2	54.3	65.2	73.6
CRML	**82.7**	**90.0**	**93.5**	**95.5**	**33.9**	**58.1**	**68.4**	**76.5**

Table 3. Performance on cross-domain.

Method	5-way 5-shot				50-way 5-shot			
	mAP	R@1	R@2	R@4	mAP	R@1	R@2	R@4
DML	36.2	57.5	73.2	86.1	6.1	19.6	29.0	41.1
FSML$_{SFT}$	49.4	65.2	77.1	86.0	9.8	24.1	35.2	47.8
FSML$_{MAML}$	51.5	67.0	78.8	87.2	10.0	23.7	35.4	48.8
FSML$_{MTL}$	40.3	62.7	28.7	41.1	6.8	20.0	29.9	42.8
CRML	**56.4**	**71.0**	**81.5**	**88.9**	**10.9**	**27.0**	**38.3**	**51.1**

Table 4. Performance on *mini*Deep Fashion.

Method	5-way 5-shot				20-way 5-shot			
	mAP	R@1	R@2	R@4	mAP	R@1	R@2	R@4
DML	31.8	50.3	65.8	80.2	11.3	26.1	37.3	50.2
FSML$_{SFT}$	35.2	51.3	66.1	79.8	12.5	26.4	37.8	50.4
FSML$_{MAML}$	38.2	**53.5**	**67.6**	80.2	**13.3**	27.7	**39.1**	**52.1**
FSML$_{MTL}$	35.2	52.2	66.8	79.9	12.3	26.8	38.0	50.6
CRML	**38.3**	50.7	66.3	**80.2**	13.0	**27.8**	**39.1**	52.0

retrieval. The result also shows that only five shots for each class is enough to boost the image retrieval quality regardless of the number of classes to retrieve. Such improvement is clear not only in Recall@k, *i.e.*, the measurement of top-k nearest neighbors, but also in mAP, *i.e.*, the quality of all distance ranks. More importantly, the proposed CRML outperforms all baselines in the most settings, improving over MAML and SFT baselines by a large margin on the *mini*ImageNet and the cross domain setting. CRML is trained to *rectify* the base feature maps by learning a small set of channel scaling and shifting parameters and thus effectively avoids overfitting to the few-shot support set from an unseen domain. In contrast, both the MAML and the SFT baselines update all parameters in the embedding functions online, thus being vulnerable to overfitting to the small number of support set. Note that the worst model is the MTL baseline, which fine-tunes the last fully-connected layer while all the other layers frozen, suggesting that fine-tuning only a single output layer in the embedding function is insufficient for online adaptation. Note that simple fine-tuning (SFT) often performs comparable to meta-learning baselines (MAML and MTL) as recently reported in transfer learning based few-shot learning work [5].

Few-shot Metric Learning is Effective on Continuous-Label Benchmarks. We also evaluate few-shot metric learning on the human pose retrieval on MPII to demonstrate its applications and show its effectiveness. Given a human image with a certain pose, the goal of the human pose retrieval is to retrieve the most similar image of a human pose, where the supervisions in MPII consist of a continuous value on a pair-wise similarity. Since such labels are expensive to collect, few-shot learning is a practical solution for this problem, while the standard image classification approach is unlikely to applied due to the pair-based form of supervisions. Table 5 summarizes the retrieval performances with increasing numbers of the pair-wise supervisions. The retrieval performance gradually improves as the number of labels increase, although the

Table 5. Performance on MPII [1]. For nDCG$_1$, the higher the better, and for mPD$_1$, the lower the better. † and ‡ denotes the performances from [23] and [20].

Metric	#(pair labels used)							
	0	25	50	100	200	300	76M†	76M‡
nDCG$_1$	40.4	41.2	41.5	41.9	42.4	43.0	70.8	74.2
mPD$_1$	31.5	30.9	30.8	30.7	30.3	29.9	17.5	16.5

Table 6. Adaptation growth rate[1] (%) of CRMLon three datasets. The models are trained and evaluated in the 5-way 5-shot setting on each dataset.

Metric →	mAP (%)			recall@1 (%)		
Dataset →	CUB	mini	cross	CUB	mini	cross
before adapt	72.2	58.1	41.8	87.5	80.7	65.3
after adapt	82.7	69.2	56.4	90.0	83.2	71.0
growth rate (%)	14.5	19.0	**34.8**	2.8	3.1	**8.7**

source classes (object classes in ImageNet [22]) used for pre-training deviate from the target classes of human poses. Few-shot adaptation achieves 7.7% of the state-of-the-art performance [20] trained with full supervisions using only 0.00039% of supervisions.

Please refer to the supplementary material for qualitative visualizations of our method and more experiments about effectiveness of few-shot metric learning with 1) 10 shots, 2) additional metric learning losses, 3) qualitative results.

5.3   Influence of Domain Gap Between Source and Target

To verify the influence of the domain gap between the source and the target classes, we conduct cross-domain experiments. Table 3 shows the results of the 5-way and 50-way 5-shot experiments on the cross-domain setting. Due to the substantial domain gap between the source and target classes, the performances are much lower than those on CUB experiment in Table 2. However, the performance improvement is between 1.5 and 2 times higher than that of CUB. We observe that CRML results in remarkable improvement, which implies CRML is learned to properly rectify the base features adapted to given a support set.

We investigate the correlation between domain gap and effects of few-shot adaptation. For *mini*ImageNet, we randomly sample 60 instances from each target class in to match the prediction set size equal to that of CUB for a fair comparison. Note that the CUB is fine-grained thus has the smallest domain gap, while the cross-domain setting has the biggest. For each dataset, we measure the ratio of performance improvement from online adaptation and refer to it as adaptation growth rate. As shown in Table 6, the growth rate increases as the domain gap arises. It implies that few-shot metric learning is more effective when the target classes diverge more from the source classes.

5.4   Few-shot Metric Learning vs. Few-shot Classification

To verify the differences between few-shot metric learning and few-shot classification, we compare them both in image retrieval and classification. We evaluate different types of few-shot classification methods for comparison: transfer-based (Baseline [5], Baseline++ [5]), optimization-based (MAML [10]), and metric-based (MatchingNet [39], ProtoNet [33], FEAT [45]) methods. Note that as mentioned earlier, unlike the transfer-based and optimization-based methods, the metric-based ones in their original forms do not use online adaptation to

Table 7. Classification and image retrieval performances of few-shot classification and few-shot metric learning methods on $mini$ImageNet in a 5-way 5-shot setting. TTA stands for test-time adaptation via gradient descents.

Method	Classification Accuracy	Image retrieval Recall@1	mAP
Transfer-based few-shot classification			
Baseline [5]	62.21	74.04	53.14
Baseline++ [5]	74.88	78.97	65.22
Metric-based few-shot classification			
MatchingNet [39]	67.42	69.70	51.10
+TTA	71.50	71.05	52.51
ProtoNet [33]	74.46	71.04	51.73
+TTA	77.15	71.89	51.78
FEAT [45]	80.37	79.15	49.73
+TTA	**80.59**	79.31	51.72
Optimization-based few-shot classification			
MAML [10]	68.80	75.81	57.03
Few-shot metric learning			
FSML$_{\text{SFT}}$	69.92	79.14	65.22
FSML$_{\text{MAML}}$	72.69	79.77	65.86
FSML$_{\text{MTL}}$	70.34	77.19	56.50
CRML	76.64	**83.22**	**69.15**

the given support on test time. For a fair comparison, we thus perform add-on online adaptation, which is denoted by TTA, for the metric-based methods by a few steps of gradient descent using the support set.

We observe that there is little correlation between the classification accuracy and the image retrieval performances as shown in Table 7. All the few-shot metric learning methods outperform few-shot classification methods on image retrieval in terms of Recall@1 and mAP, while their classification accuracies are lower than those of few-shot classification methods. Interestingly, only Baseline++ shows competitive results with few-shot metric learning on image retrieval; we believe it is because the learned vectors in Baseline++ behave similarly to proxies in proxy-based metric learning methods. The results imply that few-shot classification learning and few-shot metric learning are distinct and result in different effects indeed. Note that even metric-based few-shot classification is not adequate for organizing the overall metric space, and additional test-time adaptation contributes insignificantly to improving the image retrieval performances.

5.5   Results on $mini$DeepFashion

Our design principle for the $mini$DeepFashion is to make a binary (similar/dissimilar) relationship between two instances inconsistent across the non-target and target classes. We thus split six attributes, each of which indicates a semantic perspective that categorizes instance into (meta-)training,

Table 8. The split of *mini*Deep Fashion.

Splits	attribute (#classes)	#instances
C^{mtr}	fabric (99)	27,079
	part (95)	
	style (127)	
C^{mvl}	shape (76)	7,600
C	category (37)	8,685
	texture (57)	
	total	33,841

Belted trench coat	Open-shoulder shift dress	Cami trapeze dress	Dotted cami dress
attribute:class	attribute:class	attribute:class	attribute:class
part:belted	part:belted	shape:cami	shape:cami
category:coat	category:dress	texture:watercolor	texture:dots

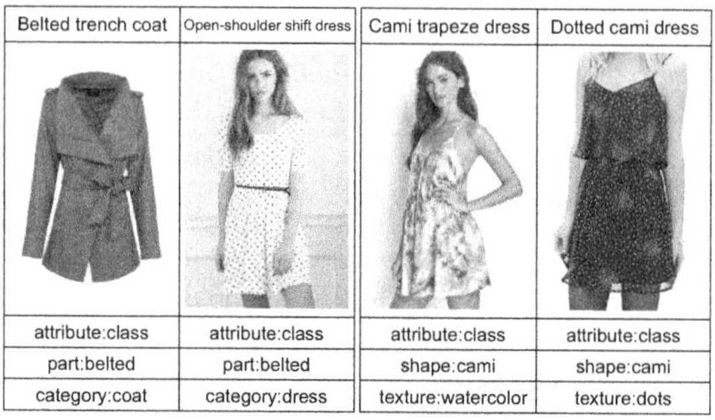

Fig. 3. Two example pairs sharing attributes. Blue rows denote positive pairs on one attribute, and red rows denote negative pairs on the other. (Color figure online)

(meta-)validation, and (meta-)testing as shown in Table 8. For example, a trench coat and a shift dress are either a positive pair in terms of parts they share, or a negative pair in terms of their categories as shown in Fig. 3. As existing few-shot learning datasets [30,40] do not assume such semantic switch, one embedding space is enough for a global information. In contrast, the assumption is no longer valid on *mini*DeepFashion, thus online adaptation is inevitable; this feature is well-aligned with the goal of few-shot metric learning. *mini*DeepFashion is built on DeepFashion [26] dataset, which is a multi-label classification dataset with six fashion attributes. We construct *mini*DeepFashion by randomly sampling 100 instances from each attribute class. The number of classes in each attribute and the number of instances in each split is shown in Table 8. Also, the class configuration for each attribute is in the supplementary materials.

Table 4 shows the results on *mini*DeepFashion. We observe that it is exceptionally challenging to reexamine distance ranking between instances online when the context of target class similarity switches from that of (meta-)training class. CRML and the baselines result in moderate performance growth from DML, opening the door for future work. Figure 4 shows the qualitative retrieval results of DML and CRML on the *mini*DeepFashion. The leftmost images are queries and the right eight images are top-eight nearest neighbors. As shown in Fig. 4, DML is misled by similar colors or shapes without adapting to target attributes, texture and category, only retrieving images of common patterns. In contrast, CRML adapts to attribute-specific data, thus retrieving correct images. For example, when the query is blue chinos, while DML only retrieves the blue pants regardless of the category, CRML retrieves the chinos successfully irrespective of the color (Fig. 4 (a)). Note that *mini*DeepFashion benchmark has an original characteristics that requires online adaptation to a certain attribute of given a support set, thus this benchmark makes more sense for evaluating few-shot metric learning in comparison to prevalent few-shot classification benchmarks.

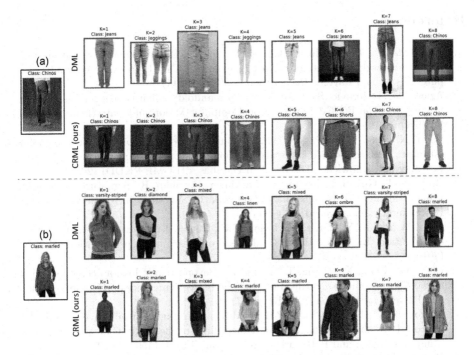

Fig. 4. Retrieval results on the (a) texture and (b) category attribute in *mini-* Deep-Fashion. The leftmost images are queries, and the right images are top-eight nearest neighbors. Green and red boxes are positive and negative images.

6   Conclusion

We have presented a few-shot metric learning framework and proposed an effective method, CRML, as well as other baseline methods. All the few-shot metric learning methods consistently outperform the conventional metric learning approach, demonstrating that they effectively adapt the learned embedding using a few annotations from target classes. Moreover, few-shot metric learning is more effective than classification approaches on relational tasks such as learning with continuous labels and multi-attribute image retrieval tasks. For this direction, we have introduced a challenging multi-attribute image retrieval dataset, *mini*DeepFashion. We believe few-shot metric learning is a new promising direction for metric learning, which effectively bridges the generalization gap of conventional metric learning.

Acknowledgements. This work was supported by Samsung Electronics Co., Ltd. (IO201208-07822-01), the IITP grants (2022-0-00959: Few-Shot Learning of Causal Inference in Vision and Language (30%), 2019-0-01906: AI Graduate School Program at POSTECH (20%)) and NRF grant (NRF-2021R1A2C3012728 (50%)) funded by the Korea government (MSIT).

References

1. Andriluka, M., Pishchulin, L., Gehler, P., Schiele, B.: 2d human pose estimation: New benchmark and state of the art analysis. In: The IEEE Conference on Computer Vision and Pattern Recognition (CVPR) (2014)
2. Aziere, N., Todorovic, S.: Ensemble deep manifold similarity learning using hard proxies. In: The IEEE Conference on Computer Vision and Pattern Recognition (CVPR) (2019)
3. Boyd, K., Eng, K.H., Page, C.D.: Area under the precision-recall curve: point estimates and confidence intervals. In: Joint European Conference on Machine Learning and Knowledge Discovery in Databases (ECML PKDD) (2013)
4. Bromley, J., Guyon, I., LeCun, Y., Säckinger, E., Shah, R.: Signature verification using a" siamese" time delay neural network. In: Advances in Neural Information Processing Systems (NeruIPS) (1994)
5. Chen, W.Y., Liu, Y.C., Kira, Z., Wang, Y.C., Huang, J.B.: A closer look at few-shot classification. In: International Conference on Learning Representations (ICLR) (2019)
6. Chen, W., Chen, X., Zhang, J., Huang, K.: Beyond triplet loss: a deep quadruplet network for person re-identification. In: The IEEE Conference on Computer Vision and Pattern Recognition (CVPR) (2017)
7. Chikontwe, P., Kim, S., Park, S.H.: Cad: Co-adapting discriminative features for improved few-shot classification. In: Proceedings of the IEEE/CVF Conference on Computer Vision and Pattern Recognition, pp. 14554–14563 (2022)
8. Chopra, S., Hadsell, R., LeCun, Y.: Learning a similarity metric discriminatively, with application to face verification. In: The IEEE Conference on Computer Vision and Pattern Recognition (CVPR) (2005)
9. Chowdhury, A., Jiang, M., Chaudhuri, S., Jermaine, C.: Few-shot image classification: Just use a library of pre-trained feature extractors and a simple classifier. In: Proceedings of the IEEE/CVF International Conference on Computer Vision, pp. 9445–9454 (2021)
10. Finn, C., Abbeel, P., Levine, S.: Model-agnostic meta-learning for fast adaptation of deep networks. In: International Conference on Machine Learning (ICML) (2017)
11. Gordo, A., Larlus, D.: Beyond instance-level image retrieval: Leveraging captions to learn a global visual representation for semantic retrieval. In: The IEEE conference on computer vision and pattern recognition (CVPR) (2017)
12. Hadsell, R., Chopra, S., LeCun, Y.: Dimensionality reduction by learning an invariant mapping. In: The IEEE Conference on Computer Vision and Pattern Recognition (CVPR) (2006)
13. He, K., Zhang, X., Ren, S., Sun, J.: Deep residual learning for image recognition. In: The IEEE Conference on Computer Vision and Pattern Recognition (CVPR) (2016)
14. Hilliard, N., Phillips, L., Howland, S., Yankov, A., Corley, C.D., Hodas, N.O.: Few-shot learning with metric-agnostic conditional embeddings. arXiv preprint arXiv:1802.04376 (2018)
15. Hoffer, E., Ailon, N.: Deep metric learning using triplet network. In: International Workshop on Similarity-Based Pattern Recognition (SIMBAD) (2015)
16. Hu, S.X., Li, D., Stühmer, J., Kim, M., Hospedales, T.M.: Pushing the limits of simple pipelines for few-shot learning: External data and fine-tuning make a difference. In: Proceedings of the IEEE/CVF Conference on Computer Vision and Pattern Recognition, pp. 9068–9077 (2022)

17. Jegou, H., Douze, M., Schmid, C.: Product quantization for nearest neighbor search. IEEE Trans. Pattern Anal. Mach. Intell. (TPAMI) **33**(1), 117–128 (2010)
18. Kang, D., Kwon, H., Min, J., Cho, M.: Relational embedding for few-shot classification. In: The IEEE International Conference on Computer Vision (ICCV) (2021)
19. Kim, S., Kim, D., Cho, M., Kwak, S.: Proxy anchor loss for deep metric learning. In: Proceedings of the IEEE/CVF Conference on Computer Vision and Pattern Recognition, pp. 3238–3247 (2020)
20. Kim, S., Seo, M., Laptev, I., Cho, M., Kwak, S.: Deep metric learning beyond binary supervision. In: The IEEE Conference on Computer Vision and Pattern Recognition (CVPR) (2019)
21. Koch, G., Zemel, R., Salakhutdinov, R.: Siamese neural networks for one-shot image recognition. In: International Conference on Machine Learning (ICML) Workshop on Deep Learning (2015)
22. Krizhevsky, A., Sutskever, I., Hinton, G.E.: Imagenet classification with deep convolutional neural networks. In: Advances in Neural Information Processing Systems (NeurIPS) (2012)
23. Kwak, S., Cho, M., Laptev, I.: Thin-slicing for pose: Learning to understand pose without explicit pose estimation. In: The IEEE Conference on Computer Vision and Pattern Recognition (CVPR) (2016)
24. Li, W.H., Liu, X., Bilen, H.: Universal representation learning from multiple domains for few-shot classification. In: Proceedings of the IEEE/CVF International Conference on Computer Vision, pp. 9526–9535 (2021)
25. Liu, B., et al.: Negative margin matters: Understanding margin in few-shot classification (2020)
26. Liu, Z., Luo, P., Qiu, S., Wang, X., Tang, X.: Deepfashion: Powering robust clothes recognition and retrieval with rich annotations. In: The IEEE Conference on Computer Vision and Pattern Recognition (CVPR) (2016)
27. Milbich, T., Roth, K., Sinha, S., Schmidt, L., Ghassemi, M., Ommer, B.: Characterizing generalization under out-of-distribution shifts in deep metric learning. Adv. Neural. Inf. Process. Syst. **34**, 25006–25018 (2021)
28. Movshovitz-Attias, Y., Toshev, A., Leung, T.K., Ioffe, S., Singh, S.: No fuss distance metric learning using proxies. In: The IEEE International Conference on Computer Vision (ICCV) (2017)
29. Qian, Q., Shang, L., Sun, B., Hu, J., Li, H., Jin, R.: Softtriple loss: Deep metric learning without triplet sampling. In: The IEEE International Conference on Computer Vision (ICCV) (2019)
30. Ravi, S., Larochelle, H.: Optimization as a model for few-shot learning. In: International Conference on Learning Representations (ICLR) (2017)
31. Schroff, F., Kalenichenko, D., Philbin, J.: Facenet: A unified embedding for face recognition and clustering. In: The IEEE Conference on Computer Vision and Pattern Recognition (CVPR) (2015)
32. Short, R., Fukunaga, K.: The optimal distance measure for nearest neighbor classification. IEEE Trans. Inform. Theor. (TIT) **27**(5), 622–627 (1981)
33. Snell, J., Swersky, K., Zemel, R.: Prototypical networks for few-shot learning. In: Advances in Neural Information Processing Systems (NeurIPS) (2017)
34. Son, J., Baek, M., Cho, M., Han, B.: Multi-object tracking with quadruplet convolutional neural networks. In: The IEEE Conference on Computer Vision and Pattern Recognition (CVPR) (2017)

35. Song, H.O., Xiang, Y., Jegelka, S., Savarese, S.: Deep metric learning via lifted structured feature embedding. In: The IEEE Conference on Computer Vision and Pattern Recognition (CVPR) (2016)
36. Sumer, O., Dencker, T., Ommer, B.: Self-supervised learning of pose embeddings from spatiotemporal relations in videos. In: The IEEE International Conference on Computer Vision (ICCV) (2017)
37. Sun, Q., Liu, Y., Chua, T.S., Schiele, B.: Meta-transfer learning for few-shot learning. In: The IEEE Conference on Computer Vision and Pattern Recognition (CVPR) (2019)
38. Triantafillou, E., Zemel, R., Urtasun, R.: Few-shot learning through an information retrieval lens. In: Advances in Neural Information Processing Systems, vol. 30 (2017)
39. Vinyals, O., Blundell, C., Lillicrap, T., Wierstra, D., et al.: Matching networks for one shot learning. In: Advances in Neural Information Processing Systems (NeurIPS) (2016)
40. Wah, C., Branson, S., Welinder, P., Perona, P., Belongie, S.: The caltech-ucsd birds-200-2011 dataset (2011)
41. Wang, J., et al.: Learning fine-grained image similarity with deep ranking. In: The IEEE Conference on Computer Vision and Pattern Recognition (CVPR) (2014)
42. Wang, X., Huang, T.E., Darrell, T., Gonzalez, J.E., Yu, F.: Frustratingly simple few-shot object detection (July 2020)
43. Wang, X., Han, X., Huang, W., Dong, D., Scott, M.R.: Multi-similarity loss with general pair weighting for deep metric learning. In: The IEEE Conference on Computer Vision and Pattern Recognition (CVPR) (2019)
44. Wohlhart, P., Lepetit, V.: Learning descriptors for object recognition and 3d pose estimation. In: The IEEE Conference on Computer Vision and Pattern Recognition (CVPR) (2015)
45. Ye, H.J., Hu, H., Zhan, D.C., Sha, F.: Few-shot learning via embedding adaptation with set-to-set functions. In: Proceedings of the IEEE/CVF Conference on Computer Vision and Pattern Recognition, pp. 8808–8817 (2020)
46. Ye, H.J., Hu, H., Zhan, D.C., Sha, F.: Few-shot learning via embedding adaptation with set-to-set functions. In: The IEEE Conference on Computer Vision and Pattern Recognition (CVPR) (2020)
47. Zhang, C., Cai, Y., Lin, G., Shen, C.: Deepemd: Few-shot image classification with differentiable earth mover's distance and structured classifiers. In: The IEEE Conference on Computer Vision and Pattern Recognition (CVPR) (2020)

3D Shape Temporal Aggregation for Video-Based Clothing-Change Person Re-identification

Ke Han[1,2], Yan Huang[1,2(✉)], Shaogang Gong[3], Yan Huang[1,2], Liang Wang[1,2], and Tieniu Tan[1,2]

[1] Center for Research on Intelligent Perception and Computing (CRIPAC), Institute of Automation, Chinese Academy of Sciences (CASIA), Beijing, China
`{ke.han,yan.huang}@cripac.ia.ac.cn`, `{yhuang,tnt}@nlpr.ia.ac.cn`
[2] University of Chinese Academy of Sciences (UCAS), Beijing, China
[3] Queen Mary University of London (QMUL), London, England
`s.gong@qmul.ac.uk`

Abstract. 3D shape of human body can be both discriminative and clothing-independent information in video-based clothing-change person re-identification (Re-ID). However, existing Re-ID methods usually generate 3D body shapes without considering identity modelling, which severely weakens the discriminability of 3D human shapes. In addition, different video frames provide highly similar 3D shapes, but existing methods cannot capture the differences among 3D shapes over time. They are thus insensitive to the unique and discriminative 3D shape information of each frame and ineffectively aggregate many redundant framewise shapes in a videowise representation for Re-ID. To address these problems, we propose a 3D Shape Temporal Aggregation (3STA) model for video-based clothing-change Re-ID. To generate the discriminative 3D shape for each frame, we first introduce an identity-aware 3D shape generation module. It embeds the identity information into the generation of 3D shapes by the joint learning of shape estimation and identity recognition. Second, a difference-aware shape aggregation module is designed to measure inter-frame 3D human shape differences and automatically select the unique 3D shape information of each frame. This helps minimise redundancy and maximise complementarity in temporal shape aggregation. We further construct a Video-based Clothing-Change Re-ID (VCCR) dataset to address the lack of publicly available datasets for video-based clothing-change Re-ID. Extensive experiments on the VCCR dataset demonstrate the effectiveness of the proposed 3STA model. The dataset is available at https://vhank.github.io/vccr.github.io.

Keywords: Clothing-change person re-identification · 3D body shape · Temporal aggregation

1 Introduction

Person re-identification (Re-ID) aims to match the same person across non-overlapping cameras. Short-term Re-ID methods [8,14,16,30] consider the

L. Wang et al. (Eds.): ACCV 2022, LNCS 13845, pp. 71–88, 2023.
https://doi.org/10.1007/978-3-031-26348-4_5

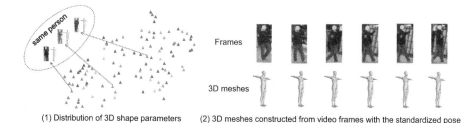

(1) Distribution of 3D shape parameters   (2) 3D meshes constructed from video frames with the standardized pose

Fig. 1. The motivation of this paper. (1) 3D shape parameters are not discriminative, as shown by randomly sampled 100 images of 10 different persons (color indicated) on the VCCR dataset, with their corresponding 3D shape parameters estimated by a 3D human estimation model [21], visualized by t-SNE. (2) The 3D shapes captured in successive video frames are usually very similar and redundant, as shown in the 3D mesh examples by SMPL [29] constructed with standardized pose parameters.

problem within a short time period assuming no changes of clothing, and therefore are mostly clothing-dependent. In practice, Re-ID over a longer time period in a more general setting, *e.g.*, over several days, probably includes clothing changes. In certain situations, a suspect may even deliberately change clothes to avoid being found. To this end, a number of methods have been proposed to address the challenge of clothing-change person Re-ID (CC Re-ID) [15,26,34,38,44].

Since clothing is less reliable in CC Re-ID, it is necessary to explore other clothing-independent characteristics, *e.g.*, body shape. Some methods consider 2D body shape features by human contours/silhouettes extraction [13,44], keypoints detection [34] or body shape disentanglement [26]. However, human body actually exists in a 3D space. 2D body shape information is not only view-dependent but also lacking of 3D depth information, which has been shown to be discriminative in Re-ID [37,42]. Some works [1,49] therefore have explored 3D shape information for Re-ID, but they suffer from two major limitations.

First, these methods [1,49] generate 3D human shapes without considering discriminative identity modelling. They usually directly employ a 3D human estimation model to generate 3D shape parameters, which are then used to construct 3D meshes by SMPL [29] for discriminating different persons. As illustrated in Fig. 1 (1), due to lack of identity modelling, the 3D shape parameters of the same person can be dispersive (especially when the same person wears different styles of clothes), while those of different persons can be close, making such 3D shape information not discriminative enough for Re-ID. **Second**, as shown in Fig. 1 (2), although a video provides richer information a single frame, the 3D shapes captured by successive frames are mostly highly similar. Existing temporal aggregation models [5,41,47] in Re-ID are usually designed for appearance instead of 3D shape information. They are insensitive to the differences among 3D shapes over time, and cannot select the unique shape information of each frame. Consequently, many redundant shapes from different frames are aggregated in a videowise shape representation, while some unique and discriminative shape information of each frame is suppressed.

To solve these problems, in this work we propose a 3D Shape Temporal Aggregation (3STA) Re-ID model for video-based CC Re-ID. Our 3STA model consists of three main modules: Identity-aware 3D Shape Generation (ISG), Difference-aware Shape Aggregation (DSA) and Appearance and Shape Fusion (ASF). 1) In order to generate the discriminative 3D shape for each video frame, the ISG embeds identity information into the 3D shape generation. This is realized by combing shape supervision of an auxiliary 3D human dataset with identity supervision of a Re-ID dataset in a joint learning framework. 2) The DSA is formulated to adaptively aggregate videowise shape representations from frames by referring to the intra-frame and inter-frame shape differences. The intra-frame shape difference enables our model to compare the changes of all the shape parameters in each frame for framewise spatial attention learning. The inter-frame shape difference is used to capture the change of each shape parameter over time to learn temporal attention. By considering both differences, DSA is sensitive to the unique and discriminative shape information in each frame and selectively aggregates into videowise shape representations with suppressed redundancy and enhanced complementarity. 3) We also exploit appearance information to model visual similarities unaffected by clothing changes, which can complement shape information especially when the target person only partially changes clothes. The ASF module is presented to fuse appearance and shape information adaptively into the final identity representation.

Another significant challenge to video-based CC Re-ID is that there is no **publicly available** dataset. For this purpose, we introduce a Video-based Clothing-Change Re-ID (VCCR) dataset in this work. Built on the attribute recognition dataset RAP [24] collected in a large indoor shopping mall, VCCR covers rich variations in clothing, cameras and viewpoints. To our best knowledge, it is currently the largest video-based CC Re-ID dataset with 4,384 tracklets of 232 clothing-change persons and 160 distractors, compared to the other dataset [3].

Our contributions are summarized as follows. 1) To our best knowledge, our 3D Shape Temporal Aggregation (3STA) model is the first attempt to explore temporal 3D shape information for video-based CC Re-ID. 2) To generate discriminative 3D shapes for Re-ID, we introduce Identity-aware 3D Shape Generation (ISG) that enforces 3D shape parameters to be person-specific. 3) The proposed Difference-aware Shape Aggregation (DSA) can be sensitive to temporal shape differences, and help minimise the redundancy of shape aggregation. 4) We construct a VCCR dataset for video-based CC Re-ID research. Extensive comparative evaluations show the effectiveness of our method against the state-of-the-art methods.

2 Related Work

Short-Term Re-ID. Short-term person Re-ID includes image-based [9,10, 17,31,32] and video-based [20,25,33] Re-ID. This research primarily relies on clothing information for discriminative person representation learning. Compared with image-based Re-ID methods, video-based Re-ID methods can leverage temporal information in video sequences to explore richer identity features.

However, the performance of these short-term Re-ID methods suffers a sharp drop when a person changes clothes.

Imaged-Based Clothing-Change Re-Id. To handle clothing variations for Re-ID, many methods have been proposed to learn clothing-independent shape information, which can be categorized into 2D and 3D shape based methods. For 2D shape learning, Qian et al. [34] present a clothing-elimination and shape-distillation model for structural representation learning from human keypoints. Yang et al. [44] directly learn feature transformation from human contour sketches. Hong et al. [13] transfer shape knowledge learned from human silhouettes to appearance features by interactive mutual learning. Li et al. [26] remove clothing colors and patterns from identity features by adversarial learning and shape disentanglement. Shu et al. [35] force the model to learn clothing-irrelevant features automatically by randomly exchanging clothes among persons.

In contrast to 2D shape that is confined to a plane, 3D shape can introduce human depth information to facilitate Re-ID. A few works have attempted to employ 3D human estimation models to construct 3D meshes for Re-ID. Zheng et al. [49] learn shape features directly with these 3D meshes instead of RGB images as inputs. However, due to the lack of consideration of identity information, the discriminability of constructed 3D meshes is not guaranteed. Chen et al. [1] propose an end-to-end framework to recover 3D meshes from original images. This method is supervised in a 2D manner by reprojecting the recovered 3D meshes back into a 2D plane again, which is lack of supervision of 3D shapes. Unlike them, our method unifies ground-truth 3D shape signals from a 3D human dataset with identity signals from a Re-ID dataset in a joint learning framework. In this way, our method can generate reliable and discriminative 3D shapes to boost shape learning for Re-ID.

Video-Based Clothing-Change Re-Id. Compared with image-based CC Re-ID, video-based CC Re-ID is rarely studied and still in the initial stage. Zhang et al. [46] make the first attempt on video-based CC Re-ID based on hand-crafted motion features from optical flow, assuming that people have constant walking patterns. Fan et al. [3] study video-based CC Re-ID with radio frequency signals reflected from human body instead of RGB color signals, thus completely removing clothing information. Different from them, we take advantage of temporal 3D shape information as a discriminative cue, which is more stable than walking patterns and easier to obtain than radio frequency signals.

Single-View 3D Human Estimation. Single-view 3D human estimation aims to construct human 3D meshes, including 3D shape and pose, from a single image. Current methods [21,23] typically predict shape and pose parameters with the supervision of 3D ground truths, and then construct 3D meshes by the SMPL model [29]. However, the 3D shape parameters estimated by these models are usually not discriminative enough and cannot fully reflect the differences of body shapes among persons, making these methods not well applied to Re-ID. To this end, our proposed ISG module combines 3D shape estimation and Re-ID in a joint learning framework to generate more discriminative shape parameters.

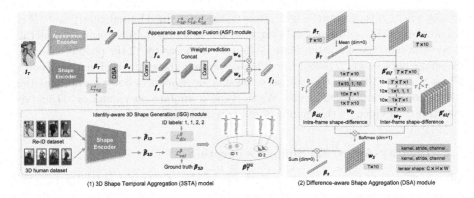

Fig. 2. An overview of the proposed 3D Shape Temporal Aggregation (3STA) model.

3   Method

In this paper, we propose a 3D Shape Temporal Aggregation (3STA) Re-ID model to learn temporal 3D shape representations for video-based CC Re-ID. As shown in Fig. 2, our model includes three main modules. The Identity-aware 3D Shape Generation (ISG) is first performed to generate discriminative 3D shape parameters for each video frame. The Difference-aware Shape Aggregation (DSA) exploits the differences across intra-frame and inter-frame 3D shape parameters to aggregate videowise shape parameters. The Appearance and Shape Fusion (ASF) further exploits appearance information to complement shape and adaptively fuses them into final representations for CC Re-ID. Let us start with an introduction to parametric 3D human estimation, which is the basis for our discriminative 3D shape learning.

3.1   Parametric 3D Human Estimation

3D human estimation models [21, 23, 29] usually parameterize 3D human body by shape parameters and pose parameters that are irrelevant to each other. Typically, SMPL [29] is modeled as a function of the pose parameters $\theta \in \mathbb{R}^{24 \times 3}$ representing the rotation vectors of 24 human joints and shape parameters $\beta \in \mathbb{R}^{10}$. Given the two parameters, SMPL can construct the 3D mesh with the corresponding pose and shape. Since pose is not person-specific, we focus only on shape parameters β in our modelling.

The SMPL model predefines a human shape template, and uses shape parameters to formulate the shape offset to the template by principal component analysis (PCA). Then the body shape is represented by a linear function B_S

$$B_S = \sum_{n=1}^{|\beta|} \beta^n \boldsymbol{S}^n, \tag{1}$$

where the shape parameters $\beta = [\beta^1, \ldots, \beta^{|\beta|}]^{\mathrm{T}}$, and $|\beta|$ is the number of parameters ($|\beta| = 10$). $\boldsymbol{S}^n \in \mathbb{R}^{3N}$ represents orthonormal principal components of shape offsets, where N is the number of vertices on the predefined human mesh, and $3N$ is the number of 3D coordinates of N vertices. The function B_S thus formulates all the shape offsets to the shape template. When S^n is shared by all the people's shapes, the shape parameters β reflect the difference among these shapes. Each parameter β^n ($n = 1, \cdots, 10$) usually controls some specific aspects of body shape, *e.g.*, the body size, waistline or leg length. We thus can formulate the change of 3D shapes over time in a video by the change of shape parameters.

3.2   Identity-Aware 3D Shape Generation

One challenge of learning 3D shape information for Re-ID is that Re-ID datasets do not contain annotations of 3D shape parameters β. In fact, it is very difficult and has to rely on special equipments to collect individual 3D shape parameters in real-world scenarios. Existing CC Re-ID methods [1,49] directly utilize 3D human estimation models to estimate shape parameters which are nevertheless not discriminative. To overcome this problem, we introduce an Identity-aware 3D Shape Generation (ISG) module that embeds identity information into the generation of shape parameters.

In the ISG module (Fig. 2 (1)), a shape encoder is modeled as a function: $\mathbb{R}^{C \times H \times W} \rightarrow \mathbb{R}^{10}$, to predict 10D shape parameters for a given image, where C, H and W are the number of channels, height and width of the image, respectively. The generated 3D shape parameters need to satisfy two requirements. (1) Validity: shape parameters are valid and close to the ground truths so that they can formulate true 3D body shape. (2) Discriminability: shape parameters of the same person are close while those of different persons are away from each other in the parameter space.

To meet the requirement (1), we introduce a 3D human dataset [19] as an auxiliary dataset with ground truths of shape parameters. Images from both the Re-ID and 3D datasets are input into the shape encoder to estimate shape parameters $\hat{\beta}_{ID}$ and $\hat{\beta}_{3D}$, respectively. We introduce a shape validity loss $\mathcal{L}_{val}^{\beta}$:

$$\mathcal{L}_{val}^{\beta} = \|\hat{\beta}_{3D} - \beta_{3D}\|^2, \tag{2}$$

where β_{3D} is ground-truth 3D shape parameters from the 3D human dataset. To meet the requirement (2), we further introduce a shape discrimination loss $\mathcal{L}_{dis}^{\beta}$ on $\hat{\beta}_{ID}$.

$$\mathcal{L}_{dis}^{\beta} = \mathcal{L}_{ce}^{\beta} + \mathcal{L}_{tri}^{\beta}, \tag{3}$$

where $\mathcal{L}_{ce}^{\beta}$ and $\mathcal{L}_{tri}^{\beta}$ are the cross-entropy and triplet losses, respectively, which are enforced by pairwise positive and negative identity labels from the Re-ID dataset. The total loss for ISG is

$$\mathcal{L}_{ISG} = \mathcal{L}_{val}^{\beta} + \alpha \mathcal{L}_{dis}^{\beta}, \tag{4}$$

where α is a weight factor. After performing ISG, the generated shape parameters for the Re-ID dataset are kept as pseudo labels (β_T^{ISG}) to train the 3STA model.

When we train the 3STA model, the shape encoder is retrained from scratch on the Re-ID dataset and does not share the weights with the shape encoder in ISG. This can decrease the interference of the 3D human data distribution bias in training the shape encoder. We denote a random tracklet composed of T frames in the Re-ID dataset as $\boldsymbol{I}_T$. The shape encoder generates shape parameters $\boldsymbol{\beta}_T \in \mathbb{R}^{T \times 10}$ and is optimised by a shape regression loss $\mathcal{L}_{reg}^s$:

$$\mathcal{L}_{reg}^s = \|\boldsymbol{\beta}_T - \boldsymbol{\beta}_T^{ISG}\|^2, \tag{5}$$

where $\boldsymbol{\beta}_T^{ISG}$ is the corresponding shape parameters generated by ISG. In this way, the shape encoder learns to regress discriminative 3D shape parameters for each video frame.

3.3 Difference-Aware Shape Aggregation

Existing temporal aggregation methods [5, 41, 47] are usually proposed for aggregating appearance information, and insensitive to the shape differences over time in a video. They thus aggregate much redundant shape information of different frames and suppress valuable unique shape information of each frame. To this end, we propose a Difference-aware 3D Shape Aggregation (DSA) module, which takes advantage of the shape differences among frames to drive the shape aggregation with suppressed redundancy and enhanced complementarity.

To make our method more sensitive to inter-frame shape differences over time, we use relative shape instead of absolute shape per frame in video shape aggregation. As shown in Section. 3.1, each shape parameter β^d controls some specific aspects of body shape, so we can formulate the subtle difference of 3D shapes among frames by the difference values of shape parameters. As shown in Fig. 2. (2), we first compute the mean shape parameters $\bar{\boldsymbol{\beta}}_T$ of a tracklet as a reference, and then obtain the shape-difference map $\boldsymbol{\beta}_{dif} = \boldsymbol{\beta}_T - \bar{\boldsymbol{\beta}}_T$ ($\boldsymbol{\beta}_{dif} \in \mathbb{R}^{T \times 10}$). We denote the value at the coordinate (t, d) as $(\boldsymbol{\beta}_{dif})_t^d$ indicating the shape difference between the d-th shape parameter of the t-th frame and the corresponding mean parameter. We introduce the intra-frame and inter-frame shape-difference references to jointly decide the weight for each shape parameter of each frame.

Intra-Frame Shape-Difference Reference. We consider a shape parameter with a larger difference to the mean parameter to be more informative than other parameters in that frame. But if most shape parameters have larger differences, their importance should be balanced because it is possibly caused by the body occlusion or shape estimation error. Therefore we introduce an intra-frame shape-difference reference $\boldsymbol{w}_D \in \mathbb{R}^{T \times 10}$ to consider all the shape parameters within a frame to balance the weight of each one. $\boldsymbol{w}_D$ for the t-th frame is formulated as

$$(\boldsymbol{w}_D)_t^1, \cdots (\boldsymbol{w}_D)_t^{10} = \text{Sigmoid}(\text{Conv}[(\boldsymbol{\beta}_{dif})_t^1, \cdots (\boldsymbol{\beta}_{dif})_t^{10}]), \tag{6}$$

where Sigmoid is the Sigmoid function, Conv is a convolutional layer, of which the kernel size is 1×10 to span all of 10 shape parameters of a frame. The output channel is 10 for 10 different parameters. Details are illustrated in Fig. 2. (2).

Inter-Frame Shape-Difference Reference. To reduce the temporal redundant information, we also introduce an inter-frame shape-difference reference $\boldsymbol{w}_T \in \mathbb{R}^{T \times 10}$ to compare each shape parameter across frames to assign temporal attention. Concretely, we compute an inter-frame shape-difference map $\boldsymbol{\beta}'_{dif} \in \mathbb{R}^{T \times T \times 10}$, on which the value at the coordinate (t_1, t_2, d) is formulated as

$$(\boldsymbol{\beta}'_{dif})^d_{t_1, t_2} = (\boldsymbol{\beta}_{dif})^d_{t_1} - (\boldsymbol{\beta}_{dif})^d_{t_2}, \tag{7}$$

where $t_1, t_2 = 1, \cdots, T$; $d = 1, \cdots, 10$. $\boldsymbol{\beta}'_{dif}$ indicates the shape difference of the d-th shape parameter between the t_1-th and t_2-th frames. $\boldsymbol{w}_T$ is then formulated by 10 convolutional layers as

$$(\boldsymbol{w}_T)^d_t = \text{Sigmoid}(\text{Conv}_d([(\boldsymbol{\beta}'_{dif})^d_{t,1}, \cdots (\boldsymbol{\beta}'_{dif})^d_{t,T}])), \tag{8}$$

where Conv_d is the d-th convolutional layer with a kernel size of 1×1, which considers all of T frames on the d-th parameter to determine the reference weight.

The reference weights $\boldsymbol{w}_D$ and $\boldsymbol{w}_T$ make our model sensitive to the shape changes both in respect to each shape parameter over time and all shape parameters in each frame at a time. They thus impose selective aggregation by minimising redundant spatial-temporal shape information across frames in a video. The final weight $\boldsymbol{w}_S = \boldsymbol{w}_T \odot \boldsymbol{w}_D$, where $\odot$ is the elementwise product, and then is normalized by the Softmax function. The aggregated videowise shape parameters $\boldsymbol{\beta}_s$ are the sum of $\boldsymbol{\beta}_T$ weighted by $\boldsymbol{w}_S$, where $\boldsymbol{\beta}_s$ is optimised by a shape-based identity loss $\mathcal{L}^s_{id}$, same as in Eq.(3), *i.e.*, the sum of the cross-entropy loss and triplet loss to learn discriminative videowise shape parameters.

3.4 Appearance and Shape Fusion

Appearance remains useful in complementing some visual similarities to shape for Re-ID, *e.g.*, when a person only changes partial clothes and/or with certain aspects of appearance unaffected by clothing changes, such as gender, age, etc. To that end, we formulate a joint appearance and shape fused representation that is adaptively learned in model training.

The appearance encoder extracts videowise appearance features $\boldsymbol{f}_a$, to be combined with videowise shape parameters $\boldsymbol{\beta}_s$. A fusion module includes two steps, *i.e.*, feature transformation and weight prediction. The feature transformation projects two feature vectors into a common feature space, defined as

$$\boldsymbol{f}_a \leftarrow \text{Sigmoid}(\text{Conv}_a(L_2(\boldsymbol{f}_a))), \boldsymbol{f}_s \leftarrow \text{Sigmoid}(\text{Conv}_s(L_2(\boldsymbol{\beta}_s))), \tag{9}$$

where L_2 is L_2 normalization, Conv_a and Conv_s are two independent convolutional layers with the kernel size of 1×1.

The weight prediction aims to estimate the weights for the two feature vectors by making them refer to each other and jointly optimise the weight for each one. This process is defined as

$$\boldsymbol{w}_a = \text{Conv}_a([\boldsymbol{f}_a, \boldsymbol{f}_s]), \quad \boldsymbol{w}_s = \text{Conv}_s([\boldsymbol{f}_a, \boldsymbol{f}_s]), \tag{10}$$

where $\boldsymbol{f}_a$ and $\boldsymbol{f}_s$ are concatenated as a tensor, which then separately goes forward through two convolutional layers Conv_a and Conv_s. They both have the kernel

Table 1. Comparison among CC Re-ID datasets. Some data are unclear because of being publicly unavailable.

Dataset	NKUP [40]	LTCC [34]	PRCC [44]	COCAS [45]	RRD-Campus [3]	Motion-ReID [46]	VCCR (Ours)
Type	image	image	image	image	radio frequency	video	video
CC IDs\Distractors	107\0	91\61	221\0	5,266\0	100\0	30\0	232\160
Cameras	15	12	3	30	5	2	23
Cropping	Detection	Detection	Manual	Manual	Detection	Manual	Manual
Tracklets	–	–	–	–	863	240	4,384
Images (Frames)	9,738	17,119	33,698	62,383	Unclear	24,480	152,610
Clothes/ID	2~3	2~14	2	2~3	Unclear	Unclear	2~10
Publicly Available	Y	Y	Y	N	N	N	Y (to be released)

size of 1×2 and thus output two weight vectors $\boldsymbol{w}_a$ and $\boldsymbol{w}_s$. A fused joint feature vector $\boldsymbol{f}_j = \boldsymbol{w}_a \odot \boldsymbol{f}_a + \boldsymbol{w}_s \odot \boldsymbol{f}_s$, where $\odot$ is element-wise product, with $\boldsymbol{f}_a$ and $\boldsymbol{f}_j$ being optimised by the appearance-based loss $\mathcal{L}_{id}^a$ and fusion-based loss $\mathcal{L}_{id}^j$. Each of them is the sum of a cross-entropy loss and a triplet loss as in Eq.(3). The overall 3STA model is jointly trained by an overall loss

$$\mathcal{L}_{all} = \mathcal{L}_{reg}^\beta + \lambda_1 \mathcal{L}_{id}^s + \lambda_2 \mathcal{L}_{id}^a + \lambda_3 \mathcal{L}_{id}^j, \tag{11}$$

where λ_1, λ_2 and λ_3 are weight factors. Discriminative appearance and shape representations, which are optimised by $\mathcal{L}_{id}^a$, $\mathcal{L}_{id}^s$ and $\mathcal{L}_{reg}^\beta$, are the foundations of contributing to a more discriminative joint representation optimised by $\mathcal{L}_{id}^j$.

4   VCCR Dataset

Given that there is no publicly available dataset for video-based CC Re-ID model learning and evaluation, we introduce a new Video-based Clothing-Change Re-ID (VCCR) dataset to be released for open access to the research community.

4.1   Collection and Labelling

We collect data from the Richly Annotated Pedestrian (RAP) dataset [24] for reducing the collection and annotation cost. Moreover, this does not lead to additional privacy issues by not introducing new data. RAP contains person images captured in an indoor shopping mall over 3 months. It was originally built for attribute recognition and annotated with identity labels. We select 232 persons that change clothes and 160 distractors that do not change clothes from the RAP dataset, with access to the corresponding videos given the permission and consent by the authors. Then we manually crop person patches from video frames. Apart from the identity and camera labels from RAP, we additionally annotate each tracklet with a clothing label. Two tracklets with the same identity label are given two different clothing labels only if there is a visible clothing change. A change of carrying items, such as bottles, books and boxes, does not affect the clothing label.

4.2   Statistics and Comparison

We compare VCCR with other CC Re-ID datasets in Table 1 in four aspects.

1) **Type**. Most existing datasets, *e.g.*, NKUP [40], LTCC [34], PRCC [44] and COCAS [45], are image-based. RRD-Campus [3] collects radio frequency signals. Only Motion-ReID [46] is video-based, but not publicly available.

2) **Scale**. Motion-ReID includes 240 tracklets, while VCCR has 4,087 tracklets of 232 clothing-change persons and 297 tracklets of 160 distractors. Each tracklet has 5 to 130 frames with an average of 35. VCCR is thus the currently largest video-based CC Re-ID dataset.

3) **Number of persons**. Since it is much more challenging to collect and label clothing-change data in videos than images, VCCR has a smaller number of clothing-change persons (232) than the image-based COCAS (5,266). But VCCR still contains more clothing-change persons than the video-based Motion-ReID (30), and also competitive compared with most image-based datasets, such as NKUP (107), LTCC (91) and PRCC (221).

4) **4) Number of clothes**. In contrast to NKUP, PRCC and COCAS with 2~3 clothes/ID, VCCR contains 2~10 clothes/ID with an average of 3.3.

4.3   Protocol

The training set includes 2,873 tracklets of 150 clothing-change persons. For test, 496 tracklets of 82 clothing-change persons constitute the query set, while 718 tracklets of these 82 persons along with 297 tracklets of 160 distractors build the gallery set. We make sure that the training and test sets have close statistics and diversity in samples. We adopt two test modes like [34], *i.e.*, the cloth-changing (CC) and standard modes, to evaluate the performance of CC Re-ID models. In the **clothing-change (CC)** mode, all the ground-truth gallery tracklets have different clothing labels to the query. In the **standard** mode, the ground-truth gallery tracklets can have either same or different clothing labels to the query. When evaluating Re-ID performances, we use the average cumulative match characteristic and report results at ranks 1, 5 and 10.

5   Experiments

5.1   Implementation Details

The appearance encoder adopts the Resnet-50 [11] backbone pretrained on ImageNet [2] to extract framewise appearance features and average pooling to produce 2048D videowise features f_a. The shape encoder is composed of a pretrained Resnet-50 backbone and three fully-connected layers of 1024, 1024 and 10 dimensions, respectively. All images are scaled to 224×112 and randomly flipped.

We first run the ISG module with the VCCR dataset and 3D human dataset Human3.6M [19]. All the tracklets of VCCR are broken into 152,610 images in total. We randomly sample 16 persons with 4 images per person from VCCR, and 64 random images from Human3.6M in each training batch. ISG is performed for 20,000 iterations with the Adam optimiser [22] (β_1=0.9 and β_2=0.999). The learning rate is set to 0.00001 and the weight factor α is set to 500. After training ISG, we keep the generated shape parameters β_T^{ISG} for VCCR to train the overall 3STA model.

The overall 3STA model is trained on VCCR. We randomly choose 8 different persons, 4 tracklets for each person and 8 successive frames for each tracklet in each training batch. We also use the Adam optimiser, with the learning rates of the shape encoder and other modules initialized at 0.00001 and 0.0001, respectively, and decayed by 0.1 after 20,000 iterations. The 3STA model is jointly trained over 30,000 epochs. The weight factor $\lambda_1=1$, $\lambda_2=10$, $\lambda_3=0.05$, and the margin parameters of all the used triplet losses are set to 0.3. The dimension of the projected feature space in the appearance and shape fusion module is 2048, i.e., $\boldsymbol{f}_a, \boldsymbol{f}_s, \boldsymbol{f}_j \in \mathbb{R}^{2048}$.

5.2 Evaluation on CC Re-Id Datasets

We compare our 3STA model and four types of state-of-the-art methods on the VCCR dataset in Table 2. In terms of the deep learning based methods, the results show a general trend that the performance is incrementally improved from image-based short-term, image-based CC, video-based short-term to video-based CC Re-ID methods. Specifically, first, **image-based short-term Re-ID** methods have the lowest accuracies, because they primarily make use of clothing information to discriminate persons and inevitably lose some discriminability under clothing-change situations. Second, **image-based CC Re-ID** methods reduce the reliance on clothing by exchanging clothes among persons [35] or using vector-neuron capsules to perceive clothing change of the same person [18]. Third, **video-based short-term** Re-ID methods have more robust Re-ID capacities due to exploiting temporal information, but they are still sensitive to clothing changes. Overall, our **video-based CC Re-ID** model 3STA achieves the highest accuracies in both CC and standard modes. The reasons are two-fold. 1) Discriminative temporal 3D shape information in videos is modelled as clothing-independent person characteristics. 2) Complementary appearance information is jointly modelled with 3D shape, resulting in the joint representation more robust to both clothing-change and clothing-consistent situations.

For completeness, we also list the released results of STFV3D [27], DynFV [6] and FITD [46] on the Motion-ReID dataset [46]. All of these methods are based on hand-crafted features. We are unable to compare them with other methods on Motion-ReID because the dataset is not publicly available, but we include a comparison on the video-based short-term Re-ID dataset PRID.

5.3 Evaluation on Short-Term Re-Id Datasets

We also conduct evaluations on the video-based short-term Re-ID datasets MARS [48] and PRID [12]. MARS is a large-scale dataset containing 1,261 persons with 20,715 tracklets. PRID includes 200 persons captured by two cameras, and only 178 persons with more than 25 frames are used, following the previous work [46].

Our 3STA model can perform clearly better than **image-based short-term and CC Re-ID** methods on both MARS and PRID, benefiting from modelling temporal 3D shape apart from clothing information. But the **video-based**

Table 2. Results on video-based CC Re-ID datasets VCCR and Motion-ReID, and short-term Re-ID datasets MARS and PRID (%). Image-based methods produce videowise features by average pooling on framewise features. Appe., Shape and Joint denote that appearance features $\boldsymbol{f}_a$, shape parameters $\boldsymbol{\beta}_s$ and joint features $\boldsymbol{f}_j$ in the 3STA model are used for evaluation (the same below).

Method Type	Methods	Features	VCCR (CC Mode)			VCCR (Standard Mode)			Motion-ReID	MARS	PRID
			MAP	Rank1	Rank5	MAP	Rank1	Rank5	Rank1	Rank1	Rank1
Image-Based Short-Term Re-ID	PCB [36]	Deep Learning	15.6	18.8	38.6	36.6	55.6	75.2	–	85.2	89.1
	MGN [39]	Deep Learning	22.6	23.6	44.9	42.7	64.4	81.9	–	86.4	90.6
	HPM [4]	Deep Learning	19.4	23.1	42.9	39.5	58.3	78.7	–	87.9	90.3
Video-Based Short-Term Re-ID	STFV3D [27]	Hand-Crafted	–	–	–	–	–	–	29.1	–	42.1
	DynFV [6]	Hand-Crafted	–	–	–	–	–	–	32.3	–	17.6
	MGH [43]	Deep Learning	30.7	34.6	54.5	51.6	76.3	87.2	–	90.0	94.8
	AP3D [7]	Deep Learning	31.6	35.9	55.8	52.1	78.0	88.4	–	**90.1**	94.6
	GRL [28]	Deep Learning	31.8	35.7	55.3	51.4	76.9	88.2	–	89.8	**95.1**
Image-Based Clothing-Change Re-ID	ReIDCaps [18]	Deep Learning	29.9	33.4	53.6	48.4	75.1	86.3	–	83.2	88.0
	SPS [35]	Deep Learning	30.5	34.5	54.1	50.6	76.5	85.5	–	82.8	87.4
Video-Based Clothing-Change Re-ID	FITD [46]	Hand-Crafted	–	–	–	–	–	–	**43.8**	–	58.7
	Appe. (3STA)	Deep Learning	29.3	32.8	52.0	46.7	74.3	84.5	–	83.7	87.8
	Shape (3STA)	Deep Learning	20.6	21.3	36.9	39.2	62.8	82.4	–	74.0	76.3
	Joint (3STA)	Deep Learning	**36.2**	**40.7**	**58.7**	**54.3**	**80.5**	**90.2**	–	89.1	93.4

short-term Re-ID methods can surpass our 3STA model, due to enhancing clothing based temporal information for better discriminating clothing-consistent persons. For the two **video-based CC Re-ID** models, 3STA significantly outperforms FITD on PRID. FITD utilizes motion cues for Re-ID with the assumption that people keep constant motion patterns, which does not always hold in practice. In contrast, 3STA explores discriminative 3D shape together with appearance, which is more stable and robust than motion cues.

5.4   Ablation Study

Appearance vs. Shape vs. Joint Representations. We compare the performance of appearance, 3D shape and joint representations in Table 2. The results show two phenomenons that deserve the attention. **1)** Appearance can achieve higher performance than 3D shape in both test modes, due to two reasons. First, when people do not change or just slightly change clothes, appearance remains more competitive than 3D shape by exploiting visual similarities for Re-ID. Second, 3D shape parameters only have 10 dimensions and they have a limited capacity of modelling complex body shape. Overall, 3D shape is best complemented with appearance instead of being used alone. **2)** The joint representations outperform both appearance and 3D shape by a significant margin. This demonstrates that our model can exploit the complementarity of two information to adaptively fuse more discriminative information, which can adapt to both cloth-changing and clothing-consistent situations better.

Identity-Aware 3D Shape Generation (ISG). The ISG module ensures the validity and discriminability of the generated 3D shape parameters by the loss

Table 3. Rank 1 accuracy on the VCCR dataset for the ablation study of losses.

ISG module		3STA model				CC Mode			Standard Mode		
$\mathcal{L}_{val}^{\beta}$	$\mathcal{L}_{dis}^{\beta}$	$\mathcal{L}_{reg}^{s}$	$\mathcal{L}_{id}^{a}$	$\mathcal{L}_{id}^{s}$	$\mathcal{L}_{id}^{j}$	Appe.	Shape	Joint	Appe.	Shape	Joint
✗	✔	✔	✔	✔	✔	32.2	12.7	31.5	73.5	57.4	72.6
✔	✗	✔	✔	✔	✔	32.7	6.3	24.1	73.9	18.7	67.3
✔	✔	✗	✔	✔	✔	31.9	14.6	26.3	73.6	44.7	64.3
✔	✔	✔	✗	✔	✔	14.5	19.7	17.6	46.2	62.5	55.7
✔	✔	✔	✔	✗	✔	32.6	15.8	29.8	73.4	57.6	72.3
✔	✔	✔	✔	✔	✗	**33.5**	**21.6**	30.3	**75.2**	62.6	73.5
✔	✔	✔	✔	✔	✔	32.8	21.3	**40.7**	74.3	**62.8**	**80.5**

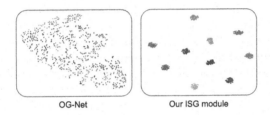

OG-Net   Our ISG module

Fig. 3. Comparing 3D shape parameters generated by the OG-Net [49] and our ISG module. The visualisation shows 700 images of 10 persons from the VCCR dataset.

$\mathcal{L}_{val}^{\beta}$ and $\mathcal{L}_{dis}^{\beta}$ in Eq. (4), respectively. We remove either of the two losses during performing ISG and show the results in Table 3 (top two lines). **1)** Removing $\mathcal{L}_{val}^{\beta}$ degenerates the accuracy of shape representations from 21.3%/62.8% to 12.7%/57.4% in the CC/starndard mode. Losing the supervision from the 3D human dataset in validity makes the shape parameters not formulate true 3D body shapes. Only using identity supervision from the Re-ID dataset for training, the model implicitly relies on appearance instead of 3D shape in minimising the loss $\mathcal{L}_{dis}^{\beta}$. **2)** Removing $\mathcal{L}_{dis}^{\beta}$ decreases the rank 1 of shape representations to 6.3%/18.7% in the CC/standard mode. In Fig. 3, we visualize the shape parameters generated by the ISG and the OG-Net [49]. OG-Net does not embed identity information, similar to removing $\mathcal{L}_{dis}^{\beta}$ from ISG. ISG enables 3D shape parameters to be separable for different persons, attributed to introducing $\mathcal{L}_{dis}^{\beta}$ to significantly improve discriminative 3D shape learning for Re-ID.

Difference-Aware 3D Shape Aggregation (DSA). 1) Intra-frame and inter-frame shape-difference references. The weight w_S in DSA is decided jointly by the intra-frame and inter-frame shape-difference references w_D and w_T. As shown in the 2nd and 3rd rows in Table 4, using w_T alone degenerates the rank 1 accuracy of shape representations from 21.3%/62.8% to 18.2%/60.7% in the CC/standard mode. The joint representations are affected similarly. The 1st and 3rd rows suggest that using w_D alone degrades the rank 1 accuracy of

Table 4. Rank 1 accuracy of different temporal aggregation methods on the VCCR dataset. $\beta_{dif} = \beta_T - \bar{\beta}_T$.

Methods	Shape	Weight	CC Mode		Standard Mode	
			Shape	Joint	Shape	Joint
Ours	β_{dif}	w_D	17.9	36.6	58.8	78.4
	β_{dif}	w_T	18.2	39.0	60.7	79.7
	β_{dif}	$w_D \odot w_T$	**21.3**	**40.7**	**62.8**	**80.5**
	β_T	w_D	16.2	35.8	58.5	77.3
	β_T	$w_D \odot w_T$	17.5	36.4	59.4	77.8

shape representations by 3.4%/4.0% in the CC/standard mode. The joint use of w_D and w_T makes DSA sensitive to the changes of spatial and temporal shape information and reduces the redundancy of shape aggregation over time.

2) Shape differences. DSA uses the shape differences among frames ($\beta_{dif} = \beta_T - \bar{\beta}_T$) to guide the weight prediction. To validate the effectiveness, we also show the result of directly using β_T to replace β_{dif} when computing w_D by Eq. (6). Comparing the 1st and 4th, or 3rd and 5th rows in Table 4, we can observe using β_{dif} improves the rank 1 accuracy obviously over β_T. The advantage of β_{dif} lies in helping the DSA module explicitly perceive the subtle 3D shape differences among frames in a form of relative shape, and reduce the reuse of redundant shape information better.

Losses. We perform the ablation study on the losses $\mathcal{L}_{reg}^s$, $\mathcal{L}_{id}^a$, $\mathcal{L}_{id}^s$ and $\mathcal{L}_{id}^j$ in training the 3STA model and show the results in Table 3. Taking off $\mathcal{L}_{reg}^s$ greatly decreases the accuracy of shape representations from 21.3%/62.8% to 14.6%/44.7% in the CC/standard mode, and also affects adversely the joint representations in a similar way. This is because $\mathcal{L}_{reg}^s$ can enable effective framewise shape learning, which is the basis of temporal shape aggregation. Our model also suffers from performance degradation in different degrees if trained without $\mathcal{L}_{id}^a$, $\mathcal{L}_{id}^s$ or $\mathcal{L}_{id}^j$. This reveals that the discriminative joint representations have to be built on discriminative appearance and shape representations.

6   Conclusion

To our best knowledge, for the first time this paper has formulated a model to learn discriminative temporal 3D shape information for video-based CC Re-ID. First, our proposed 3STA model has included an ISG module, which considers identity modelling to generate the discriminative 3D shape for each frame. Then, a DSA module that is sensitive to the shape differences among frames has been proposed to aggregate framewise shape representations into videowise ones. query Kindly provide the page range for Refs. [31].It selectively exploits the unique shape information of each frame to reduce the redundancy of shape aggregation. Moreover, we have also contributed a VCCR dataset for the video-based CC Re-ID research community.

Acknowledgements. This work was jointly supported by National Key Research and Development Program of China Grant No. 2018AAA0100400, National Natural Science Foundation of China (62236010, 62276261, 61721004, and U1803261), Key Research Program of Frontier Sciences CAS Grant No. ZDBS-LY- JSC032, Beijing Nova Program (Z201100006820079), CAS-AIR, the fellowship of China postdoctoral science foundation (2022T150698), China Scholarship Council, Vision Semantics Limited, and the Alan Turing Institute Turing Fellowship.

References

1. Chen, J., et al.: Learning 3D shape feature for texture-insensitive person re-identification. In: Proceedings of the IEEE/CVF Conference on Computer Vision and Pattern Recognition (CVPR), pp. 8146–8155 (2021)
2. Deng, J., Dong, W., Socher, R., Li, L.J., Li, K., Fei-Fei, L.: ImageNet: a large-scale hierarchical image database. In: IEEE conference on computer vision and pattern recognition (CVPR), pp. 248–255 (2009)
3. Fan, L., Li, T., Fang, R., Hristov, R., Yuan, Y., Katabi, D.: Learning longterm representations for person re-identification using radio signals. In: Proceedings of the IEEE/CVF conference on computer vision and pattern recognition (CVPR), pp. 10699–10709 (2020)
4. Fu, Y., et al.: Horizontal pyramid matching for person re-identification. In: AAAI, pp. 8295–8302 (2019)
5. Gao, J., Nevatia, R.: Revisiting temporal modeling for video-based person reid. arXiv preprint arXiv:1805.02104 (2018)
6. Gou, M., Zhang, X., Rates-Borras, A., Asghari-Esfeden, S., Sznaier, M., Camps, O.: Person re-identification in appearance impaired scenarios. arXiv preprint arXiv:1604.00367 (2016)
7. Gu, X., Chang, H., Ma, B., Zhang, H., Chen, X.: Appearance-Preserving 3D Convolution for Video-Based Person Re-identification. In: Vedaldi, A., Bischof, H., Brox, T., Frahm, J.-M. (eds.) ECCV 2020. LNCS, vol. 12347, pp. 228–243. Springer, Cham (2020). https://doi.org/10.1007/978-3-030-58536-5_14
8. Han, K., Huang, Y., Chen, Z., Wang, L., Tan, T.: Prediction and Recovery for Adaptive Low-Resolution Person Re-Identification. In: Vedaldi, A., Bischof, H., Brox, T., Frahm, J.-M. (eds.) ECCV 2020. LNCS, vol. 12371, pp. 193–209. Springer, Cham (2020). https://doi.org/10.1007/978-3-030-58574-7_12
9. Han, K., Huang, Y., Song, C., Wang, L., Tan, T.: Adaptive super-resolution for person re-identification with low-resolution images. Pattern Recogn. **114**, 107682 (2021)
10. Han, K., Si, C., Huang, Y., Wang, L., Tan, T.: Generalizable person re-identification via self-supervised batch norm test-time adaption **36**, pp. 817–825 (2022)
11. He, K., Zhang, X., Ren, S., Sun, J.: Deep residual learning for image recognition. In: Proceedings of the IEEE conference on computer vision and pattern recognition (CVPR), pp. 770–778 (2016)

12. Hirzer, M., Beleznai, C., Roth, P.M., Bischof, H.: Person Re-identification by Descriptive and Discriminative Classification. In: Heyden, A., Kahl, F. (eds.) SCIA 2011. LNCS, vol. 6688, pp. 91–102. Springer, Heidelberg (2011). https://doi.org/10.1007/978-3-642-21227-7_9

13. Hong, P., Wu, T., Wu, A., Han, X., Zheng, W.: Fine-grained shape-appearance mutual learning for cloth-changing person re-identification. In: Proceedings of the IEEE/CVF conference on computer vision and pattern recognition (CVPR), pp. 10513–10522 (2021)

14. Huang, Y., Wu, Q., Xu, J., Zhong, Y.: SBSGAN: suppression of inter-domain background shift for person re-identification. In: Proceedings of the IEEE/CVF International Conference on Computer Vision (ICCV), pp. 9527–9536 (2019)

15. Huang, Y., Wu, Q., Xu, J., Zhong, Y., Zhang, Z.: Clothing status awareness for long-term person re-identification. In: Proceedings of the IEEE/CVF International Conference on Computer Vision (ICCV), pp. 11895–11904 (2021)

16. Huang, Y., Wu, Q., Xu, J., Zhong, Y., Zhang, Z.: Unsupervised domain adaptation with background shift mitigating for person re-identification. Int. J. Comput. Vis. **129**(7), 2244–2263 (2021). https://doi.org/10.1007/s11263-021-01474-8

17. Huang, Y., Xu, J., Wu, Q., Zheng, Z., Zhang, Z., Zhang, J.: Multi-pseudo regularized label for generated data in person re-identification. Trans. Image Process. **28**(3), 1391–1403 (2018)

18. Huang, Y., Xu, J., Wu, Q., Zhong, Y., Zhang, P., Zhang, Z.: Beyond scalar neuron: adopting vector-neuron capsules for long-term person re-identification. TCSVT **30**(10), 3459–3471 (2019)

19. Ionescu, C., Papava, D., Olaru, V., Sminchisescu, C.: Human3.6m: large scale datasets and predictive methods for 3D human sensing in natural environments. TPAMI 36(7), 1325–1339 (2013)

20. Isobe, T., Zhu, F., Wang, S.: Revisiting temporal modeling for video super-resolution. arXiv preprint arXiv:2008.05765 (2020)

21. Kanazawa, A., Black, M.J., Jacobs, D.W., Malik, J.: End-to-end recovery of human shape and pose. In: CVPR, pp. 7122–7131(2018)

22. Kingma, D.P., Ba, J.: Adam: a method for stochastic optimization. arXiv preprint arXiv:1412.6980 (2014)

23. Kolotouros, N., Pavlakos, G., Black, M.J., Daniilidis, K.: Learning to reconstruct 3d human pose and shape via model-fitting in the loop. In: Proceedings of the IEEE/CVF international conference on computer vision (ICCV), pp. 2252–2261 (2019)

24. Li, D., Zhang, Z., Chen, X., Huang, K.: A richly annotated pedestrian dataset for person retrieval in real surveillance scenarios. Trans. Image Process. **28**(4), pp. 1575–1590 (2018)

25. Li, J., Wang, J., Tian, Q., Gao, W., Zhang, S.: Global-local temporal representations for video person re-identification. In: Proceedings of the IEEE/CVF international conference on computer vision (CVPR), pp. 3958–3967 (2019)

26. Li, Y.J., Luo, Z., Weng, X., Kitani, K.M.: Learning shape representations for clothing variations in person re-identification. arXiv preprint arXiv:2003.07340 (2020)

27. Liu, K., Ma, B., Zhang, W., Huang, R.: A spatio-temporal appearance representation for viceo-based pedestrian re-identification. In: Proceedings of the IEEE international conference on computer vision (ICCV), pp. 3810–3818 (2015)

28. Liu, X., Zhang, P., Yu, C., Lu, H., Yang, X.: Watching you: global-guided reciprocal learning for video-based person re-identification. In: Proceedings of the IEEE/CVF Conference on Computer Vision and Pattern Recognition (CVPR), pp. 13334–13343 (2021)

29. Loper, M., Mahmood, N., Romero, J., Pons-Moll, G., Black, M.J.: SMPL: a skinned multi-person linear model. ACM Trans. Graph. (TOG), 34(6), 1–16 (2015)
30. Niu, K., Huang, Y., Ouyang, W., Wang, L.: Improving description-based person re-identification by multi-granularity image-text alignments. Trans. Image Process. **29**,5542–5556 (2020)
31. Niu, K., Huang, Y., Wang, L.: Fusing two directions in cross-domain adaption for real life person search by language. In: Proceedings of the IEEE/CVF International Conference on Computer Vision Workshops (ICCV) Workshops (2019)
32. Niu, K., Huang, Y., Wang, L.: Textual dependency embedding for person search by language. In: Proceedings of the 28th ACM International Conference on Multimedia (ACMMM), pp. 4032–4040 (2020)
33. Pathak, P., Eshratifar, A.E., Gormish, M.: Video person re-id: fantastic techniques and where to find them. arXiv preprint arXiv:1912.05295 (2019)
34. Qian, X., et al.: long-term cloth-changing person re-identification. arXiv preprint arXiv:2005.12633 (2020)
35. Shu, X., Li, G., Wang, X., Ruan, W., Tian, Q.: Semantic-guided pixel sampling for cloth-changing person re-identification. IJIS **28**, 1365–1369 (2021)
36. Sun, Y., Zheng, L., Yang, Y., Tian, Q., Wang, S.: Beyond Part Models: Person Retrieval with Refined Part Pooling (and A Strong Convolutional Baseline). In: Ferrari, V., Hebert, M., Sminchisescu, C., Weiss, Y. (eds.) ECCV 2018. LNCS, vol. 11208, pp. 501–518. Springer, Cham (2018). https://doi.org/10.1007/978-3-030-01225-0_30
37. Uddin, M.K., Lam, A., Fukuda, H., Kobayashi, Y., Kuno, Y.: Fusion in dissimilarity space for RGB-d person re-identification. Array **12**, 100089 (2021)
38. Wan, F., Wu, Y., Qian, X., Chen, Y., Fu, Y.: When person re-identification meets changing clothes. In: Proceedings of the IEEE/CVF Conference on Computer Vision and Pattern Recognition Workshops (CVPRW), pp. 830–831 (2020)
39. Wang, G., Yuan, Y., Chen, X., Li, J., Zhou, X.: Learning discriminative features with multiple granularities for person re-identification. In: Proceedings of the 26th ACM international conference on Multimedia (ACMMM), pp. 274–282 (2018)
40. Wang, K., Ma, Z., Chen, S., Yang, J., Zhou, K., Li, T.: A benchmark for clothes variation in person re-identification. Int. J. Intell. Syst. **35**(12), 1881–1898 (2020)
41. Wang, Y., Zhang, P., Gao, S., Geng, X., Lu, H., Wang, D.: Pyramid spatial-temporal aggregation for video-based person re-identification. In: Proceedings of the IEEE/CVF International Conference on Computer Vision (ICCV), pp. 12026–12035 (2021)
42. Wu, A., Zheng, W.S., Lai, J.H.: Robust depth-based person re-identification. IEEE Trans. Image Process. **26**(6), 2588–2603 (2017)
43. Yan, Y., et al.: Learning multi-granular hypergraphs for video-based person re-identification. In: Proceedings of the IEEE/CVF conference on computer vision and pattern recognition (CVPR), pp. 2899–2908 (2020)
44. Yang, Q., Wu, A., Zheng, W.S.: Person re-identification by contour sketch under moderate clothing change. IEEE Trans. Pattern Anal. Mach. Intell. **43**(6), 2029–2046 (2019)
45. Yu, S., Li, S., Chen, D., Zhao, R., Yan, J., Qiao, Y.: Cocas: a large-scale clothes changing person dataset for re-identification. In: Proceedings of the IEEE/CVF Conference on Computer Vision and Pattern Recognition (CVPR), pp. 3400–3409 (2020)
46. Zhang, P., Wu, Q., Xu, J., Zhang, J.: Long-term person re-identification using true motion from videos. In: 2018 IEEE Winter Conference on Applications of Computer Vision (WACV), pp. 494–502 (2018)

47. Zhang, Z., Lan, C., Zeng, W., Chen, Z.: Multi-granularity reference-aided attentive feature aggregation for video-based person re-identification. In: Proceedings of the IEEE/CVF conference on computer vision and pattern recognition (CVPR), pp. 10407–10416 (2020)
48. Zheng, L., Bie, Z., Sun, Y., Wang, J., Su, C., Wang, S., Tian, Q.: MARS: A Video Benchmark for Large-Scale Person Re-Identification. In: Leibe, B., Matas, J., Sebe, N., Welling, M. (eds.) ECCV 2016. LNCS, vol. 9910, pp. 868–884. Springer, Cham (2016). https://doi.org/10.1007/978-3-319-46466-4_52
49. Zheng, Z., Zheng, N., Yang, Y.: Parameter-efficient person re-identification in the 3D space. arXiv preprint arXiv:2006.04569 (2020)

Robustizing Object Detection Networks Using Augmented Feature Pooling

Takashi Shibata[1][✉], Masayuki Tanaka[2], and Masatoshi Okutomi[2]

[1] NTT Corporation, Kanagawa, Japan
t.shibata@ieee.org
[2] Tokyo Institute of Technology, Tokyo, Japan

Abstract. This paper presents a framework to robustize object detection networks against large geometric transformation. Deep neural networks rapidly and dramatically have improved object detection performance. Nevertheless, modern detection algorithms are still sensitive to large geometric transformation. Aiming at improving the robustness of the modern detection algorithms against the large geometric transformation, we propose a new feature extraction called augmented feature pooling. The key is to integrate the augmented feature maps obtained from the transformed images before feeding it to the detection head without changing the original network architecture. In this paper, we focus on rotation as a simple-yet-influential case of geometric transformation, while our framework is applicable to any geometric transformations. It is noteworthy that, with only adding a few lines of code from the original implementation of the modern object detection algorithms and applying simple fine-tuning, we can improve the rotation robustness of these original detection algorithms while inheriting modern network architectures' strengths. Our framework overwhelmingly outperforms typical geometric data augmentation and its variants used to improve robustness against appearance changes due to rotation. We construct a dataset based on MS COCO to evaluate the robustness of the rotation, called COCO-Rot. Extensive experiments on three datasets, including our COCO-Rot, demonstrate that our method can improve the rotation robustness of state-of-the-art algorithms.

1 Introduction

There has been remarkable progress in object detection by modern network architectures [3,32,45], large image datasets with accurate annotations [16,17,33], and sophisticated open-sources [1,5,41,54]. Despite these successes, a significant issue still remains; it is sensitive to unexpected appearance changes in the wild, such as geometric transformation, occlusions, and image degradation. Particularly, rotation robustness is simple yet significant for object detection. In such cases as first-person vision, drone-mounted cameras, and robots in accidents and disasters, images are taken with unexpected camera poses and often contain large rotations.

Supplementary Information The online version contains supplementary material available at https://doi.org/10.1007/978-3-031-26348-4_6.

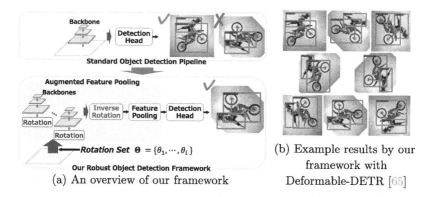

(a) An overview of our framework

(b) Example results by our framework with Deformable-DETR [65]

Fig. 1. An overview of our proposed framework (a) and example results by our framework (b). The rotation robustness of a modern object detection network can be substantially improved by our framework based on augmented feature pooling.

A typical approach to improve robustness to geometric transformation is data augmentation (DA) [10,11,30,46] and test-time augmentation (TTA [21,47]). Although the DA and TTA are essential learning protocols for image classification, they are powerless for rotation transformation in object detection. The reason for this is that a bounding box for an augmented image with rotation becomes much looser than the originally annotated bounding box, as we will describe in details in Sect. 3. The loosened bounding box includes a large area of background, which dramatically harm training and inference performance. This loosened bounding box problem is a common and significant challenge in object detection with the large geometric transformations such as rotation. As we will show later, DA and TTA cannot overcome the loosened bounding box problem.

A further challenge in improving the robustness of the object detection network to geometric transformation is the orientation bias of backbone feature extraction. The weight of the common backbones for object detection networks, e.g. ResNet [21] and Swin Transformer [37], are optimized for the frontal direction due to the orientation bias of the training data. Those standard backbones cannot be directly applicable to object detection tasks with arbitrary rotations. There is a strong demand for a general framework that can easily inherit the strengths of highly expressive backbones and modern object detection architectures while improving robustness to geometric transformations.

We propose a robust feature extraction framework for large geometric transformation based on augmented feature pooling. This paper focuses on rotation as a simple-yet-influential case of geometric transformation, while our framework is applicable to any kind of geometric transformations. The key is to integrate feature maps obtained from geometrically augmented images before feeding it to a detection head. It can be achieved by adding two processes: inverse rotation and feature pooling as shown in Fig. 1 (a). Examples of bounding boxes detected by our framework with Deformable DETR [65] are shown in Fig. 1 (b). We aim to improve robustness without additional annotation cost, so we only use the already annotated bounding boxes. Despite that, our proposed augmented feature pooling can substantially improve the robustness against rotation by simply adding a few lines of code to the original implementation of the existing method, and fine-tuning the model while freezing the backbone parameters. It can be easily applied to highly expressive backbones with many parameters, e.g. Swin-Transformer [37], because the proposed method does not require backbone optimization.

Our main contributions are summarized as follows[1] 1) We propose a rotation robust feature extraction framework based on augmented feature pooling, which is applicable to various modern object detection networks. 2) We conducted preliminary experiments for investigating the problems of object detection networks on rotation robustness. 3) We constructed an object detection dataset with arbitrary rotations using MS COCO [33] for evaluating the robustness against the rotation. 4) Extensive experiments on three datasets demonstrate that our method can significantly improve the performance of state-of-the-art methods.

2   Related Works

Object Detection. The architectures of recent object detection consist of three components: backbone, neck, and detection head. Based on the detection head's architecture, the existing object detection algorithms can be classified into single-stage detectors [6,19,32,35,42–44] and two-stage detectors [2,4,22,39,40,45,49, 53]. While anchors are widely used, anchor-free approaches [29,48,59,64] and keypoint-based approaches [28,62] have been proposed. Beyond those CNN-based methods, Transformers have also been employed in detection networks, combining a transformer-based architecture [3,13,65] with a CNN-based backbone [12,18,21,36,56] or using a transformer-based backbone [37]. Those methods implicitly assume that the target objects are facing in the front.

Data Augmentation and Test Time Augmentation. Data augmentation (DA) has become essential for training protocol. Learnable data augmentation algorithms by reinforcement learning and random search have been proposed [10,11,30]. Data augmentation is also effective at inference, which is called Test-Time Augmentation (TTA) [21,47]. If those DA and TTA for the classification task are naively applied to the rotated-object detection task, the detection performance will be significantly degraded because the augmentation makes the bounding box loose. Recently, a DA algorithm [25] to handle that problem has been proposed to approximate the bounding box with an inscribed ellipse to improve the robustness for small rotation. In contrast, our proposed method can be more robust to larger rotations.

Rotation-Invariant CNNs and Datasets. Rotation invariance is a fundamental challenge in pattern recognition, and many approaches have been proposed. Aiming at extracting features invariant to affine transformations including rotation, various network architectures have been proposed [7–9,26,38,50,52, 57,60,61,63]. Alignment-based approaches [23,24,51] have also been presented.

---

[1] Our code of will be available at http://www.ok.sc.e.titech.ac.jp/res/DL/index.html.

Those methods are not directly applicable to the state-of-the-art object detection algorithms because those methods do not support the latest advantages including transformer-based approaches [37,65].

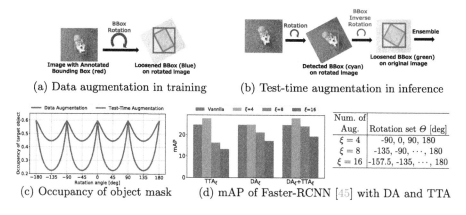

(a) Data augmentation in training   (b) Test-time augmentation in inference

(c) Occupancy of object mask   (d) mAP of Faster-RCNN [45] with DA and TTA

Fig. 2. Loosened bounding-box problem of DA and TTA.

For specific applications such as remote sensing focusing on rotation robustness, network architectures and image datasets have been designed based on an oriented bounding box [14,15,20,34,55,58], where its rotation angle information is annotated in addition to the center position, width, and height. While the oriented bounding box is practical for these specific applications, it is not applicable to standard object detection datasets [16,17,33]. It is also difficult to share the advantages of modern object detection developed for the standard datasets.

3   Two Challenges of Object Detection for Rotation Robustness

We discuss two challenges of object detection for rotation: the loosened bounding boxes and the sensitivity of backbone feature extraction on object detection task.

Loosened Bounding Box. When we apply geometrical data augmentation, we need to generate a new bounding box for the rotated image. If there is no segmented mask along object boundary, we have to generate a new bounding box from the originally annotated bounding box. The bounding rectangle of the rotated bounding box is commonly used as the new bounding box.

Consequently, geometrical transformation of the data augmentation (DA) and the test time augmentation (TTA) generates loosened bounding boxes, as shown in Fig. 2 (a) and (b). To evaluate the looseness of the generated bounding box, we measured the occupancy of the target object in the generated bounding box for each rotation angle during training with DA and inference with TTA[2]

---

[2] Note that the TTA curve assumes that each inference before ensemble is ideal, and thus this occupancy is the upper bound.

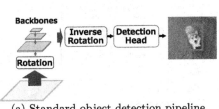

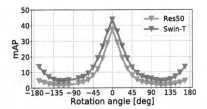

(a) Standard object detection pipeline with rotation and inverse rotation

(b) Detection accuracy (mAP) of (a) with respect to the rotation angle.

Fig. 3. Rotation sensitivity for backbone feature extraction.

using MS COCO [33] as shown in Fig. 2 (c). These analysis show that (i) the loosened bounding box problem can only be avoided at integer multiples of $90°C$, where the occupancy does not decrease, (ii) the looseness is dramatically increased where there is a deviation from those four angles.

To demonstrate the harm of the loosened bounding box for DA and TTA, we measured the mAP of Faster-RCNN [45] with DA and TTA on the rotated version of MS COCO val dataset (we will describe details in Sect. 5.1) as shown in Fig. 2(d). For DA and TTA, we evaluated mAP using three sets of rotation angles for data augmentations denoted by $\xi=4$, 8, and 16, where ξ is the number of augmentation. Here, angles are assigned at equal intervals from all directions, depending on ξ. More specifically, those three rotation sets are given by $\Theta=[-180, -90, 0, 90]$ for $\xi=4$, $\Theta=[-180, -135, \cdots, 135]$ for $\xi=8$, and $\Theta=[-180, -157.5, \cdots, 157.5]$ for $\xi=16$, respectively. As shown in Fig. 2(d), DA, TTA and those combination are only effective for $\xi = 4$ because the loosened bounding box degrades the training and inference performances for any angle except for the integer multiples of $90°C$. The performance of DA and TTA (except for $\xi = 4$) become worse than naive fine tuning where the detection head is naively refined using training data containing arbitrary rotation.

Sensitivity of Backbone Feature Extraction. Backbone feature extraction of the object detection network is also sensitive to rotation transformations, i.e. the backbone feature extraction is not rotation invariant. When the rotated image is used as the input image, as shown in Fig. 3(a), the feature map obtained by the backbone feature map is also rotated. Even though the rotated feature map is aligned by inverse rotation, the feature map still deviates significantly from the feature map extracted from the original input image. As a result, the rotation of the input image dramatically reduces the detection accuracy (see Fig. 3(b)). Surprisingly, the sensitivity of the backbone to rotation is a common challenge, not only in commonly used backbones like the ResNet50 [21], but also in modern transformers like the Swin-Transformer [37].

4 Proposed Rotation Robust Object Detection Framework

Our proposed method aims to improve the robustness to large geometric transformations such as rotation while inheriting the strengths and weights of existing

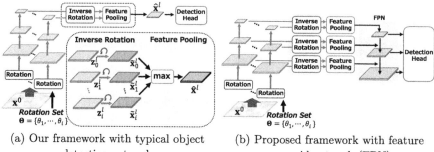

(a) Our framework with typical object detection network

(b) Proposed framework with feature pyramid network (FPN)

Fig. 4. Proposed augmented feature pooling architecture. Feature maps obtained from the rotated images are inversely rotated and integrated by feature pooling. Then, the integrated features are fed into a detection head or FPN.

detection networks, avoiding the loosened bounding box problem and the sensitivity of backbone feature extraction. The key is to introduce augmented feature pooling, which integrates the set of the feature maps obtained from the rotated images before feeding it to the detection head. In the following, we describe our augmented feature pooling and its extension to Feature Pyramid Network [31]. Then, we explain how to extend our framework to transformer-based backbones such as Swin Transformer [37], and discuss its application to modern object detection networks.

4.1 Architecture of Augmented Feature Pooling

Figure 4 shows an overview of our proposed augmented feature pooling. Our augmented feature pooling is a simple architecture that inserts the inverse rotation and the feature pooing between the backbone and the detection head. Let $\mathbf{x}^0$, $\mathbf{x}^l$, and F^l be an input image, the l-th stage's feature map, and the backbone of l-th stage, respectively. The l-th stage's feature map $\mathbf{x}^l$ is obtained from $(l{-}1)$-th stage's feature map $\mathbf{x}^{l-1}$ and l-th stage's backbones F^l as follows:

$$\mathbf{x}^l = F^l \circ \mathbf{x}^{l-1} = F^l(\mathbf{x}^{l-1}), \tag{1}$$

where "$\circ$" is composing operator.

We start our discussion with our proposed augmented feature pooling with a standard detection architecture shown in Fig. 4(a). We generate a set of augmented feature maps by the rotation angle θ defined by the rotation set $\Theta = \{\theta_1, \cdots, \theta_i\}$. To obtain this set of the augmented feature maps, we first generate the set of rotated images $Z^0 = \{\mathbf{z}_0^0, \cdots \mathbf{z}_i^0\}$, where the i-th rotated image $\mathbf{z}_i^0$ is generated by $\mathbf{z}_i^0 = R_{\theta_i}(\mathbf{x}^0)$ using the rotation operator R_θ. The rotation operator R_θ represents the rotation within the image plane by angle theta around the image center. Each of these rotated images $\mathbf{z}_i^0$ is fed to the backbone $F = F^l \circ F^{l-1} \circ \cdots \circ F^1$, resulting in the set of rotated feature maps $Z^l = \{\mathbf{z}_0^l, \cdots \mathbf{z}_i^l\}$.

Inverse Rotation. Alignment of feature maps is essential for feature pooling from the augmented feature maps because the object detection task simultaneously estimates the bounding box location with the class label. A set of the aligned feature maps $\tilde{X}^l$ corresponding to each augmentation is obtained by the inverse rotation $R_{-\theta_i}(\cdot)$ as follows:

$$\tilde{X}^l = \{\tilde{\mathbf{x}}_0^l, \cdots, \tilde{\mathbf{x}}_i^l\} = \{R_{-\theta_0}(\tilde{\mathbf{z}}_0^l), \cdots, R_{-\theta_i}(\tilde{\mathbf{z}}_i^l)\}. \tag{2}$$

Feature Pooling. Our proposed feature pooling performs a element-wise max pooling from the set of aligned feature maps as

$$\left(\hat{\mathbf{x}}^l\right)_k = \max_i \left(\tilde{\mathbf{x}}_i^l\right)_k, \tag{3}$$

where k is an index of an element of the feature map, and $(\hat{\mathbf{x}}^l)_k$ and $(\tilde{\mathbf{x}}_i^l)_k$ are the k-th element of $\hat{\mathbf{x}}^l$ and $\tilde{\mathbf{x}}^l$, respectively. From Eqs. (1), (2) and (3), our augmented feature pooling with the rotation set Θ is formally given by

$$\left(\hat{\mathbf{x}}^l\right)_k = \max_{\theta \in \Theta} \left(R_{-\theta}(F^l \circ \cdots \circ F^1 \circ R_\theta(\mathbf{x}))\right)_k. \tag{4}$$

In a typical object detection task, the extracted raw feature map $\mathbf{x}^l$ is used as an input for the detection head. Our proposed method feeds the pooled feature map $\hat{\mathbf{x}}^l$ to the detection head instead of the raw feature map $\mathbf{x}^{l3}$.

Extension to Feature Pyramid Network. Our proposed framework can be easily extended to Feature Pyramid Networks (FPN) [31] as shown in Fig. 4(b). The inverse rotation and the feature pooling are applied to the augmented feature maps for each stage, and those pooled feature maps are fed to the FPN module. The set of pooled feature maps is denoted as $\{\hat{\mathbf{x}}^0 \cdots \hat{\mathbf{x}}^m \cdots \hat{\mathbf{x}}^l\}$. Using Eq. (4), the m-th stage's pooled feature map $\hat{\mathbf{x}}^m$ is formally represented as follows:

$$\left(\hat{\mathbf{x}}^m\right)_k = \max_{\theta \in \Theta} \left(R_{-\theta}(F^m \circ \cdots \circ F^1 \circ R_\theta(\mathbf{x}))\right)_k. \tag{5}$$

Rotation Set Designs. By designing the rotation set Θ, our proposed method can control whether to focus on a specific angle range or robust to arbitrary rotation. For example, if the rotation set Θ is uniform and dense, the robustness against arbitrary angles is improved, which is the main focus of this paper. On the other hand, if the rotation set Θ is intensively sampled around a target angle, e.g. 0 [deg] for the case where the target object is approximately facing in the front, the robustness of object detection accuracy around the target angle is improved. We will discuss the effectiveness of the rotation set designs in Sect. 5.2.

Beyond CNNs: Transformer-Based Backbone. Our proposed method can also be applied to transformer-based backbones with spatial structures such as Swin Transformer [37]. Figure 5 shows the details of the CNN-based and the Swin Transformer-based architectures. When FPN is used together with

---

[3] The dimensions of feature map $\mathbf{x}^l$ are the same as the original backbones.

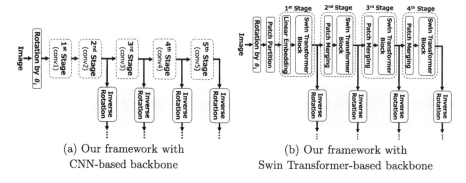

(a) Our framework with   (b) Our framework with
CNN-based backbone   Swin Transformer-based backbone

Fig. 5. Extension to transformer-based backbone.

CNN-based backbones, e.g. ResNet [21], ResNeXt [56], our augmented feature pooling is applied to the feature maps obtained from Stages-2, 3, 4, and 5. On the other hand, for Swin Transformer, our augmented feature pooling is applied to the feature map immediately after the *Swin-Transformer Block* of each stage, i.e. just before the *Patch Merging*. The proposed method can be easily applied to the transformer-based backbone with spatial structure and thus inherits their rich feature representation and pre-trained weight.

4.2 Applying to Object Detection Networks

Our proposed framework is applicable to various types of detection heads such as single-stage detectors [32,42], two-stage detectors [45], and transformer-based detectors [65] without any changes in those detection head's architectures. Our framework aims to improve the robustness of rotation-sensitive detectors while taking advantage of the weight of the pre-trained backbones. We consider this robustness improvement as a downstream task, freezing the backbone and optimizing only the parameters of the detection head. Limiting the optimization parameters to the detection head allows us to quickly achieve robustness against the rotation transformation with much less computation than optimizing all parameters including the backbone.

5 Experiments

5.1 Setting

Datasets and Evaluation Measures. While MNIST-Rot12k [27], which is a rotated version of the original MNIST in any direction, is widely used for the classification task, there is no common dataset with this kind for the generic object detection task. Therefore, we constructed a new dataset[4] containing arbitrary rotation using MS COCO [33], called *COCO-Rot*, and evaluated the performance of our augmented feature pooling. Our *COCO-Rot* is composed of

---

[4] The details of our dataset are described in our supplemental.

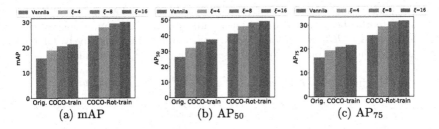

(a) mAP   (b) AP$_{50}$ (c) AP$_{75}$

Fig. 6. Performance on *COCO-Rot-val* by the proposed method with various number of augmentations ξ. The mAP, AP$_{50}$, and AP$_{75}$ with MS COCO-train [33] and *COCO-Rot-train* as training data are shown, respectively.

COCO-Rot-train and *COCO-Rot-val*, which were generated from the original MS COCO training and validation data, respectively. We automatically annotated the bounding box for *COCO-Rot-train* and *COCO-Rot-val* based on rotated ground truth segmentation mask instead of manual annotation. The numbers of images for *COCO-Rot-train* and *COCO-Rot-val* are 118K and 5K, respectively. For training, we used the original MS COCO-train [33] or *COCO-Rot-train* as training data, respectively. In addition, we also demonstrated our performance on two publicly available datasets, PASCAL VOC [17] and Synthetic Fruit Dataset [25]. We used MS COCO detection evaluation measures [33], i.e. the mean Average Precision (mAP), AP$_{50}$, and AP$_{75}$.

Implementation Details. We implemented our code based on MMDetection [5] with PyTorch [41]. The default training protocol in MMDetection [5] was employed unless otherwise noted. SGD was used for optimization, and the training schedule is 1x (i.e. 12 epochs with warmup and step decay, the learning rate is set to 2.0×10^{-2} to 2.0×10^{-4} for Faster-RCNN [45]). NVIDIA A100, P100 and K80 GPUs were used for our experiments. For evaluation of our framework and existing framework, we only trained the heads while fixing the feature extraction backbones. Unless otherwise noted, we used the pre-trained model using the original MS COCO [33] as initial weights for training. Resnet 50 was used in most of our experiments. We set the batch size to 16. The detailed training protocols are described in our supplementary material.

5.2 Effectiveness of Augmented Feature Pooling

We demonstrate the effectiveness of our proposed method in terms of the effectiveness by increasing the number of augmentations, the comparison with DA and TTA, the applicability to various backbones, and the effectiveness of our rotation set design using Faster-RCNN [45]. Applicability to various detection heads will be described in Sec. 5.3.

Improvement by Increasing the Number of Augmentations. We first demonstrate the effectiveness of our proposed method by increasing the number of augmentations. Subscript ξ represents the number of augmentations as Ours$_\xi$. For example, ours with the four augmentations is denoted as Ours$_4$. We used three

Table 1. Performance of mAP on *COCO-Rot-val* by DA, TTA, and ours. Bold and italic indicate the best and second best results for each column, respectively. MS COCO-train and *COCO-Rot-train* are used as training data, respectively. Green and red characters show the increase or decrease in performance from vanilla.

(a) Original MS COCO-train.

Method	mAP
Vanilla	15.8
+ TTA$_4$	24.7 (+8.9)
+ DA$_4$	21.2 (+5.4)
+ DA$_4$+TTA$_4$	24.3 (+8.5)
Ours$_{16}$	21.3 (+5.5)
+ TTA$_4$	26.3 (+10.5)
+ DA$_4$	**26.7 (+10.9)**
+ DA$_4$ + TTA$_4$	*26.5 (+10.7)*

(b) *COCO-Rot-train.*

Method	mAP
Vanilla	24.6
+ TTA$_4$	27.6 (+3.0)
+ DA$_4$	24.8 (+0.2)
+ DA$_4$+TTA$_4$	27.8 (+3.2)
+ *Oracle DA$_4$*	24.8 (+0.2)
+ *Oracle DA$_4$*+TTA$_4$	27.7 (+3.1)
Ours$_{16}$	**30.0 (+5.4)**
+ TTA$_4$	**30.0(+5.4)**

rotation sets with different numbers of augmentations $\xi = 4$, 8, and 16, where ξ is the number of augmentation for ours. Angles are assigned at equal intervals from all directions, depending on ξ. We only train the head of the detection network using *COCO-Rot-train* or the original MS COCO-train [33] while fixing the feature extraction backbones. The performance on *COCO-Rot-val* were evaluated. We also evaluated the performances of the original backbone feature extraction without our augmented feature pooling, which we call *vanilla* in the following. In vanilla, the original backbone is also frozen during training for their detection heads using *COCO-Rot-train* or original MS COCO-train, respectively.

Figure 6 shows mAP, AP$_{50}$ and AP$_{75}$ on *COCO-Rot-val* by the proposed method with various number of augmentations ξ. In our method, the performance is steadily improved as ξ is increased because we can avoid the loose bounding box problem. Note that naive DA, TTA and those combinations are only effective for $\xi = 4$ as shown in Sect. 3 (see Fig. 2(d)). In the following, unless otherwise noted, for the number of augmentations in the following experiments, we set $\xi = 16$ for our proposed method, which was highest mAPs for our method. On the other hand, we fix $\xi = 4$ for DA, TTA as DA$_4$ and TTA$_4$ because DA, TTA and those combination are only effective for $\xi = 4$, i.e. $\Theta=[-180, -90, 0, 90]$ (Table 1).

Comparison with DA and TTA. To demonstrate the superiority of our augmented feature pooling, we evaluated the performance of our method, DA, and TTA. For fair comparison, we also used *COCO-Rot-train* to train the heads for vanilla, DA, and TTA. We also compared our approach with a tightened bounding box using the instance segmentation mask label when perform rotation augmentation, called as *Oracle DA$_4$*. Table 1 shows mAP of our method, vanilla, DA, *Oracle DA*, and TTA. The values in parentheses are the increase (green) from mAP of the vanilla backbone feature extraction. We can clearly see that our proposed method with DA (Table 1(a)) or TTA (Table 1 (b)) can achieve the highest mAP compared to naive DA and TTA[5]. Note that our proposed method

---

[5] As shown in our supplemental, AP$_{50}$ and AP$_{75}$ are also the highest in the proposed method as well as mAP.

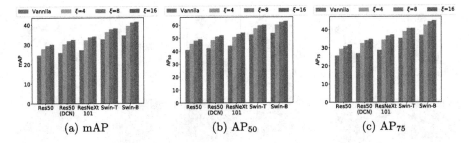

Fig. 7. Performance on *COCO-Rot-val* by the proposed method with various backbones. The proposed method is applicable to both CNN-based and transformer-based backbones. The robustness is improved by increasing the number of augmentations ξ.

also outperforms the tight box-based DA, called *Oracle DA*. This is because our augmented feature pooling can simultaneously solve both the rotation-bias problem of backbone feature extraction and the loosened bounding box problem as discussed in Sect. 3.

Applicability to Various Backbones. Our proposed method is effective for both CNN-based and transformer-based backbones. We evaluated the applicability of our framework using the following five major backbones with FPN [31], Resnet50 [21], Resnet50 [21] with DCN [12], Resnet101 [21], ResNeXt101 [56], Swin-T [37], and Swin-S [37]. The performance was evaluated on *COCO-Rot-val*. Again, the rotation set Θ was defined by uniformly assigning from all directions with equal intervals according to the number of augmentations ξ. Figure 7 shows mAP, AP_{50} and AP_{75} of our proposed framework with various numbers of augmentations and that of vanilla. Our framework substantially improves the performance for all backbones compared with the vanilla backbone feature extraction. We can see that the mAP, AP_{50} and AP_{75} are improved by increasing the number of augmentations ξ for all backbones. Note that our proposed method can further improve the performance of DCN [12] designed to compensate for the positional deformation. Our proposed method is applicable to such geometrical-transformation-based backbones. Figure 8 shows the visual comparisons between our proposed method and vanilla. There are many false positives in the vanilla backbone feature extraction (blue arrows in the first and the second rows) and false negatives (green arrows in the third row). Specifically, skateboards and people are falsely detected (blue arrows) in the first row, and fire hydrants (green arrows) can not be detected in the third row. In contrast, our proposed method can successfully detect those objects, even when using the same training dataset *COCO-Rot-train*. More visual comparisons are shown in our supplemental.

Effectiveness of Rotation Set Design. As described in Sect. 4.1, our framework can control whether to focus on the robustness to arbitrary angle range or a specific angle range by designing the rotation set Θ. To demonstrate this, we evaluated mAP, AP_{50}, and AP_{75} for the three rotation set designs denoted by Set 1, Set 2, and Set 3 shown in Fig. 9. Here, Set 1 has only a single angle at 0 [deg], Set 2 has the five angles equally sampled among ±45-° range, and Set 3

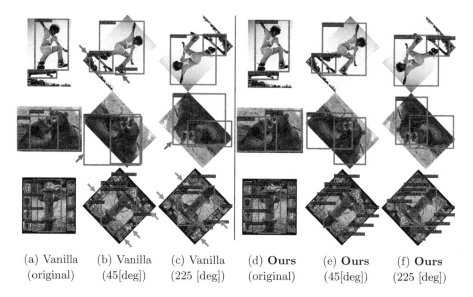

(a) Vanilla	(b) Vanilla	(c) Vanilla	(d) **Ours**	(e) **Ours**	(f) **Ours**
(original)	(45[deg])	(225 [deg])	(original)	(45[deg])	(225 [deg])

Fig. 8. Example results by our proposed method and vanilla using Faster-RCNN [45] with Swin-Transformer [37]. In the results of vanilla, the green and the blue arrows, i.e. ↖ and ↖, indicate **false negatives** and **false positives**, respectively. Contrary to vanilla backbone feature extraction, our proposed method can detect target objects with accurate bounding boxes for various rotation angles.

has 16 angles equally sampled among ± 180-° ranges. Figure 9 shows mAP, AP_{50}, and AP_{75} for each rotation angle for those three rotation set designs. Compared to Set 1 (blue line), mAP, AP_{50} and AP_{75} for Set 2 (green line) are improved in the wide-angle range centered on 0 [deg]. Furthermore, in Set 3 (red line), mAP, AP_{50}, and AP_{75} are improved on average for all rotation angles. From these results, we can see that our proposed method enables us to improve the robustness for arbitrary angles, and at the same time, it can also improve the robustness for a specific angle range by designing the rotation set.

Other Datasets. We compared the performance of our proposed method with the state-of-the-art method [25] focusing rotation augmentation for object detection using the PASCAL VOC and Synthetic Fruit datasets. For a fair comparison, the backbone and the detection head were both optimized as in [25]. We used ResNet50, which has the smallest expressive power of the backbone, in our implementation. Here, we set the rotation set Θ as seven angles sampled at equal intervals from a range of ± 15 [deg] as in [25]. Table 2 (a) shows AP_{50} and AP_{75} of the proposed and the existing methods. Note that the value for the existing method is taken from [25]. As shown in Table 2, our proposed method achieves substantially higher performance in both AP_{50} and AP_{75}. In contrast to [25], our framework can improve the robustness over a broader range by designing the rotation set Θ as mentioned in Sect. 4.1. In this sense, our framework is a more general and versatile framework that encompasses the existing method [25].

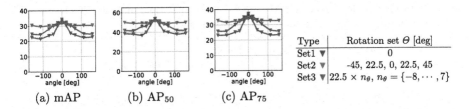

(a) mAP   (b) AP_{50} (c) AP_{75}

Type	Rotation set Θ [deg]
Set1 ▼	0
Set2 ▼	-45, 22.5, 0, 22.5, 45
Set3 ▼	$22.5 \times n_\theta, n_\theta = \{-8, \cdots, 7\}$

Fig. 9. Performance comparisons with various rotation set. Our framework can control whether to focus on robust to arbitrary rotation or a specific angle range by designing the rotation set Θ. (Color figure online)

Table 2. Result on other datasets. Bold indicates the most accurate methods. (a) Comparison of AP_{50} and AP_{75} on PASCAL VOC and Synthetic Fruit. (b) Results of AP_{50} on *PASCAL VOC-Rot*.

(a) PASCAL VOC and Synthetic Fruit

Datasets	PASCAL VOC [17]		Synth. Fruit [25]	
Methods	AP_{50}	AP_{75}	AP_{50}	AP_{75}
Ellipse+RU [25]	81.6	58.0	95.8	93.2
Ours	**89.6**	**69.4**	**96.7**	**93.6**

(b) *PASCAL VOC-Rot*

Method	AP_{50}	AP_{75}
Vanilla	64.9	35.6
+ TTA_4	72.3 (+7.4)	40.5 (+4.9)
$Ours_{16}$	77.3 (+12.4)	**45.7 (+10.1)**
+ TTA_4	**78.8 (+13.9)**	44.2 (+8.6)

In many practical scenarios, we cannot obtain an annotated segmented mask along object boundary due to the high cost of annotation. Even in such a case, the proposed method works better than naive TTA. To demonstrate this, we constructed a new dataset, called *PASCAL VOC-Rot*, by rotating the image and the originally annotated bounding box of the original PASCAL VOC in an arbitrary rotation. Table 2 (b) shows AP_{50} on *PASCAL VOC-Rot*[6]. Note that the bounding boxes for those evaluation and training datasets are loose because there are no segmented masks in PASCAL VOC. As shown in Table 2, our proposed method is relatively more effective than naive TTA (Table 2).

5.3 Applicability to Modern Object Detection Architectures

The proposed method is applicable to various types of object detection networks including single-stage, two-stage, and transformer-based architectures. To demonstrate the versatility of our proposed method, the following widely used and state-of-the-art object detection networks were used for our evaluation: Faster-RCNN [45] (two-stage), Retinanet [32] (single-stage), YOLOF [6] (single-stage), FSAF [64] (anchor-free), ATSS [59] (anchor-free), and Deformable-DETR [65] (transformer-based).

Tables 3(a) and (b) show mAP on *COCO-Rot-val* of our proposed method, vanilla with DA_4, our proposed method with TTA_4 and vanilla with TTA_4, respectively[7]. We can clearly see that our proposed method substantially

---

[6] Note that, in PASCAL VOC, the standard evaluation metric is AP_{50}.

[7] We also show AP_{50} and AP_{75} in our supplementary material.

Table 3. Overall performance of mAP on *COCO-Rot-val*. Bold indicates the best result for each column. *COCO-Rot-train* is used for training.

(a) Our method and DA with various object detection networks.

Baseline	Backbone/Neck	Vanilla	Vanilla+DA$_4$	Ours$_{16}$
Faster-RCNN [45]	ResNet50 w/ FPN	24.6	24.8 (+0.2)	**30.0 (+5.4)**
RetinaNet [32]	ResNet50 w/ FPN	24.1	24.3 (+0.2)	**29.6 (+5.5)**
FSAF [64]	ResNet50 w/ FPN	24.6	24.9 (+0.3)	**30.1 (+5.5)**
ATSS [59]	ResNet50 w/ FPN	26.6	27.0 (+0.4)	**32.2 (+5.6)**
YOLOF [6]	ResNet50	24.4	24.6 (+0.2)	**28.1 (+3.7)**
D-DETR (++ two-stage) [65]	ResNet50	35.9	37.3 (+1.4)	**39.5 (+3.6)**

(b) Our method and TTA with various object detection networks.

Baseline	Backbone/Neck	Vanilla	Vanilla+TTA$_4$	Ours$_{16}$+TTA$_4$
Faster-RCNN [45]	ResNet50 w/ FPN	24.6	27.6 (+3.0)	**30.0 (+5.4)**
RetinaNet [32]	ResNet50 w/ FPN	24.1	27.3 (+3.2)	**29.5 (+5.4)**
FSAF [64]	ResNet50 w/ FPN	24.6	27.3 (+2.7)	**30.1 (+5.5)**
ATSS [59]	ResNet50 w/ FPN	26.6	29.3 (+2.7)	**32.0 (+5.4)**
YOLOF [6]	ResNet50	24.4	26.9 (+2.5)	**28.2 (+3.8)**
D-DETR (++ two-stage) [65]	ResNet50	35.9	37.6 (+1.7)	**39.2 (+3.3)**

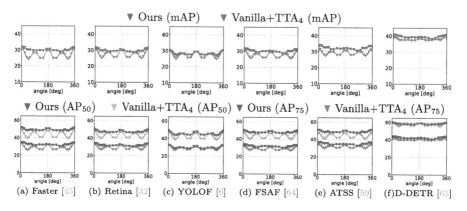

Fig. 10. Comparison of mAP, AP$_{50}$ and AP$_{75}$ for each rotation angle between our method and vanilla with TTA$_4$. The combination of various object detection with our method improves the robustness against rotation compared to vanilla with TTA$_4$.

improves the mAP for all the detection architectures than DA and TTA. Finally, we evaluated mAP, AP$_{50}$, and AP$_{75}$ of our proposed method and vanilla with TTA$_4$ for each rotation angle as shown in Fig. 10. For mAP and AP$_{75}$, our proposed method outperforms vanilla with TTA$_4$ for almost all angles. In AP$_{50}$, the proposed method is comparable to vanilla with TTA$_4$ only in [65], and our method outperforms vanilla with TTA$_4$ in other detection networks. These results show that our method is applicable to various detection networks.

6   Conclusions

We have proposed the rotation robust feature extraction framework using augmented feature pooling. The key is to integrate the augmented feature maps obtained from the rotated images before feeding it to the detection head without changing the original network architecture. We can obtain robustness against rotation using the proposed framework by freezing the backbone and fine-tuning detection head. Extensive experiments on three datasets demonstrated that our method improves the robustness of state-of-the-art algorithms. Unlike TTA and DA, the performance of the proposed method improves as the number of augmentations is increased.

References

1. Abadi, M., et al.: Tensorflow: a system for large-scale machine learning. In: 12th USENIX Symposium on Operating Systems Design and Implementation, OSDI 2016, pp. 265–283 (2016)
2. Cai, Z., Vasconcelos, N.: Cascade r-cnn: high quality object detection and instance segmentation. IEEE Trans. Pattern Anal. Mach. Intell. (TPAMI) (2019). https://doi.org/10.1109/tpami.2019.2956516
3. Carion, N., Massa, F., Synnaeve, G., Usunier, N., Kirillov, A., Zagoruyko, S.: End-to-end object detection with transformers. In: Vedaldi, A., Bischof, H., Brox, T., Frahm, J.-M. (eds.) ECCV 2020. LNCS, vol. 12346, pp. 213–229. Springer, Cham (2020). https://doi.org/10.1007/978-3-030-58452-8_13
4. Chen, K., et al.: Hybrid task cascade for instance segmentation. In: IEEE Conference Computer on Vision Pattern Recognition (CVPR) (2019)
5. Chen, K., et al.: MMDetection: open mmlab detection toolbox and benchmark. arXiv preprint arXiv:1906.07155 (2019)
6. Chen, Q., Wang, Y., Yang, T., Zhang, X., Cheng, J., Sun, J.: You only look one-level feature. In: IEEE Conference on Computer Vision Pattern Recognition (CVPR) (2021)
7. Cheng, G., Han, J., Zhou, P., Xu, D.: Learning rotation-invariant and fisher discriminative convolutional neural networks for object detection. IEEE Trans. Image Process. **28**(1), 265–278 (2018)
8. Cheng, G., Zhou, P., Han, J.: RIFD-CNN: rotation-invariant and fisher discriminative convolutional neural networks for object detection. In: IEEE Conference on Computer Vision Pattern Recognition (CVPR), pp. 2884–2893 (2016)
9. Cohen, T., Welling, M.: Group equivariant convolutional networks. In: International Conference on Machine Learning (ICML) (2016)
10. Cubuk, E.D., Zoph, B., Mane, D., Vasudevan, V., Le, Q.V.: Autoaugment: learning augmentation policies from data. In: IEEE Conference on Computer Vision Pattern Recognition (CVPR) (2019)
11. Cubuk, E.D., Zoph, B., Shlens, J., Le, Q.V.: Randaugment: practical automated data augmentation with a reduced search space. In: IEEE Conference on Computer Vision Pattern Recognition Workshop (CVPRW) (2020)
12. Dai, J., et al.: Deformable convolutional networks. In: International Conference on Computer Vision (ICCV), pp. 764–773 (2017)

13. Dai, Z., Cai, B., Lin, Y., Chen, J.: Up-detr: unsupervised pre-training for object detection with transformers. In: IEEE Conference on Computer Vision Pattern Recognition (CVPR), pp. 1601–1610 (2021)

14. Ding, J., Xue, N., Long, Y., Xia, G.S., Lu, Q.: Learning ROI transformer for oriented object detection in aerial images. In: IEEE Conference on Computer Vision Pattern Recognition (CVPR), pp. 2849–2858 (2019)

15. Ding, J., et al.: Object detection in aerial images: a large-scale benchmark and challenges (2021)

16. Everingham, M., Eslami, S.A., Van Gool, L., Williams, C.K., Winn, J., Zisserman, A.: The pascal visual object classes challenge: a retrospective. Int. J. Comput. Vis. (IJCV) **111**(1), 98–136 (2015)

17. Everingham, M., Van Gool, L., Williams, C.K., Winn, J., Zisserman, A.: The pascal visual object classes (voc) challenge. Int. J. Comput. Vis. (IJCV) **88**(2), 303–338 (2010)

18. Gao, S., Cheng, M.M., Zhao, K., Zhang, X.Y., Yang, M.H., Torr, P.H.: Res2net: a new multi-scale backbone architecture. IEEE Trans. Pattern Anal. Mach. Intell. (TPAMI) **43**, 652–662 (2019)

19. Ghiasi, G., Lin, T.Y., Le, Q.V.: Nas-fpn: learning scalable feature pyramid architecture for object detection. In: IEEE Conference on Computer Vision Pattern Recognition (CVPR), pp. 7036–7045 (2019)

20. Han, J., Ding, J., Xue, N., Xia, G.S.: Redet: a rotation-equivariant detector for aerial object detection. In: IEEE Conference on Computer Vision Pattern Recognition (CVPR) (2021)

21. He, K., Zhang, X., Ren, S., Sun, J.: Deep residual learning for image recognition. In: IEEE Conference on Computer Vision Pattern Recognition (CVPR) (2016)

22. He, K., Gkioxari, G., Dollar, P., Girshick, R.: Mask r-cnn. In: 2017 IEEE International Conference on Computer Vision (ICCV) (2017)

23. Jaderberg, M., Simonyan, K., Zisserman, A., Kavukcuoglu, K.: Spatial transformer networks. Adv. Neural Inform. Process. Syst. (NeurIPS) **28**, 1–9 (2015)

24. Jeon, Y., Kim, J.: Active convolution: learning the shape of convolution for image classification. In: IEEE Conference on Computer Vision Pattern Recognition (CVPR), pp. 4201–4209 (2017)

25. Kalra, A., Stoppi, G., Brown, B., Agarwal, R., Kadambi, A.: Towards rotation invariance in object detection. In: International Conference on Computer Vision (ICCV) (2021)

26. Laptev, D., Savinov, N., Buhmann, J.M., Pollefeys, M.: Ti-pooling: transformation-invariant pooling for feature learning in convolutional neural networks. In: IEEE Conference on Computer Vision Pattern Recognition (CVPR) (2016)

27. Larochelle, H., Erhan, D., Courville, A., Bergstra, J., Bengio, Y.: An empirical evaluation of deep architectures on problems with many factors of variation. In: International Conference on Machine Learning (ICML) (2007)

28. Law, H., Deng, J.: Cornernet: detecting objects as paired keypoints. In: European Conference Computer Vision (ECCV), pp. 765–781. Springer, Heidelberg (2018)

29. Li, X.: Generalized focal loss: learning qualified and distributed bounding boxes for dense object detection. Adv. Neural Inform. Process. Syst. (NeurIPS) **33**, 21002–21012 (2020)

30. Lim, S., Kim, I., Kim, T., Kim, C., Kim, S.: Fast autoaugment. Adv. Neural Inform. Process. Syst. (NeurIPS) **32**, 1–11 (2019)

31. Lin, T.Y., Dollár, P., Girshick, R., He, K., Hariharan, B., Belongie, S.: Feature pyramid networks for object detection. In: IEEE Conference on Computer Vision Pattern Recognition (CVPR), pp. 2117–2125 (2017)

32. Lin, T.Y., Goyal, P., Girshick, R., He, K., Dollár, P.: Focal loss for dense object detection. In: International Conference on Computer Vision (ICCV), pp. 2980–2988 (2017)
33. Lin, T.-Y., et al.: Microsoft COCO: common objects in context. In: Fleet, D., Pajdla, T., Schiele, B., Tuytelaars, T. (eds.) ECCV 2014. LNCS, vol. 8693, pp. 740–755. Springer, Cham (2014). https://doi.org/10.1007/978-3-319-10602-1_48
34. Liu, L., Pan, Z., Lei, B.: Learning a rotation invariant detector with rotatable bounding box. arXiv preprint arXiv:1711.09405 (2017)
35. Liu, W., Anguelov, D., Erhan, D., Szegedy, C., Reed, S., Fu, C.-Y., Berg, A.C.: SSD: single shot multibox detector. In: Leibe, B., Matas, J., Sebe, N., Welling, M. (eds.) ECCV 2016. LNCS, vol. 9905, pp. 21–37. Springer, Cham (2016). https://doi.org/10.1007/978-3-319-46448-0_2
36. Liu, Y., Wang, Y., Wang, S., Liang, T., Zhao, Q., Tang, Z., Ling, H.: Cbnet: a novel composite backbone network architecture for object detection. In: Proceedings of the AAAI Conference on Artificial Intelligence (AAAI), pp. 11653–11660 (2020)
37. Liu, Z., Lin, Y., Cao, Y., Hu, H., Wei, Y., Zhang, Z., Lin, S., Guo, B.: Swin transformer: hierarchical vision transformer using shifted windows. In: International Conference on Computer Vision (ICCV) (2021)
38. Marcos, D., Volpi, M., Komodakis, N., Tuia, D.: Rotation equivariant vector field networks. In: International Conference on Computer Vision (ICCV) (2017)
39. Pang, J., et al.: Towards balanced learning for instance recognition. Int. J. Comput. Vis. (IJCV) 129(5), 1376–1393 (2021)
40. Pang, J., Chen, K., Shi, J., Feng, H., Ouyang, W., Lin, D.: Libra r-cnn: towards balanced learning for object detection. In: IEEE Conference on Computer Vision Pattern Recognition (CVPR) (2019)
41. Paszke, A., et al.: Pytorch: an imperative style, high-performance deep learning library. Adv. Neural Inform. Process. Syst. (NeurIPS) 32, 8026–8037 (2019)
42. Redmon, J., Divvala, S., Girshick, R., Farhadi, A.: You only look once: unified, real-time object detection. In: IEEE Conference on Computer Vision Pattern Recognition (CVPR), pp. 779–788 (2016)
43. Redmon, J., Farhadi, A.: Yolo9000: better, faster, stronger. In: IEEE Conference Computer Vision Pattern Recognition (CVPR), pp. 7263–7271 (2017)
44. Redmon, J., Farhadi, A.: Yolov3: an incremental improvement (2018)
45. Ren, S., He, K., Girshick, R., Sun, J.: Faster r-cnn: towards real-time object detection with region proposal networks. Adv. Neural Inform. Process. Syst. (NeurIPS) 28, 1–9 (2015)
46. Shorten, C., Khoshgoftaar, T.M.: A survey on image data augmentation for deep learning. J. Big Data 6(1), 1–48 (2019)
47. Simonyan, K., Zisserman, A.: Very deep convolutional networks for large-scale image recognition. In: International Conference on Learning Representation (ICLR) (2015)
48. Tian, Z., Shen, C., Chen, H., He, T.: Fcos: fully convolutional one-stage object detection. arXiv preprint arXiv:1904.01355 (2019)
49. Vu, T., Jang, H., Pham, T.X., Yoo, C.D.: Cascade rpn: delving into high-quality region proposal network with adaptive convolution. Adv. Neural Inform. Process. Syst. (NeurIPS) 32, 1–11 (2019)
50. Wang, Z., et al.: Orientation invariant feature embedding and spatial temporal regularization for vehicle re-identification. In: International Conference on Computer Vision (ICCV), pp. 379–387 (2017)

51. Weng, X., Wu, S., Beainy, F., Kitani, K.M.: Rotational rectification network: enabling pedestrian detection for mobile vision. In: Winter Conference on Applications of Computer Vision (WACV), pp. 1084–1092. IEEE (2018)
52. Worrall, D.E., Garbin, S.J., Turmukhambetov, D., Brostow, G.J.: Harmonic networks: deep translation and rotation equivariance. In: IEEE Conference on Computer Vision Pattern Recognition (CVPR) (2017)
53. Wu, Y., et al.: Rethinking classification and localization for object detection. arXiv (2019)
54. Wu, Y., Kirillov, A., Massa, F., Lo, W.Y., Girshick, R.: Detectron2 (2019). https://github.com/facebookresearch/detectron2
55. Xia, G.S., et al.: Dota: a large-scale dataset for object detection in aerial images. In: The IEEE Conference on Computer Vision Pattern Recognition (CVPR) (2018)
56. Xie, S., Girshick, R., Dollár, P., Tu, Z., He, K.: Aggregated residual transformations for deep neural networks. In: IEEE Conference on Computer Vision Pattern Recognition (CVPR), pp. 1492–1500 (2017)
57. Xu, W., Wang, G., Sullivan, A., Zhang, Z.: Towards learning affine-invariant representations via data-efficient cnns. In: Winter Conference on Applications of Computer Vision (WACV) (2020)
58. Yang, S., Pei, Z., Zhou, F., Wang, G.: Rotated faster r-cnn for oriented object detection in aerial images. In: Proceedings of ICRSA (2020)
59. Zhang, S., Chi, C., Yao, Y., Lei, Z., Li, S.Z.: Bridging the gap between anchor-based and anchor-free detection via adaptive training sample selection. arXiv preprint arXiv:1912.02424 (2019)
60. Zhang, Z., Jiang, R., Mei, S., Zhang, S., Zhang, Y.: Rotation-invariant feature learning for object detection in vhr optical remote sensing images by double-net. IEEE Access **8**, 20818–20827 (2019)
61. Zhang, Z., Chen, X., Liu, J., Zhou, K.: Rotated feature network for multi-orientation object detection. arXiv preprint arXiv:1903.09839 (2019)
62. Zhou, X., Wang, D., Krähenbühl, P.: Objects as points. arXiv preprint arXiv:1904.07850 (2019)
63. Zhou, Y., Ye, Q., Qiu, Q., Jiao, J.: Oriented response networks. In: IEEE Conference on Computer Vision Pattern Recognition (CVPR) (2017)
64. Zhu, C., He, Y., Savvides, M.: Feature selective anchor-free module for single-shot object detection. In: IEEE Conference Computer Vision Pattern Recognition (CVPR), pp. 840–849 (2019)
65. Zhu, X., Su, W., Lu, L., Li, B., Wang, X., Dai, J.: Deformable detr: deformable transformers for end-to-end object detection. In: International Conference on Learning Representation (2021). https://openreview.net/forum?id=gZ9hCDWe6ke

Reading Arbitrary-Shaped Scene Text from Images Through Spline Regression and Rectification

Long Chen, Feng Su$^{(\boxtimes)}$ ⓘ, Jiahao Shi, and Ye Qian

State Key Laboratory for Novel Software Technology, Nanjing University,
Nanjing 210023, China
suf@nju.edu.cn

Abstract. Scene text in natural images contains a wealth of valuable semantic information. To read scene text from the image, various text spotting techniques that jointly detect and recognize scene text have been proposed in recent years. In this paper, we present a novel end-to-end text spotting network SPRNet for arbitrary-shaped scene text. We propose a parametric B-spline centerline-based representation model to describe the distinctive global shape characteristics of the text, which helps to effectively deal with interferences such as local connection and tight spacing of text and other object, and a text is detected by regressing its shape parameters. Further, exploiting the text's shape cues, we employ adaptive projection transformations to rectify the feature representation of an irregular text, which improves the accuracy of the subsequent text recognition network. Our method achieves competitive text spotting performance on standard benchmarks through a simple architecture equipped with the proposed text representation and rectification mechanism, which demonstrates the effectiveness of the method in detecting and recognizing scene text with arbitrary shapes.

Keywords: Scene text spotting · Spline · Regression · Rectification

1 Introduction

Scene text in natural images carries a wealth of semantic information, which is of great importance in various real-world applications. To read the scene text from the image, text spotting methods first localize text regions in the image and then recognize the character sequences contained in them. Due to the complex and varied appearance of text, scene text spotting has been a challenging task and attracted increasing research attention in recent years.

Most of recent scene text spotting methods [5,18,23,31,42,46] integrated text detection and recognition into an end-to-end framework to exploit the complementarity of these two tasks to effectively improve the performance of the

Supplementary Information The online version contains supplementary material available at https://doi.org/10.1007/978-3-031-26348-4_7.

L. Wang et al. (Eds.): ACCV 2022, LNCS 13845, pp. 107–123, 2023.
https://doi.org/10.1007/978-3-031-26348-4_7

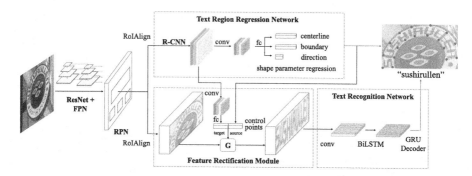

Fig. 1. Illustration of the architecture of the proposed text spotting network SPRNet. The network first detects text with arbitrary shapes in the image by a spline-based text region representation and regression model. An adaptive spatial rectification module is then employed to transform the text's feature representation to a regular shape to facilitate the subsequent text recognition. 'G' denotes the grid sampling operation for feature deformation based on predicted control points.

whole spotting model. Meanwhile, to alleviate the difficulties that the irregular shape of a scene text causes to a text recognition network, variant techniques like shape rectification [23] and spatial attention [31] have been employed in recent text spotting methods for generating appropriate features for text recognition.

Despite the great progress in enhancing scene text spotting performance, most of existing text spotting methods employed a text region localization mechanism based on either segmentation [29,31,46] or regression of discrete boundary points [39], which did not capture the text's shape characteristics as a whole (e.g., via a global shape model) and sometimes required some post-processing like grouping or fitting to obtain the final text region.

In this paper, we propose a novel end-to-end scene text spotting network SPRNet, which integrates a spline-based parametric shape regression network for localizing arbitrary-shaped text region, an adaptive text feature rectification module, and a light-weight text recognition network. Figure 1 shows the overall architecture of the proposed text spotting network. The key contributions of our work are summarized as follows:

– We propose a spline-based representation and regression model for detecting arbitrary-shaped text. The model geometrically describes the global shape of a text and its intrinsic smoothness and regularity with a parametric B-spline centerline and associated boundary cues. Compared with the segmentation- or boundary-based text representations employed in previous text spotting methods, our parametric, centerline-based representation of text is less susceptible to interferences such as local connection and tight spacing of text and other object due to its modeling and constraints on the overall shape and regularity of the text. Moreover, the model obtains directly the complete boundary of the text as the localization result, eliminating the need for post-processing that segmentation-based methods usually rely on to obtain the final text boundary.

- We integrate a shape rectification mechanism with the text spotting model for recognizing text with arbitrary shapes. The rectification module exploits adaptive projection transformations and the shape cues of an irregular text obtained by the detection module to regularize the text's feature representation, which effectively improves the accuracy of the subsequent text recognition network.
- Our text spotting method achieves competitive performance on several scene text benchmarks.

2   Related Work

Scene Text Detection. Most of recent scene text detection methods can be generally categorized into two schemes: segmentation-based and regression-based. Segmentation-based methods [27,41,45,49] localize text regions by predicting a text/non-text label for every image pixel using some fully convolutional networks [26]. Accordingly, a text region is usually modeled as a connected set of text pixels in these methods, and some of them [5,27,49] further model a text's centerline region as a shrunk mask of the whole text area consisting of a set of points on the text's central axis associated with local geometric attributes such as centerline/character orientations and boundary offsets, and certain post-processing is often required to generate the final boundary of the text. Regression-based methods [20,44,50] predict text candidates by regressing their bounding box parameters based on generated proposals or from dense features directly, while a text region is usually depicted by its polygonal boundary with discrete vertices.

Note both pixel-based and boundary-based text representations employed in most previous work capture only local constraints such as connectedness or offset between individual pixels or boundary points, lacking accurate description of a text's global shape characteristics. Comparatively, our method geometrically and holistically depicts the text shape with a parametric representation based on B-spline.

Scene Text Recognition. Recent text recognition methods usually employ some sequence models like RNN to recognize the character sequence in an image region as a whole, avoiding error-prone segmentation of individual characters. Particularly, the encoder-decoder framework has often been employed in text recognition, with the encoder encoding the text region into a feature sequence and the decoder predicting a sequence of most probable character labels corresponding to the features with connectionist temporal classification (CTC) [6,33] or attention mechanisms [3,16]. To cope with text in irregular shapes, some recent methods further proposed rectification [28,34,35,47,48] and 2D attention [17] techniques for obtaining appropriate text features for recognition. For example, in [47], a text's shape was characterized by a point-based centerline associated with local geometric attributes similar to [27], which was used to generate the fiducial points of a TPS transformation for rectifying the feature maps of an irregular text.

Scene Text Spotting. Earlier scene text spotting methods [12,19,40] often employed a two-stage pipeline that performed text detection and recognition in separate steps. Due to the complementarity of text detection and recognition tasks, however, it is difficult for these two-stage spotting methods to attain holistically optimal performance.

Most of recent scene text spotting methods [5,18,23,31,42,46] employed an end-to-end detection and recognition pipeline for improved spotting performance. Particularly, to handle arbitrary-shaped scene text, some methods introduced spatial rectification measures [5,23] to help obtain regularized representations of the text or spatial attention mechanisms [18,31] to adaptively align features with characters for recognition. For example, in [31], a Mask R-CNN based instance segmentation model was combined with a seq2seq recognition model and a spatial attention mechanism for text spotting. On the other hand, ABCNet [23] first localized the text boundary depicted by two Bezier curves, and then exploited the BezierAlign operation to generate rectified features of the text for recognition. Our method differs from previous work in two main aspects—the text region representation and regression model and the text feature rectification mechanism, which are described in detail in following respective sections.

3 Methodology

We propose an effective scene text spotting network SPRNet. As shown in Fig. 1, the network localizes arbitrary-shaped text regions in the image with a spline-based text shape representation and regression model, and then adaptively rectifies the feature representation of an irregular text for subsequent text recognition.

3.1 Text Localization via Spline-Based Shape Regression

Different from most previous segmentation-based and boundary point-based text region representation schemes used for scene text detection, which lack precise description and effective constraint for the global shape of one text, we propose a parametric, geometric text region modeling and regression scheme, which captures the holistic shape characteristics of a text to improve the text region localization accuracy. Specifically, as shown in Fig. 2, a text region is modeled by a n-order B-spline centerline describing the global layout of the text and a series of boundary cues capturing its local shape details.

The B-spline centerline is formulated as:

$$B(t) = \sum_{i=0}^{m} P_i \mathcal{N}_{i,n}(t) \tag{1}$$

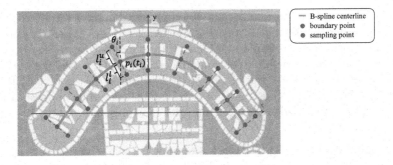

Fig. 2. Illustration of the spline-based representation of a text region.

where $P_{i=0,\cdots,m}$ denote the $m+1$ control points of the B-spline, and $\mathcal{N}_{i,n}(t)$ is the basis function associated with P_i, which is defined recursively with a predesignated knot vector $[\bar{t}_0, \bar{t}_1, \cdots, \bar{t}_{m+n}]$ as follows:

$$\mathcal{N}_{i,1}(t) = \begin{cases} 1, \bar{t}_i \leq t < \bar{t}_{i+1} \\ 0, otherwise \end{cases}$$

$$\mathcal{N}_{i,n}(t) = \frac{t - \bar{t}_i}{\bar{t}_{i+n-1} - \bar{t}_i} \mathcal{N}_{i,n-1}(t) + \frac{\bar{t}_{i+n} - t}{\bar{t}_{i+n} - \bar{t}_{i+1}} \mathcal{N}_{i+1,n-1}(t) \qquad (2)$$

In addition to the centerline, we further depict the contour of a text region with two sets of boundary points $\{v_i\}_{i=1..w}^u$ and $\{v_i\}_{i=1..w}^l$ on the upper and lower boundaries of the text region respectively as shown in Fig. 2. Each pair of two corresponding boundary points (v_i^u, v_i^l) are connected by a line segment L_i, and its length above and below the centerline are described by a pair of parameters l_i^u and l_i^l, the angle between L_i and the coordinate axis is described by a parameter θ_i, and the intersection point between L_i and the centerline (called a *sampling point*) is represented by its corresponding spline variable value t_i. Accordingly, a text region is geometrically described by the control points $P_{i=0,\cdots,m}$ of the B-spline centerline and the parameters $\{t_i, l_i^u, l_i^l, \theta_i\}$ of the boundary points.

Our spline-based, geometric text region representation model differs essentially from the segmentation-based representations employed by previous scene text detection and spotting methods [5,41,43,49]. The explicit parametric modeling of the global centerline provides effective shape constraints for robustly and accurately localizing text in cluttered scenes such as closely spaced or partially overlapping text instances, which are often challenging for segmentation-based detection methods. Moreover, compared to previous text representations that modeled a text region by its upper and lower boundaries like in [23], our centerline-based representation of the text region is usually less affected by variations of text geometry and style such as nonuniform sizes of characters in a text which often cause more significant changes to the boundary of the text region than to its centerline, and better captures the smoothness of the overall shape of a text.

We generate training labels for the parameters of the text region representation model in a similar manner to that adopted in [36] on the basis of the common polygonal annotations of text region provided in most scene text bench-marks. Particularly, different from ABCNet which requires generating ground-truth labels for the control points of Bezier curve boundaries, we do not gener-ate annotations for the B-spline centerline's control points. Instead, we generate ground-truth labels for a series of k *path points* located on the text centerline, which act as more direct constraints on the B-spline centerline and are easier to be inferred from text region features than the control points.

Text Region Regression Network. To infer the shape parameters of a text candidate in an image, as shown in Fig. 1, the text region regression network takes a text region proposal's feature maps generated by the ResNet50 [10], FPN [21], and RPN [32] backbone as input, and employs a Cascade R-CNN [2] to refine the proposal's position and assign it a text/non-text score. Next, the network employs three branches, each comprising several convolution, pooling, and full-connected layers, to predict the parameters of the B-spline centerline, the boundary points, and the text direction respectively. The detailed configuration of the network is given in the supplementary material.

Localization Loss. We employ a multitask text region localization loss $\mathcal{L}_{loc}$ on each text region proposal, which integrates a RPN loss L_{rpn} [32], a Cascade R-CNN loss L_{rcnn} [2], and a text region regression loss L_{reg}:

$$\mathcal{L}_{loc} = \lambda_1 L_{rpn} + \lambda_2 L_{rcnn} + \lambda_3 L_{reg} \qquad (3)$$

where λ_1, λ_2, and λ_3 are set to 1.0.

The text region regression loss L_{reg} measures the approximation accuracy of the predicted text region relative to the ground-truth, which is formulated as the combination of a centerline loss L_{spline}, a boundary loss L_{bound}, and a text direction loss L_{dir}:

$$L_{reg}(\boldsymbol{P}, \boldsymbol{T}_c, \boldsymbol{T}_b, \boldsymbol{\Theta}, \boldsymbol{l}, \boldsymbol{Q}^*, \boldsymbol{V}^*, \boldsymbol{\Theta}^*, \boldsymbol{l}^*, \boldsymbol{d}, \boldsymbol{d}^*) = \lambda_4 L_{spline}(\boldsymbol{P}, \boldsymbol{T}_c, \boldsymbol{Q}^*)$$
$$+ \lambda_5 L_{bound}(\boldsymbol{P}, \boldsymbol{T}_b, \boldsymbol{\Theta}, \boldsymbol{l}, \boldsymbol{V}^*, \boldsymbol{\Theta}^*, \boldsymbol{l}^*) + \lambda_6 L_{dir}(\boldsymbol{d}, \boldsymbol{d}^*) \qquad (4)$$

where $\boldsymbol{P} = \{\boldsymbol{P}_0, \cdots, \boldsymbol{P}_m\}$ are predicted control points of the B-spline centerline defined by Eq. (1). $\boldsymbol{T}_c = \{t_1^c, \cdots, t_k^c\}$ and $\boldsymbol{T}_b = \{t_1^b, \cdots, t_w^b\}$ are predicted spline variable values for the path points and the sampling points on the centerline respectively, while $\boldsymbol{Q}^*$ and $\boldsymbol{V}^*$ are the ground-truth coordinates of path points and boundary points respectively. $\boldsymbol{\Theta}, \boldsymbol{l} = [\boldsymbol{l}^u, \boldsymbol{l}^l]$ and $\boldsymbol{\Theta}^*, \boldsymbol{l}^*$ are the predicted and ground-truth angles and lengths of the lines connecting sampling points to corresponding boundary points respectively. $L_{dir}(\boldsymbol{d}, \boldsymbol{d}^*)$ is the binary cross-entropy loss between the predicted text direction probability vector $\boldsymbol{d}$ and the ground-truth one-hot direction label vector $\boldsymbol{d}^*$ which is generated for a text region based on the angle θ_t between the text's main axis (i.e. the line connecting the first and last path points) and the x axis to categorize it to horizontal if $\theta_t < 50°$ and vertical otherwise. The weights λ_4, λ_5, and λ_6 are experimentally set to 5.0, 5.0, and 0.5 respectively in this work.

The centerline loss L_{spline} measures how accurately the predicted B-spline centerline approximates the ground-truth path points $\boldsymbol{Q}^*$ and is formulated as:

$$L_{spline}(\boldsymbol{P}, \boldsymbol{T}_c, \boldsymbol{Q}^*) = smooth_{L1}(|\mathcal{F}(\boldsymbol{P}, \boldsymbol{T}_c) - \boldsymbol{Q}^*|) \tag{5}$$

where the function $\mathcal{F}(\boldsymbol{P}, \boldsymbol{T})$ computes a set of s output points corresponding to a set of spline variable values $\boldsymbol{T} = \{t_1, \cdots, t_s\}$, which are located on the B-spline defined by the control points $\boldsymbol{P} = \{\boldsymbol{P}_0, \cdots, \boldsymbol{P}_m\}$:

$$\mathcal{F}(\boldsymbol{P}, \boldsymbol{T}) = \begin{bmatrix} \boldsymbol{T}_1 \\ .. \\ \boldsymbol{T}_s \end{bmatrix} \begin{bmatrix} \boldsymbol{N}_{0,n} & \boldsymbol{N}_{1,n} & .. & \boldsymbol{N}_{m,n} \end{bmatrix} \begin{bmatrix} \boldsymbol{P}_0 \\ \boldsymbol{P}_1 \\ .. \\ \boldsymbol{P}_m \end{bmatrix} \tag{6}$$

where $\boldsymbol{N}_{i,n}$ denotes the coefficient vector of the ith basis function of B-spline, and $\boldsymbol{T}_j = [t_j^{n-1}, t_j^{n-2}, \cdots, t_j^0]$ with t_j being the spline variable value for the jth output point. Therefore, $\mathcal{F}(\boldsymbol{P}, \boldsymbol{T}_c)$ yields the set of predicted path points.

The function $smooth_{L1}(\cdot)$ is defined as:

$$smooth_{L1}(x) = \begin{cases} 0.5x^2, & if\ |x| < 1 \\ |x| - 0.5, & otherwise \end{cases} \tag{7}$$

The boundary loss L_{bound} measures the accuracy of the predicted boundary points of text region relative to the ground-truth $\boldsymbol{V}^*$ and is formulated as:

$$L_{bound}(\boldsymbol{P}, \boldsymbol{T}_b, \boldsymbol{\Theta}, \boldsymbol{l}, \boldsymbol{V}^*, \boldsymbol{\Theta}^*, \boldsymbol{l}^*) = smooth_{L1}(|\mathcal{G}(\boldsymbol{P}, \boldsymbol{T}_b, \boldsymbol{\Theta}, \boldsymbol{l}) - \boldsymbol{V}^*|)$$
$$+ smooth_{L1}(sum(|\boldsymbol{\Theta} - \boldsymbol{\Theta}^*|)) + smooth_{L1}(sum(|\boldsymbol{l} - \boldsymbol{l}^*|)) \tag{8}$$

where the function $\mathcal{G}(\boldsymbol{P}, \boldsymbol{T}_b, \boldsymbol{\Theta}, \boldsymbol{l})$ computes w pairs of boundary points based on the set of sampling points computed by $\mathcal{F}(\boldsymbol{P}, \boldsymbol{T}_b)$ and the predicted parameters $\boldsymbol{\Theta}, \boldsymbol{l}$ of lines connecting sampling and boundary points. Moreover, we maintain two separate sets of $\boldsymbol{\Theta}, \boldsymbol{l}$ parameters to better capture shape characteristics of horizontal and vertical text respectively, and compute L_{bound} on the parameter set corresponding to the direction label $\boldsymbol{d}^*$.

3.2 Spatial Rectification of Text Features

To alleviate the difficulties caused by irregular text shapes (e.g., curved or per-spectively distorted) to a text recognizer, we introduce an adaptive shape rec-tification module to spatially regularize the text's feature representation before feeding it to the recognizer for improved recognition accuracy. Different from most previous text rectification methods [30, 34, 35] which used spatial trans-form network (STN) [13] with thin-plate-spline (TPS) transformation to deform the text's shape, we employ a piecewise linear deformation model based on pro-jection transformation for feature sampling and mapping to reduce non-linear distortion to the text's shape during rectification, while keeping sufficient defor-mation flexibility for widely varied shapes of scene text.

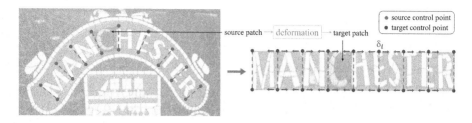

Fig. 3. Illustration of spatial rectification of irregular text. Note the deformation actually occurs on the feature maps rather than the image itself.

Specifically, given the predicted boundary points of a text region, as shown in Fig. 3, we first use the line connecting each pair of boundary points on the upper and lower text boundaries respectively to divide the feature map of the text region into a strip of adjacent quadrilateral patches (called source patches), each of which will be deformed individually.

Next, we map each source patch to a target patch in the output (rectified) feature map as shown in Fig. 3. Different from predefining a set of fixed-size target patches on the output feature map using a uniform grid as employed in previous methods [35,48], we propose a variable target grid by predicting an offset δ for each grid point to allow a target patch's boundary to deviate adaptively from the uniform grid position, which increases the model's flexibility for rectifying non-uniform distortions of text. Note the offsets of the target grid points are end-to-end learned with the recognition task without any extra supervision.

Finally, we compute the feature values in a target patch by grid-sampling features in the corresponding source patch to obtain a regular feature representation of the text region for recognition.

Feature Patch Deformation. We employ projection transformation as the mapping function between the source and target patches because of its linearity which helps keep shape characteristics of character and the fact that most scene text has a certain degree of perspective distortion resulting from the viewing process.

Using the four boundary points of a source patch as four *source* control points and the four corner points of the corresponding target patch as *target* control points, the homogeneous deformation matrix $\mathcal{H}$ of a projection transformation for the patch can be formulated as:

$$\mathcal{H} = reshp([\boldsymbol{b} \quad 1])_{3\times3} \tag{9}$$

where function $reshp(\cdot)_{3\times3}$ reshapes the input tensor to a 3×3 view, and $\boldsymbol{b}$ is a 1×8 vector computed as:

$$\boldsymbol{b} = \mathcal{A}^{-1}\boldsymbol{x} \tag{10}$$

where $\boldsymbol{x}$ is an 8×1 vector containing the coordinates of the four target control points. $\mathcal{A}$ is an 8×8 matrix formulated as follows based on the Direct Linear Transformation (DLT) algorithm [8]:

$$\mathcal{A} = \begin{bmatrix} r_x^{(0)} & r_y^{(0)} & 1 & 0 & 0 & 0 & -r_x^{(0)} * t_x^{(0)} & -r_y^{(0)} * t_x^{(0)} \\ 0 & 0 & 0 & r_x^{(0)} & r_y^{(0)} & 1 & -r_x^{(0)} * t_y^{(0)} & -r_y^{(0)} * t_y^{(0)} \\ \cdots & & & & & & & \\ r_x^{(3)} & r_y^{(3)} & 1 & 0 & 0 & 0 & -r_x^{(3)} * t_x^{(3)} & -r_y^{(3)} * t_x^{(3)} \\ 0 & 0 & 0 & r_x^{(3)} & r_y^{(3)} & 1 & -r_x^{(3)} * t_y^{(3)} & -r_y^{(3)} * t_y^{(3)} \end{bmatrix} \qquad (11)$$

where $(r_x^{(i)}, r_y^{(i)})$ and $(t_x^{(i)}, t_y^{(i)})$ are the (x, y) coordinates of the ith source and target control points respectively.

Given the deformation matrix $\mathcal{H}$, a position $\boldsymbol{p}_t$ in the target patch is mapped back to the position $\boldsymbol{p}_r = \mathcal{H}^{-1}\boldsymbol{p}_t$ in the source patch. Accordingly, we compute the feature value in the position $\boldsymbol{p}_t$ in the target patch's feature map by bilinear interpolation of feature values neighbouring to $\boldsymbol{p}_r$ in the source feature map. This grid sampling operation is represented by the symbol 'G' in Fig. 1.

3.3   Text Recognition

Given the rectified feature maps of one text region, we employ a light-weight attention-based sequence-to-sequence recognition network to recognize the text. As shown in Fig. 1, the network first employs several convolutional layers to produce a feature map of height 1, and then uses a bidirectional LSTM to encode long-range forward and backward dependencies between the column feature vectors of the feature map and outputs a sequence of features. A gated recurrent unit (GRU) decoder with Bahdanau attention is finally employed to decode the feature sequence into a character label sequence. More details about character sequence prediction with GRU can be found in [28], and the configuration of the recognition network is presented in the supplementary material.

Recognition Loss. The text recognition loss $\mathcal{L}_{rec}$ is formulated as:

$$\mathcal{L}_{rec} = -\sum_{i=1}^{N}\sum_{j=1}^{NC} \mathbb{I}(\hat{\mathbf{y}}_i^j = 1)log(\mathbf{y}_i^j) \qquad (12)$$

where N is the length of the predicted character label distribution sequence $\{\mathbf{y}_i\}$, NC is the total number of different characters, $\{\hat{\mathbf{y}}_i\}$ is the ground-truth one-hot label distribution sequence, and $\mathbb{I}(\cdot)$ is a binary function that returns 1 if its input is evaluated as true and returns 0 otherwise.

3.4   Text Spotting Loss

The total loss of the text spotting model is a combination of the text region localization loss $\mathcal{L}_{loc}$ and the text recognition loss $\mathcal{L}_{rec}$:

$$\mathcal{L} = \lambda_l \mathcal{L}_{loc} + \lambda_r \mathcal{L}_{rec} \qquad (13)$$

where the weights λ_l and λ_r are set to 1.0 and 0.2 respectively in this work.

4   Experiments

4.1   Datasets

We evaluate our scene text spotting method on three challenging benchmarks: TotalText, CTW1500, and ICDAR2015. **TotalText** [4] is composed of 1255 and 300 images for training and testing respectively and contains large numbers of curved text instances, each annotated by a polygonal boundary of 10 vertices. **CTW1500** [24] contains 1000 training images and 500 testing images with many challenging long curved text, each annotated by a polygonal boundary of 14 vertices. **ICDAR2015** [14] consists of 1000 training images and 500 testing images with multi-oriented accidental scene text instances, each annotated by a quadrilateral bounding box. We employ precision P, recall R, and f-measure F to evaluate text spotting performance.

4.2   Implementation Details

We implement the proposed text spotting network on the basis of the PyTorch framework and conduct the experiments on a NVIDIA Tesla V100 GPU. We depict the text centerline by a cubic B-spline (order $n = 4$) with 5 control points ($m = 4$) and an open uniform knot vector, and approximate the centerline with $k = 17$ path points. We employ $w = 9$ pairs of boundary points on the upper and lower boundaries of a text region.

The spotting network is optimized using stochastic gradient descent with a weight decay of 0.0001 and a momentum of 0.9. The network is first pre-trained on a combined dataset similar to that used in [29] for 90K iterations with the learning rate starting from 0.01 and reduced to 0.001 for the last 20K iterations. The combined dataset contains training samples of SynthText [7], ICDAR 2013 [15], ICDAR 2015 [14], COCO-Text [38], and Total-Text [4] datasets, with a sampling ratio $2 : 2 : 2 : 2 : 1$ among these datasets for generating a mini-batch of 10. Next, we fine-tune separate spotting models for different test datasets using their own training sets. For TotalText and CTW1500 curved text datasets, the learning rate is initialized to 0.001 for the first 40K training iterations and is reduced to 0.0001 for further 20K iterations. For ICDAR2015 dataset, a learning rate of 0.001 is used during 40K training iterations of the network.

4.3   Ablation Study

Effectiveness of Spline-Based Text Region Regression. We verify the effectiveness of the proposed spline-based text region representation and regression model by comparing the text detection performance of some variants of the text region regression network in Table 1. The model 'Baseline' uses the Cascade R-CNN backbone to predict the bounding boxes of text instances in the image. The model 'Mask' replaces the shape parameter regression branches with the mask branch in Mask R-CNN [9] for text detection.

Table 1. Text detection performance of the proposed spline-based text region regression model and two variant models

Model	TotalText			ICDAR2015		
	P	R	F	P	R	F
Baseline	73.3	72.7	73.0	79.5	74.8	77.1
Mask	85.0	82.2	83.6	89.7	80.9	85.1
Proposed	**85.7**	**85.1**	**85.4**	**91.1**	**85.4**	**88.1**

Table 2. Text detection performance using variant number of control points for the B-spline centerline of a text region

Num	TotalText			ICDAR2015		
	P	R	F	P	R	F
4	**85.8**	84.5	85.1	89.0	**86.9**	87.9
5	85.7	**85.1**	**85.4**	**91.1**	85.4	**88.1**
6	85.5	85.0	85.3	90.2	85.6	87.8
7	85.1	84.6	84.8	88.4	86.5	87.4

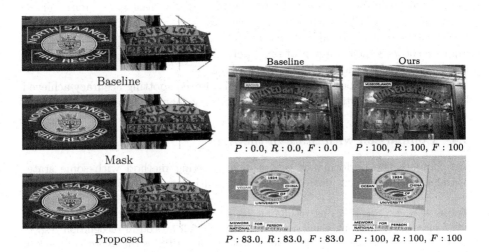

Baseline   Ours

P : 0.0, R : 0.0, F : 0.0   P : 100, R : 100, F : 100

P : 83.0, R : 83.0, F : 83.0   P : 100, R : 100, F : 100

Fig. 4. Text detection results obtained by variant models in Table 1.

Fig. 5. Text spotting results obtained by the baseline model (left) and our model (right) in Table 3. Detected text instances are marked with green boxes. Incorrect recognition results are shown with red text.

Compared to the baseline, the proposed spline-based text region regression model substantially improves the text detection performance through more accurate and flexible modeling of the text region. It also achieves higher detection f-measure than the mask mechanism [9], showing the advantages of the proposed parameterized, geometric representation of the text over the pixel-level representation in accurately describing the shape of the text. Figure 4 presents some text detection results obtained by variant models in Table 1. The proposed model yields more accurate text region boundaries than the others.

We further inspect the impact of using different numbers of control points for the B-spline centerline of a text region on the detection performance. As shown in Table 2, a B-spline with 5 control points is usually sufficient to cope with the different shape complexities of most scene text.

Effectiveness of Text Feature Rectification for Text Spotting. We verify the effectiveness of our text feature rectification mechanism in scene text spotting. Table 3 compares the spotting performance with our rectification model, a STN-based rectification model similar to [34] for adaptive text shape deformation, and a baseline model that removes the rectification module from the spotting network (i.e., feeding features of a text region directly to the text recognition module).

As shown in Table 3, introducing adaptive rectification of text features ahead of recognition significantly enhances the text spotting performance owing to the rectified, more regular representation of the text, especially on benchmarks with curved/irregular text instances like TotalText as expected. Figure 5 shows some examples of text spotting results obtained by the baseline model and our

Table 3. Text spotting performance with variant rectification models

Model	TotalText			ICDAR2015		
	P	R	F	P	R	F
Baseline	75.8	73.0	74.4	69.9	69.0	69.5
STN	80.8	72.8	76.6	71.7	69.0	70.3
Ours	**81.2**	**78.9**	**80.0**	**72.2**	**69.3**	**70.7**

rectification-based model respectively. The improved spotting accuracy achieved by our model shows its effectiveness for arbitrary-shaped scene text spotting.

4.4   Comparison with State-of-the-Arts

We compare the performance of our text spotting method with some state-of-the-art methods on both curved and multi-oriented text benchmarks in

Table 4. Scene text spotting results on TotalText and CTW1500. 'None' and 'Full' are f-measure of spotting using no lexicon and the full lexicon in recognition respectively. 'Det' is the f-measure of text detection results. 'FPS' is the inference speed on TotalText. In each column, the best result is shown in bold and the second best result is shown with underline. Methods marked with * exploited additional character-level labels besides the common word-level labels in training and are not included in ranking.

Method	TotalText			CTW1500			FPS
	Det	None	Full	Det	None	Full	
TextNet [37]	63.5	54.0	–	–	–	–	–
FOTS [22]	–	–	–	62.8	21.1	39.7	–
Qin et al.[31]	83.3	67.8	–	–	–	–	4.8
TextDragon [5]	80.3	48.8	74.8	83.6	39.7	72.4	–
ABCNet [23]	–	64.2	75.7	–	45.2	74.1	17.9
Text Perceptron [30]	85.2	<u>69.7</u>	78.3	84.6	57.0	–	–
PAN++ [42]	86.0	68.6	<u>78.6</u>	–	–	–	21.1
ABCNet v2 [25]	**87.0**	**70.4**	78.1	<u>84.7</u>	<u>57.5</u>	**77.2**	10
Mask TextSpotter [29] *	83.9	52.9	71.8	–	–	–	4.8
CharNet [46] *	85.6	66.6	–	–	–	–	–
Mask TextSpotter v3 [18] *	–	71.2	78.4	–	–	–	–
Ours	<u>86.6</u>	67.8	**80.0**	**84.9**	59.6	<u>75.0</u>	8.6

Table 5. Scene text spotting results on ICDAR2015. 'S', 'W', and 'G' are f-measure of spotting using the strong (100 words), weak (1000+ words), and generic (90K words) lexicons respectively. Methods marked with * exploited additional character-level labels besides the common word-level labels in training and are not included in ranking.

Method	Word Spotting			End-to-End Recognition			FPS
	S	W	G	S	W	G	
Deep TextSpotter [1]	58.0	53.0	51.0	54.0	51.0	47.0	9.0
TextBoxes++ [19]	76.5	69.0	54.4	73.3	65.9	51.9	–
FOTS [22]	84.7	79.3	63.3	81.1	75.9	60.8	7.5
He *et al.*[11]	<u>85.0</u>	<u>80.0</u>	65.0	82.0	77.0	63.0	–
TextDragon [5]	**86.2**	**81.6**	<u>68.0</u>	<u>82.5</u>	<u>78.3</u>	65.2	2.6
Text Perceptron [30]	84.1	79.4	67.9	80.5	76.6	65.1	–
PAN++ [42]	–	–	–	82.7	78.2	69.2	13.8
ABCNet v2 [25]	–	–	–	**82.7**	**78.5**	**73.0**	10
Mask TextSpotter [29] *	79.3	74.5	64.2	79.3	73.0	62.4	2.6
CharNet [46] *	–	–	–	83.1	79.2	69.1	–
Mask TextSpotter v3 [18] *	83.1	79.1	75.1	83.3	78.1	74.2	2.5
Ours	82.7	77.0	**70.7**	**82.7**	76.6	<u>70.6</u>	6.2

Tables 4 and 5. Note that, besides the word-level annotations of text, some methods (marked with *) further exploited external character-level annotations as extra supervision information, which are not available in the benchmark datasets.

Curved Text Spotting. Table 4 shows that our method achieves the best results in two text spotting and one text detection tasks on TotalText and CTW1500 curved text datasets and comparable results in the rest of detection/spotting tasks, which demonstrate the method's capability to accurately localize and recognize various curved text in natural images.

Particularly, compared to Text Perceptron which combined a TPS-based feature rectification module with a focusing attention recognizer [3] and ABCNet which employed a Bezier curve-based feature sampling mechanism for recognizing irregular text, our rectification and spotting model achieves higher performance on most evaluation metrics on the two curved text benchmarks. ABCNet v2 further extended the ABCNet's backbone (e.g. introducing the BiFPN and CoordConv modules) and its training mechanism for enhanced performance. When the ResNet+FPN backbone of ABCNet is used, which is similar to that employed in our model, it achieves a text spotting f-measure of 67.4 on Total-Text and 54.6 on CTW1500 using no lexicon [25]. On the other hand, unlike our method employing common word-level annotations of text as supervision information, Mask TextSpotter v3 exploited both word-level and character-level annotations (e.g. bounding boxes and category indices of characters) for training the model and employed a combinatory text recognition strategy integrating character-level pixel voting and spatial attention mechanisms.

Multi-oriented Text Spotting. On ICDAR2015 which consists of multi-oriented but mostly straight text instances, as shown in Table 5, our method also achieves comparable text spotting performance among the methods that similarly exploit only word-level annotations of text and common training datasets. The good results of our method on the curved and multi-oriented text benchmarks demonstrate its effectiveness in spotting scene text in arbitrary shapes.

4.5   Qualitative Results

Figure 6 shows some text spotting results of our method. The proposed spotting network robustly detects and recognizes various scene text with largely varied appearances and qualities. More examples of scene text spotting results and discussions of limitations can be found in the supplementary material.

Fig. 6. Examples of text spotting results. Detected text instances are marked with green boxes, with corresponding recognition results shown nearby. (Color figure online)

5   Conclusions

We present a method for accurately spotting arbitrary-shaped scene text in natural images. A parametric text representation and regression model based on the spline centerline is proposed to capture the distinctive global shape characteristics of text for robustly localizing text instances with varied appearances. The method further spatially rectifies the feature representation of an irregularly shaped text with an adaptive deformation model before feeding it to the text recognition network, which effectively improves the text spotting accuracy. In the future work, we will explore integrating effective language models with the recognition network and further improving the collaboration between detection and recognition modules for enhancing the performance of the method.

References

1. Busta, M., Neumann, L., Matas, J.: Deep TextSpotter: an end-to-end trainable scene text localization and recognition framework. In: 2017 IEEE International Conference on Computer Vision (ICCV), pp. 2223–2231 (2017)
2. Cai, Z., Vasconcelos, N.: Cascade R-CNN: delving into high quality object detection. In: CVPR, pp. 6154–6162 (2018)
3. Cheng, Z., Bai, F., Xu, Y., Zheng, G., Pu, S., Zhou, S.: Focusing attention: towards accurate text recognition in natural images. In: 2017 IEEE International Conference on Computer Vision (ICCV), pp. 5086–5094 (Oct 2017)
4. Chng, C.K., Chan, C.S.: Total-text: a comprehensive dataset for scene text detection and recognition. In: ICDAR, pp. 935–942 (2017)
5. Feng, W., He, W., Yin, F., Zhang, X., Liu, C.: TextDragon: an end-to-end framework for arbitrary shaped text spotting. In: ICCV, pp. 9075–9084 (2019)
6. Graves, A., Fernández, S., Gomez, F., Schmidhuber, J.: Connectionist temporal classification. In: Proceedings of the 23rd International Conference on Machine Learning - ICML 2006, pp. 369–376 (2006)
7. Gupta, A., Vedaldi, A., Zisserman, A.: Synthetic data for text localisation in natural images. In: CVPR, pp. 2315–2324 (2016)
8. Hartley, R., Zisserman, A.: Multiple View Geometry in Computer Vision. Cambridge University Press, Cambridge (2003)
9. He, K., Gkioxari, G., Dollár, P., Girshick, R.B.: Mask R-CNN. In: ICCV, pp. 2980–2988 (2017)
10. He, K., Zhang, X., Ren, S., Sun, J.: Deep residual learning for image recognition. In: CVPR, pp. 770–778 (2016)
11. He, T., Tian, Z., Huang, W., Shen, C., Qiao, Y., Sun, C.: An end-to-end textspotter with explicit alignment and attention. In: CVPR, pp. 5020–5029 (2018)
12. Jaderberg, M., Simonyan, K., Vedaldi, A., Zisserman, A.: Reading text in the wild with convolutional neural networks. IJCV 116(1), 1–20 (2016)
13. Jaderberg, M., Simonyan, K., Zisserman, A., Kavukcuoglu, K.: Spatial transformer networks. In: NIPS, pp. 2017–2025 (2015)
14. Karatzas, D., et al.: ICDAR 2015 competition on robust reading. In: ICDAR, pp. 1156–1160 (2015)
15. Karatzas, D., et al.: ICDAR 2013 robust reading competition. In: ICDAR, pp. 1484–1493 (2013)
16. Lee, C.Y., Osindero, S.: Recursive recurrent nets with attention modeling for OCR in the wild. In: 2016 IEEE Conference on Computer Vision and Pattern Recognition (CVPR), pp. 2231–2239 (2016)
17. Li, H., Wang, P., Shen, C., Zhang, G.: Show, attend and read: a simple and strong baseline for irregular text recognition. In: AAAI Conference on Artificial Intelligence, vol. 33, pp. 8610–8617 (2019)
18. Liao, M., Pang, G., Huang, J., Hassner, T., Bai, X.: Mask TextSpotter v3: segmentation proposal network for robust scene text spotting. In: Vedaldi, A., Bischof, H., Brox, T., Frahm, J.-M. (eds.) ECCV 2020. LNCS, vol. 12356, pp. 706–722. Springer, Cham (2020). https://doi.org/10.1007/978-3-030-58621-8_41
19. Liao, M., Shi, B., Bai, X.: Textboxes++: a single-shot oriented scene text detector. TIP 27(8), 3676–3690 (2018)
20. Liao, M., Shi, B., Bai, X., Wang, X., Liu, W.: TextBoxes: a fast text detector with a single deep neural network. In: AAAI, pp. 4161–4167 (2017)

21. Lin, T., Dollár, P., Girshick, R.B., He, K., Hariharan, B., Belongie, S.J.: Feature pyramid networks for object detection. In: CVPR, pp. 936–944 (2017)
22. Liu, X., Liang, D., Yan, S., Chen, D., Qiao, Y., Yan, J.: FOTS: fast oriented text spotting with a unified network. In: CVPR, pp. 5676–5685 (2018)
23. Liu, Y., Chen, H., Shen, C., He, T., Jin, L., Wang, L.: ABCNet: real-time scene text spotting with adaptive bezier-curve network. In: CVPR, pp. 9806–9815 (2020)
24. Liu, Y., Jin, L., Zhang, S., Zhang, S.: Detecting curve text in the wild: new dataset and new solution. CoRR abs/1712.02170 (2017). https://arxiv.org/abs/1712.02170
25. Liu, Y., et al.: Abcnet v2: adaptive Bezier-curve network for real-time end-to-end text spotting. IEEE Trans. Pattern Anal. Mach. Intell. **44**(11), 8048–8064 (2022)
26. Long, J., Shelhamer, E., Darrell, T.: Fully convolutional networks for semantic segmentation. In: CVPR, pp. 3431–3440 (2015)
27. Long, S., Ruan, J., Zhang, W., He, X., Wu, W., Yao, C.: TextSnake: a flexible representation for detecting text of arbitrary shapes. In: Ferrari, V., Hebert, M., Sminchisescu, C., Weiss, Y. (eds.) ECCV 2018. LNCS, vol. 11206, pp. 19–35. Springer, Cham (2018). https://doi.org/10.1007/978-3-030-01216-8_2
28. Luo, C., Jin, L., Sun, Z.: MORAN: a multi-object rectified attention network for scene text recognition. Pattern Recogn. **90**, 109–118 (2019)
29. Lyu, P., Liao, M., Yao, C., Wu, W., Bai, X.: Mask TextSpotter: an end-to-end trainable neural network for spotting text with arbitrary shapes. In: ECCV, pp. 71–88 (2018)
30. Qiao, L., et al.: Text perceptron: towards end-to-end arbitrary-shaped text spotting. In: Proceedings of the Thirty-Fourth AAAI Conference on Artificial Intelligence, pp. 11899–11907 (2020)
31. Qin, S., Bissacco, A., Raptis, M., Fujii, Y., Xiao, Y.: Towards unconstrained end-to-end text spotting. In: ICCV, pp. 4703–4713 (2019)
32. Ren, S., He, K., Girshick, R.B., Sun, J.: Faster R-CNN: towards real-time object detection with region proposal networks. In: NIPS, pp. 91–99 (2015)
33. Shi, B., Bai, X., Yao, C.: An end-to-end trainable neural network for image-based sequence recognition and its application to scene text recognition. IEEE Trans. Pattern Anal. Mach. Intell. **39**(11), 2298–2304 (2017)
34. Shi, B., Wang, X., Lyu, P., Yao, C., Bai, X.: Robust scene text recognition with automatic rectification. In: 2016 IEEE Conference on Computer Vision and Pattern Recognition (CVPR), pp. 4168–4176 (Jun 2016)
35. Shi, B., Yang, M., Wang, X., Lyu, P., Yao, C., Bai, X.: ASTER: an attentional scene text recognizer with flexible rectification. IEEE Trans. Pattern Anal. Mach. Intell. **41**(9), 2035–2048 (2019)
36. Shi, J., Chen, L., Su, F.: Accurate arbitrary-shaped scene text detection via iterative polynomial parameter regression. In: Computer Vision - ACCV 2020, pp. 241–256 (2021)
37. Sun, Y., Zhang, C., Huang, Z., Liu, J., Han, J., Ding, E.: TextNet: irregular text reading from images with an end-to-end trainable network. In: Jawahar, C.V., Li, H., Mori, G., Schindler, K. (eds.) ACCV 2018. LNCS, vol. 11363, pp. 83–99. Springer, Cham (2019). https://doi.org/10.1007/978-3-030-20893-6_6
38. Veit, A., Matera, T., Neumann, L., Matas, J., Belongie, S.J.: Coco-text: dataset and benchmark for text detection and recognition in natural images. CoRR abs/1601.07140 (2016)
39. Wang, H., et al.: All you need is boundary: toward arbitrary-shaped text spotting. In: AAAI, pp. 12160–12167 (2020)

40. Wang, T., Wu, D.J., Coates, A., Ng, A.Y.: End-to-end text recognition with convolutional neural networks. In: Proceedings of the 21st International Conference on Pattern Recognition, pp. 3304–3308 (2012)
41. Wang, W., et al.: Shape robust text detection with progressive scale expansion network. In: CVPR, pp. 9336–9345 (2019)
42. Wang, W., et al.: Pan++: towards efficient and accurate end-to-end spotting of arbitrarily-shaped text. IEEE Trans. Pattern Anal. Mach. Intell. **44**(9), 5349–5367 (2022)
43. Wang, W., et al.: Efficient and accurate arbitrary-shaped text detection with pixel aggregation network. In: ICCV, pp. 8439–8448 (2019)
44. Wang, X., Jiang, Y., Luo, Z., Liu, C., Choi, H., Kim, S.: Arbitrary shape scene text detection with adaptive text region representation. In: CVPR, pp. 6449–6458 (2019)
45. Wu, Y., Natarajan, P.: Self-organized text detection with minimal post-processing via border learning. In: ICCV, pp. 5010–5019 (2017)
46. Xing, L., Tian, Z., Huang, W., Scott, M.R.: Convolutional character networks. In: ICCV, pp. 9125–9135 (2019)
47. Yang, M., et al.: Symmetry-constrained rectification network for scene text recognition. In: 2019 IEEE/CVF International Conference on Computer Vision (ICCV), pp. 9146–9155 (2019)
48. Zhan, F., Lu, S.: ESIR: end-to-end scene text recognition via iterative image rectification. In: CVPR, pp. 2054–2063 (2019)
49. Zhang, C., et al.: Look more than once: an accurate detector for text of arbitrary shapes. In: CVPR, pp. 10544–10553 (2019)
50. Zhou, X., et al.: EAST: an efficient and accurate scene text detector. In: CVPR, pp. 2642–2651 (2017)

IoU-Enhanced Attention for End-to-End Task Specific Object Detection

Jing Zhao[1], Shengjian Wu[1], Li Sun[1,2(✉)], and Qingli Li[1]

[1] Shanghai Key Laboratory of Multidimensional Information Processing,
Shanghai, China
[2] Key Laboratory of Advanced Theory and Application in Statistics and Data
Science, East China Normal University, Shanghai, China
sunli@ee.ecnu.edu.cn

Abstract. Without densely tiled anchor boxes or grid points in the image, sparse R-CNN achieves promising results through a set of object queries and proposal boxes updated in the cascaded training manner. However, due to the sparse nature and the one-to-one relation between the query and its attending region, it heavily depends on the self attention, which is usually inaccurate in the early training stage. Moreover, in a scene of dense objects, the object query interacts with many irrelevant ones, reducing its uniqueness and harming the performance. This paper proposes to use IoU between different boxes as a prior for the value routing in self attention. The original attention matrix multiplies the same size matrix computed from the IoU of proposal boxes, and they determine the routing scheme so that the irrelevant features can be suppressed. Furthermore, to accurately extract features for both classification and regression, we add two lightweight projection heads to provide the dynamic channel masks based on object query, and they multiply with the output from dynamic convs, making the results suitable for the two different tasks. We validate the proposed scheme on different datasets, including MS-COCO and CrowdHuman, showing that it significantly improves the performance and increases the model convergence speed. Codes are available at https://github.com/bravezzzzzz/IoU-Enhanced-Attention.

1 Introduction

Object detection is a fundamental task in computer vision, which aims to locate and categorize semantic regions with bounding boxes. Traditionally, there are two-stage [13,19,33] methods based on the densely tiled anchors or one-stage [25, 28,31,32,41] methods built on either anchors or grid points. However, they are both complained for handcrafted designs, *e.g.* the anchor shapes, the standard for positive and negative training samples assignment, and the extra post processing step, like Non-Maximum Suppression (NMS).

This work is supported by the Science and Technology Commission of Shanghai Municipality under Grant No. 22511105800, 19511120800 and 22DZ2229004.

Supplementary Information The online version contains supplementary material available at https://doi.org/10.1007/978-3-031-26348-4_8.

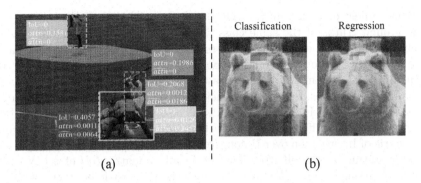

(a)   (b)

Fig. 1. Two ways to enhance sparse R-CNN. (a) focuses on a particular proposal box (in yellow) and its corresponding attention matrix in sparse R-CNN ($attn$) and our model ($\widehat{attn}$). It shows that IoU-ESA can effectively restrict the query attending region, hence keeping the uniqueness of object query. In (b), two channel masks, dynamically predicted from object queries, help to highlight different regions on the features for classification and regression tasks. (Color figure online)

DETR [3] is a simpler model that utilizes the structure of transformer. In its decoder, a sparse set of learnable queries absorb object contents from the image by cross attention and also from each other by self attention. Then bipartite matching connected the updated queries and ground truths, assigning only one query to a ground truth box for the classification and regression loss. The one-to-one label assignment rule prevents redundant boxes output from the model. Therefore, NMS is no longer needed. However, DETR still suffers from the slow training convergence. TSP [40] selects features from the encoder to initialize the query set, and SMCA [11] directly predicts a spatial mask to weight the attention matrix. Both of them can accelerate the training speed.

The above works have a sparse query set but must attend to the complete image densely. Deformable DETR [50] changes the standard attention into a deformable one, constraining each query to attend to a small set of sampling points around the reference point. It significantly reduces the calculations of attention patterns and improves convergence speed. Sparse R-CNN [39] utilizes RoI align [17] to make local object embedding for loss heads. It has paired object query q_i and proposal box b_i, and q_i only interacts with the corresponding image feature within box b_i through dynamic convs, where $i = 1, 2, \cdots, N$ and N is the total number of the query slots and boxes. Apart from that, it still performs self attention among the query set so that one q_i knows about the others. Sparse R-CNN is a cascaded structure with multiple stages, and b_i is refined in each stage to approach the ground truth box progressively.

Despite the simple designs, sparse R-CNN relies on self attention to model the relationship between different object queries. Since q_i provides kernels for each b_i, self attention indirectly enlarges the receptive field of the dynamic convs within each box b_i. However, q_i is summarized into a vector without spatial dimensions, and the self attention ignores the spatial coverage of b_i. In other words, q_i may be intensively influenced by other q_j simply because they have similar query and key vectors, without considering the locations of b_i and b_j.

For dense object detection in MS-COCO and CrowdHuman datasets, there are many ground truths of the same class within an image. The dynamic routing in self attention can be ineffective due to the lack of guidance from the spatial prior, especially in the early training epochs.

In this paper, we intend to enhance the self attention for object queries in sparse R-CNN by considering the geometry relation between two corresponding boxes, as is shown in Fig. 1a. Similar idea is also adopted in Relation Net [19]. However, we take a different strategy without adding any parameters. The common metric of Intersection over Union (IoU) is directly utilized as a natural prior for value routing during self attention. Specifically, a matrix IoU of size $N \times N$ is computed among N proposal boxes, in which the IoU metric between two boxes b_i and b_j determines each element IoU_{ij}. Note that IoU is in the same size with the attention matrix $attn$, so the element-wise multiplication can combine them, and the result is used for weighting the value matrix to update object queries.

To further enhance the feature for classification and localization in the dual loss heads, we design a dynamic channel weighting (DCW) module to produce different and suitable features for two tasks. The idea is to add a connection between object query and RoI feature to achieve dynamic cross weighting. It employs the object query q_i to predict two channel masks with the exact dimension of q_i. Then the two masks are applied to weight the output feature given by the dynamic convs. As a result, channels can be highlighted differently for the two tasks, focusing on the relevant regions for classification and regression, as shown in Fig. 1b. Similar to the dynamic conv, there are two light-weight heads in the DCW module, and both have the sigmoid function at the end, like the SE-Net [20], constraining the value range from 0 to 1. After the DCW, features for the two tasks are no longer shared, potentially benefiting both.

The summary of our contributions is listed as follows:

- We propose an IoU-enhanced self attention module in sparse R-CNN, which brings in the spatial relation prior to the attention matrix. Here the IoU of the two proposal boxes is utilized as a metric to evaluate the similarities of different object queries and to guide the value routing in self attention.
- We design an inserted DCW module with the cross weighting scheme, which outputs two types of channel masks based on object queries, and then uses them to weight the feature from the original dynamic conv. The DCW separates the RoI features for the classification and regression heads.
- Extensive experiments on two datasets of CrowdHuman and MS-COCO are carried out, which validate the proposed method effectively boosts the quality of detection results without significantly increasing the model size and the amount of the calculations.

2   Related Work

Object detection [7,43] has been intensively investigated in computer vision, particularly in the framework of a deep neural network [47]. Recently, due to the great success of transformer [42] and its application on image classification

[10,29,44], the performance of detection has been significantly improved due to strong backbones pre-trained on Imagenet [8]. Here we only give a brief review on models, particularly for object detection, primarily including two types which are dense and sparse detectors.

2.1 Dense Detector

Typical dense methods are either two-stage or one-stage. From R-CNN [14], Fast R-CNN [13] to Faster R-CNN [33], two-stage [19,36] or even multi-stage [2] methods become tightly fitted into CNN, achieving the end-to-end training manner. In these methods, object proposals, which are the roughly located object regions, are first obtained. Then, RoI align [17] extracts these proposal regions from the entire image, making the second stage only observes the RoIs and focuses on improving their qualities. One-stage methods detect objects in a single shot without RoI align. SSD [28] and YOLO [31,32] mimic the first stage in Faster R-CNN, and try to give the final locations and the belonging classes directly . However, the detection quality of them significantly lags behind their two-stage competitors. RetinaNet [25] attributes the inferiority of one-stage methods to the extreme imbalance between the positive and negative training samples and designs a focal loss to deal with it, which down weights the easy negative during training. Based on feature pyramid network (FPN) [24] and sample assigning schemes, e.g., ATSS [46], PAA [22] or OTA [12], one-stage method RetinaNet achieves competitive performance.

Except for the early versions of YOLO [31,32], most of the above works have densely tiled initial boxes in the image, known as anchors. However, their sizes and ratios are important but challenging to choose. Recent works demonstrate that single-stage anchor-free methods achieve promising results. Without tiled anchors, these methods directly classify each grid point and regress bounding boxes for positive samples. Different strategies have been adopted to locate a box from a grid. FCOS [41] outputs distances to four borders of the ground truth box. At the same time, it also predicts centerness as a supplement quality metric for NMS. RepPoints [45] utilizes deformable conv [6] to spread irregular points around the object center, and then collects all point positions for locating the ground truth box. CornerNet [23] adopts a bottom-up scheme, which estimates Gaussian heatmaps for the top-left and bottom-right corners of a ground truth, and tries to match them from the same ground truth through an embedding feature. A similar idea is also used to predict heatmaps for object centers [48]. RepPoints-v2 [5] integrates the idea of CornerNet by adding a branch to predict corner heatmaps, which also helps refine the predicted box from irregular points during inference.

2.2 Sparse Detector

Although anchor-free methods greatly alleviate the complexity, they still need to predict on the densely tiled grids. Since a ground truth is potentially assigned to many candidates as positive, NMS becomes the inevitable post processing step to reduce the redundant predictions. However, NMS is complained for its heuristic design, and many works try to improve [1,21,27] or even eliminate it [38]. But these works often bring in complex modules. DETR [3] gives a simple NMS-free solution under the transformer structure [42]. It assumes a sparse set of learnable queries updated repeatedly by both self and cross attentions. The query is finally employed to predict its matching ground truth. Note that one ground truth box is only assigned to its belonging query through bipartite matching. Therefore NMS is not required anymore. Despite its simple setting, DETR takes a long time to converge. TSP [40] finds that the unstable behavior of cross attention in the early training epochs is the leading cause of the slow convergence. Hence it removes cross attention and only performs self attention among selected RoIs. Using FPN as backbone, TSP-FCOS and TSP-RCNN can achieve satisfactory results with 36 epochs. SMCA [11] incorporates the spatial prior predicted directly from query to modulate the attention matrix and increases convergence speed. Deformable DETR [50] uses the idea in deformable conv-v2 [49] into transformer's encoder and decoder. Different from traditional non-local attention operation, each query in deformable attention straightly samples the features at irregular positions around the reference point and predicts the dynamic weight matrix to combine them. The weighted feature updates to query for the next stage until it feeds into the loss heads. Sparse R-CNN [39] can be regarded as a variant of DETR and assumes an even simpler setting. It has a set of learnable object queries and proposal boxes, and kernels given by query slots process only regions within proposal boxes. Details are provided in Sect. 3.1. We believe that sparse R-CNN can be easily enhanced without breaking its full sparse assumption and increasing many calculations and parameters.

3   Method

We now introduce the proposed method, mainly including IoU-enhanced self attention (IoU-ESA) and dynamic channel weighting (DCW). The overview of the method is shown in Fig. 2. Before illustrating the details of the designed modules, we first review and analyze the preliminary works of sparse R-CNN.

3.1   Preliminaries and Analysis on Sparse R-CNN

Basic Formulation. Details about sparse RCNN can also be found in Fig. 2. It has a set of the learnable object queries q to provide one-to-one dynamic interactions within their corresponding learnable proposal boxes b. These q are expected to encode the content of the ground truth boxes. The proposal boxes b represent the rectangles. They locate on features of the whole image for regional RoIs r by RoI align. Dynamic convs, with the two sets of parameters (Params1-2) generated from the object queries q by a linear projection, is performed on RoI features r. Then, object features o for dual loss heads can be obtained. Note that the dynamic convs directly link the object queries q with their corresponding RoIs r. Sparse R-CNN is in an iterative structure with multiple stages.

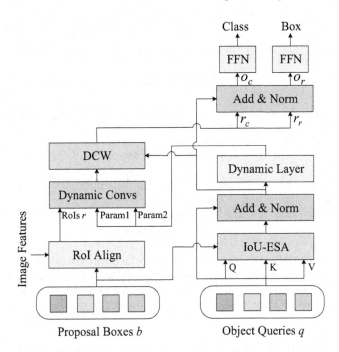

Fig. 2. An overview illustration of our approach. Besides the original dynamic convs module in sparse R-CNN, two inserted modules, IoU-Enhanced Self Attention (IoU-ESA) and Dynamic Channel Weighting (DCW), are designed. The former enhances the self attention among object queries by IoU between proposal boxes. The latter outputs disentangled features for the two tasks, classification and regression, in object detection.

Particularly, both q and b are updated in the intermediate stage and potentially used for the next one. In the initial stage, q and b are model parameters which can be tuned by back-propagation during training.

Sparse R-CNN is similar to DETR in the following aspects: (i) They both have learnable object queries, which intends to describe and capture the ground truths. In sparse R-CNN, it is the proposal feature, while in DETR, it is the positional encoding. (ii) In these two methods, self attention is performed among object queries to aggregate information from the entire set. (iii) Dynamic convs and cross attention are comparable. Both perform interactions between object queries and image feature, refining the object embeddings used for classification and regression. In summary, sparse R-CNN can be regarded as a variant of DETR depending on RoI align, with a different way to generate object embeddings for loss heads.

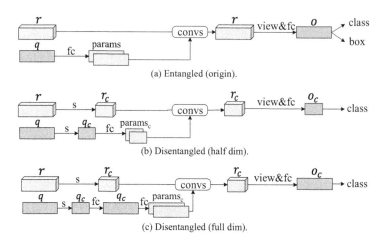

Fig. 3. Analysis on feature disentanglement. (a)–(c) indicate the original sparse RCNN, disentangled dynamic conv with half dimension for each branch, and disentangled dynamic conv with a projection head to recover full dimension, respectively. Note that, we only show the classification branch in (b) and (c).

Analysis on Self Attention. The unique property of sparse R-CNN is its sparse-in sparse-out paradigm throughout the processing steps. Neither dense candidates nor interacting with global features exists in the pipeline. However, we find that the self attention plays an important role, since it models the relation among q and enlarges the receptive field of it. Although q is a summarized vector without spatial coverage, the geometry relation is also expected to be discovered in the self attention and to be considered in value routing. To validate our assumption, we train one model without self attention and the other in which the self attention matrix is replaced by the IoU matrix computed directly between proposal boxes. Their performances are compared with the original sparse R-CNN in Table 1. These experiments are conducted on MS-COCO. It can be seen that the worst performance, with the AP of 38.5, is from the model w/o MSA without any self attention among object queries. When IoU matrix is used as the attention matrix, the results from IoU-MSA have been greatly recovered. However, the original self attention still achieves the optimal performance. Thus, we intend to enhance the self attention in sparse R-CNN by utilizing the IoU as a geometry prior to guide the routing scheme in self attention. Note that, all the self attentions are in the form of multi-head. The analyze result on CrowdHuman will be presented in the supplementary material.

Analysis on Feature Disentanglement. As is described in [36], the two tasks in detection concentrate on different parts of objects. The classification branch concerns about the region of semantics to support its class type, while the regression branch is more related to object contours for accurately locating it. Therefore, it is beneficial to disentangle the features for them. Note that sparse R-CNN uses the same entangled feature for the two tasks, which can be potentially improved.

We train two models which have separate features and structures for the classification and regression, as shown in Fig. 3. Both of them use the same dimension (e.g., 256) setting on object queries q and RoI features r. The first one in Fig. 3b allocates half dimensions (e.g., 128) for each task. Therefore the dynamic conv and its later layers become fully independent. Note that this setting saves the model parameters and calculations. The other in Fig. 3c has a similar intention but considers to compensating the dimension lost before giving the feature to dual heads. The number of channels is recovered to the original dimension as in sparse R-CNN, which is realized by an extra projection layer. One thing that needs to mention is that although the newly added layer brings along more model parameters and calculations, it is still efficient compared with the original model, mainly due to the significant dimension reduction during dynamic convs. Their performances on MS-COCO are listed in Table 2. Compared to the original sparse R-CNN with entangled features and structures for both tasks, the half dim model metric does not obviously degrade, but the parameters and Flops are apparently less. The full dim model is slightly better than the original and is still slightly efficient. These results clearly demonstrate the effectiveness of task disentanglement in sparse R-CNN. However, we think the simple strategy to divide channels for different tasks does not fully consider the dynamic weighting scheme from the object query. Based on this observation, we intend to enhance the feature for classification and localization by a dynamic channel weighting (DCW) module, which gives a better disentanglement for the two tasks at the cost of only a slight increase in model parameters.

Table 1. Analysis on self attention. Three different models, including origin sparse R-CNN with multi-head self attention (MSA), without self attention and attention matrix replaced by IoU matrix (IoU-MSA), are trained on MS-COCO.

Method	AP	Params	FLOPs
MSA	42.8	106.2M	133.8G
w/o MSA	38.5	104.7M	133.6G
IoU-MSA	41.8	105.4M	133.7G

Table 2. Analysis on feature disentanglement on MS-COCO. AP, the number of parameters, and the amount of calculations of the three models are listed. 'En.' and 'Dis.' indicate entangled and disentangled feature dimension.

Method	AP	Params	FLOPs
En. (original)	42.8	106.2M	133.8G
Dis. (half dim)	42.7	97.0M	132.9G
Dis. (full dim)	43.1	103.7M	133.5G

3.2   IoU-Enhanced Self Attention (IoU-ESA)

As is depicted in Sect. 3.1, multi-head self attention (MSA) is applied on the object queries q to obtain the global information from each other. Here we further

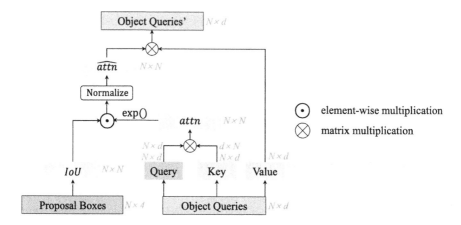

Fig. 4. Details of IoU-ESA module. Normal self attention is still executed among object queries. The attention matrix *attn*, calculated from Q and K, is further modulated by *IoU* matrix through element-wise multiplication. *IoU* matrix reflects similarity of proposal boxes, and it also determines the value routing scheme.

illustrate its mechanism as in Eq. (1),

$$\text{MSA}(\mathbf{Q}, \mathbf{K}, \mathbf{V}) = [\text{Attn}\,(\mathbf{Q}_i, \mathbf{K}_i, \mathbf{V}_i)]_{i=1:H}\mathbf{W}^o \tag{1}$$

$$\text{Attn}\,(\mathbf{Q}_i, \mathbf{K}_i, \mathbf{V}_i) = \text{softmax}\left(\frac{\mathbf{Q}_i\mathbf{K}_i^{\text{T}}}{\sqrt{d}}\right)\mathbf{V}_i \tag{2}$$

where $\mathbf{Q}, \mathbf{K}, \mathbf{V} \in \mathbb{R}^{N \times d}$ are the input query, key and value matrices, projected from object queries q by parallel layers with different parameters. d is the hidden dimension after the projection, and H is the total number of heads. $\text{Attn}\,(\mathbf{Q}, \mathbf{K}, \mathbf{V})$ denotes the standard attention function with $\mathbf{Q}$, $\mathbf{K}$, and $\mathbf{V}$ as its input, and it is further clarified in Eq. (2). $\mathbf{W}^o$ is a set of learnable parameters used to combine multiple heads, and the subscripts specify the index of the head. Note that each head independently routes its own value $\mathbf{V}$ according to the similarity between $\mathbf{Q}$ and $\mathbf{K}$.

We intend to enhance the self attention by considering the geometry relation between two corresponding boxes, and the details are given in Fig. 4. Basically, IoU-ESA utilizes another way to measure the similarity between $\mathbf{Q}$ and $\mathbf{K}$. Compared with exhaustively measure between any slot pair through $attn = \mathbf{Q}\mathbf{K}^{\text{T}}/\sqrt{d} \in \mathbb{R}^{N \times N}$, IoU is a good prior and it is easier to compute. IoU-ESA takes advantage of it without adding any parameters. Specifically, each element of *attn* before normalize is multiplied by its corresponding element in *IoU* matrix, which is computed from the pairs of proposal boxes. The IoU-enhanced self attention is formulated as:

$$\widehat{attn}_{ij} = \frac{\exp\,(attn_{ij}) \cdot IoU_{ij}}{\sum_{k=1}^{N} \exp\,(attn_{ik}) \cdot IoU_{ik}} \tag{3}$$

where i, j are indexed from 1 to N, $\widehat{attn} \in \mathbb{R}^{N \times N}$ is the enhanced attention matrix, and $IoU \in [0, 1]^{N \times N}$ is measured among N proposal boxes based on the IoU metric.

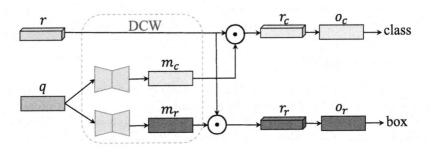

Fig. 5. Details of DCW module and its way to be applied on RoI feature. Two channel masks m_c and m_r are generated from object queries q by the lightweight bottlenecks, and they enhance the RoI features r output from dynamic convs.

3.3 Dynamic Channel Weighting (DCW)

As is analyzed in Sect. 3.1, it is better to use different features for the two loss heads in detection. However, sparse R-CNN sends the same features into both heads, which is obviously not optimal. Although the simple strategy, to separate object queries q and RoI features r along the channel dimension, can reduce the parameters and calculations without degrading the performance, a better dynamic scheme, which fully utilizes q, is still needed.

Here we propose a dynamic channel weighting (DCW) module to separate and strengthen the two features. As shown in Fig. 5, two channel masks m_c and $m_r \in \mathbb{R}^{N \times d}$ are generated by linear projection layers in parallel with the dynamic layer for the parameters in dynamic convs. The sigmoid is used as the activation, constraining each element in m_c and m_r between 0–1. Different from SE-Net, the DCW is actually a dynamic cross weighting scheme. Based on the object queries q, they generate m_c and m_r with the same number of channels as the RoI features r, and the value in them shares for each spatial position in r. These two channel masks highlight different areas in the proposal box, hence producing more relevant features for classification and regression branches. The DCW module can be formulated as follows:

$$r' = r \odot \sigma \left(g_m \left(q \right) \right) \tag{4}$$

where q and r indicate object queries and RoI features after dynamic conv. g_m is a bottleneck including two connected linear projection layers, predicting two masks based on q. $\sigma(\cdot)$ represents the sigmoid activation. The result r' is the enhanced RoI features for different tasks, and it can be r_c or r_r as is shown in

Fig. 5. Note that the separate r_c and r_r go through independent layers, which project them into object embeddings o_c and o_r. These features are finally given to the two loss heads to compute different types of penalties.

Since the original sparse R-CNN is built in cascaded structure, the object queries q need to be updated for the next stage. To keep it simple, in our implementation, o_c and o_r are added together to form the new q in the next stage.

3.4   Training Objectives

Following [3,37,39,50], we take the Hungarian algorithm to make the prediction and ground truth pairs. The bipartite matching manner avoids post processing like NMS. The matching cost is the same as the training loss, $\mathcal{L} = \lambda_{cls} \cdot \mathcal{L}_{cls} + \lambda_{L1} \cdot \mathcal{L}_{L1} + \lambda_{giou} \cdot \mathcal{L}_{giou}$. Here, $\mathcal{L}_{cls}$ is focal loss [25] used for classification. $\mathcal{L}_{L1}$ and $\mathcal{L}_{giou}$ representing the L1 loss and generalized IoU loss [34] are employed for box regression. λ_{cls}, λ_{L1} and λ_{giou} are coefficients balancing the impact of each term.

4   Experiment

Datasets. We perform extensive experiments with variant backbones on the pedestrian detection benchmark CrowdHuman [35] and 80-category MS-COCO 2017 detection dataset [26].

For CrowdHuman, training is performed on ∼15k train images and testing is evaluated on the ∼4k validation images. There are a total of 470k human instances from train and validation subsets and 23 persons per image on average, with various kinds of occlusions in the dataset. Each human instance is annotated with a head bounding-box, human visible-region bounding-box and human full-body bounding-box. We only use the full-body bounding-box for training. For evaluation, we follow the standard Caltech [9] evaluation metric mMR, which stands for the Log-Average Missing Rate over false positives per image (FPPI) ranging in $[10^{-2}, 100]$. Average Precision (AP) and Recall are also provided to better evaluate our methods.

For COCO dataset, training is performed on train2017 split (∼118k images) and testing is evaluated on val2017 set (5k images). Each image is annotated with bounding-boxes. On average, there are 7 instances per image, up to 63 instances in a single image in the training set, ranging from small to large on the same image. If not specified, we report AP as bbox AP, the integral metric over multiple thresholds. We report validation AP at the last training epoch.

Implementation Details. All experiments are implemented based on Detectron2 codebase [15]. We initialize backbone from the pre-trained model on ImageNet [8], and other newly added layers with Xavier [16]. By default, we treat ResNet-50 [18] as the backbone, and use FPN [24] to deal with different sizes of ground truth boxes. AdamW [30] is employed to optimize the training loss with 10^{-4} weight decay. The initial learning rate is 2.5×10^{-5}, divided by 10 at

40th epoch for CrowdHuman and by 10 at 27th and 33rd epochs for COCO. All models are trained on 4 RTX 3090 GPUs with batch size of 16, for 50 epochs on CrowdHuman, while 36 epochs on COCO. Images are resized such that the shorter edge is at most 800 pixels while the longer side is 1,333. The loss weights are set as $\lambda_{cls} = 2$, $\lambda_{L1} = 5$ and $\lambda_{giou} = 2$. Following sparse R-CNN, the default number of object queries and proposal boxes are 500 for CrowdHuman, 100 for COCO. All models have 6 iterative stages. The gradients on proposal boxes are detached in the next stage for stable training. There is no post processing step like NMS at the evaluation.

4.1   Ablation Study

Ablation Study on CrowdHuman. CrowdHuman consists of images with highly overlapped pedestrians. IoU-ESA restricts the attending range of the object queries q during self attention, according to Eq. (3), so q can intentionally consider the geometry relation for updating itself. On the other hand, DCW gives the independent feature representations for classification and regression, which is also conducive to improve the performance. Table 3 shows the detailed ablations on them. It clearly indicates that both modules can significantly improve the performance. When the two modules are combined together, they can bring further enhancements.

Ablation Study on COCO. We further carry out a similar experiment on the richer and larger dataset MS-COCO. The results are shown in Table 4. Note that compared to CrowdHuman, COCO has variety of classes and dense objects at the same time. It can be seen that both IoU-ESA and DCW modules can also perform well on COCO.

Table 3. Ablation study results on CrowdHuman. All experiments are based on ResNet-50, and number of object queries and proposal boxes are 500.

IoU-ESA	DCW	AP ↑	mMR ↓	Recall ↑
		89.2	48.3	95.9
✓		90.4	47.8	96.7
	✓	89.9	47.4	96.3
✓	✓	**90.9**	**47.2**	**96.9**

Table 4. Ablation study results on COCO. All experiments are based on ResNet-50, and number of object queries and proposal boxes are set to 100.

IoU-ESA	DCW	AP	AP$_{50}$	AP$_{75}$
		42.8	61.2	45.7
✓		43.9	63.0	47.6
	✓	43.6	62.3	47.3
✓	✓	**44.4**	**63.4**	**48.3**

4.2   Main Results

Main Results on CrowdHuman. To thoroughly evaluate the performance of our proposed methods, we do plenty of experiments based on the same backbone (ResNet-50) for different methods. The results are shown in Table 5. It obviously shows that our work outperforms the sparse R-CNN and other works especially the sparse detector, including DETR and deformable DETR.

Note that we increase all metrics, including AP, mMR and Recall, at the same time. Particularly, the enhanced version can achieve +1.7, −1.1, +1.0 gains on AP, mMR and Recall, respectively, compared with sparse R-CNN.

Table 5. Performance comparisons on CrowdHuman.

Method	NMS	AP ↑	mMR ↓	Recall ↑
Faster R-CNN	✓	85.0	50.4	90.2
RetinaNet	✓	81.7	57.6	88.6
FCOS	✓	86.1	55.2	94.3
AdaptiveNMS	✓	84.7	49.7	91.3
DETR	○	66.1	80.6	–
Deformable DETR	○	86.7	54.0	92.5
Sparse R-CNN	○	89.2	48.3	95.9
Ours	○	**90.9**	**47.2**	**96.9**

Main Results on COCO. Table 6 shows the performances of our proposed methods with two standard backbones, ResNet-50 and ResNet-101, and it also compares with some mainstream object detectors or DETR-series. It can be seen that the sparse R-CNN outperforms well-established detectors, such as Faster R-CNN and RetinaNet. Our enhanced version can further improve its detection accuracy. Our work reaches 44.4 AP, with 100 query slots based on ResNet-50, which gains 4.2 and 1.6 AP from Faster R-CNN and sparse R-CNN. Note that our model only slightly increases the calculations from 134 to 137 GFlops. However, it is still efficient compared with other models like deformable DETR, TSP and SMCA. With 300 query slots, we can achieve 46.4 on AP, which is not only superior to 45.0 from sparse R-CNN, but also outperforms other methods with even bigger backbone like ResNet-101. Our method consistently boosts the performance under ResNet-101. Particularly, it gives the AP at 45.6 and 47.5 with 100 and 300 query slots, respectively, compared with 44.1 and 46.4 from sparse R-CNN of the same settings. More surprisingly, our method reaches 49.6 on the Swin-T backbone, surpassing sparse R-CNN 1.7 AP. The results of our method on COCO test-dev set are reported in Table 7. With the help of test-time augmentation, the performance can be further improved.

Table 6. Performances of different detectors on COCO 2017 val set. Here '*' indicates that the model is with 300 object queries and random crop training augmentation, 'Res' indicates ResNet. The top two sections show results from Detectron2 [15], mmdetection [4] or original papers [3,11,39,50].

Method	Backbone	Epoch	Params	FLOPs	AP	AP_{50}	AP_{75}	AP_S	AP_M	AP_L
RetinaNet [15]	Res-50	36	38M	206G	38.7	58.0	41.5	23.3	42.3	50.3
RetinaNet [15]	Res-101	36	57M	273G	40.4	60.3	43.2	24.0	44.3	52.2
Faster R-CNN [15]	Res-50	36	42M	180G	40.2	61.0	43.8	24.2	43.5	52.0
Faster R-CNN [15]	Res-101	36	61M	232G	42.0	62.5	45.9	25.2	45.6	54.6
Cascade R-CNN [4]	Res-50	20	69M	235G	41.0	59.4	44.4	22.7	44.4	54.3
DETR [3]	Res-50	500	41M	86G	42.0	62.4	44.2	20.5	45.8	61.1
DETR [3]	Res-101	500	60M	152G	43.5	63.8	46.4	21.9	48.0	61.8
Deformable DETR [50]	Res-50	50	40M	173G	43.8	62.6	47.7	26.4	47.1	58.0
TSP-FCOS [40]	Res-50	36	52M	189G	43.1	62.3	47.0	26.6	46.8	55.9
TSP-RCNN [40]	Res-50	36	64M	188G	43.8	63.3	48.3	28.6	46.9	55.7
TSP-FCOS [40]	Res-101	36	–	255G	44.4	63.8	48.2	27.7	48.6	57.3
TSP-RCNN [40]	Res-101	36	–	254G	44.8	63.8	49.2	29.0	47.9	57.1
SMCA [11]	Res-50	50	40M	152G	43.7	63.6	47.2	24.2	47.0	60.4
SMCA [11]	Res-101	50	58M	218G	44.4	65.2	48.0	24.3	48.5	61.0
Sparse R-CNN [39]	Res-50	36	106M	134G	42.8	61.2	45.7	26.7	44.6	57.6
Sparse R-CNN [39]	Res-101	36	125M	201G	44.1	62.1	47.2	26.1	46.3	59.7
Sparse R-CNN* [39]	Res-50	36	106M	152G	45.0	63.4	48.2	26.9	47.2	59.5
Sparse R-CNN* [39]	Res-101	36	125M	219G	46.4	64.6	49.5	28.3	48.3	61.6
Sparse R-CNN* [39]	Swin-T	36	110M	158G	47.9	67.3	52.3	–	–	–
Ours	Res-50	36	133M	137G	44.4	63.4	48.3	27.5	47.1	58.2
Ours	Res-101	36	152M	203G	45.6	64.4	49.8	27.5	48.5	60.2
Ours*	Res-50	36	133M	160G	46.4	65.8	50.7	29.9	49.0	60.5
Ours*	Res-101	36	152M	227G	47.5	66.8	51.8	30.4	50.7	62.5
Ours*	Swin-T	36	136M	167G	49.6	69.4	54.4	33.8	51.9	64.4

Table 7. Performances of different detectors on COCO test-dev set. 'ReX' indicates ResNeXt, and 'TTA' indicates test-time augmentations, following the settings in [46].

Method	Backbone	TTA	AP	AP_{50}	AP_{75}	AP_S	AP_M	AP_L
Faster RCNN	Res-101		36.2	59.1	39.0	18.2	39.0	48.2
RetinaNet	Res-101		39.1	59.1	42.3	21.8	42.7	50.2
Cascade R-CNN	Res-101		42.8	62.1	46.3	23.7	45.5	55.2
TSD	Res-101		43.2	64.0	46.9	24.0	46.3	55.8
TSP-RCNN	Res-101		46.6	66.2	51.3	28.4	49.0	58.5
Sparse R-CNN*	ReX-101		46.9	66.3	51.2	28.6	49.2	58.7
Ours	Res-101		45.6	64.9	49.6	26.9	47.9	58.4
Ours*	Res-101		47.8	67.1	52.3	29.0	50.1	60.4
Ours	Res-101	✓	49.5	68.7	54.9	32.4	51.5	61.9
Ours*	Res-101	✓	50.8	70.2	56.8	34.1	52.7	63.1
Ours*	Swin-T	✓	52.2	71.9	58.4	35.4	53.7	65.0

5 Conclusion

This paper enhances sparse R-CNN and makes it more suitable for dense object detection. We design two modules, including IoU-ESA and DCW. The former strengthens the original self attention performed on object queries. By employing IoU of two corresponding proposal boxes, the object query is updated by other relevant ones under the guide from spatial relations. The latter allows the object query to provide two extra dynamic channel weights for classification and regression, so image feature within proposal boxes can be highlighted in different ways to meet the requirement of the two detection tasks. Both two modules effectively boost the performance of spare R-CNN. They can significantly increase the metrics on CrowdHuman and MS-COCO based on different backbones. Considering the fact that sparse R-CNN has relatively more parameters than other methods, particularly in its multi-stage dynamic heads, we will explore the ways to reduce model size and to increase the efficiency.

References

1. Bodla, N., Singh, B., Chellappa, R., Davis, L.S.: Soft-nms-improving object detection with one line of code. In: Proceedings of the IEEE International Conference on Computer Vision, pp. 5561–5569 (2017)
2. Cai, Z., Vasconcelos, N.: Cascade r-cnn: delving into high quality object detection. In: Proceedings of the IEEE Conference on Computer Vision and Pattern Recognition, pp. 6154–6162 (2018)
3. Carion, N., Massa, F., Synnaeve, G., Usunier, N., Kirillov, A., Zagoruyko, S.: End-to-end object detection with transformers. In: Vedaldi, A., Bischof, H., Brox, T., Frahm, J.-M. (eds.) ECCV 2020. LNCS, vol. 12346, pp. 213–229. Springer, Cham (2020). https://doi.org/10.1007/978-3-030-58452-8_13
4. Chen, K., et al.: Mmdetection: Open mmlab detection toolbox and benchmark. arXiv preprint arXiv:1906.07155 (2019)
5. Chen, Y., et al.: Reppoints v2: verification meets regression for object detection. Adv. Neural Inf. Process. Syst. **33** (2020)
6. Dai, J., et al.: Deformable convolutional networks. In: Proceedings of the IEEE International Conference on Computer Vision, pp. 764–773 (2017)
7. Dalal, N., Triggs, B.: Histograms of oriented gradients for human detection. In: 2005 IEEE Computer Society Conference on Computer Vision and Pattern Recognition (CVPR 2005), vol. 1, pp. 886–893. IEEE (2005)
8. Deng, J., Dong, W., Socher, R., Li, L.J., Li, K., Fei-Fei, L.: Imagenet: a large-scale hierarchical image database. In: 2009 IEEE Conference on Computer Vision and Pattern Recognition, pp. 248–255. IEEE (2009)
9. Dollar, P., Wojek, C., Schiele, B., Perona, P.: Pedestrian detection: an evaluation of the state of the art. IEEE Trans. Pattern Anal. Mach. Intelli. **34**(4), 743–761 (2011)
10. Dosovitskiy, A., et al.: An image is worth 16×16 words: transformers for image recognition at scale. arXiv preprint arXiv:2010.11929 (2020)
11. Gao, P., Zheng, M., Wang, X., Dai, J., Li, H.: Fast convergence of detr with spatially modulated co-attention. arXiv preprint arXiv:2101.07448 (2021)

12. Ge, Z., Liu, S., Li, Z., Yoshie, O., Sun, J.: Ota: optimal transport assignment for object detection. In: Proceedings of the IEEE/CVF Conference on Computer Vision and Pattern Recognition, pp. 303–312 (2021)
13. Girshick, R.: Fast r-cnn. In: Proceedings of the IEEE International Conference on Computer Vision, pp. 1440–1448 (2015)
14. Girshick, R., Donahue, J., Darrell, T., Malik, J.: Rich feature hierarchies for accurate object detection and semantic segmentation. In: Proceedings of the IEEE Conference on Computer Vision and Pattern Recognition, pp. 580–587 (2014)
15. Girshick, R., Radosavovic, I., Gkioxari, G., Dollár, P., He, K.: Detectron (2018). https://github.com/facebookresearch/detectron
16. Glorot, X., Bengio, Y.: Understanding the difficulty of training deep feedforward neural networks. In: Proceedings of the Thirteenth International Conference on Artificial Intelligence and Statistics, pp. 249–256. JMLR Workshop and Conference Proceedings (2010)
17. He, K., Gkioxari, G., Dollár, P., Girshick, R.: Mask r-cnn. In: Proceedings of the IEEE International Conference on Computer Vision, pp. 2961–2969 (2017)
18. He, K., Zhang, X., Ren, S., Sun, J.: Deep residual learning for image recognition. In: Proceedings of the IEEE Conference on Computer Vision and Pattern Recognition, pp. 770–778 (2016)
19. Hu, H., Gu, J., Zhang, Z., Dai, J., Wei, Y.: Relation networks for object detection. In: Proceedings of the IEEE Conference on Computer Vision and Pattern Recognition, pp. 3588–3597 (2018)
20. Hu, J., Shen, L., Sun, G.: Squeeze-and-excitation networks. In: Proceedings of the IEEE Conference on Computer Vision and Pattern Recognition, pp. 7132–7141 (2018)
21. Huang, X., Ge, Z., Jie, Z., Yoshie, O.: NMS by representative region: towards crowded pedestrian detection by proposal pairing. In: Proceedings of the IEEE/CVF Conference on Computer Vision and Pattern Recognition, pp. 10750–10759 (2020)
22. Kim, K., Lee, H.S.: Probabilistic anchor assignment with IoU prediction for object detection. In: Vedaldi, A., Bischof, H., Brox, T., Frahm, J.-M. (eds.) ECCV 2020. LNCS, vol. 12370, pp. 355–371. Springer, Cham (2020). https://doi.org/10.1007/978-3-030-58595-2_22
23. Law, H., Deng, J.: Cornernet: detecting objects as paired keypoints. In: Proceedings of the European Conference on Computer Vision (ECCV), pp. 734–750 (2018)
24. Lin, T.Y., Dollár, P., Girshick, R., He, K., Hariharan, B., Belongie, S.: Feature pyramid networks for object detection. In: Proceedings of the IEEE Conference on Computer Vision and Pattern Recognition, pp. 2117–2125 (2017)
25. Lin, T.Y., Goyal, P., Girshick, R., He, K., Dollár, P.: Focal loss for dense object detection. In: Proceedings of the IEEE International Conference on Computer Vision, pp. 2980–2988 (2017)
26. Lin, T.-Y., et al.: Microsoft COCO: common objects in context. In: Fleet, D., Pajdla, T., Schiele, B., Tuytelaars, T. (eds.) ECCV 2014. LNCS, vol. 8693, pp. 740–755. Springer, Cham (2014). https://doi.org/10.1007/978-3-319-10602-1_48
27. Liu, S., Huang, D., Wang, Y.: Adaptive NMS: refining pedestrian detection in a crowd. In: Proceedings of the IEEE/CVF Conference on Computer Vision and Pattern Recognition, pp. 6459–6468 (2019)
28. Liu, W., et al.: SSD: single shot multibox detector. In: Leibe, B., Matas, J., Sebe, N., Welling, M. (eds.) ECCV 2016. LNCS, vol. 9905, pp. 21–37. Springer, Cham (2016). https://doi.org/10.1007/978-3-319-46448-0_2

29. Liu, Z., et al.: Swin transformer: hierarchical vision transformer using shifted windows. arXiv preprint arXiv:2103.14030 (2021)
30. Loshchilov, I., Hutter, F.: Decoupled weight decay regularization. In: International Conference on Learning Representations (2018)
31. Redmon, J., Divvala, S., Girshick, R., Farhadi, A.: You only look once: unified, real-time object detection. In: Proceedings of the IEEE Conference on Computer Vision and Pattern Recognition, pp. 779–788 (2016)
32. Redmon, J., Farhadi, A.: Yolo9000: better, faster, stronger. In: Proceedings of the IEEE Conference on Computer Vision and Pattern Recognition, pp. 7263–7271 (2017)
33. Ren, S., He, K., Girshick, R., Sun, J.: Faster r-cnn: towards real-time object detection with region proposal networks. Adv. Neural Inf. Process. Syst. **28**, 91–99 (2015)
34. Rezatofighi, H., Tsoi, N., Gwak, J., Sadeghian, A., Reid, I., Savarese, S.: Generalized intersection over union: a metric and a loss for bounding box regression. In: Proceedings of the IEEE/CVF Conference on Computer Vision and Pattern Recognition, pp. 658–666 (2019)
35. Shao, S., et al.: Crowdhuman: a benchmark for detecting human in a crowd. arXiv preprint arXiv:1805.00123 (2018)
36. Song, G., Liu, Y., Wang, X.: Revisiting the sibling head in object detector. In: Proceedings of the IEEE/CVF Conference on Computer Vision and Pattern Recognition, pp. 11563–11572 (2020)
37. Stewart, R., Andriluka, M., Ng, A.Y.: End-to-end people detection in crowded scenes. In: Proceedings of the IEEE Conference on Computer Vision and Pattern Recognition, pp. 2325–2333 (2016)
38. Sun, P., Jiang, Y., Xie, E., Yuan, Z., Wang, C., Luo, P.: Onenet: towards end-to-end one-stage object detection. arXiv preprint arXiv:2012.05780 (2020)
39. Sun, P., et al.: Sparse r-cnn: end-to-end object detection with learnable proposals. In: Proceedings of the IEEE/CVF Conference on Computer Vision and Pattern Recognition, pp. 14454–14463 (2021)
40. Sun, Z., Cao, S., Yang, Y., Kitani, K.: Rethinking transformer-based set prediction for object detection. arXiv preprint arXiv:2011.10881 (2020)
41. Tian, Z., Shen, C., Chen, H., He, T.: FCOS: fully convolutional one-stage object detection. In: Proceedings of the IEEE/CVF International Conference on Computer Vision, pp. 9627–9636 (2019)
42. Vaswani, A., et al.: Attention is all you need. Adv. Neural Inf. Process. Syst., 5998–6008 (2017)
43. Viola, P., Jones, M.: Rapid object detection using a boosted cascade of simple features. In: Proceedings of the 2001 IEEE Computer Society Conference on Computer Vision and Pattern Recognition, CVPR 2001. vol. 1. IEEE (2001)
44. Wang, W., et al.: Pyramid vision transformer: a versatile backbone for dense prediction without convolutions. arXiv preprint arXiv:2102.12122 (2021)
45. Yang, Z., Liu, S., Hu, H., Wang, L., Lin, S.: Reppoints: point set representation for object detection. In: Proceedings of the IEEE/CVF International Conference on Computer Vision, pp. 9657–9666 (2019)
46. Zhang, S., Chi, C., Yao, Y., Lei, Z., Li, S.Z.: Bridging the gap between anchor-based and anchor-free detection via adaptive training sample selection. In: Proceedings of the IEEE/CVF Conference on Computer Vision and Pattern Recognition, pp. 9759–9768 (2020)
47. Zhao, Z.Q., Zheng, P., Xu, S.T., Wu, X.: Object detection with deep learning: a review. IEEE Trans. Neural Netw. Learn. Syst. **30**(11), 3212–3232 (2019)

48. Zhou, X., Wang, D., Krähenbühl, P.: Objects as points. arXiv preprint arXiv:1904.07850 (2019)
49. Zhu, X., Hu, H., Lin, S., Dai, J.: Deformable convnets v2: more deformable, better results. In: Proceedings of the IEEE/CVF Conference on Computer Vision and Pattern Recognition, pp. 9308–9316 (2019)
50. Zhu, X., Su, W., Lu, L., Li, B., Wang, X., Dai, J.: Deformable DETR: deformable transformers for end-to-end object detection. In: International Conference on Learning Representations (2020)

HAZE-Net: High-Frequency Attentive Super-Resolved Gaze Estimation in Low-Resolution Face Images

Jun-Seok Yun[1], Youngju Na[1], Hee Hyeon Kim[1], Hyung-Il Kim[2], and Seok Bong Yoo[1(✉)]

[1] Department of Artificial Intelligence Convergence, Chonnam National University, Gwangju, Korea
sbyoo@jnu.ac.kr
[2] Electronics and Telecommunications Research Institute, Daejeon, Korea

Abstract. Although gaze estimation methods have been developed with deep learning techniques, there has been no such approach as aim to attain accurate performance in low-resolution face images with a pixel width of 50 pixels or less. To solve a limitation under the challenging low-resolution conditions, we propose a high-frequency attentive super-resolved gaze estimation network, i.e., HAZE-Net. Our network improves the resolution of the input image and enhances the eye features and those boundaries via a proposed super-resolution module based on a high-frequency attention block. In addition, our gaze estimation module utilizes high-frequency components of the eye as well as the global appearance map. We also utilize the structural location information of faces to approximate head pose. The experimental results indicate that the proposed method exhibits robust gaze estimation performance even in low-resolution face images with 28×28 pixels. The source code of this work is available at https://github.com/dbseorms16/HAZE_Net/.

1 Introduction

Human gaze information provides principal guidance on a person's attention and is significant in the prediction of human behaviors and speculative intentions. Accordingly, it has been widely used in various applications, such as human-computer interaction [1,2], autonomous driving [3], gaze target detection [4], and virtual reality [5]. Most of the existing methods for estimating the human gaze by utilizing particular equipment (e.g., eye-tracking glasses and virtual reality/augmented reality devices) are not suitable for real-world applications [6–8]. Recently, to solve this problem, face image's appearance-based gaze estimation methods that learn a direct mapping function from facial appearance or eyes to human gaze are considered. To accurately estimate the human gaze, an image

Supplementary Information The online version contains supplementary material available at https://doi.org/10.1007/978-3-031-26348-4_9.

with well-preserved eye features (e.g., the shape of the pupil) and well-separated boundaries (e.g., the boundary between the iris and the eyelids) is crucial.

Recent studies [9,10] show reliable performance for the gaze estimation with high-resolution (HR) face images from 448 × 448 to 6000 × 4000 pixels including abundant eye-related features. However, in the real world, even the face region detected from the HR image may have a low-resolution (LR) depending on the distance between the camera and the subject, as shown in the upper-left of Fig. 1. Most of the existing gaze estimation methods use fixed-size images. Thus, when the distance between the camera and the subject is large, it leads to the severe degradation of gaze estimation due to the lack of resolution of the eye patches, as shown in the first row of Fig. 1.

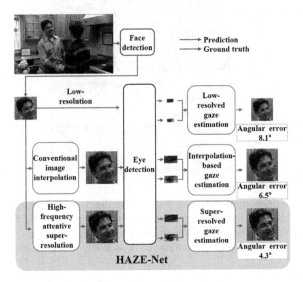

Fig. 1. Examples of gaze estimation approaches in the real world. HAZE-Net introduced a high-frequency attentive super-resolved gaze estimation that outperforms conventional methods by a large margin.

To deal with the problem, the conventional image interpolation approach can be adopted so that the image resolution for eye regions can be enhanced as shown in the second row of Fig. 1. However, since it only works based on the limited relationship between the surrounding pixels, this method cannot resolve the degradation of gaze estimation performance. As an alternative, image super-resolution (SR) methods [11–18] have been considered. These methods are performed to restore HR images from the LR images. Accordingly, the SR module learns how to reconstruct the LR images to HR images. However, this is an ill-posed problem with various possible answers. This indicates that conventional SR modules do not stably enhance eye features and boundaries, which are essential for gaze estimation. In other words, although the SR approach may help improve the quality of the image, it does not guarantee an ideal mapping for the optimal performance of the gaze estimation. The results of each image upscaling method for an LR face image with 28×28 pixels are shown in Fig. 2. A severe degradation problem occurs in the up-sampled image when bicubic interpolation is applied to the LR image, as shown in bicubic results. On the other hand, the conventional SR module shows higher quality, as shown in DRN [12] results. Nevertheless, it can be seen that the boundary between the iris and the pupil is not distinctly differentiated. This demonstrates that the conventional SR method does not provide an optimal mapping guideline for gaze estimation.

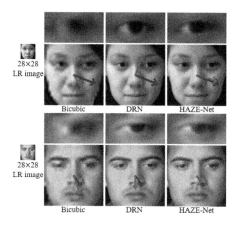

Fig. 2. Visual comparison of different 4× up-sampling methods for LR face images with 28 × 28 pixels. The first row for each LR image shows the enlarged eye images using different methods. The second row for each LR image represents the ground truth (blue arrow) and predicted gaze (red arrow), respectively. (Color figure online)

In this paper, we propose a high-frequency attentive super-resolved gaze estimation network, so-called HAZE-Net, which is mainly comprised of two modules: 1) SR module based on a high-frequency attention block (HFAB) and 2) global-local gaze estimation module. To deal with the limitations in the conventional SR methods, we reinforce the high-frequency information inspired by the observation that the contour of eye features and their boundaries correspond to high-frequency in the frequency domain. Through the proposed SR module, we observe that it preserves the shape of the pupil well and distinctly differentiates the boundary between the iris and the pupil. In addition to the SR module, we devise a global-local gaze estimation module. Based on the super-resolved face images and corresponding global-local (face-eye) appearance maps are used to improve the gaze estimation performance. In addition, we use the coordinates of five landmarks (e.g., eyes, nose, both corners of the mouth) containing the structural location information of the face to provide an appropriate guide to the head pose. Moreover, the devised two modules are collaboratively trained via the proposed alternative learning strategy. In this process, we add a constraint on the SR module to produce a face image that is favorable to gaze estimation. It contributes to improving gaze estimation performance, as shown in HAZE-Net results in Fig. 2. We test gaze estimation performance under LR conditions using MPIIFaceGaze [19] and EyeDiap [20] datasets. The proposed method effectively estimates the gaze under challenging LR conditions (e.g., 28 × 28 pixels face image). The major contributions of the paper are as follows:

- The HFAB proposed in our SR module strengthens the high-frequency information including eye regions, which is crucial for gaze estimation. With the contribution of HFAB, the LR face image can be enhanced to be suitable for gaze estimation.
- Our gaze estimation module utilizes the global-local appearance map obtained via high-frequency extraction. It improves the performance to be robust to person-specific appearance and illumination changes.
- HAZE-Net performs favorably against typical gaze estimation models under challenging LR conditions.

2   Related Work

Appearance-Based Gaze Estimation. Gaze estimation can be divided into two methods: model-based methods [21–23] and appearance-based methods [24–32]. Model-based methods estimate human gazes from the shape of the pupil and boundaries between the iris and the pupil by handcrafted features. However, recently, appearance-based methods have been in the spotlight owing to large datasets and the advancement of deep learning techniques. These methods learn how to extract embedded features for gaze estimation. As one of the early-stage methods, GazeNet [24] takes a grayscale eye patch as input and estimates the gaze vector. It shows a better gaze estimation performance by additionally using a head pose vector. As an extended version, the performance was further improved by using the VGG network. Spatial-Weights CNN [19] utilizes not only the eye region but also full-face images. Spatial weights are used to encode important positions in the face image. This method weights regions of the face that are useful for gaze estimation. Through this, more weight is assigned to a specific area in the face image. Furthermore, iTracker [26] receives two eyes, face, and face positions as input and predicts the gaze vector. Dilated-Net [27] utilizes dilated-convolutions to extract high-level features without reducing spatial resolution. It is to capture such small changes in eye appearance. Kang Wang et al. [28] point out the difficulty of generalizing gaze estimation because of appearance variations, head pose variations, and over-fitting issues with point estimation. To deal with these issues, they introduced adversarial learning and Bayesian framework in their network so that it can be practically used in real-world applications. Focusing on the fact that over-parameterized neural networks can be quickly over-fitted, disentangling transforming encoder-decoder (DT-ED) [29] performs few-shot adaptive gaze estimation for learning person-specific gaze networks with very few calibration samples. Coarse-to-fine adaptive network (CA-Net) [25] extracts coarse-grained features from face images to estimate basic gaze direction and fine-grained features from eye images to estimate gaze. In addition, according to the above strategy, they design a bigram model that connects two gaze directions and a framework that introducing attention components to adaptively acquire appropriate subdivided functions. Additionally, there is an attempt to support gaze estimation by utilizing semantic segmentation which identifies different regions of the eyes such as the pupils and iris pupils. RITnet [31] exploits boundary-aware loss functions with a loss scheduling strategy to distinguish coherent regions with crisp region boundaries. PureGaze [32] purifies unnecessary features for gaze estimation (e.g., illumination, personal appearance, and facial expression) through adversarial training with a gaze estimation network and a reconstruction network. Nevertheless, the appearance-based gaze estimation can have a high variance in performance depending on the person-specific appearance (e.g., colors of pupil and skin). In this paper, we devise the global-local appearance map for the gaze estimation to be robust to person-specific appearance. Also, our gaze estimation module effectively learns high-frequency features to be robust to illumination and resolution changes.

Unconstrained Gaze Estimation. Despite the emergence of appearance-based methods for gaze estimation, there are limitations on estimating gaze from real-world images owing to various head poses, occlusion, illumination changes, and challenging LR conditions. According to the wide range of head pose, obtaining both eyes in the occluded or illuminated image is difficult. Park et al. [10] proposed a model specifically designed for the task of gaze estimation from single eye input. FARE-Net [9] is inspired by a condition where the two eyes of the same person appear asymmetric because of illumination. It optimizes the gaze estimation results by considering the difference between the left and right eyes. The model consists of FAR-Net and E-net. FAR-Net predicts 3D gaze directions for both eyes and is trained with an asymmetric mechanism. The asymmetric mechanism is to sum the loss generated by the asymmetric weight and the gaze direction of both eyes. E-Net learns the reliability of both eyes to balance symmetrical and asymmetrical mechanisms. To solve this problem in a different approach, region selection network (RSN) [33] learns to select regions for effective gaze estimation. RSN utilizes GAZE-Net as an evaluator to train the selection network. To effectively train and evaluate the above methods, unconstrained datasets which are collected in real-world settings have been emerged [34–38]. Recently, some studies have introduced self-supervised or unsupervised learning to solve the problem of the lack of quantitative real-world datasets [39–41]. In addition, GAN-aided methods [42,43] can be applied to solve the lack of datasets problem. The above studies have been conducted to solve various constraints, but studies in unconstrained resolutions are insufficient. When the recently proposed gaze estimation [24,27,32,44] modules are applied to the LR environment, it is experimentally shown that the performance of these modules is not satisfactory. To deal with this, we propose HAZE-Net which shows an acceptable gaze accuracy under challenging LR conditions.

3   Method

This section describes the architecture of the proposed high-frequency attentive super-resolved gaze estimation, that so-called HAZE-Net. The first module for the proposed method is the SR module based on HFABs that is a key component to strengthen the high-frequency component of LR face images. The second module is the global-local gaze estimation module, where discriminative eye features are learned. Note that two modules are collaboratively learned. The overall architecture of HAZE-Net is shown in Fig. 3.

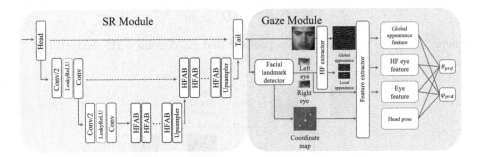

Fig. 3. HAZE-Net architecture. In the first module (yellow panel), given the face image of an LR image, we employ our SR module based on the HFAB. The input image goes through one or two down samples according to the scale factor. In the second module (violet panel), our gaze estimation module utilizes eye patches and the global-local appearance map. We feed four features to the final fully connected layer to obtain the estimated gaze angle $(\theta_{prd}, \phi_{prd})$. (Color figure online)

3.1   Super-Resolution Module

Our SR module is mainly composed of HFABs to exaggerate the high-frequency components that are highly related to gaze estimation performance. Figure 4 shows the high-frequency extractor (HF extractor) for extracting high-frequency components from the input. We use the DCT principle that indicates that the more directed from the top-left to the bottom-right in the zigzag direction, the higher is the frequency component. 2D-DCT denoted by $\mathcal{F}$ transforms input $\boldsymbol{I}$ into the DCT spectral domain $\boldsymbol{D}$ for each channel:

$$\boldsymbol{D}^{(i)} = \mathcal{F}(\boldsymbol{I}^{(i)}), \; i = 1, 2, ... n, \tag{1}$$

where i is a channel index, and n is the number of channels. We create a binary mask $\boldsymbol{m}$ by using a hyper-parameter λ which decides the masking point as follows:

$$\boldsymbol{m} = \begin{cases} 0, & y < -x + 2\lambda h \\ 1, & \text{otherwise} \end{cases}, \tag{2}$$

where h denotes the height of I, and x, y denote the horizontal and vertical coordinates of m, respectively. The size of m equals I. The hyper-parameter λ ranges from 0 to 1. If the λ is too small, overfitting occurs because finer features with low-frequency are emphasized and used for learning. On the other hand, if the λ is too large, most of the useful information for gaze estimation such as the shape of the pupil and the boundaries between the iris and the eyelids is lost, preventing performance improvement. The high-frequency can be separated by element-wise product of D and m. The high-frequency features in the DCT spectral domain are transformed into the spatial domain through 2D-IDCT denoted by $\mathcal{F}^{-1}$:

$$E_h(I) = \mathcal{F}^{-1}(D \otimes m), \tag{3}$$

where $\otimes$ denotes the element-wise product and E_h denotes a HF extractor.

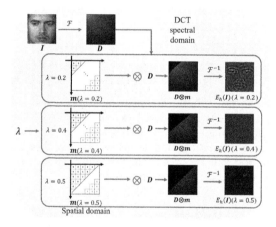

Fig. 4. Architecture of HF extractor. Given the spatial domain image or feature map, we map the spatial domain into DCT spectral domain. λ is a hyper-parameter that indexes the high-frequency component to be extracted from the top-left to the bottom-right in the zigzag direction. The mask determined by λ is multiplied by the feature of the DCT spectral domain. We finally get the high-frequency spatial domain image through 2D-IDCT.

The HFAB utilizes the residual channel attention block (RCAB) [14] structure, as shown in Fig. 5. RCAB extracts informative feature components to learn the channel statistic. The high-frequency feature map extracted by the HF extractor and the original feature map are assigned to the HFAB as input, as shown in Fig. 5. The original feature map is reinforced by a residual group stacked RCABs for image restoration. To exaggerate the insufficient high-frequency in the original process, the high-frequency feature map is input through a module consisting of two RCABs. Two enhanced results are added to obtain a high-frequency exaggerated feature map. This result becomes a feature in which the outline of the face and the boundary between the elements of the eyes are emphasized. Our SR module is composed of the HFABs, $\mathcal{H}$, and the HF extractors, E_h. The architecture of our SR module is given on the left side of Fig. 3. The module takes an LR image as input and magnifies it to a target size through bicubic interpolation.

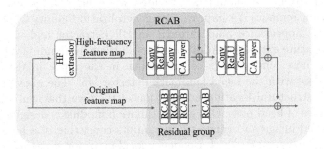

Fig. 5. Architecture of HFAB. Given a feature map, we separate the image into the high-frequency feature map and the original feature map. Both features are fed through independent RCABs. We empirically employ two RCABs for the high-frequency feature map and five RCABs for the original feature map. CA layer allows RCAB to learn useful channel-wise features and enhance discriminative learning ability.

After the head layer extracts the features from the magnified input, the feature size is reduced through the down block consisting of two convolution layers and a LeakyReLU activation layer. The original features and high-frequency features extracted by E_h are passed through the group of HFABs. The high-frequency enhanced feature through the HFAB is upscaled to the target size through up-block, U, consisting of a pixel shuffle layer and two convolution layers. Finally, this extended feature is concatenated with the feature extracted from the head layer and converted into the super-resolved RGB image I_{SR} through the tail layer as follows:

$$I_{SR} = U(\mathcal{H}(E_h(f_d), f_d)) \oplus f_b), \qquad (4)$$

where f_b is the feature extracted from a bicubic-upsampled image, f_d is the feature reduced in size by the downblock and $\oplus$ is the concatenation operation.

3.2   Gaze Estimation Module

The performance of the appearance-based gaze estimation module depends on the resolution of the image received as an input. In general, the proportion of a face in the image is usually small and variable. Thus, resizing the LR face image to a larger size causes severe loss of information that is important to gaze estimation. Therefore, in this paper, we propose the super-resolved gaze estimation module that is robust under LR conditions. As our gaze estimation module adaptively learns through super-resolved images with exaggerated high-frequency, it preserves information that helps estimate gaze under the LR environment. Our module secures stable input by adding additional high-frequency components that are insensitive to this environment. The gaze estimation module has a high variance in performance depending on the appearance of a person. This is because face images contain redundant low-frequency information. Thus, unnecessary information should be excluded while high-frequencies that help gaze estimation remain. We improve our generalization ability by obtaining a high-frequency appearance map through an HF extractor and using it as an input of the feature extractor. Additionally, we utilize five landmark coordinates such as eyes, nose, and corners of the mouth in the input image during the training process, with the facial landmark detector [45]. The above five coordinates are the structural location information of the face that can be used as a proper guidance of head pose. Our gaze estimation module is designed to receive a super-resolved image and generates five inputs consisting of a high-frequency global appearance map, two high-frequency local maps for each eye, and two eye images. The global appearance map refers to the features that only leave high-frequency features extracted from the face and facial landmarks. Meanwhile, the local appearance map is only extracted from eye patches by the same procedure. It utilizes Resnet-18 as the backbone to extract features from each input. The five features extracted from each input are concatenated into a vector and put into a fully connected layer of size 512. A two-dimensional head-pose vector is used to train our gaze estimation module that predicts a gaze angle $(\theta_{prd}, \phi_{prd})$.

3.3   Loss Function

HAZE-Net employs two loss functions for the SR and gaze estimation modules. The two loss functions are appropriately combined according to the proposed alternative learning strategy.

SR Loss. Our module uses L1 Loss, which is commonly used in SR tasks. The loss function minimizes the difference in pixel values between the original image and the SR result image as follows:

$$L_{SR} = \frac{1}{N}\|F_{SR}(I_{LR}) - I_{HR}\|_1, \tag{5}$$

where N is the image batch size, I_{LR} is the LR image taken as the input, and I_{HR} is the original HR image. In addition, F_{SR} is our high-frequency attention SR module.

Gaze Estimation Loss. The proposed gaze estimation module predicts (θ, ϕ) which represents pitch and yaw, respectively. The predicted $(\theta_{prd}, \phi_{prd})$ are compared with the ground truth (θ_{gt}, ϕ_{gt}) and the mean squared error as the loss function. This is the loss function of our gaze estimation module:

$$L_{GE} = \frac{1}{N}\sum_{i=1}^{N}((\theta_{prd}{}^{i} - \theta_{gt}{}^{i})^2 + (\phi_{prd}{}^{i} - \phi_{gt}{}^{i})^2). \qquad (6)$$

Total Loss. The total loss function is a combination of the SR loss and the gaze estimation loss. Therefore, the total loss is defined as follows:

$$L_{Total} = L_{SR} + \alpha L_{GE}, \qquad (7)$$

where α is a hyper-parameter that scales gaze estimation loss. If the loss scale is focused on one side, it tends to diverge. Thus, it should be appropriately tuned according to the purpose of each phase. The detailed hyper-parameters according to the phase are introduced in Sect. 3.4.

3.4   Alternative End-to-End Learning

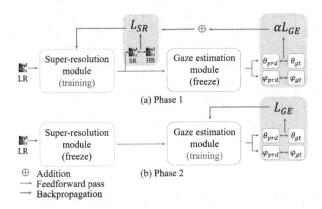

Fig. 6. Flowchart of HAZE-Net's alternative end-to-end learning architecture. (a) Phase 1: SR module training while freezing gaze estimation module. (b) Phase 2: Gaze estimation module training while freezing SR module.

This section describes the learning strategies for the proposed HAZE-Net. It is not simply a structural combination of the SR module and the gaze estimation module but a complementary combination through the proposed alternative end-to-end learning. We initialize each module with pre-trained parameters for each task. To train the end-to-end model stably, we combine two modules and apply different losses at each phase, as shown in Fig. 6. We found that using training on our module is more efficient and effective than training from scratch. In phase 1, the SR module is trained while the gaze estimation module is frozen. We use the weighted sum of the SR loss and gaze estimation loss, L_{Total}, as shown in Fig. 6(a). We combine two loss functions for our modules to learn complementarily. α is used to perform training by scaling the magnitude of SR loss and gaze estimation loss. Given the scale difference between the two losses, we found that setting the α to 0.1 is the best empirical choice. In phase 2, the gaze estimation module is trained while the SR module is frozen, as shown in Fig. 6(b). We use only gaze loss L_{GE} in phase 2. Although training both modules without freezing is a possible option, we found that the performance was poor compared with our strategy. The SR images produced by our end-to-end trained SR module generally show clear boundaries between the components of the eyes and the clear shape of the pupil. Although our module may not guarantee a better peak signal-to-noise ratio (PSNR) score, it performs better in the gaze estimation task than simply combining the two separated state-of-the-art (SOTA) modules.

4   Experiments

4.1   Datasets and Evaluation Metrics

Datasets. Based on the datasets used in recent studies [9,29,44], we accordingly evaluate our module on the MPIIFaceGaze [19] and EyeDiap [20] datasets. To simulate LR conditions, we set the HR size to 112×112 and set LR size according to the scale factor (e.g., 56×56, 37×37, 28×28). For example, if the scale factor is $2\times$, the resolution of the LR image is 56×56. If the scale factor is $4\times$, the resolution of the LR image is 28×28. The MPIIFaceGaze contains a total of 45,000 images from 15 subjects. We split 9,000 images of three subjects (p00, p02, p14) for validation, and others are used for the training set. The EyeDiap contains a total of 94 video clips from 16 subjects. We prepare 16,665 images as in [25]. We split 2,384 images consisting of two subjects (p15, p16) and others are utilized for the training set. When generating LR images, we utilize the built-in resize function of MATLAB.

Environment. Our module is implemented in PyTorch 1.8.0, and experiments for comparison with other modules are conducted in the same environment. We train each module for 100 epochs with a batch size of 80 and use the Adam optimizer. In addition, we empirically set the hyper-parameter λ to 0.2

Evaluation Metric. We compare the proposed SR module with SOTA SR methods. For qualitative comparison, we compared the PSNR and structural similarity index measure (SSIM) [46] values of the different methods for scale factors ($2\times$, $3\times$, and $4\times$). Also, we compare the proposed gaze estimation module with other modules. We compute the angular error between the predicted gaze vector and the ground-truth gaze vector, and represent the performance of the module as an angular error to numerically show the performance.

4.2   Performance Comparison by Module

Comparison of SR Modules. We compare our SR module with SOTA SR modules [11–13] in terms of both quantitative results and visual results. All SR modules are trained according to their losses and training methods on MPI-IFaceGaze and Eyediap datasets from scratch. Note that gaze datasets are rescaled to simulate low-resolution constraint settings. Therefore, the SR losses are calculated between the SR result and HR image. We present a comparison in terms of high-frequency restoration. As shown in Fig 7, the proposed module enhances the lines, which are high-frequency components, better than DBPN [11] and DRN [12]. SwinIR [13] is comparable to our SR module. As shown in Table 1, the HAZE-Net shows a lower tendency in terms of PSNR and SSIM than the SOTA SR modules. However, as shown in Fig. 7, the proposed HAZE-Net can adequately enhance high-frequency components to be suitable for gaze estimation task that requires clear boundaries. To prove the superiority of our SR module, we measure angular errors on each SR result with the baseline gaze estimation module consisting of ResNet-18 and fully connected layers. As shown in Table 1, our SR module provides the lowest angular error compared with other SOTA SR modules. The HFAB proposed in our SR module strengthens the high-frequency information such as eye features (e.g., the shape of the pupil) and boundaries (e.g., the boundary between the iris and the eyelids). It leads improvement of gaze estimation performance. Moreover, our SR module can restore clean HR image robust to noise of the input image even while maintaining the high-frequency information.

Table 1. Performance comparison with SOTA SR modules for 2×, 3×, and 4×. The best and the second-best results are highlighted in red and blue colors, respectively.

SR module	Scale	MPIIFaceGaze		EyeDiap	
		PSNR/SSIM	Angular error	PSNR/SSIM	Angular error
Bicubic	2	30.83/0.8367	7.23	35.62/0.9436	5.96
DBPN [11]		34.35/0.8882	6.64	39.61/0.9716	5.64
DRN [12]		33.73/0.8128	6.46	38.70/0.9228	5.44
SwinIR [13]		34.40/0.8911	6.51	40.36/0.9735	6.47
Our SR module		34.28/0.8263	6.23	39.65/0.9041	4.68
Bicubic	3	26.23/0.6939	7.73	31.46/0.8722	5.64
DBPN [11]		31.43/0.8257	6.69	37.02/0.9447	5.18
DRN [12]		31.59/0.8279	8.52	36.19/0.9165	6.55
SwinIR [13]		31.67/0.9086	6.62	36.93/0.9657	5.32
Our SR module		31.33/0.8219	6.49	36.82/0.9392	4.96
Bicubic	4	25.84/0.6429	9.32	29.58/0.8066	6.22
DBPN [11]		29.69/0.7704	7.06	34.96/0.9128	5.83
DRN [12]		29.77/0.7735	6.85	33.42/0.8516	5.82
SwinIR [13]		30.26/0.8723	7.54	34.02/0.9452	5.79
Our SR module		29.59/0.7769	6.60	32.73/0.8934	5.54

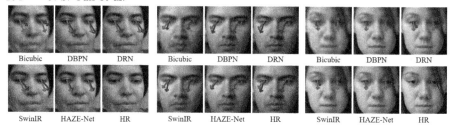

Bicubic   DBPN   DRN   Bicubic   DBPN   DRN   Bicubic   DBPN   DRN

SwinIR   HAZE-Net   HR   SwinIR   HAZE-Net   HR   SwinIR   HAZE-Net   HR

Fig. 7. Qualitative comparison of our SR module with SOTA modules on 4× SR.

Comparison of Gaze Estimation Module. In this section, we compare the results of our gaze estimation module with other gaze estimation modules. We select a gaze estimation module to compare with our module. Among the recent gaze estimation modules, we exclude modules that use few data [39], unlabeled [40,41]. All gaze estimation module is trained using an image of size 112 × 112 from scratch. For a fair comparison, we commonly use ResNet-18 as a backbone of all gaze estimation modules. As presented in Table 2, our method shows the best gaze angular error on the MPIIFaceGaze dataset and the second-best gaze angular error on the EyeDiap dataset. It indicates the superiorty of our gaze estimation module due to the global-local appearance map. As shown in Fig. 8, our module shows robust performance under challenging illumination conditions.

Table 2. Performance comparison with gaze estimation modules for 112 × 112 HR images. The best and the second-best results are highlighted in red and blue colors, respectively.

Gaze estimation module	MPIIFaceGaze angular error	EyeDiap angular error
GazeNet [24]	5.88	4.25
RT-GENE [44]	5.52	4.65
DilatedNet [27]	5.03	4.53
PureGaze [32]	5.71	3.88
Our gaze estimation module	4.95	4.12

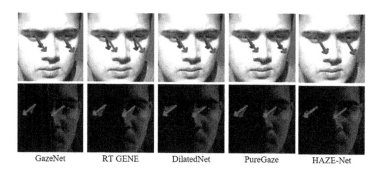

GazeNet   RT GENE   DilatedNet   PureGaze   HAZE-Net

Fig. 8. Visualization of gaze estimation results under challenging illumination conditions. Blue and red arrows represent ground truth and predicted gaze, respectively. (Color figure online)

4.3 Comparison Under LR Conditions

To verify performance under the LR conditions, we compare HAZE-Net, and the combination of SR module and gaze estimation module. Each module is trained with gaze datasets accordingly Tables 1 and 2. For fair comparison under the LR conditions, each gaze estimation modules are fine-tuned with the results of SR modules. Moreover, we set the gaze estimation baselines that are trained with LR image. In Sect. 4.2, the results show that our module presents lower PSNR and SSIM than those of SwinIR. In contrast, HAZE-Net exhibits the lowest angular error, as presented in Table 3. This is because HAZE-Net successfully enhances high-frequency components, which are critical for gaze estimation performance, compared to other SR modules.

Table 3. Performance comparison with SOTA SR modules combined with gaze estimation modules under LR conditions. The best and the second-best results are highlighted in red and blue colors, respectively.

SR module	Gaze Estimation Module	MPII FaceGaze Angular error	EyeDiap Angular Error	SR module	Gaze Estimation Module	MPII FaceGaze Angular error	EyeDiap Angular errorr
LR (56×56)	GazeNet	9.42	9.17	LR (37×37)	GazeNet	9.82	9.87
	RT-GENE	10.13	10.33		RT-GENE	10.54	11.41
	DilatedNet	15.47	17.21		DilatedNet	16.59	18.23
Bicubic 2× (112×112)	GazeNet	6.89	4.46	Bicubic 3× (111×111)	GazeNet	8.00	5.02
	RT-GENE	6.23	4.96		RT-GENE	7.33	5.64
	DilatedNet	5.59	4.55		DilatedNet	6.85	5.93
	PureGaze	6.71	4.25		PureGaze	7.92	4.91
DBPN 2× (112×112)	GazeNet	6.07	4.05	DBPN 3× (111×111)	GazeNet	6.55	4.15
	RT-GENE	6.59	4.82		RT-GENE	5.65	4.79
	DilatedNet	5.21	4.58		DilatedNet	5.60	5.22
	PureGaze	5.94	3.98		PureGaze	6.38	3.92
DRN 2× (112×112)	GazeNet	6.17	4.36	DRN 3× (111×111)	GazeNet	6.59	4.52
	RT-GENE	5.76	4.69		RT-GENE	6.52	5.38
	DilatedNet	5.14	5.04		DilatedNet	5.73	5.18
	PureGaze	6.01	5.51		PureGaze	6.21	5.72
SwinIR 2× (112×112)	GazeNet	6.47	4.21	SwinIR 3× (111×111)	GazeNet	7.25	4.51
	RT-GENE	5.54	4.76		RT-GENE	6.46	5.02
	DilatedNet	5.03	4.39		DilatedNet	5.56	4.47
	PureGaze	5.77	4.33		PureGaze	7.09	4.39
HAZE-Net 2× (112×112)		4.93	3.90	HAZE-Net 3× (111×111)		5.14	3.74

SR module	Gaze estimation module	MPIIFaceGaze angular error	EyeDiap angular error
LR (28×28)	GazeNet	10.45	11.53
	RT-GENE	10.76	12.69
	DilatedNet	17.89	19.23
Bicubic 4× (112×112)	GazeNet	9.23	5.57
	RT-GENE	9.32	6.22
	DilatedNet	7.52	6.14
	PureGaze	9.17	4.96
DBPN 4× (112×112)	GazeNet	7.10	4.81
	RT-GENE	7.03	5.70
	DilatedNet	5.89	4.51
	PureGaze	6.94	4.45
DRN 4× (112×112)	GazeNet	7.04	4.90
	RT-GENE	7.05	5.65
	DilatedNet	5.82	5.39
	PureGaze	6.88	4.30
SwinIR 4× (112×112)	GazeNet	8.14	5.21
	RT-GENE	7.44	5.50
	DilatedNet	6.38	4.67
	PureGaze	7.97	4.56
HAZE-Net 4×(112×112)		5.56	4.02

Table 4. Quantitative results for evaluating the effects of the HF extractor on the MPIIFaceGaze dataset. The experiment is conducted for 4×.

HF extractor	PSNR	SSIM	Angular error
×	30.80	0.8397	7.27
O	29.59	0.7769	6.60

Table 5. Quantitative results for evaluating the effects of global and local appearance maps. The experiments are conducted under HR (112×112) conditions.

Global map	Local map	RGB eye patch	MPIIFaceGaze angular error
×	×	O	5.56
×	O	O	5.27
O	×	O	5.23
O	O	O	4.95

Table 6. Gaze estimation performance of different levels of α and λ. The best results are highlighted in red.

MPIIFaceGaze angular error			
α	0	0.1	1
HAZE-Net 4×	5.47	4.95	5.13
λ	0.2	0.4	0.5
HAZE-Net 4×	4.95	5.19	5.71

4.4 Ablation Study

Effect of HF Extractor. We investigate the impact of the HF extractor in order to verify the element performance of our first contribution introduced in Sect. 1. We measure the angular error by using our gaze estimation module trained with HR images. As presented in Table 4, although the HF extractor shows a lower PSNR and SSIM, it exhibits better gaze estimation performance. This indicates that the proposed module enhanced high-frequency components that are suitable for gaze estimation tasks.

Global-Local Appearance Map. In this section, we demonstrate the effectiveness of the global-local appearance map introduced as our second contribution. Table 5 shows that both global and local appearance maps help to improve gaze estimation performance. In particular, using the global-local appearance map provides 0.61° lower angular error than using only RGB eye images.

Hyper-parameters. We clarify and specify how we decided the value of the hyper-parameters α, and λ. As present in Table 6, when α is 0.1, λ is 0.2, it provides the best performance in terms of gaze estimation. $\alpha = 0.1$ means effectiveness of end-to-end learning. $\lambda = 0.2$ is used to improve the generalization performance of HAZE-Net. If the λ is too large (e.g., 0.4,0.5), most of the useful information for gaze estimation is lost. We determined the hyper-parameters according to these results.

Limitations. Our module may still be somewhat limited in two aspects in its practical application. In the first aspect, the inference time of our module in 112×112 HR resolution that is measured in $2\times$ is 46ms, $3\times$ is 41ms, and $4\times$ is 103ms with NVIDIA RTX 3080 GPU. Therefore, it is slightly difficult to apply in an environment that requires real-time. Second, as our experiment assumes only the bicubic kernel, there is a possibility that the performance will deteriorate in a real environment where the blur kernel is blinded.

5   Conclusion

In this paper, we propose a high-frequency attentive super-resolved gaze estimation network. In the SR module, we introduce the HFAB to effectively exaggerate high-frequency components for gaze estimation. In the gaze estimation module, we introduce the global-local high-frequency appearance map. Furthermore, alternative end-to-end learning is performed to effectively train our module. With the contribution of techniques described above, HAZE-Net significantly improves the performance of the gaze estimation module under LR conditions. Extensive experiments including ablation studies demonstrate the superiority of our method over the existing methods.

Acknowledgements. This work was supported by the IITP grant (No.2020-0-00004), the NRF grant (NRF-2020R1A4A1019191) funded by the Korea government (MSIT), and the Industrial Fundamental Technology Develop Program (No.20018699) funded by the Korea government (MOTIE).

References

1. Massé, B., Ba, S., Horaud, R.: Tracking gaze and visual focus of attention of people involved in social interaction. IEEE Trans. Pattern Anal. Mach. Intell. **40**(11), 2711–2724 (2017)
2. Schauerte, B., Stiefelhagen, R.: "look at this!" learning to guide visual saliency in human-robot interaction. In: 2014 IEEE/RSJ International Conference on Intelligent Robots and Systems, pp. 995–1002. IEEE (2014)
3. Fridman, L., Reimer, B., Mehler, B., Freeman, W.T.: Cognitive load estimation in the wild. In: Proceedings of the 2018 CHI Conference on Human Factors in Computing Systems, pp. 1–9 (2018)
4. Fang, Y., et al.: Dual attention guided gaze target detection in the wild. In: Proceedings of the IEEE/CVF Conference on Computer Vision and Pattern Recognition, pp. 11390–11399 (2021)

5. Patney, A., et al.: Perceptually-based foveated virtual reality. In: ACM SIGGRAPH 2016 Emerging Technologies, pp. 1–2 (2016)
6. Fathi, A., Li, Y., Rehg, J.M.: Learning to recognize daily actions using gaze. In: Fitzgibbon, A., Lazebnik, S., Perona, P., Sato, Y., Schmid, C. (eds.) ECCV 2012. LNCS, vol. 7572, pp. 314–327. Springer, Heidelberg (2012). https://doi.org/10.1007/978-3-642-33718-5_23
7. Ishikawa, T.: Passive driver gaze tracking with active appearance models (2004)
8. Yamazoe, H., Utsumi, A., Yonezawa, T., Abe, S.: Remote gaze estimation with a single camera based on facial-feature tracking without special calibration actions. In: Proceedings of the 2008 Symposium on Eye Tracking Research and Applications, pp. 245–250 (2008)
9. Cheng, Y., Zhang, X., Feng, L., Sato, Y.: Gaze estimation by exploring two-eye asymmetry. IEEE Trans. Image Process. **29**, 5259–5272 (2020)
10. Park, S., Spurr, A., Hilliges, O.: Deep pictorial gaze estimation. In: Proceedings of the European Conference on Computer Vision, pp. 721–738 (2018)
11. Haris, M., Shakhnarovich, G., Ukita, N.: Deep back-projection networks for super-resolution. In: Proceedings of the IEEE Conference on Computer Vision and Pattern Recognition, pp. 1664–1673 (2018)
12. Guo, Y., et al.: Closed-loop matters: dual regression networks for single image super-resolution. In: Proceedings of the IEEE/CVF Conference on Computer Vision and Pattern Recognition, pp. 5407–5416 (2020)
13. Liang, J., Cao, J., Sun, G., Zhang, K., Van Gool, L., Timofte, R.: Swinir: image restoration using swin transformer. In: Proceedings of the IEEE/CVF International Conference on Computer Vision, pp. 1833–1844 (2021)
14. Zhang, Y., Li, K., Li, K., Wang, L., Zhong, B., Fu, Y.: Image super-resolution using very deep residual channel attention networks. In: Proceedings of the European Conference on Computer Vision, pp. 286–301 (2018)
15. Dai, T., Cai, J., Zhang, Y., Xia, S.T., Zhang, L.: Second-order attention network for single image super-resolution. In: Proceedings of the IEEE/CVF Conference on Computer Vision and Pattern Recognition, pp. 11065–11074 (2019)
16. Mei, Y., Fan, Y., Zhou, Y., Huang, L., Huang, T.S., Shi, H.: Image super-resolution with cross-scale non-local attention and exhaustive self-exemplars mining. In Proceedings of the IEEE/CVF Conference on Computer Vision and Pattern Recognition, pp. 5690–5699 (2020)
17. Niu, B., et al.: Single image super-resolution via a holistic attention network. In: Vedaldi, A., Bischof, H., Brox, T., Frahm, J.-M. (eds.) ECCV 2020. LNCS, vol. 12357, pp. 191–207. Springer, Cham (2020). https://doi.org/10.1007/978-3-030-58610-2_12
18. Zhang, Y., Tian, Y., Kong, Y., Zhong, B., Fu, Y.: Residual dense network for image super-resolution. In: Proceedings of the IEEE Conference on Computer Vision and Pattern Recognition, pp. 2472–2481 (2018)
19. Zhang, X., Sugano, Y., Fritz, M., Bulling, A.: It's written all over your face: full-face appearance-based gaze estimation. In: Proceedings of the IEEE Conference on Computer Vision and Pattern Recognition Workshops, pp. 51–60 (2017)
20. Mora, K.A.F., Monay, F., Odobez, J.M.: Eyediap: a database for the development and evaluation of gaze estimation algorithms from rgb and rgb-d cameras. In: Proceedings of the Symposium on Eye Tracking Research and Applications, pp. 255–258 (2014)
21. Baltrusaitis, T., Zadeh, A., Lim, Y.C., Morency, L.P.: Openface 2.0: facial behavior analysis toolkit. In: 2018 13th IEEE International Conference on Automatic Face and Gesture Recognition, pp. 59–66. IEEE (2018)

22. Park, S., Zhang, X., Bulling, A., Hilliges, O.: Learning to find eye region landmarks for remote gaze estimation in unconstrained settings. In: Proceedings of the 2018 ACM Symposium on Eye Tracking Research and Applications, pp. 1–10 (2018)
23. Valenti, R., Sebe, N., Gevers, T.: Combining head pose and eye location information for gaze estimation. IEEE Trans. Image Process. **21**(2), 802–815 (2011)
24. Zhang, X., Sugano, Y., Fritz, M., Bulling, A.: Appearance-based gaze estimation in the wild. In: Proceedings of the IEEE Conference on Computer Vision and Pattern Recognition, pp. 4511–4520 (2015)
25. Cheng, Y., Huang, S., Wang, F., Qian, C., Feng, L.: A coarse-to-fine adaptive network for appearance-based gaze estimation. In: Proceedings of the AAAI Conference on Artificial Intelligence, vol. 34, pp. 10623–10630 (2020)
26. Krafka, K., et al.: Eye tracking for everyone. In: Proceedings of the IEEE Conference on Computer Vision and Pattern Recognition, pp. 2176–2184 (2016)
27. Chen, Z., Shi, B.E.: Appearance-based gaze estimation using dilated-convolutions. In: Jawahar, C.V., Li, H., Mori, G., Schindler, K. (eds.) ACCV 2018. LNCS, vol. 11366, pp. 309–324. Springer, Cham (2019). https://doi.org/10.1007/978-3-030-20876-9_20
28. Wang, K., Zhao, R., Su, H., Ji, Q.: Generalizing eye tracking with bayesian adversarial learning. In: Proceedings of the IEEE/CVF Conference on Computer Vision and Pattern Recognition, pp. 11907–11916 (2019)
29. Park, S., De Mello, S., Molchanov, P., Iqbal, U., Hilliges, O., Kautz, J.: Few-shot adaptive gaze estimation. In: Proceedings of the IEEE/CVF International Conference on Computer Vision, pp. 9368–9377 (2019)
30. Nie, X., Feng, J., Zhang, J., Yan, S.: Single-stage multi-person pose machines. In: Proceedings of the IEEE/CVF International Conference on Computer Vision, pp. 6951–6960 (2019)
31. Chaudhary, A.K., et al.: Ritnet: real-time semantic segmentation of the eye for gaze tracking. In: 2019 IEEE/CVF International Conference on Computer Vision Workshop, pp. 3698–3702. IEEE (2019)
32. Cheng, Y., Bao, Y., Feng, L.: Puregaze: purifying gaze feature for generalizable gaze estimation. In: Proceedings of the AAAI Conference on Artificial Intelligence, vol. 36, pp. 436–443 (2022)
33. Zhang, X., Sugano, Y., Bulling, A., Hilliges, O.: Learning-based region selection for end-to-end gaze estimation. In: 31st British Machine Vision Conference, p. 86. British Machine Vision Association (2020)
34. Kellnhofer, P., Recasens, A., Stent, S., Matusik, W., Torralba, A.: Gaze360: physically unconstrained gaze estimation in the wild. In: Proceedings of the IEEE/CVF International Conference on Computer Vision, pp. 6912–6921 (2019)
35. Marin-Jimenez, M.J., Kalogeiton, V., Medina-Suarez, P., Zisserman, A.: Laeo-net: revisiting people looking at each other in videos. In: Proceedings of the IEEE/CVF Conference on Computer Vision and Pattern Recognition, pp. 3477–3485 (2019)
36. Recasens, A., Khosla, A., Vondrick, C., Torralba, A.: Where are they looking? Adv. Neural Inf. Process. Syst. **28** (2015)
37. Tomas, H., et al.: Goo: a dataset for gaze object prediction in retail environments. In: Proceedings of the IEEE/CVF Conference on Computer Vision and Pattern Recognition, pp. 3125–3133 (2021)
38. Zhang, X., Park, S., Beeler, T., Bradley, D., Tang, S., Hilliges, O.: ETH-XGaze: a large scale dataset for gaze estimation under extreme head pose and gaze variation. In: Vedaldi, A., Bischof, H., Brox, T., Frahm, J.-M. (eds.) ECCV 2020. LNCS, vol. 12350, pp. 365–381. Springer, Cham (2020). https://doi.org/10.1007/978-3-030-58558-7_22

39. Kothari, R., De Mello, S., Iqbal, U., Byeon, W., Park, S., Kautz, J.: Weakly-supervised physically unconstrained gaze estimation. In: Proceedings of the IEEE/CVF Conference on Computer Vision and Pattern Recognition, pp. 9980–9989 (2021)

40. Sun, Y., Zeng, J., Shan, S., Chen, X.: Cross-encoder for unsupervised gaze representation learning. In: Proceedings of the IEEE/CVF International Conference on Computer Vision, pp. 3702–3711 (2021)

41. Yu, Y., Odobez, J.M.: Unsupervised representation learning for gaze estimation. In: Proceedings of the IEEE/CVF Conference on Computer Vision and Pattern Recognition, pp. 7314–7324 (2020)

42. Shrivastava, A., Pfister, T., Tuzel, O., Susskind, J., Wang, W., Webb, R.: Learning from simulated and unsupervised images through adversarial training. In: Proceedings of the IEEE Conference on Computer Vision and Pattern Recognition, pp. 2107–2116 (2017)

43. Zheng, Z., Zheng, L., Yang, Y.: Unlabeled samples generated by gan improve the person re-identification baseline in vitro. In: Proceedings of the IEEE International Conference on Computer Vision, pp. 3754–3762 (2017)

44. Fischer, T., Chang, H.J., Demiris, Y.: Rt-gene: real-time eye gaze estimation in natural environments. In: Proceedings of the European Conference on Computer Vision, pp. 334–352 (2018)

45. Deng, J., Guo, J., Ververas, E., Kotsia, I., Zafeiriou, S.: Retinaface: single-shot multi-level face localisation in the wild. In: Proceedings of the IEEE/CVF Conference on Computer Vision and Pattern Recognition, pp. 5203–5212 (2020)

46. Wang, Z., Bovik, A.C., Sheikh, H.R., Simoncelli, E.P.: Image quality assessment: from error visibility to structural similarity. IEEE Trans. Image Process. **13**(4), 600–612 (2004)

LatentGaze: Cross-Domain Gaze Estimation Through Gaze-Aware Analytic Latent Code Manipulation

Isack Lee, Jun-Seok Yun, Hee Hyeon Kim, Youngju Na, and Seok Bong Yoo[✉]

Department of Artificial Intelligence Convergence, Chonnam National University,
Gwangju, Korea
sbyoo@jnu.ac.kr

Abstract. Although recent gaze estimation methods lay great emphasis on attentively extracting gaze-relevant features from facial or eye images, how to define features that include gaze-relevant components has been ambiguous. This obscurity makes the model learn not only gaze-relevant features but also irrelevant ones. In particular, it is fatal for the cross-dataset performance. To overcome this challenging issue, we propose a gaze-aware analytic manipulation method, based on a data-driven approach with generative adversarial network inversion's disentanglement characteristics, to selectively utilize gaze-relevant features in a latent code. Furthermore, by utilizing GAN-based encoder-generator process, we shift the input image from the target domain to the source domain image, which a gaze estimator is sufficiently aware. In addition, we propose gaze distortion loss in the encoder that prevents the distortion of gaze information. The experimental results demonstrate that our method achieves state-of-the-art gaze estimation accuracy in a cross-domain gaze estimation tasks. This code is available at https://github.com/leeisack/LatentGaze/.

1 Introduction

Human gaze information is essential in modern applications, such as human computer interaction [1,2], autonomous driving [3], and robot interaction [4]. With the development of deep-learning techniques, leveraging convolution neural networks (CNNs), appearance-based gaze estimation has led to significant improvements in gaze estimation. Recently, various unconstrained datasets with wide gaze range, head pose range are proposed. Although these allow a gaze estimator to learn a broader range of gaze and head pose, the improvement is

---

I. Lee and J.-S. Yun—These authors contributed equally to this paper as co-first authors.

Supplementary Information The online version contains supplementary material available at https://doi.org/10.1007/978-3-031-26348-4_10.

L. Wang et al. (Eds.): ACCV 2022, LNCS 13845, pp. 161–178, 2023.
https://doi.org/10.1007/978-3-031-26348-4_10

limited to the fixed environment. Because real-world datasets contain unseen conditions, such as the various personal appearances, illuminations, and background environments, the performance of the gaze estimator be degraded significantly. It suggests that gaze-irrelevant factors cause unexpected mapping relations, so called overfitting. It makes a model barely handle various unexpected factors in a changing environment. Therefore, to handle this issue, the domain adaptation method, which uses a small number of target samples in training, and the domain generalization method, which only utilizes source domain data, have received much attention recently. However, in the real world, the former approach does not always hold in practice because the target data for adaptation are usually unavailable. Consequently, the latter approach is highly effective because of its practicability. Especially in gaze estimation, these problems are more because the dimensions of gaze-irrelevant features are much larger than those of gazerelevant features, which are essential to gaze estimation. This problem must be considered because it affects the learning of the network in an unexpected way. Although a few studies have suggested handling this issue in gaze estimation tasks, it remains a challenge.

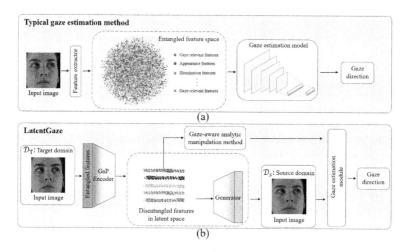

Fig. 1. (a) Typical gaze estimation method which utilizes entangled features. (b) LatentGaze which utilizes latent code with domain shifted image.

In this paper, to solve this issue, we propose a gaze-aware latent code manipulation module based on the data-driven approach. In this process, we focus on the following two characteristics of Generative Adversarial Networks (GAN) inversion for downstream tasks: reconstruction and editability. The former characteristic assures that the attributes of the input images can be mapped into a latent code without losing principal information. The latter enables manipulation or editing concerning a specific attribute, allowing the latent code to be selectively used by excluding or remaining the information in downstream tasks.

Based on these, our proposed method utilizes statistical analysis in the latent domain to find elements that have a high correlation with the gaze information. To this end, we split the dataset into binary-labeled groups, which are a left-staring group and a right-staring group. We then perform two-sample t-tests to see whether the difference between the distributions of two groups' latent codes is statistically significant. Therefore, we could find which elements are related to gaze. In addition, the specific information of the latent code is attentively used to train a gaze estimator. As shown in Fig. 1, typical gaze estimation methods are vulnerable to domain shifts because they use an image with entangled attributes. However, GAN inversion separates intertwined characteristics within an image into semantic units (e.g., illumination, background, and head pose). We exploit this advantage by using the GAN inversion encoder as a backbone for the two modules in our framework. We propose a target-to-source domain shift module with a GAN-based encoder-decoder. This module maps the unseen target image into the source distribution, where the gaze estimator is sufficiently aware, by replacing the attributes of the image with the attributes of the source domain. To this end, we first use the GAN inversion method, encoder, to extract the input image's attributes in a semantically disentangled manner. Nevertheless, while coarse information is generally maintained after inversion, finer information such as gaze angle and skin tone are subject to distortion. Thus, we introduce our proposed gaze distortion loss in the training process of the encoder to minimize the distortion of gaze between the input and the generated images. Our two modules complementarily improve performance during gaze estimation because they encompass both the image and the latent domains. The contributions of this study can be summarized as follows.

- We propose an intuitive and understandable gaze-aware latent code manipulation module to select semantically useful information for gaze estimation based on a statistical data-driven approach.
- We propose a target-to-source domain shift module that maps the target image into the source domain image space with a GAN-based encoder-decoder. It significantly improves the cross-dataset gaze estimation performance.
- We demonstrate the correlation between manipulated latent features and gaze estimation performance through visualizations and qualitative experiments. As a result, our framework outperforms the state-of-the-art (SOTA) methods in both fixed and cross-dataset evaluations.

2   Related Work

2.1   Latent Space Embedding and GAN Inversion

Recent studies have shown that GANs effectively encode various semantic information in latent space as a result of image generation. Diverse manipulation approaches have been proposed to extract and control image attributes. At an early stage, Mirza et al. [5] trained to create a conditional image which enables the control of a specific attribute of an image. Subsequently, Abdal et al. [6] analyze three semantic editing operations that can be applied on vectors in the latent space. Shen et al. [7] adopted a data-driven approach and used principal component analysis to learn the most important directions. Park et al. [8] presented a simple yet effective approach to conditional continuous normalizing flows in the GAN latent space conditioned by attribute features. In the present study, we propose an intuitive and understandable manipulation method to find the direction that correlates with gaze estimation performance. In addition, we prove the effectiveness and necessity of latent code manipulation, particularly by showing cross-domain evaluations. However, such manipulations in the latent space are only applicable to images generated from pre-trained GANs rather than to any given real image. GAN inversion maps a real image into a latent space using a pre-trained generator. Inversion not only considers reconstruction performance but must also be semantically meaningful to perform editing. To this end, Zhu et al. [9] introduced a domain-guided encoder and domain regularized optimization to enable semantically significant inversion. Tov et al. [10] investigated the characteristics of high-quality inversion in terms of distortion, perception, and editability and showed the innate tradeoffs among them. Typically, encoders learn to reduce distortion, which represents the similarity between the input and target images, both in the RGB and feature domains. Zhe et al. [11] proposed an architecture that learns to disentangle and encode these irrelevant variations in a self-learned manner. Zheng et al. [12] proposed a method to relieve distortion of the eye region by leveraging GAN training to synthesize an eye image adjusted on a target gaze direction. However, in this study, the distortion is redefined from the gaze estimation perspective. By utilizing GAN inversion as a backbone for two proposed modules, we effectively improve generalization ability of our framework. First, we take the editability of GAN inversion to extract gaze-relevant features and utilize them in gaze-aware analytic selection manipulation module. Second, we improve the estimation performance for unseen datasets by shifting target domain images into source domain images utilizing GAN inversion and generator. It reduces the bias between the distributions of the two independent image spaces. These help the gaze estimator improve cross-domain performance without touching the target samples.

2.2   Domain Adaptation and Generalization

Most machine learning methods commonly encounter the domain shift problem in practice owing to the distribution shift between the source and target domain datasets. Consequently, a model suffers significantly from performance degradation in the target domain. Several research fields have been explored extensively to address this problem. Domain adaptation utilizes a few target samples during the training [13]. Recently, Bao et al. [14] performed the domain adaptation under the restriction of rotation consistency and proposed rotation-enhanced unsupervised domain adaptation (RUDA) for cross-domain gaze estimation. Wang et al. [15] proposed a gaze adaptation method, namely contrastive regression gaze adaptation (CRGA), for generalizing gaze estimation on the target domain in an unsupervised approach. Although such methods show performance improvement, they do not always hold in practice as target data are difficult to obtain or are unknown. In contrast, domain generalization aims to generalize a model to perform well in any domain, using only the source domain dataset. Unsupervised domain generalization is primarily based on two methodologies: self-supervised learning and adversarial learning. An advantage of self-supervised learning is that the model learns generic features while solving a pretext task. This makes the model less prone to overfitting to the source domain [16]. Adversarial learning allows a generative model to learn the disentangled representations. Here, we introduce GAN inversion to utilize its out-of-distribution generalization ability.

2.3   Gaze Estimation

Several appearance-based gaze estimation methods have been introduced [17–19]. Recent studies have shown significant improvements in gaze estimation performance using various public datasets [20]. However, cross-person and cross-dataset problems remain in gaze estimation tasks. This stems from variance across a large number of subjects with different appearances. To improve the cross-person performance, Park et al. [8] utilized an auto-encoder to handle person-specific gaze and a few calibration samples. Yu et al. [21] improved the person-specific gaze model adaptation by generating additional training samples through the synthesis of gaze-redirected eye images. Liu et al. [22] proposed an outlier-guided collaborative learning for generalizing gaze in the target domain. Kellnhofer et al. [23] utilized a discriminator to adapt the model to the target domain. Cheng et al. [24] introduced domain generalization method through gaze feature purification. This method commonly assumes that gaze-irrelevant features make the model overfit in the source domain and perform poorly in the target domain. Although these methods have improved cross-person performance, they cannot always be applicable as the target data are difficult to obtain or even unavailable in the real world. In consideration of this, generalizing the gaze estimation across datasets is required. However, since facial features are tightly intertwined with illumination, face color, facial expression, it is difficult to separate gaze-relevant features from them. To address this, we improve the generalization ability by analyzing latent codes and selectively utilize the attributes that are favorable to the gaze estimator.

3   LatentGaze

3.1   Preliminaries

Gaze estimation models which use full-faces learn a biased mapping on the image-specific information such as facial expressions, personal appearance, and illumination rather than learn only the gaze-relevant information of the source dataset.

While it leads to a strong mapping ability in a fixed domain, it still remains as a challenging problem as it causes serious performance degradation in cross-domain gaze estimation. Therefore, this paper aims to dodge the bias-variance tradeoff dilemma by shifting the domain of image space from target dataset to source dataset. In this section, we introduce some notations, objective definitions and condition.

Conditions. While the majority of generalization works have been dedicated to the multi-source setting, our work only assumes a single-source domain setting for two motivations. First, it is easier to expand from a single-source domain generalization to a multi-source domain generalization. Second, our study specifically attempts to solve the out-of-distribution problem when the training data are roughly homogeneous.

Definition of Latent Code Manipulation. The first application of GAN inversion is to selectively utilize gaze-related attributes in the input image. We invert a target domain image into the latent space because it allows us to manipulate the inverted image in the latent space by discovering the desired code with gaze-related attributes. This technique is usually referred to as latent code manipulation. In this paper, we will interchangeably use the terms "latent code" and "latent vector". Our goal is to ensure that the manipulated latent vector and gaze vector are highly correlated. Therefore, we need to find the manipulation operator H that determines which elements are correlated to gaze information in the latent vector. This objective can be formulated as:

$$H^* = \max_{H} \nabla f(H(X)), \tag{1}$$

$$\nabla f = [\frac{\partial f}{\partial x_1}, \frac{\partial f}{\partial x_2}, \dots, \frac{\partial f}{\partial x_d}]^T, \tag{2}$$

where f denotes an optimal gaze estimator, X denotes latent vector. x denotes an element of X, and d denotes the dimension of latent space.

Definition of Domain Shifting Process. Given labeled source domain D_S, the goal is to learn a shifting model f that maps an unseen image from target domain D_T to D_S by only accessing source data from D_S. We take GAN inversion method into consideration to utilize its out-of-distribution generalization ability. It can support inverting the real-world images, that are not generated by the same process of the source domain training data. Our objective can be formulated as follows:

$$S^* = \underset{s}{\text{argmax}} \|F(S(D_T)) - F(D_S)\|, \tag{3}$$

where F denotes the general feature extractor, and S^* denotes the domain shifter. Note that domain the generalization process should not include the target sample during the training.

3.2   The LatentGaze Framework Overview

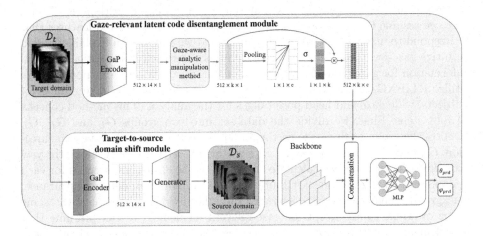

Fig. 2. Overview of the LatentGaze framework. Our framework is mainly composed of two modules. The target-to-source domain shift module maps the unseen target image into the source distribution where the gaze estimator is sufficiently aware. The gaze-aware analytic selection manipulation module finds the gaze-relevant feature through statistical analysis in the latent domain. Thereafter, the gaze estimate model utilizes each feature to estimate the gaze vector (θ_{prd}, ϕ_{prd}).

Figure 2 shows the overall architecture of the proposed framework. It consists of two complementary modules. The first module, i.e., the target-to-source domain shift module, maps a target domain image to a source domain image. The encoder extracts an image into a latent code. To reduce the possible gaze distortion in the encoder, we propose gaze distortion loss to preserve the gaze information. The generator generates an attributes-persevered image in the source domain, based on the extracted latent code. The second module, i.e., the gaze-aware latent code manipulation module, improves the generalization performance by manipulating features that are highly related to gaze in the latent code through statistical hypothesis test. Subsequently, the latent vector and the image are concatenated. And through two fc-layers, the gaze direction is predicted as θ_{prd} and ϕ_{prd}. These two modules effectively utilize the property that latent code passed through inversion is semantically disentangled. These modules achieve complementary benefits in our framework by jointly reinforcing cross-domain generalization. Qualitative and quantitative experiments and visualizations show that our framework achieves robust performance, on par with the state-of-the-art approaches.

3.3   Statistical Latent Editing

Recently, extensive research has been conducted on the discovery of semantically meaningful latent manipulations. Radford et al. [25] investigated the vector arithmetic property in the latent space of a pre-trained generator and demonstrated that averaging the latent code of samples for visual concepts showed consistent and stable generations that semantically obeyed arithmetic. Therefore, we assume that it is possible to statistically find gaze-related features that correspond to a specific attribute. In this study, we propose an explainable and intuitive gaze-aware analytic manipulation method to select semantically useful information for gaze estimation based on a statistical data-driven approach. We utilize ETH-XGaze [26] and MPIIFaceGaze [27] which have a large number of subjects, wide gaze and head pose ranges for the analysis. This method consists of four steps. First, we divide the data set into two groups G_L and G_R. G_L denotes the group having gaze vectors of 30°C to 90°C along the yaw axis direction. G_R denotes the group having gaze vectors of –30°C to –90°C along the yaw axis direction. Second, we calculate the mean of the entire latent codes for each group. Since gaze-irrelevant attributes such as illumination, person-appearance, background environments are averaged out, the divided groups have the same value of latent codes except for gaze-relevant elements. We can presume that the elements with a large difference between the group's values are significantly related to gaze features. From then on, since adjacent tensors represent similar features, we consider 16 tensors as a single chunk and define it as a unit of statistical operations as follows.

$$C^i = \frac{1}{16} \sum_{j=1}^{16} C^{ij}, \ i = 1, 2, \ldots, k, \tag{4}$$

where C^i denotes i-th chunk, k denotes the number of chunks in a latent code, and j represents the number of elements in a chunk. C_L^i and C_R^i each denotes the chunk of the group C_L^i and C_R^i . Third, we sort the chunks in a descending order to select the most gaze-related chunks. Fourth, we perform paired t-tests for two expected values of each chunk from two groups.

Paired t-test for Two Expected Values. We perform the t-test and the size of the samples in the two groups is sufficiently large, so the distribution of the difference between the sample means approximates to Gaussian distribution by the central limit theorem, and each sample variance approximates to the population variance. The hypotheses and test statistic are formulated as:

$$H_0 : \mu_L^i = \mu_R^i, \ H_1 : \mu_L^i \neq \mu_R^i, \ i = 1, 2, \ldots d, \tag{5}$$

$$T = \frac{\overline{X_L^i} - \overline{X_R^i}}{\sqrt{\frac{\sigma(X_L^i)}{n_L} + \frac{\sigma(X_R^i)}{n_R}}}, \tag{6}$$

where H_0 denotes null hypothesis, H_1 denotes alternative hypothesis, μ_L^i and μ_R^i denote the population average of both groups for the i-th chunk, d denotes the dimension of the latent space, T denotes test statistic, X_L^i and X_R^i each denotes sample statistic of C_L^i and C_R^i, n_L and n_R each denotes to the number of G_L and G_R, and σ denotes standard deviation. The above procedure can be found in Fig. 3. We found that a number of chunks belonging to the critical region are in the channels 4 and 5. However, as the indices of the gaze-related chunks subject to slightly differ depending on the dataset, we utilize a channel attention layer (CA-layer) [28] to improve generalization performance in cross-domain. We verify the efficacy of our proposed manipulation method through visualizations and experiments. Furthermore, we show the effectiveness of our gaze-aware analytic manipulation method. As shown in Fig. 4, it shows the chunks extracted from our manipulation method are related to gaze information The Fig. 4(c) shows the generated images from the latent code that are replaced only gaze-relevant chunks from the other group. They are similar to the appearance of the Fig. 4(b) while preserving the gaze direction of the Fig. 4(a). Since it is not completely disentangled, it does not completely maintain the appearance, illumination of Fig. 4(b), but it does not affect the performance of gaze estimation.

3.4 Domain Shift Module

The goal of this module is to properly map a target image to the source latent space by an out-of-distribution generalization of GAN inversion. Latent code should be mapped to a semantically aligned code based on the knowledge that emerged in the latent space. However, due to the large bias of distributions between the source and target datasets, the mapping ability of the encoder network hardly covers the target domain. Furthermore, general extraction of

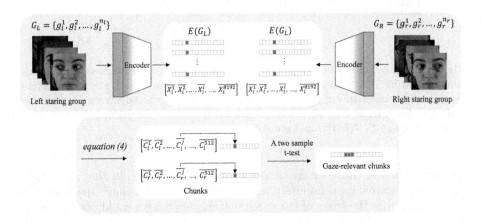

Fig. 3. Overview of the statistical latent editing. The person-paired gaze datasets are divided into the left staring group (G_L) and right staring group (G_R). Each image is embedded into latent code by the GaP-encoder. After that, the latent code values of each group averaged and subtracted the mean values of G_L and G_R. The larger values of the result have a high correlation with the gaze information. Due to this statistical hypothesis test, we classify the gaze-relevant and gaze-irrelevant.

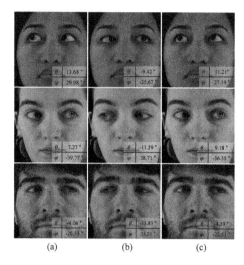

(a)   (b)   (c)

Fig. 4. Visualizations of latent code manipulation. (a) shows images staring one side. (b) shows images staring the other side in the same subjects. (c) shows the generated images with the latent codes that combines the gaze-irrelevant chunks of (b) and the gaze-relevant chunks of (a).

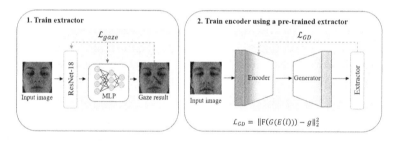

Fig. 5. Training process of GaP-encoder that preserves gaze-relevant feature during domain shifting. The training process consists of two phases. First, the gaze-relevant extractor is trained to estimate the gaze vector from the facial image. Second, the GaP-encoder is trained by $\mathcal{L}_{GD}$ that is difference between the estimated result from generated image $F(G(E(l)))$ and ground truth g.

features from the entire image makes it extremely difficult in that it requires extracting features from infinitely large unseen real-world image space. Therefore, to minimize the space of the space that unseen images could possibly have, we capture only face with RetinaFace [29] and crop images. This could drastically reduce the unexpected input from different environments. Since the values of a latent code is set along one specific direction to the corresponding attribute, the latent code is meaningful beyond simply compressing the facial information. The second goal of the generator is to perform image reconstruction at the pixel level. The generator is pre-trained with the source dataset. This makes it possible to utilize dataset-specific mapping ability as the generated images in the space where the gaze estimator is fully capable. Especially, the label space of the gaze estimation task remains the same as it can be represented with two angles of azimuth and elevation. In general, since the generatable space of each attribute trained with the source domain is larger, the generator can reconstruct an image domain from the latent domain with attributes aligned in a semantical way. To obtain latent codes, we benchmarked e4e-encoder [10] which embed images into latent code with disentanglement characters. However, e4e-encoder cannot preserve gaze-relevant features, since the gaze-relevant feature is a relatively finer attribute. To tackle this issue, we propose novel gaze distortion loss L_{GD} that preserves gaze features in the encoding process. We call the encoder with this loss the GaP-encoder. As shown in Fig. 5, training process of encoder consists of two phases. First, we train extractor F that is combined with ResNet-18 and MLP to estimate gaze vector from face images. Second, we utilize F to calculate L_{GD} from the generated image $G(E(I))$. E is an encoder, G is a generator, I is input image, respectively. Therefore, L_{GD} is denoted as:

$$\mathcal{L}_{GD} = \|F(G(E(I))) - g\|_2^2, \tag{7}$$

where g denotes 3D ground-truth gaze angle. The encoder is trained with our proposed $\mathcal{L}_{GD}$ on top of three losses used in [10] that consists of $\mathcal{L}_2$. $\mathcal{L}_{LPIPS}$ [30], $\mathcal{L}_{sim}$ [10] and total loss is formulated as follows:

$$\mathcal{L}_{total}(I) = \lambda_{l2}\mathcal{L}_2(I) + \lambda_{LPIPS}\mathcal{L}_{LPIPS}(I) + \lambda_{sim}\mathcal{L}_{sim}(I) + \lambda_{GD}\mathcal{L}_{GD}(I), \tag{8}$$

where λ_{l2} , λ_{LPIPS} , λ_{sim} , λ_{GD} are the weights of each loss. In our method, the extractor F may have gaze angle error. Since our method performs end-to-end training with encoder and extractor F during training process, F can be optimized to generate the image which preserved gaze relevant features.

4   Experiments

4.1   Datasets and Setting

Datasets. Based on the datasets used in recent gaze estimation studies, we used four gaze datasets, ETH-XGaze [26], MPIIFaceGaze [27], and EyeDiap [31]. The ETH-XGaze dataset provides 80 subjects (i.e., 756,540 images) that consist of various illumination conditions, gaze and head pose ranges. The MPIIFaceGaze provides 45000 facial images from 15 subjects. The EyeDiap dataset provides 16,674 facial images from 14 participants. The smaller the number of subjects and various gaze direction datasets, the better to extract gaze-relevant features through statistical methods. For this reason, datasets such as ETH-XGaze and MPIIFaceGaze are useful for extracting gaze-relevant features.

Cross-Domain Evaluation Environment. We used ETH-XGaze that has wider gaze and head pose ranges than the others. To evaluate cross-dataset performance of gaze estimation, MPIIFaceGaze and EyeDiap are used. All facial images are resized to 256×256 for stable encoder-decoder training. We conducted two cross-dataset evaluation tasks, E(ETH-XGaze) → M (MPIIGaze), E → D (EyeDiap).

Dataset Preparation. We utilized RetinaFace [30] to detect the face region of an image and consistently cropped according to the region for all datasets to induce the generator to generate equal size of faces.

Evaluation Metric. To numerically represent the performance of the gaze estimation models, we computed the angular error between the predicted gaze vector $\hat{g}$ and 3D ground-truth gaze angle g, and the angular error can be computed as:

$$\mathcal{L}_{angular} = \frac{\hat{g} \cdot g}{\|\hat{g}\|\|g\|} \tag{9}$$

where $\hat{g}$ and g each denotes predicted 3D gaze angle and the ground truth angle, and $\|g\|$ denotes absolute value of g, and $\cdot$ is the dot product.

4.2 Comparison with Gaze Estimation Models on Single Domain

An experiment was conducted to verify the performance of the gaze estimation model in a single dataset. We achieved favorable performance against other gaze estimation methods on single-domain as well as cross-domain evaluations. We utilized the ETH-XGaze and MPIIFaceGaze dataset which provides a standard gaze estimation protocol on this experiment. ETH-XGaze is divided into 15 validation sets and the rest of the training sets. MPIIFaceGaze is divided into 36,000 training sets, 9,000 validation sets based on each individual. To compare with the SOTA gaze estimation model, we tested RT-Gene, Dilated-Net models using eye images, and ResNet-18, Pure-Gaze using face images. As shown in Table 1, the proposed LatentGaze shows a lower angular error than most others, indicating the superiority of our framework. In addition, as shown in Fig. 6, personal appearances are converted into favorable conditions for gaze estimation. Since it helps the gaze estimation model extract high-quality gaze-relevant features, the model can provide good gaze estimation performance in single domain. In addition to the SOTA performance of our model on the single domain evaluation, we propose a specialized model to cover the cross-domain evaluation.

Table 1. Performance comparison with gaze estimation models for ETH-XGaze and MPIIFaceGaze dataset.

Method	ETH-XGaze angular error (°)	MPIIFaceGaze angular error (°)
ResNet-18	4.71	5.14
Dilated-Net	4.79	4.82
Pure-Gaze	4.52	5.51
RT-Gene	4.21	4.31
LatentGaze	3.94	3.63

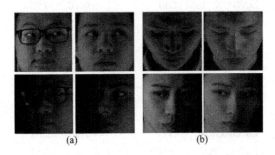

Fig. 6. Visual results of LatentGaze in single domain. (a) The glasses are eliminated without blurring and geometric deformation. (b) The illumination is changed favorable to the gaze estimator.

4.3   Comparison with Gaze Estimation Models on Cross Domain

We compare LatentGaze with the SOTA gaze estimation methods on cross-domain evaluation. In general, in a cross-dataset environment, gaze estimation methods show significant performance degradation due to overfitting to the source domain. However, our model solves the overfitting of gaze-irrelevant features by enabling selective latent codes utilization through gaze-aware analytic manipulation. In addition, the target-to-source domain shift module maps the input image to an image space where the gaze estimator is sufficiently aware. As shown in Table 2, our method shows the best or favorable gaze angular error in two cross-domain environments. Here, the performance of E→D is poor compared to E→M. The reason is that the difference in resolution between the source domain and the target domain used for generator training is extreme. However, we can solve this through preprocessing through super-resolution and denoising. In this experiment, no more than 100 target samples for adoption were randomly selected. It demonstrates that our framework is suitable for a real-world application where the distribution of target domain is usually unknown. Moreover, as shown in Fig. 7, it shows that the proposed framework can shift the target domain to the source domain while preserving gaze-relevant features. Since it helps the gaze estimation module maintains gaze estimation performance, our framework provides the robustness of cross-domain evaluation. Even if small illumination change is generated during the domain shift process as shown in Fig. 7, it does not affect the gaze estimation performance as shown in Table 2.

Table 2. Performance comparison with SOTA gaze estimation models on cross-dataset evaluation. The best and the second-best results are highlighted in red and blue colors, respectively.

Method	Target samples	E→M	E→D
Gaze360 [23]	>100	5.97	7.84
GazeAdv [32]	500	6.75	8.10
DAGEN [16]	500	6.16	9.73
ADDA [33]	1000	6.33	7.90
UMA [34]	100	7.52	12.37
PnP-GA [22]	<100	5.53	6.42
PureGaze [24]	<100	5.68	7.26
CSA [15]	<100	5.87	5.95
RUDA [14]	<100	5.70	6.29
LatentGaze	N/A	7.98	9.81
LatentGaze	<100	5.21	7.81

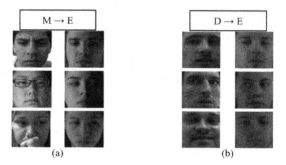

(a)   (b)

Fig. 7. Visualizations of generated images by target-to-source domain shift module. The results show that our module maps the unseen target images into the source distribution by replacing the image's attributes with the source domain's. (a) MPIIFaceGaze dataset is mapped into the domain of the ETH-XGaze dataset. (b) EyeDiap dataset is mapped into the domain of the ETH-XGaze dataset.

4.4   Ablation Study

In this section, we conduct experiments about how two modules affect the gaze estimation performance. We also show the necessity of our $\mathcal{L}_{GD}$ in the encoder training process by comparing the generated images with and without the loss. Finally, to show the effectiveness of the gaze-aware analytic manipulation method, we first find out the gaze-related index in latent codes through the proposed method and replace one group's gaze-related elements from the other group's ones. And we visualize the generated images from the replaced latent codes to show the gaze of the images are replaced accordingly.

Effect of Gaze-aware Analytic Manipulation. In this section, we demonstrate the effectiveness of the gaze-aware analytic manipulation. We used both an image and the manipulated latent code for the MPIIFaceGaze dataset. We used 448 (all), 256 (56%), and 64 (14%) chunks in the latent code, respectively. We removed gaze-irrelevant chunks found by our gaze-aware analytic manipulation method. As presented in Table 3, the angular error using gaze-aware analytic manipulation is lower than that when using many chunks. It indicates that our manipulation method correctly separates gaze-relevant features from tightly intertwined ones. Consequently, the gaze estimation model does not have to consider features that impede the gaze estimation performance.

Table 3. Quantitative results for evaluating the effects of gaze-aware analytic manipulation.

# of Chucks	MPIIFaceGaze angular error ($°$)
None (only Resnet-18)	5.14
All	11.34
256	7.21
64	3.63

Effect of GaP-Encoder. To verify the effectiveness of the Gap-encoder, we compare the e4e-encoder benchmarked in this paper with the GaP-encoder. To fair comparison, we train each encoder with 30 epochs, and the angular error is measured by the same gaze estimation module. Also, each image result is generated by the same generator. As presented in Table 4, the GaP-encoder exhibits better gaze estimation performance because our model preserves gaze-relevant features after the domain shifting. As shown in Fig. 8, since our GaP-encoder effectively embed the image into latent code with preserving gaze-relevant features, the generated image shows the same pupil position which is highly related to gaze estimation performance.

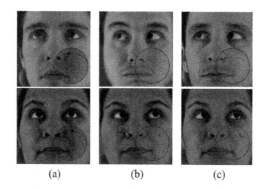

Fig. 8. Visual comparison with e4e-encoder and GaP-encoder. All results are generated by the same generator. (a) Images generated by e4e-encoder's latent codes. (b) The original images, (c) Images generated by GaP-encoder's latent codes. Red and blue dots represent the predicted gaze vectors. Green dot represents ground-truth gaze vector. (Color figure online)

Table 4. Performance comparison with encoder models. This is the result of using only the latent code from each encoder.

# of Chucks	MPIIFaceGaze angular error (°)
e4e-encoder	28.51
GaP-encoder	3.54

5   Conclusion

In this paper, we presented the first practical application of GAN inversion to solve the cross-dataset problem in gaze estimation with latent code. Our proposed LatentGaze framework consists of a target-to-source domain shift module, which maps the target image into the source domain image space, and a gaze-aware analytic selection manipulation module, which selectively manipulates gaze-relevant features by a statistical data-driven approach. Furthermore, we propose gaze distortion loss that prevents the distortion of gaze information caused by inversion. Our quantitative and qualitative experiments and visualizations show our approach performs favorably against the SOTA methods.

Acknowledgements. This work was supported by the NRF grant (NRF-2020R1A4A1019191), the IITP grant (No.2021-0-02068), the funded by the Korea government (MSIT), and the Industrial Fundamental Technology Develop Program (No.20018699) funded by the Korea government (MOTIE).

References

1. Massé, B., Ba, S., Horaud, R.: Tracking gaze and visual focus of attention of people involved in social interaction. IEEE Trans. Pattern Anal. Mach. Intell. **40**(11), 2711–2724 (2017)
2. Schauerte, B., Stiefelhagen, R.: "look at this!" learning to guide visual saliency in human-robot interaction. In: 2014 IEEE/RSJ International Conference on Intelligent Robots and Systems, pp. 995–1002. IEEE (2014)
3. Yuan, G., Wang, Y., Yan, H., Fu, X.: Self-calibrated driver gaze estimation via gaze pattern learning. Knowl.-Based Syst. **235**, 107630 (2022)
4. Lanillos, P., Ferreira, J.F., Dias, J.: A Bayesian hierarchy for robust gaze estimation in human-robot interaction. Int. J. Approx. Reason. **87**, 1–22 (2017)
5. Mirza, M., Osindero, S.: Conditional generative adversarial nets (2014)
6. Abdal, R., Qin, Y., Wonka, P.: Image2stylegan: How to embed images into the stylegan latent space? In: Proceedings of the IEEE/CVF International Conference on Computer Vision, pp. 4432–4441 (2019)
7. Shen, Y., Gu, J., Tang, X., Zhou, B.: Interpreting the latent space of gans for semantic face editing, pp. 9243–9252 (2020)
8. Park, S., Mello, S. D., Molchanov, P., Iqbal, U., Hilliges, O., Kautz, J.: Few-shot adaptive gaze estimation. In: Proceedings of the IEEE/CVF International Conference on Computer Vision, pp. 9368–9377 (2019)
9. Zhu, J., Shen, Y., Zhao, D., Zhou, B.: In-domain gan inversion for real image editing, pp. 592–608 (2020)
10. Tov, O., Alaluf, Y., Nitzan, Y., Patashnik, O., Cohen-Or, D.: Designing an encoder for stylegan image manipulation. ACM Trans. Graph. (TOG) **40**(4), 1–14 (2021)
11. He, Z., Spurr, A., Zhang, X., Hilliges, O.: Photo-realistic monocular gaze redirection using generative adversarial networks. In: Proceedings of the IEEE/CVF International Conference on Computer Vision, pp. 6932–6941 (2019)
12. Zheng, Y., Park, S., Zhang, X., De Mello, S., Hilliges, O.: Self-learning transformations for improving gaze and head redirection, vol. 33, pp. 13127–13138 (2020)
13. Cui, S., Wang, S., Zhuo, J., Su, C., Huang, Q., Tian, Q.: Gradually vanishing bridge for adversarial domain adaptation. In: Proceedings of the IEEE/CVF Conference on Computer Vision and Pattern Recognition, pp. 12455–12464 (2020)
14. Bao, Y., Liu, Y., Wang, H., Lu, F.: Generalizing gaze estimation with rotation consistency. In: Proceedings of the IEEE/CVF Conference on Computer Vision and Pattern Recognition, pp. 4207–4216 (2022)
15. Wang, K., Zhao, R., Su, H., Ji, Q.: Generalizing eye tracking with bayesian adversarial learning. In: Proceedings of the IEEE/CVF Conference on Computer Vision and Pattern Recognition, pp. 11907–11916 (2019)
16. Guo, Z., Yuan, Z., Zhang, C., Chi, W., Ling, Y., Zhang, S.: Domain adaptation gaze estimation by embedding with prediction consistency. In: Proceedings of the Asian Conference on Computer Vision (2020)
17. Tzeng, E., Hoffman, J., Saenko, K., Darrell, T.: Adversarial discriminative domain adaptation. In: Proceedings of the IEEE Conference on Computer Vision and Pattern Recognition, pp. 7167–7176 (2017)
18. Krafka, K., et al.: Eye tracking for everyone. In: Proceedings of the IEEE Conference on Computer Vision and Pattern Recognition, pp. 2176–2184 (2016)
19. Zhang, X., Sugano, Y., Fritz, M., Bulling, A.: It's written all over your face: full-face appearance-based gaze estimation. In: Proceedings of the IEEE Conference on Computer Vision and Pattern Recognition Workshops, pp. 51–60 (2017)

20. Fischer, T., Chang, H.J., Demiris, Y.: Rt-gene: real-time eye gaze estimation in natural environments. In: Proceedings of the European Conference on Computer Vision, pp. 334–352 (2018)
21. Yu, Y., Liu, G., Odobez, J.M.: Improving few-shot user-specific gaze adaptation via gaze redirection synthesis. In: Proceedings of the IEEE/CVF Conference on Computer Vision and Pattern Recognition, pp. 11937–11946 (2019)
22. Liu, Y., Liu, R., Wang, H., Lu, F.: Generalizing gaze estimation with outlier-guided collaborative adaptation. In: Proceedings of the IEEE/CVF International Conference on Computer Vision, pp. 3835–3844 (2021)
23. Kellnhofer, P., Recasens, A., Stent, S., Matusik, W., Torralba, A.: Gaze360: physically unconstrained gaze estimation in the wild, pp. 6912–6921 (2019)
24. Cheng, Y., Bao, Y., Feng, L.: Puregaze: purifying gaze feature for generalizable gaze estimation. In: Proceedings of the AAAI Conference on Artificial Intelligence, vol. 36, pp. 436–443 (2022)
25. Radford, A., Metz, L., Chintala, S.: Unsupervised representation learning with deep convolutional generative adversarial networks (2015)
26. Zhang, X., Park, S., Beeler, T., Bradley, D., Tang, S., Hilliges, O.: ETH-XGaze: a large scale dataset for gaze estimation under extreme head pose and gaze variation. In: Vedaldi, A., Bischof, H., Brox, T., Frahm, J.-M. (eds.) ECCV 2020. LNCS, vol. 12350, pp. 365–381. Springer, Cham (2020). https://doi.org/10.1007/978-3-030-58558-7_22
27. Zhang, X., Sugano, Y., Fritz, M., Bulling, A.: Mpiigaze: real-world dataset and deep appearance-based gaze estimation, vol. 41, pp. 162–175. IEEE (2017)
28. Zhang, Y., Li, K., Li, K., Wang, L., Zhong, B., Fu, Y.: Image super-resolution using very deep residual channel attention networks. In: Proceedings of the European Conference on Computer Vision, pp. 286–301 (2018)
29. Deng, J., Guo, J., Ververas, E., Kotsia, I., Zafeiriou, S.: Retinaface: single-shot multi-level face localisation in the wild. In: Proceedings of the IEEE/CVF Conference on Computer Vision and Pattern Recognition, pp. 5203–5212 (2020)
30. Zhang, R., Isola, P., Efros, A. A., Shechtman, E., Wang, O.: The unreasonable effectiveness of deep features as a perceptual metric. In: Proceedings of the IEEE Conference on Computer Vision and Pattern Recognition, pp. 586–595 (2018)
31. Mora, K.A.F., Monay, F., Odobez, J.M.: Eyediap: a database for the development and evaluation of gaze estimation algorithms from rgb and rgb-d cameras. In: Proceedings of the Symposium on Eye Tracking Research and Applications, pp. 255–258 (2014)
32. Wang, K., Zhao, R., Su, H., Ji, Q.: Generalizing eye tracking with bayesian adversarial learning. In: Proceedings of the IEEE/CVF Conference on Computer Vision and Pattern Recognition, pp. 11907–11916 (2019)
33. Tzeng, E., Hoffman, J., Saenko, K., Darrell, T.: Adversarial discriminative domain adaptation. In: Proceedings of the IEEE Conference on Computer Vision and Pattern Recognition, pp. 7167–7176 (2017)
34. Cai, M., Lu, F., Sato, Y.: Generalizing hand segmentation in egocentric videos with uncertainty-guided model adaptation. In: Proceedings of the IEEE/CVF Conference on Computer Vision and Pattern Recognition, pp. 14392–14401 (2020)

Cross-Architecture Knowledge Distillation

Yufan Liu[1,2], Jiajiong Cao[5], Bing Li[1,4(✉)], Weiming Hu[1,2,3], Jingting Ding[5], and Liang Li[5]

[1] National Laboratory of Pattern Recognition, Institute of Automation, Chinese Academy of Sciences, Beijing, China
`bli@nlpr.ia.ac.cn`
[2] School of Artificial Intelligence, University of Chinese Academy of Sciences, Beijing, China
[3] CAS Center for Excellence in Brain Science and Intelligence Technology, Beijing, China
[4] PeopleAI, Inc., Beijing, China
[5] Ant Financial Service Group, Beijing, China

Abstract. Transformer attracts much attention because of its ability to learn global relations and superior performance. In order to achieve higher performance, it is natural to distill complementary knowledge from Transformer to convolutional neural network (CNN). However, most existing knowledge distillation methods only consider homologous-architecture distillation, such as distilling knowledge from CNN to CNN. They may not be suitable when applying to cross-architecture scenarios, such as from Transformer to CNN. To deal with this problem, a novel cross-architecture knowledge distillation method is proposed. Specifically, instead of directly mimicking output/intermediate features of the teacher, partially cross attention projector and group-wise linear projector are introduced to align the student features with the teacher's in two projected feature spaces. And a multi-view robust training scheme is further presented to improve the robustness and stability of the framework. Extensive experiments show that the proposed method outperforms 14 state-of-the-arts on both small-scale and large-scale datasets.

Keywords: Knowledge distillation · Cross architecture · Model compression

1 Introduction

Knowledge distillation (KD) has become a fundamental topic for model performance promotion. It has been successfully applied to various applications including model compression [1] and knowledge transfer [2]. KD usually adopts a teacher-student framework, where the student model is trained under the guidance of the teacher's knowledge. The knowledge is usually defined by soft outputs or intermediate features of the teacher model.

L. Wang et al. (Eds.): ACCV 2022, LNCS 13845, pp. 179–195, 2023.
https://doi.org/10.1007/978-3-031-26348-4_11

Existing KD methods focus on convolutional neural network (CNN). However, there recently emerge many new networks such as Transformer. It shows superior on different computer vision tasks including image classification [3] and detection [4], while its huge computation and limited platform acceleration support limits the application of Transformer, especially for edge devices. On the other hand, with several years of development, there are sufficient acceleration libraries including CUDA [5], TensorRT [6] and NCNN [7], making CNN hardware friendly on both servers and edge devices. To this end, it is a natural idea to distill the knowledge from high-performance Transformer to compact CNN. However, there is a large gap between the two architectures. As shown in Fig. 1-(a), Transformer consists of self-attention-based transformer blocks while CNN contains a sequence of convolutional blocks. Further, the features are arranged in a totally different way. The intermediate outputs of CNNs are formed with c channels of $h' \times w'$ feature maps. Different from CNN, the features of Transformer consist of N feature vectors with $3hw$ elements, where N refers to the patch number.

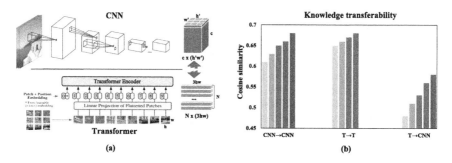

Fig. 1. (a) The comparison of CNN and Transformer. The formation of the features are absolutely different. (b) The cosine similarity between features from different models on ImageNet. Note that the features are mapped into the same dimension by a linear projection. For "CNN→CNN", the bars represent the similarities between CNN ResNet152 and CNNs {ResNet18, ResNet32, ResNet50, ResNet101, ResNet152}; For "T→T", the bars represent the similarities between Transformer ViT-L/16 and Transformers {ViT-B/32, ViT-B/16, ViT-L/32, ViT-L/16}; For "T→CNN", the bars represent the similarities between Transformer ViT-L/16 and CNNs {ResNet18, ResNet32, ResNet50, ResNet101, ResNet152}.

Unfortunately, existing methods focus on homologous-architecture KD such as CNN →CNN and Transformer→Transformer, which are not suitable for the cross-architecture scenarios. As shown Fig. 1-(b), the knowledge "transferability" is defined quantitatively. In particular, the output feature of the student is aligned to the feature space of the teacher, and then, the cosine similarity of the aligned student feature vector and the teacher feature vector is computed. For homologous-architecture cases, the transferability is between $0.6 - 0.7$, while it is much lower, typically lower than 0.55, on the cross-architecture condition. Consequently, it is more difficult to distill knowledge across different architectures and a new KD framework should be designed to deal with it.

In this work, a novel cross-architecture knowledge distillation method is proposed to bridge the large gap between Transformer and CNN. With the help of the proposed framework, the knowledge from Transformer is efficiently transferred to the student CNN network and the knowledge transferability is significantly improved via this method. It encourages the student to learn both local spatial features (with the original CNN model) and the complementary global features (from the transformer teacher model). In particular, two projectors including a partially cross attention (PCA) projector and a group-wise linear (GL) projector, are designed. Instead of directly mimicking the output of the teacher, these two projectors align the intermediate student feature into two different feature spaces and knowledge distillation is further operated in the two feature spaces. The PCA projector maps the student feature into the Transformer attention space of the teacher. This projector encourages the student to learn the global relation from the Transformer teacher. The GL projector maps the student feature into the Transformer feature space in a pixel-by-pixel manner. This projector directly alleviates the feature formation differences between the teacher and the student. In addition, to alleviate the instability caused by the diversity in the cross-architecture framework, we propose a cross-view robust training scheme. Multi-view samples are generated to disturb the student network. And a multi-view adversarial discriminator is constructed to distinguish the teacher features and the disturbed student features, while the student is trained to confuse the discriminator. After convergence, the student can be more robust and stable.

Extensive experiments are conducted on both large-scale datasets and small-scale datasets, including ImageNet [8] and CIFAR [9]. The experimental results of different teacher-student pairs demonstrate that the proposed method stably performs better than 14 state-of-the-arts. In summary, the main contributions of our work are three-fold:

- We propose a cross-architecture knowledge distillation framework to distill excellent Transformer knowledge to guide CNN. In this framework, partially cross attention (PCA) projector and group-wise linear (GL) projector are designed to align the student feature space and promote the transferability between teacher features and student features.
- We propose a multi-view robust training scheme to improve the stability and robustness of the student network.
- Experimental results show that the proposed method is effective and outperforms 14 state-of-the-arts on both large-scale datasets and small-scale datasets.

2   Related Work

Hinton et al. [10] proposes the concept of knowledge distillation, using the soft output of teacher to guide the learning of student. Recently, it has been applied

mainly to model compression [1] and knowledge transfer [2]. Different forma-
tions of distilled knowledge are explored to better guide the student network,
including final output [10,11] and hint layer knowledge [12–19]. For hint layer
knowledge, many endeavors have been taken to match the student hint layers and the teacher-guided layers. For example, AT [12] defines single-channel
attention maps as knowledge. However, the computation of the attention maps
causes channel-dimension information loss. FitNet [13] directly distills the features from intermediate layers without information loss. However, this restriction
is somewhat hard and not all the information is beneficial. Liu *et al.* [17] distill
the knowledge called instance relationship graph (IRG), which contains instance
feature, instance feature relationship and feature space transformation. It is not
limited by the dimension mismatch between the teacher and the student.

The methods above all focus on convolutional neural network (CNN).
Recently, Transformer becomes increasingly popular because of its impressive
performance. However, due to the totally different architecture, many previous
KD methods can not be directly applied to Transformers. There are some works
[20–22] studying knowledge distillation between Transformers. DeiT [20] proposes a distillation token similar to the class token, to make the student Transformer learn the hard label from the teacher and ground truth (GT). MINILM
[21] focuses on the attention mechanisms in Transformer and distills the corresponding self-attention information. IR [22] distills the internal representations
(*e.g.*, self-attention map) from the teacher Transformer to the student Transformer.

In summary, existing methods usually present a transformation to match the
teacher's features and the student's features. However, nearly all of them require
similar or even the same architecture between teacher and student. To deal
with the cross-architecture knowledge distillation problem, we carefully design
projectors to match the teacher and the student in the same feature space.
Consequently, a compact student CNN model can well learn the global feature
from a teacher Transformer model despite the big gap in the architectures.

3   The Proposed Method

In this section, the framework of the proposed method is first introduced. Then,
two key components of the framework including cross-architecture projectors
and a cross-view robust training scheme are presented. The former is constructed
to alleviate the feature mismatch for cross-architecture scenarios and help the
student learn the global relation of the features, while the latter is adopted to
improve the robustness and stability of the student. Finally, the loss function
and training procedure are described.

3.1   Framework

The overall framework of the proposed method is depicted in Fig. 2. In this
figure, the upper pink network represents the teacher network, while the lower

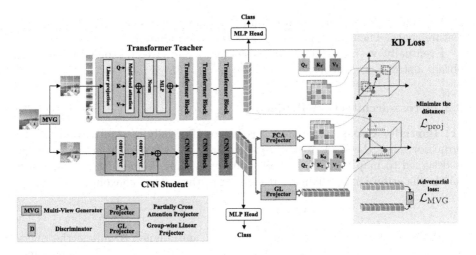

Fig. 2. Overall framework of the proposed method.

blue network is the student network. For the transformer teacher Θ^{T}, the input sample $\mathbf{x} \in \mathbb{R}^{3 \times H \times W}$ is divided into $(N = \frac{HW}{hw})$ patches $\{x_n \in \mathbb{R}^{3 \times h \times w}\}_{n=1}^{N}$. After the inference of several transformer blocks, the feature $\mathbf{h}_{\mathrm{T}} \in \mathbb{R}^{N \times (3hw)}$ is generated. And the final predicted possibility is then computed via a multi layer perceptron (MLP) head as shown in Fig. 2. For the CNN student Θ^{S}, it receives the whole image without patch-wise partition as input. Similarly, after the inference of several CNN blocks, the final student feature $\mathbf{h}_{\mathrm{S}} \in \mathbb{R}^{c \times (h'w')}$ can be obtained. Note that c is the channel number and $h'w' = \frac{HW}{2^{2s}}$. The s denotes the number of CNN stages (usually equals 4). It is then used to predict the class.

Due to the differences of the design principles and architectures between transformers and CNNs, it is hard to make the student features directly mimic the teacher features using the existing KD methods. To solve this problem, we propose a cross-architecture projector which consists of a partially cross attention (PCA) projector and a group-wise linear (GL) projector. The PCA projector maps the student features into the transformer attention space. By mapping the CNN feature space to this attention space, it is easier for the student to learn the global relationship among different regions by minimizing the distances between the student attention maps and the teacher attention maps. The GL projector maps the student features into the transformer feature space. In this transformer feature space, the student is guided to mimic the global transformer features in a pixel-by-pixel manner.

To improve the robustness and stability of the student, a cross-view robust training scheme is proposed. Multi-view samples are generated by a multi-view generator which randomly conducts some transformations and generates mask and noise adding to the inputs. Fed with the multi-view inputs, the student generates different features. A multi-view adversarial discriminator is constructed to distinguish the teacher features and the student features in the transformer feature space. Then the goal is to puzzle the discriminator.

Eventually, we integrate the proposed losses and give end-to-end training to obtain a strong student network.

3.2   Cross-architecture Projector

(1) Partially Cross Attention Projector Partially cross attention (PCA) projector maps the student feature space into transformer attention space. It is designed to map the CNN features to Query, Key, Value matrices and then mimic the attention mechanism. It consists of three 3×3 convolutional layers:

$$\{Q_S, K_S, V_S\} = \text{Proj}_1(\mathbf{h_S}), \tag{1}$$

where the matrixes Q_S, K_S, V_S are computed and aligned to mimic the query Q_T, the key K_T and the value V_T of the Transformer teacher. In the transformer attention space, the self-attention of the student is calculated as:

$$\text{Attn}_S = softmax(\frac{Q_S(K_S)^T}{\sqrt{d}})V_S, \tag{2}$$

in which d is the query size. The calculation of Attn_T is similar. Hence, we can minimize the distance between the attention maps of the teacher and those of the student to guide the student network. To further improve the robustness of the student, we construct the partially cross attention of the student to replace the original Attn_S:

$$\text{PCAttn}_S = softmax(\frac{g(Q_S)(g(K_S))^T}{\sqrt{d}})g(V_S),$$

$$\text{s.t.} \quad g(M_S(i,j)) = \begin{cases} M_T(i,j), & p \geq 0.5 \\ M_S(i,j), & p < 0.5 \end{cases}, (M = Q, K, V). \tag{3}$$

Note that (i, j) denotes the matrix element index of M. The function $g(\cdot)$ replaces the Q_S, K_S, V_S matrixes of the student by the corresponding matrixes of the teacher, with the probability p subject to uniform distribution. In this manner, the loss is constructed:

$$\mathcal{L}_{\text{proj}_1} = ||\text{Attn}_T - \text{PCAttn}_S||_2^2 + ||\frac{V_T \cdot V_T}{\sqrt{d}} - \frac{V_S \cdot V_S}{\sqrt{d}}||_2^2, \tag{4}$$

to make the student mimic the teacher in the attention space.

(2) Group-wise Linear Projector Group-wise linear (GL) projector maps the student feature into transformer feature space. It consists of several shared-weight fully-connected (FC) layers:

$$\mathbf{h'_S} = \text{Proj}_2(\mathbf{h_S}), \tag{5}$$

where $\mathbf{h}'_S \in \mathbb{R}^{N \times (3hw)}$ is aligned to have the same dimension with teacher feature $\mathbf{h}_T$. Specifically, for the regular image input with size of 224×224, the dimensions are $\mathbf{h}_S \in \mathbb{R}^{256 \times 196}$ and $\mathbf{h}'_S \in \mathbb{R}^{196 \times 768}$. In order to realize a pixel-by-pixel mapping manner, the projector needs at least 196 FC layers with 256×768 parameters. each of them maps the pixel from the original feature space to the corresponding "pixel" in the transformer space. A large number of FC layers may cause huge computation. In order to obtain a compact projector, we propose the **group-wise** linear projector where a 4×4 neighborhood shares an FC layer. Hence, the GL projector only contains 16 FC layers. Furthermore, *drop-out* is also adopted to reduce the computation and improve the robustness. Finally, after obtaining the new aligned student feature, the loss is computed as:

$$\mathcal{L}_{\text{proj}_2} = ||\mathbf{h}_T - \mathbf{h}'_S||_2^2, \tag{6}$$

to minimize the distance between the teacher feature and the student feature in the transformer feature space.

3.3 Cross-view Robust Training

Due to the big difference between the architectures of the teacher and the student, it is not that easy for the student to learn to be robust. To improve the robustness and the stability of the student network, we proposed a cross-view robust training scheme. The proposed training scheme contains two important components, *i.e.*, a multi-view generator (MVG) and the corresponding multi-view adversarial discriminator. The MVG takes the original image as the input, and generates images with different transformations with some probability:

$$\tilde{\mathbf{x}} = \text{MVG}(\mathbf{x}) = \begin{cases} \text{Trans}(\mathbf{x}), & p \geq 0.5 \\ \mathbf{x}, & p < 0.5 \end{cases}, \tag{7}$$

in which $\text{Trans}(\cdot)$ contains the common transformations, such as color jettering, random crop, rotation, patch-wise mask, *etc.* The probability p is also subject to the uniform distribution. These transformed versions of the samples are then fed to the student network. Subsequently, the multi-view adversarial discriminator is constructed to distinguish the teacher feature $\mathbf{h}_T$ and the transformed student feature $\mathbf{h}'_S$, which is comprised of a three-FC-layer network. In this manner, the target of the cross-view robust training is to confuse the discriminator and obtain a robust student feature. The training loss of the discriminator is computed as:

$$\mathcal{L}_{\text{MAD}} = \frac{1}{m} \sum_{k=1}^{m} \left[-\log D(\mathbf{h}_T^{(k)}) - \log(1 - D(\mathbf{h}'^{(k)}_S)) \right]. \tag{8}$$

Note that $D(\cdot)$ denotes the multi-view adversarial discriminator. And m is the total number of training samples. For the student network which can be seen as the generator in the adversarial training, the loss is written as:

$$\mathcal{L}_{\text{MVG}} = \frac{1}{m} \sum_{k=1}^{m} \left[\log(1 - D(\mathbf{h}'^{(k)}_S)) \right]. \tag{9}$$

Minimizing this loss can help to generate the student feature $\mathbf{h}'_S$ which distributes similarly to that of the teacher feature $\mathbf{h}_T$.

3.4   Optimization

In this subsection, we introduce the overall optimization and the training procedure of the proposed method. In order to train the student network, the loss function can be obtained by:

$$\mathcal{L}_{\text{total}} = (\mathcal{L}_{\text{proj}_1} + \mathcal{L}_{\text{proj}_2}) + \lambda \cdot \mathcal{L}_{\text{MVG}}, \tag{10}$$

in which λ is the penalty coefficient balancing the loss terms. For the multi-view adversarial discriminator, the loss function is $\mathcal{L}_{\text{MAD}}$ in Eq. (8).

The overall training procedure of the proposed method is summarized in Algorithm 1. In detail, the cross-architecture teacher-student framework is first constructed. The PCA projector and the GL projector are then embedded in the student network to map the student features into the teacher attention space and feature space. Subsequently, a cross-view robust training scheme is adopted to train the framework. The framework main body (*i.e.*, $\mathbf{\Theta}^S$, $\text{Proj}_1(\cdot)$ and $\text{Proj}_2(\cdot)$) and the multi-view adversarial discriminator $D(\cdot)$ are alternatively updated. After convergence, the modules $\text{Proj}_1(\cdot)$, $\text{Proj}_2(\cdot)$ and $D(\cdot)$ are removed and only the compact student network $\mathbf{\Theta}^S$ is remained to carry out the inference phase.

Algorithm 1: The procedure of cross-architecture knowledge distillation.

Input: Database $\mathcal{D}^{\text{train}} = \{\mathbf{x}^{\text{train}}, \mathbf{y}^{\text{train}}\}$, $\mathbf{\Theta}^S$, $\mathbf{\Theta}^T$, $D(\cdot)$, $\text{Proj}_1(\cdot)$, $\text{Proj}_2(\cdot)$.

1 $e = 0$;
2 Initialize $\mathbf{\Theta}^S$, $\text{Proj}_1(\cdot)$, $\text{Proj}_2(\cdot)$ and $D(\cdot)$;
3 **repeat**
4   Compute the transformed features $\mathbf{h}'_S$ and $\{Q_S, K_S, V_S\}$ through $\text{Proj}_1(\cdot)$ and $\text{Proj}_2(\cdot)$, using Eq. (1) and Eq. (5);
5   Update $\mathbf{\Theta}^S$, $\text{Proj}_1(\cdot)$ and $\text{Proj}_2(\cdot)$ using Eq. (10);
6   **if** $e\%5 = 0$ **then**
7 | Update $D(\cdot)$ using Eq. (8);
8   **end**
9      $e = e + 1$;
10 **until** *done*;
11 Remove $\text{Proj}_1(\cdot)$, $\text{Proj}_2(\cdot)$ and $D(\cdot)$, and predict the label through $\mathbf{\Theta}^S$ in inference phase;
12 **return** $\mathbf{\Theta}^S$.

---

4 Experiments

4.1 Settings

Databases and Networks. We evaluate the proposed method on two databases: CIFAR [9] and ImageNet [8]. The data are augmented using the same strategies as in the PyTorch official examples [23]. For networks, we use the popular CNNs as the student network, including ResNets [24], MobileNet v2 [25], Xception [26] and EfficientNet [27]. The typical Transformers are applied as the teacher network, such as ViT [3], and Swin Transformer [28].

Implementation Details. We train all the networks from scratch. For CIFAR datasets, the total number of epochs is 200 with a standard batch size of 64. The learning rate is initialized as 0.1 and multiplied by 0.1 at epoch 100 and epoch 150. For ImageNet, the total number of epochs is 120 with a 256 batch size. The learning rate is initialized as 0.1 and multiplied by 0.1 at epoch 30, epoch 60 and epoch 90, respectively. A standard stochastic gradient descent (SGD) optimizer with 10^{-4} weight decay and 0.9 momentum is adopted. All the experiments are conducted on a platform with 8 Nvidia Tesla GPU cards and 96-core Intel(R) Xeon(R) Platinum 8163 CPU. In addition, every single setting is repeated 5 times with different random seeds on Pytorch.

4.2 Performance Comparison

We compare the performance of our method with 14 state-of-the-art knowledge distillation methods, including Logits [10], FitNet [13], AT [12], IRG [17], RKD [29], CRD [30], OFD [14], ReviewKD [31], LONDON [32], AFD [33], AB [34], FT [35], DeiT [20] and MINILM [21]. Among them, Logits, FitNet, AT, IRG, RKD, CRD, OFD, ReviewKD and LONDON are CNN-based KD methods, and DeiT and MINILM are transformer-based KD methods. There exist few related works for the Transformer-CNN framework. Consequently, several CNN-based methods including logits, RKD and IRG are adopted for cross-architecture scenarios, since these methods do not rely on the CNN architectures. Besides, for a fair comparison, we select CNNs and Transformers with similar FLoating-point OPerations (FLOPs) or similar accuracy as the teacher network or the student network.

Evaluation on CIFAR. Table 1 presents the KD results on CIFAR100. As shown in this table, three KD modes of the teacher-student frameworks, including CNN-CNN, Transformer-CNN and Transformer-Transformer, are evaluated. It can be seen that the proposed method has the best performance among all the methods, including CNN-based KD methods and transformer-based methods. For the most commonly used CNN-CNN mode, the proposed cross-architecture KD method shows superior performance. It is because the CNN student learns complementary global information from the Transformer teacher. The performance gap is even larger (usually more than 1%) when the Transformer teacher and the CNN teachers have similar FLOPs. Because under similar computation cost, Transformer teacher usually has higher accuracy than CNN teacher. For the Transformer-CNN mode, a higher performance gain (an average gain of 2.7%) is obtained compared with the CNN-CNN methods. This indicates that existing KD methods do not take full advantage of the Transformer teacher, though they can be adopted to the cross-architecture scenario. In Transformer-Transformer mode, the proposed method results mostly surpass the Transformer-based KD results. Although the Xceptionx2 model is slightly inferior to the ViT-B/16 model, the performance gain of Xceptionx2 is higher than that of ViT-B/16. This indicates that cross-architecture KD can obtain higher promotion than the conventional homologous-architecture KD. Besides, in our cross-architecture framework, it is easier to adopt and accelerate the CNN student into practical application.

Evaluation on ImageNet. Experiments are conducted on ImageNet to further verify the generalization and effectiveness of the proposed method. As shown in Table 2, our method exhibits the best performance on ImageNet. Similar to the settings of CIFAR, two homologous-architecture modes including CNN-CNN and Transformer-Transformer and one cross-architecture mode, *i.e.*, Transformer-CNN, are compared. Different from homologous-architecture methods, the proposed cross-architecture framework encourages the student to learn both local spatial features (with the original CNN model) and complementary global features (from the transformer teacher model). Consequently, the CNN student obtains higher performance. Especially, from Table 2, some CNNs (*e.g.*, ResNet50x2-80.72%) guided by Transformer even surpasses the Transformer with similar model computation (*e.g.*, ViT-B/32-78.29%), by more than 1.03% accuracy. With hardware-friendly attributes, these improved CNNs are more potential for edge device applications.

4.3   Ablation Study

(1) **Different teacher-student pairs.** In order to verify the generalization of the proposed method, we evaluate it with different cross-architecture teacher-student pairs in Table 3. It can be observed that our cross-architecture method obtains significant performance promotion across different teacher-student pairs, compared with the baseline. In addition, the accuracies of the student continue increasing as the teacher's performance becomes better. At this end, Transformer can be an excellent teacher since it usually obtains better performance with similar FLOPs compared with a CNN network. Using Transformer to guide the learning of a CNN student can be a potential direction.

(2) **Effectiveness of the proposed projector.** We analyze the effectiveness of the proposed PCA projector and GL projector. Experimental results on ImageNet in Fig. 3-(a) show great performance gain when the two projectors are

Table 1. Performance comparison on CIFAR100. Note that "x2" denotes the channel number of this network is twice of the original ResNet's. And "x3" has the analogous meaning.

Mode	Teacher	Student	Methods	Test accuracy	Teacher	Student	Methods	Test accuracy
CNN→CNN	ResNet152x2 (212.0 GFLOPs)	ResNet50 (4.1 GFLOPs)	Baseline_T	91.03%	ResNet101x3 (205.0 GFLOPs)	ResNet50x2 (15.9 GFLOPs)	Baseline_T	90.98%
			Baseline_S	85.02%			Baseline_S	88.21%
			Logits	86.53%			Logits	89.07%
			FitNet	85.37%			FitNet	88.51%
			AT	86.41%			AT	89.18%
			RKD	86.22%			RKD	89.39%
			IRG	86.87%			IRG	89.89%
			OFD	86.79%			OFD	89.62%
			CRD	86.91%			CRD	89.94%
			ReviewKD	87.03%			ReviewKD	90.04%
			LONDON	87.16%			LONDON	89.98%
	ViT-B/16	ResNet50	**Ours**	87.39%	ViT-B/16	ResNet50x2	**Ours**	90.33%
	ViT-L/16	ResNet50	**Ours**	**88.09%**	ViT-L/16	ResNet50x2	**Ours**	**90.97%**
Transformer →CNN	ViT-B/16 (55.4 GFLOPs)	ResNet50 (4.1 GFLOPs)	Baseline_T	90.92%	ViT-L/16 (190.7 GFLOPs)	ResNet50 (4.1 GFLOPs)	Baseline_T	92.46%
			Baseline_S	85.02%			Baseline_S	85.02%
			Logits	86.42%			Logits	86.69%
			RKD	86.13%			RKD	86.73%
			IRG	86.59%			IRG	86.91%
			Ours	**87.39%**			**Ours**	**88.09%**
	ViT-B/16 (55.4 GFLOPs)	ResNet50x2 (15.9 GFLOPs)	Baseline_T	90.92%	ViT-L/16 (190.7 GFLOPs)	ResNet50x2 (15.9 GFLOPs)	Baseline_T	92.46%
			Baseline_S	88.21%			Baseline_S	88.21%
			Logits	88.86%			Logits	89.28%
			RKD	89.11%			RKD	89.51%
			IRG	89.38%			IRG	89.68%
			Ours	**90.33%**			**Ours**	**90.97%**
	Swin-L (103.9 GFLOPs)	ResNet50 (4.1 GFLOPs)	Baseline_T	93.78%	Swin-L (103.9 GFLOPs)	ResNet50x2 (15.9 GFLOPs)	Baseline_T	93.78%
			Baseline_S	85.02%			Baseline_S	88.21%
			Logits	86.78%			Logits	88.93%
			RKD	86.91%			RKD	90.02%
			IRG	87.06%			IRG	89.97%
			Ours	**88.46%**			**Ours**	**91.21%**
Transformer → Transformer	ViT-L/16 (190.7 GFLOPs)	ViT-B/16 (55.4 GFLOPs)	Baseline_T	92.46%	Swin-L (103.9 GFLOPs)	ViT-B/16 (55.4 GFLOPs)	Baseline_T	93.78%
			Baseline_S	90.92%			Baseline_S	90.92%
			Logits	91.45%			Logits	91.74%
			IRG	91.59%			IRG	91.88%
			DeiT	91.57%			DeiT	91.91%
			MINILM	91.44%			MINILM	91.75%
	ViT-L/16	Xceptionx2 (57.3G/90.27%)	**Ours**	91.15%	Swin-L	Xceptionx2 (57.3G/90.27%)	**Ours**	91.36%
	ViT-L/16	ResNet101x3	**Ours**	**91.84%**	Swin-L	ResNet101x3	**Ours**	**92.07%**
	ViT-L/16 (190.7 GFLOPs)	ViT-B/32 (13.8 GFLOPs)	Baseline_T	92.46%	Swin-L (103.9 GFLOPs)	ViT-B/32 (13.8 GFLOPs)	Baseline_T	93.78%
			Baseline_S	89.46%			Baseline_S	89.46%
			Logits	90.22%			Logits	90.59%
			IRG	90.39%			IRG	90.95%
			DeiT	90.40%			DeiT	90.99%
			MINILM	90.26%			MINILM	90.62%
	ViT-L/16 (190.7 GFLOPs)	ResNet152 (11.0 G/89.57%)	**Ours**	**90.66%**	Swin-L (103.9 GFLOPs)	ResNet152 (11.0 G/89.57%)	**Ours**	**91.20%**

* Baseline_T: Baseline model of the teacher network.
* Baseline_S: Baseline model of the student network.

involved during the KD procedure. It indicates that PCA and GL projectors significantly improve the quality of the CNN feature, though they are removed during the inference phase. We further evaluate the transferability after adding these two projectors in Fig. 3-(b). The cosine similarity is increased by a large

Table 2. Performance comparison on ImageNet.

Mode	Teacher	Student	Methods	Test accuracy Top1/Top5	Teacher	Student	Methods	Test accuracy Top1/Top5
CNN→CNN	ResNet152x2 (212.0 GFLOPs)	ResNet50x2 (15.9 GFLOPs)	Baseline_T	81.95/96.02	ResNet101x3 (205.0 GFLOPs)	ResNet50x2 (15.9 GFLOPs)	Baseline_T	82.03/96.06
			Baseline_S	78.16/93.91			Baseline_S	78.16/93.91
			Logits	79.06/94.67			Logits	79.19/94.71
			AT	79.01/94.66			AT	78.92/94.63
			FT	79.12/94.69			FT	79.11/94.69
			AB	78.93/94.62			AB	79.01/94.65
			OFD	79.63/94.81			OFD	79.55/94.79
			AFD	79.38/94.76			AFD	79.45/94.78
			IRG	79.85/94.87			IRG	79.75/94.84
			ReviewKD	80.12/94.99			ReviewKD	80.08/94.97
			LONDON	80.09/94.97			LONDON	80.15/95.01
	ViT-B/16	ResNet50x2	**Ours**	80.74/95.38	ViT-B/16	ResNet50x2	**Ours**	80.72/95.38
	ViT-L/16	ResNet50x2	**Ours**	**80.92/95.43**	ViT-L/16	ResNet50x2	**Ours**	**81.01/95.46**
Transformer →CNN	ViT-B/16 (55.4 GFLOPs)	ResNet50 (4.1 GFLOPs)	Baseline_T	82.17/96.11	ViT-L/16 (190.7 GFLOPs)	ResNet50 (4.1 GFLOPs)	Baseline_T	84.20/96.93
			Baseline_S	76.28/93.03			Baseline_S	76.28/93.03
			Logits	77.02/93.40			Logits	77.45/93.57
			RKD	77.27/93.50			RKD	77.82/93.75
			IRG	77.39/93.55			IRG	77.75/93.71
			Ours	**78.34/94.06**			**Ours**	**78.85/94.31**
	ViT-B/16 (55.4 GFLOPs)	ResNet50x2 (15.9 GFLOPs)	Baseline_T	82.17/96.11	ViT-L/16 (190.7 GFLOPs)	ResNet50x2 (15.9 GFLOPs)	Baseline_T	84.20/96.93
			Baseline_S	78.16/93.91			Baseline_S	78.16/93.91
			Logits	79.02/94.62			Logits	79.31/94.72
			RKD	79.68/94.82			RKD	79.78/94.85
			IRG	79.60/94.79			IRG	79.83/94.88
			Ours	**80.72/95.38**			**Ours**	**81.01/95.46**
	Swin-L (103.9 GFLOPs)	ResNet50 (4.1 GFLOPs)	Baseline_T	87.32/98.21	Swin-L (103.9 GFLOPs)	ResNet50x2 (15.9 GFLOPs)	Baseline_T	87.32/98.21
			Baseline_S	76.28/93.03			Baseline_S	78.16/93.91
			Logits	77.60/93.64			Logits	79.68/94.83
			RKD	77.85/93.76			RKD	79.92/94.92
			IRG	77.89/93.79			IRG	80.10/94.99
			Ours	**78.96/94.42**			**Ours**	**81.39/95.64**
Transformer → Transformer	ViT-L/16 (190.7 GFLOPs)	ViT-B/16 (55.4 GFLOPs)	Baseline_T	84.20/96.93	Swin-L (103.9 GFLOPs)	ViT-B/16 (55.4 GFLOPs)	Baseline_T	87.32/98.21
			Baseline_S	82.17/96.11			Baseline_S	82.17/96.11
			Logits	83.18/96.55			Logits	83.49/96.65
			IRG	83.27/96.59			IRG	83.60/96.69
			DeiT	83.38/96.63			DeiT	83.71/96.72
			MINILM	83.17/96.55			MINILM	83.55/96.65
	ViT-L/16	Xceptionx2 (80.37%/95.24%)	**Ours**	82.56/96.34	Swin-L	Xceptionx2 (80.37%/95.24%)	**Ours**	82.98/96.45
	ViT-L/16	ResNet152x2	**Ours**	**83.62/96.74**	Swin-L	ResNet101x3	**Ours**	**84.37/96.97**
	ViT-L/16 (190.7 GFLOPs)	ViT-B/32 (13.8 GFLOPs)	Baseline_T	84.20/96.93	Swin-L (103.9 GFLOPs)	ViT-B/32 (13.8 GFLOPs)	Baseline_T	87.32/98.21
			Baseline_S	78.29/94.08			Baseline_S	78.29/94.08
			Logits	79.40/94.76			Logits	79.30/94.73
			IRG	79.20/94.64			IRG	79.10/94.60
			DeiT	79.37/94.75			DeiT	79.27/94.71
			MINILM	79.29/94.70			MINILM	79.19/94.67
	ViT-L/16	ResNet152 (78.31%/94.05%)	**Ours**	**80.47/95.29**	Swin-L	ResNet152 (78.31%/94.05%)	**Ours**	**81.09/95.52**

margin and is even higher than that of the homologous-architecture. Therefore, it is possible to increase the knowledge transferability between Transformer and CNN by carefully designed KD methods.

(3) Effectiveness of the cross-view robust training. As reported in Fig. 3-(a), for regular evaluation without noise, student networks obtain 0.2%–0.4%

Table 3. Performance results of different teacher-student pairs on ImageNet. Note that the brackets behind the networks report the FLOPs of the networks.

Teacher	Student	Teacher accuracy		Student accuracy		Ours accuracy	
		Top1	Top5	Top1	Top5	Top1	Top5
ViT-B/16 (55.4G)	ResNet50 (4.1 GFLOPs)	82.17%	96.11%	76.28%	93.03%	78.34%	94.06%
ViT-L/16 (190.7G)		84.20%	96.93%	76.28%	93.03%	78.85%	94.31%
DeiT-B (55.4G)		83.12%	96.52%	76.28%	93.03%	78.53%	94.13%
Swin-B (15.4G)		86.38%	98.01%	76.28%	93.03%	78.87%	94.29%
Swin-L (103.9G)		87.32%	98.21%	76.28%	93.03%	78.96%	94.42%
ViT-B/16	ResNet18 (1.9 GFLOPs)	82.17%	96.11%	69.76%	89.08%	71.73%	90.41%
ViT-L/16		84.20%	96.93%	69.76%	89.08%	72.02%	90.52%
DeiT-B		83.12%	96.52%	69.76%	89.08%	71.85%	90.45%
Swin-B		86.38%	98.01%	69.76%	89.08%	72.01%	90.52%
Swin-L		87.32%	98.21%	69.76%	89.08%	72.09%	90.57%
ViT-B/16	MobileNetV2 (0.3 GFLOPs)	82.17%	96.11%	71.88%	90.29%	73.34%	91.01%
ViT-L/16		84.20%	96.93%	71.88%	90.29%	73.52%	91.18%
DeiT-B		83.12%	96.52%	71.88%	90.29%	73.40%	91.06%
Swin-B		86.38%	98.01%	71.88%	90.29%	73.56%	91.21%
Swin-L		87.32%	98.21%	71.88%	90.29%	73.66%	91.25%
ViT-B/16	EfficientNetB0 (1.6 GFLOPs)	82.17%	96.11%	77.69%	93.53%	79.23%	94.50%
ViT-L/16		84.20%	96.93%	77.69%	93.53%	79.34%	94.54%
DeiT-B		83.12%	96.52%	77.69%	93.53%	79.30%	94.52%
Swin-B		86.38%	98.01%	77.69%	93.53%	79.38%	94.55%
Swin-L		87.32%	98.21%	77.69%	93.53%	79.52%	94.60%

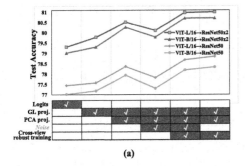
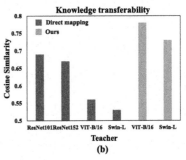

(a)   (b)

Fig. 3. (a) Performance of each component in the proposed method. (b) The cosine similarities between the features from different models. The student network is ResNet50. Among these blue bars, the features are mapped into the same dimension with the teacher features by a linear projector. All the results are obtained on ImageNet. (Color figure online)

top-1 accuracy gain on ImageNet with the cross-view robust training scheme. To further verify its effectiveness, we also report the results for noisy evaluation, where the validation dataset is augmented differently from the training augmentation. Under this protocol, the top-1 accuracy gain after adding the cross-view

Table 4. Evaluation on other visual tasks, including object detection, instance segmentation and face anti-spoofing.

Task (Dataset)	Teacher backbone	Student backbone	Method	AP	ΔAP
Object Detection (COCO)	–	ResNet50	Baseline	34.5	0
	ResNet152x2		Logits	35.0	0.5
	ViT-L/16		Logits	34.9	0.4
	ViT-L/16		**Ours**	**35.5**	**1.0**
	–	ResNet101	Baseline	37.1	0
	ResNet152x2		Logits	37.7	0.6
	ViT-L/16		Logits	37.4	0.3
	ViT-L/16		**Ours**	**38.1**	**1.0**
	–	ResNeXt101	Baseline	39.2	0
	ResNet152x2		Logits	39.8	0.6
	ViT-L/16		Logits	39.6	0.4
	ViT-L/16		**Ours**	**40.3**	**1.1**
Task (Dataset)	Teacher backbone	Student backbone	Method	AP	ΔAP
Instance Segmenta- tion (COCO)	–	ResNet50	Baseline	32.6	0
	ResNet152x2		Logits	33.3	0.7
	ViT-L/16		Logits	33.1	0.5
	ViT-L/16		**Ours**	**33.6**	**1.0**
	–	ResNet101	Baseline	33.9	0
	ResNet152x2		Logits	34.5	0.6
	ViT-L/16		Logits	34.2	0.3
	ViT-L/16		**Ours**	**34.8**	**0.9**
	–	ResNeXt101	Baseline	35.1	0
	ResNet152x2		Logits	35.5	0.4
	ViT-L/16		Logits	35.3	0.2
	ViT-L/16		**Ours**	**35.9**	**0.8**
Task (Dataset)	Teacher backbone	Student backbone	Method	EER	$-\Delta$EER
Face Anti- Spoofing (CelebA- Spoof)	–	ResNet18	Baseline	1.6	0
	ResNet152x2		Logits	1.6	0
	ViT-L/16		Logits	1.6	0
	ViT-L/16		**Ours**	**1.3**	**0.3**
	–	Inception-v3	Baseline	1.4	0
	ResNet152x2		Logits	1.3	0.1
	ViT-L/16		Logits	1.4	0
	ViT-L/16		**Ours**	**1.1**	**0.3**
	–	ResNeXt26	Baseline	1.3	0
	ResNet152x2		Logits	1.3	0
	ViT-L/16		Logits	1.3	0
	ViT-L/16		**Ours**	**0.9**	**0.4**

robust training scheme is enlarged to more than 1.0%. It demonstrates that the proposed robust training scheme enhances the noise robustness of the student network.

(4) Applications on other tasks. The proposed cross-architecture KD method also performs well on other tasks. As shown in Table 4, our method is evaluated on three visual tasks including object detection [36], instance segmentation [37] and face anti-spoofing [38].

For detection and segmentation, we follow the recent protocol of the COCO database [39] and report average precision (AP). Note that AP in segmentation is computed using mask intersection over union (IoU). The proposed method shows superiority compared with the conventional KD method in Table 4. For the conventional KD method Logits, the performance of the cross-architecture mode is even worse than the performance of the homologous-architecture mode. This further manifests that our method effectively solves the mismatching problem of cross-architecture KD. In addition, for face anti-spoofing, which is a binary classification task, we adopt ResNet18, Inception-v3 and ResNext26 as the backbones of the student. Equal Error Rate (EER) is reported as the evaluation metric. And the experiments are conducted on CelebA-Spoof [38], which is one of the largest datasets for face anti-Spoofing. It is worth mentioning that there exist few useful information of class correlation on the binary classification task. Hence, conventional KD method Logits has marginal enhancement on the student. In contrast, the proposed method also obtains a satisfactory performance from Table 4. It is interesting to notice that, though the proposed method is designed for the classification task, it has good generalization when it is directly applied to other tasks such as detection and segmentation.

5   Conclusions

In this paper, a novel cross-architecture knowledge distillation method is proposed. In particular, two projectors including a partially cross attention (PCA) projector and a group-wise Linear (GL) projector are presented The two projectors promote the knowledge transferability from teacher to student. In order to further improve the robustness and stability of the framework, a multi-view robust training scheme is proposed. Extensive experimental results show that our method outperforms 14 state-of-the-arts on both large-scale datasets and small-scale datasets.

Acknowledgements. This work was supported by the National Key Research and Development Program of China (Grant No. 2020AAA0106800), the National Natural Science Foundation of China (No. 62192785, Grant No.61902401, No. 61972071, No. U1936204, No. 62122086, No. 62036011, No. 62192782 and No. 61721004), the Beijing Natural Science Foundation No. M22005, the CAS Key Research Program of Frontier Sciences (Grant No. QYZDJ-SSW-JSC040). The work of Bing Li was also supported by the Youth Innovation Promotion Association, CAS.

References

1. Cheng, Y., Wang, D., Zhou, P., Zhang, T.: A survey of model compression and acceleration for deep neural networks. arXiv preprint arXiv:1710.09282 (2017)
2. Tan, C., Sun, F., Kong, T., Zhang, W., Yang, C., Liu, C.: A survey on deep transfer learning. In: Kůrková, V., Manolopoulos, Y., Hammer, B., Iliadis, L., Maglogiannis, I. (eds.) ICANN 2018. LNCS, vol. 11141, pp. 270–279. Springer, Cham (2018). https://doi.org/10.1007/978-3-030-01424-7_27

3. Dosovitskiy, A., et al.: An image is worth 16×16 words: transformers for image recognition at scale. arXiv preprint arXiv:2010.11929 (2020)
4. Carion, N., Massa, F., Synnaeve, G., Usunier, N., Kirillov, A., Zagoruyko, S.: End-to-end object detection with transformers. In: Vedaldi, A., Bischof, H., Brox, T., Frahm, J.-M. (eds.) ECCV 2020. LNCS, vol. 12346, pp. 213–229. Springer, Cham (2020). https://doi.org/10.1007/978-3-030-58452-8_13
5. Nvidia: Cuda. Nvidia (2007). https://developer.nvidia.com/cuda-zone
6. Nvidia: Tensorrt. Nvidia (2022). https://developer.nvidia.com/tensorrt
7. Tencent: Ncnn. Tencent (2017). https://github.com/Tencent/ncnn
8. Russakovsky, O., et al.: ImageNet large scale visual recognition challenge. Int. J. Comput. Vision **115**, 211–252 (2015)
9. Krizhevsky, A., Hinton, G.: Learning multiple layers of features from tiny images. Technical report, Citeseer (2009)
10. Hinton, G., Vinyals, O., Dean, J.: Distilling the knowledge in a neural network. arXiv preprint arXiv:1503.02531 (2015)
11. Ba, L.J., Caruana, R.: Do deep nets really need to be deep? arXiv preprint arXiv:1312.6184 (2013)
12. Zagoruyko, S., Komodakis, N.: Paying more attention to attention: improving the performance of convolutional neural networks via attention transfer. arXiv preprint arXiv:1612.03928 (2016)
13. Romero, A., Ballas, N., Kahou, S.E., Chassang, A., Gatta, C., Bengio, Y.: Fitnets: hints for thin deep nets. arXiv preprint arXiv:1412.6550 (2014)
14. Heo, B., Kim, J., Yun, S., Park, H., Kwak, N., Choi, J.Y.: A comprehensive overhaul of feature distillation. In: Proceedings of the IEEE/CVF International Conference on Computer Vision, pp. 1921–1930 (2019)
15. Huang, Z., Wang, N.: Like what you like: knowledge distill via neuron selectivity transfer. arXiv preprint arXiv:1707.01219 (2017)
16. Yim, J., Joo, D., Bae, J., Kim, J.: A gift from knowledge distillation: fast optimization, network minimization and transfer learning. In: Proceedings of the IEEE Conference on Computer Vision and Pattern Recognition, pp. 4133–4141 (2017)
17. Liu, Y., Cao, J., Li, B., Yuan, C., Hu, W., Li, Y., Duan, Y.: Knowledge distillation via instance relationship graph. In: Proceedings of the IEEE/CVF Conference on Computer Vision and Pattern Recognition, pp. 7096–7104 (2019)
18. Song, J., Chen, Y., Ye, J., Song, M.: Spot-adaptive knowledge distillation. IEEE Trans. Image Process. **31**, 3359–3370 (2022)
19. Song, J., et al.: Tree-like decision distillation. In: Proceedings of the IEEE/CVF Conference on Computer Vision and Pattern Recognition, pp. 13488–13497 (2021)
20. Touvron, H., Cord, M., Douze, M., Massa, F., Sablayrolles, A., Jégou, H.: Training data-efficient image transformers & distillation through attention. In: International Conference on Machine Learning, pp. 10347–10357. PMLR (2021)
21. Wang, W., Wei, F., Dong, L., Bao, H., Yang, N., Zhou, M.: Minilm: deep self-attention distillation for task-agnostic compression of pre-trained transformers. arXiv preprint arXiv:2002.10957 (2020)
22. Aguilar, G., Ling, Y., Zhang, Y., Yao, B., Fan, X., Guo, C.: Knowledge distillation from internal representations. In: Proceedings of the AAAI Conference on Artificial Intelligence, vol. 34, pp. 7350–7357 (2020)
23. Paszke, A., et al.: Automatic differentiation in pytorch. In: Advances in Neural Information Processing Systems Workshop (2017)
24. He, K., Zhang, X., Ren, S., Sun, J.: Deep residual learning for image recognition. In: Proceedings of the IEEE Conference on Computer Vision and Pattern Recognition, pp. 770–778 (2016)

25. Sandler, M., Howard, A., Zhu, M., Zhmoginov, A., Chen, L.C.: Mobilenetv 2: inverted residuals and linear bottlenecks. In: Proceedings of the IEEE Conference on Computer Vision and Pattern Recognition, pp. 4510–4520 (2018)

26. Chollet, F.: Xception: deep learning with depthwise separable convolutions. In: Proceedings of the IEEE Conference on Computer Vision and Pattern Recognition, pp. 1251–1258 (2017)

27. Tan, M., Le, Q.: Efficientnet: rethinking model scaling for convolutional neural networks. In: International Conference on Machine Learning, pp. 6105–6114. PMLR (2019)

28. Liu, Z., et al.: Swin transformer: hierarchical vision transformer using shifted windows. In: Proceedings of the IEEE/CVF International Conference on Computer Vision, pp. 10012–10022 (2021)

29. Park, W., Kim, D., Lu, Y., Cho, M.: Relational knowledge distillation. In: Proceedings of the IEEE/CVF Conference on Computer Vision and Pattern Recognition, pp. 3967–3976 (2019)

30. Tian, Y., Krishnan, D., Isola, P.: Contrastive representation distillation. arXiv preprint arXiv:1910.10699 (2019)

31. Chen, P., Liu, S., Zhao, H., Jia, J.: Distilling knowledge via knowledge review. In: In: Proceedings of the IEEE/CVF Conference on Computer Vision and Pattern Recognition, pp. 5008–5017 (2021)

32. Shang, Y., Duan, B., Zong, Z., Nie, L., Yan, Y.: Lipschitz continuity guided knowledge distillation. In: Proceedings of the IEEE/CVF International Conference on Computer Vision, pp. 10675–10684 (2021)

33. Wang, K., Gao, X., Zhao, Y., Li, X., Dou, D., Xu, C.Z.: Pay attention to features, transfer learn faster cnns. In: International conference on learning representations. (2019)

34. Heo, B., Lee, M., Yun, S., Choi, J.Y.: Knowledge transfer via distillation of activation boundaries formed by hidden neurons. In: Proceedings of the AAAI Conference on Artificial Intelligence, vol. 33, pp. 3779–3787 (2019)

35. Kim, J., Park, S., Kwak, N.: Paraphrasing complex network: network compression via factor transfer. Adv. Neural Inf. Process. Syst. **31**, 1–10 (2018)

36. Ren, S., He, K., Girshick, R., Sun, J.: Faster r-cnn: towards real-time object detection with region proposal networks. Adv. Neural Inf. Process. Syst. **28**, 91–99 (2015)

37. He, K., Gkioxari, G., Dollár, P., Girshick, R.: Mask r-cnn. In: Proceedings of the IEEE International Conference on Computer Vision, pp. 2961–2969 (2017)

38. Zhang, Y., et al.: CelebA-spoof: large-scale face anti-spoofing dataset with rich annotations. In: Vedaldi, A., Bischof, H., Brox, T., Frahm, J.-M. (eds.) ECCV 2020. LNCS, vol. 12357, pp. 70–85. Springer, Cham (2020). https://doi.org/10.1007/978-3-030-58610-2_5

39. Lin, T.-Y., et al.: Microsoft COCO: common objects in context. In: Fleet, D., Pajdla, T., Schiele, B., Tuytelaars, T. (eds.) ECCV 2014. LNCS, vol. 8693, pp. 740–755. Springer, Cham (2014). https://doi.org/10.1007/978-3-319-10602-1_48

Cross-Domain Local Characteristic Enhanced Deepfake Video Detection

Zihan Liu, Hanyi Wang, and Shilin Wang[✉]

School of Electronic Information and Electrical Engineering, Shanghai Jiao Tong University, Shanghai, China
{lzh123,why_820,wsl}@sjtu.edu.cn

Abstract. As ultra-realistic face forgery techniques emerge, deepfake detection has attracted increasing attention due to security concerns. Many detectors cannot achieve accurate results when detecting unseen manipulations despite excellent performance on known forgeries. In this paper, we are motivated by the observation that the discrepancies between real and fake videos are extremely subtle and localized, and inconsistencies or irregularities can exist in some critical facial regions across various information domains. To this end, we propose a novel pipeline, Cross-Domain Local Forensics (XDLF), for more general deepfake video detection. In the proposed pipeline, a specialized framework is presented to simultaneously exploit local forgery patterns from space, frequency, and time domains, thus learning cross-domain features to detect forgeries. Moreover, the framework leverages four high-level forgery-sensitive local regions of a human face to guide the model to enhance subtle artifacts and localize potential anomalies. Extensive experiments on several benchmark datasets demonstrate the impressive performance of our method, and we achieve superiority over several state-of-the-art methods on cross-dataset generalization. We also examined the factors that contribute to its performance through ablations, which suggests that exploiting cross-domain local characteristics is a noteworthy direction for developing more general deepfake detectors.

1 Introduction

Recent years have witnessed tremendous progress in face forgery techniques [13, 25,29,40], i.e., deepfake, due to the emergence of deep generative models. As such techniques can synthesize highly realistic fake videos without considerable human effort, they can easily be abused by malicious attackers to counterfeit imperceptible identities or behaviors, thereby causing severe political and social threats. To mitigate such threats, numerous automatic deepfake detection methods [6,9,20,31,34,46,53,54] have been proposed.

Most studies formulated deepfake detection as a binary classification problem with global supervision (i.e., real/fake) for training. They relied on convolutional neural networks (CNN) to extract discriminative features to detect forgeries.

© The Author(s), under exclusive license to Springer Nature Switzerland AG 2023
L. Wang et al. (Eds.): ACCV 2022, LNCS 13845, pp. 196–214, 2023.
https://doi.org/10.1007/978-3-031-26348-4_12

While these methods achieved satisfactory accuracy when the training and test sets have similar distributions, their performance significantly dropped when encountering novel manipulations. Therefore, many works [20,26,34,36] aimed at improving generalization to unseen forgeries with diverse approaches.

With the continuous refinement of face forgery methods, the discrepancies between real and fake videos are increasingly subtle and localized. Inconsistencies or irregularities can exist in some critical local regions across various information domains, e.g., space [1,43], frequency [6,31,34,42], and time [4,18,24,44] domains. However, these anomalies are so fine-grained that vanilla CNN often fails to capture them. Many detection algorithms exploited local characteristics to enhance generalization performance. However, these algorithms still had some limitations in representing local features. On the one hand, some algorithms [20] solely relied on a specific facial region to distinguish between real and fake videos while ignoring other facial regions, which restricted the detection performance. On the other hand, many algorithms [6,34,36] made insufficient use of local representation and cannot aggregate local information from various domains.

In this work, we are motivated by the above observation. It is reasonable to assume that incorporating more local regions and information domains can improve detection performance. We expect to design a specialized model to implement this idea and verify its performance through extensive experiments. We aim to guide the model to capture subtle artifacts around some high-level facial local regions that are sensitive to forgeries due to complicated natural motions. These regions are referred to as the forgery-sensitive local regions (FSLR) in this paper, which are abundant in high-level semantics that can enhance the model's generalization capability. We also consider the feasibility of simultaneously exploiting information from space, frequency, and time domains based on a 3D CNN backbone.

To this end, we propose Cross-Domain Local Forensics (XDLF), a novel pipeline specially designed for feature extraction across multiple domains and local artifacts enhancement. Four forgery-sensitive local regions (i.e., left eye, right eye, nose, and mouth) are extracted to guide the model to capture subtle artifacts around these regions. To simultaneously leverage information from space, frequency, and time domains, we design a two-stream 3D CNN based framework to learn a cross-domain dense representation for forgery detection.

To demonstrate the effectiveness of our framework, extensive experiments were conducted on several benchmark datasets, including FaceForensics++ [43], Celeb-DF [29], and DFDC [13]. Our results show the superiority of the proposed method over many state-of-the-art approaches on cross-dataset generalization.

Our main contributions are as follows:

- We leverage four forgery-sensitive local regions of a human face to guide the model to enhance subtle artifacts and localize potential anomalies around those regions. Using bounding boxes of those regions, we extract regional features as an attention to help the model focus more on those regions. We validated our design through ablations.

- We present a novel deepfake video detection pipeline that simultaneously exploits information from space, frequency, and time domains, thus learning a cross-domain dense representation for better generalization.
- We achieve impressive performance on extensive experiments, and our method outperforms several state-of-the-art methods on cross-dataset generalization.

2   Related Work

2.1   Deepfake Detection

Existing deepfake detection algorithms can fall into two categories, namely image-based methods and video-based methods, depending on whether temporal information is explicitly exploited across frames.

Image-Based Methods. Earlier image-based methods employed hand-crafted facial features to detect forgeries, e.g., steganalysis features [55], inconsistent head poses [50], and anomalous visual artifacts [37]. However, these methods underperformed on more realistic forgeries synthesized with more advanced face manipulation technologies recently. With the tremendous progress of deep learning, many works [1,43] utilized state-of-the-art convolutional neural networks (CNN), e.g., Xception [7], to extract features from facial images and perform binary classification. More recently, an increasing number of CNN-based methods have been proposed from various perspectives. They aimed at exploring the crucial discrepancies between real and fake images, continuously improving the detection performance. These methods included leveraging frequency spectrum [16,31,34,36,42], attention mechanism [10,53], extra identity information [3,9], self-supervised learning [26,28,54], etc.

Video-Based Methods. Unlike image-based methods, video-based methods distinguish real and fake videos based on a sequence of aligned frames. Most works managed to model the temporal consistency across frames, since current face manipulation techniques struggled to generate temporally coherent fake videos. These methods [19,30,36,44] utilized recurrent neural networks (RNN) or 3D CNN to extract spatio-temporal features of facial movements. They can focus on unnatural eye blinking [27], irregular mouth motion [20], inconsistent visual-auditory modalities [2,8,38,56]. In contrast, our method designs a two-stream 3D CNN based framework to mine forgery patterns from space, frequency, and time domains. We also leverage four facial forgery-sensitive local regions to enhance imperceptible artifacts for forgery defect localization.

2.2   Generalization to Unseen Forgeries

While current methods achieved excellent accuracy in the scenario where the training and test sets have similar distributions, they cannot generalize very well to unseen forgeries and tend to overfit to manipulation-specific artifacts. It is of paramount importance for deployed detectors to learn generalized representation regardless of forgery types. To this end, many works focused on improving generalization to unseen forgeries with diverse approaches. Several works [34,36] used a two-branch architecture to exploit information from the RGB domain and the frequency domain, exploiting generalized frequency patterns to expose the discrepancies. Our method has a similar idea but far different designs. Moreover, a series of self-supervised methods [26,28,54] demonstrated superior generalization. These methods relied on self-generated fake data targeted at specific patterns without the need for conventional forgery training data. The patterns can be face warping artifacts [28], blending boundary [26], source feature inconsistency [54]. Lip-Forensics [20] exhibited remarkable performance on cross-dataset generalization by pre-training a spatio-temporal network to perform lipreading and fine-tuning on a deepfake dataset. We followed its experimental settings due to similar goals.

3   Proposed Method

3.1   Overview

In this section, we first explain the motivation of our work, and then briefly introduce the pipeline of our proposed method.

Motivation. Recent studies [20,37,42,53] have shown that the discrepancies between real and fake videos contain implicitly in local subtle regions, where manipulation artifacts may exist across various information domains. Unfortunately, most deepfake datasets have no manipulation masks as local supervision. Without external location guidance of facial semantic regions that are sensitive to forgeries, it is often difficult for detectors to localize those subtle artifacts. We observe that current detection algorithms had two limitations in leveraging local information:

- Some algorithms [20,27] relied on a single facial region as the criterion to detect forgeries, while ignoring the effect of other critical local regions, which may restrict the performance. Our framework exploits four forgery-sensitive local regions (FSLR) of a human face, which are used to guide the model to enhance subtle artifacts and localize more potential anomalies based on our newly proposed FSLR-Guided Feature Enhancement.
- Many algorithms made insufficient use of local regions to detect anomalies, which can be embodied in multiple information domains, e.g., space, frequency, and time domains. To the best of our knowledge, few studies have been done to simultaneously capture local features across these three domains. We note that the Two-branch [36] method extracted spatial/frequency and temporal features at two stages with CNN and RNN, respectively, without cross reference among these features. To this end, we propose a two-stream framework, Cross-Domain Local Forensics, to simultaneously exploit local information from those three domains.

Pipeline. Motivated by the above observations, we propose a novel feature extraction framework **Cross-Domain Local Forensics (XDLF)** for more general deepfake video detection. Figure 1 illustrates the overall pipeline of XDLF.

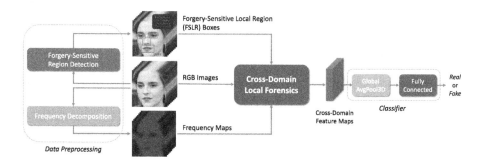

Fig. 1. Pipeline of our proposed framework XDLF. The end-to-end training consists of three stages: Data Preprocessing, Cross-Domain Local Forensics, and Classifier.

The pipeline takes as input a sequence of aligned RGB frames. First, the data preprocessing consists of two procedures. On the one hand, **Frequency Decomposition** takes as input RGB images to generate frequency maps where manipulation traces in the frequency domain are amplified, especially for those videos with high compression. On the other hand, **Forgery-Sensitive Region Detection** takes as input RGB images to extract bounding boxes of four **forgery-sensitive local regions** (**FSLR**) that are abundant in high-level defects. The four FSLRs are left eye, right eye, nose, and mouth. Then, sequences of RGB images, frequency maps, and FSLR boxes are input into **Cross-Domain Local Forensics** (**XDLF**) to learn a comprehensive and generalized cross-domain features. Finally, a classifier comprising a 3D global average pooling layer and a fully-connected layer is used to make predictions.

3.2 Data Preprocessing

Frequency Decomposition. Recent studies [31,52] observed that up-sampling is a necessary step of most existing face manipulation methods, and

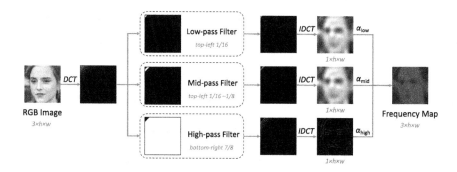

Fig. 2. Pipeline of Frequency Decomposition. This module generates frequency maps where manipulation traces in the frequency domain are amplified adaptively.

cumulative up-sampling can leave apparent anomalies in the frequency domain, which provides clues for detecting manipulated videos. Inspired by F^3-Net [42], we design Frequency Decomposition to obtain multi-band frequency maps adaptively. Figure 2 shows the pipeline of this module.

For each RGB image X in a frame sequence, we first calculate the frequency response with Discrete Cosine Transform (DCT) $\mathcal{D}$. Then, filters of low, middle, and high frequency bands f_i, $i \in \{\text{low}, \text{mid}, \text{high}\}$ are used to obtain three frequency components. We follow the settings in [42] to construct filters. Next, Inversed Discrete Cosine Transform (IDCT) $\mathcal{D}^{-1}$ is applied to three frequency components to obtain the corresponding spatial components Y_i, $i \in \{\text{low}, \text{mid}, \text{high}\}$. Finally, the three spatial components are concatenated to attain the frequency map Y. Before concatenation, each component is multiplied by a learnable weight $\alpha_i \in (0, 1)$, $i \in \{\text{low}, \text{mid}, \text{high}\}$ to enable the model to adaptively concentrate on the interested frequency band for a flexible representation of frequency features. The above can be summarized as Eq. 1, 2, where $\odot$ is the element-wise product.

$$Y_i = \mathcal{D}^{-1}\{\mathcal{D}(X) \odot f_i\}, \; i \in \{\text{low}, \text{mid}, \text{high}\} \tag{1}$$

$$Y = \text{Concat}(\alpha_{\text{low}} Y_{\text{low}}, \; \alpha_{\text{mid}} Y_{\text{mid}}, \; \alpha_{\text{high}} Y_{\text{high}}) \tag{2}$$

Forgery-Sensitive Region Detection. Current face manipulation techniques struggled to generate temporally coherent fake faces, especially in high-level semantic regions that have continual motions and thereby sensitive to forgeries. We hope to guide the model to pay more attention to these regions. Therefore, we extract bounding boxes of four forgery-sensitive local regions (FSLR): left eye, right eye, nose, and mouth. These four manually selected regions are further leveraged by **FSLR-Guided Feature Enhancement (FGFE)** as an external guidance. For each RGB image, we first compute 68 facial landmarks based on a face detector. Then, the landmarks are used to crop bounding boxes of those four regions based on preset box sizes. Each box can be expressed as a quadruple (h_1, h_2, w_1, w_2) where (h_1, w_1) is the top-left vertex and (h_2, w_2) is the bottom-right vertex. The four boxes are stacked to generate the 4×4 FSLR box matrix.

3.3   Cross-Domain Local Forensics

We propose a novel two-stream collaborative learning framework for cross-domain feature extraction, Cross-Domain Local Forensics (XDLF), which is based on a spatio-temporal convolutional backbone. As is illustrated in Fig. 3, the framework consists of two symmetric 3D CNN backbones: 3D-CNN(A) extracts spatio-temporal features of RGB images, and 3D-CNN(B) extracts frequency-temporal features of frequency maps. The features of the two modalities are cross-referenced and merged at low, middle, and high levels of the backbone, with **Cross Attention** and **Feature Fusion**, respectively. Moreover, we apply **FSLR-Guided Feature Enhancement** to the low-level features of both streams, thus enhancing the local subtle artifacts of shallow features under the

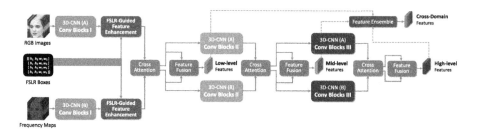

Fig. 3. Framework of Cross-Domain Local Forensics. We adopt a two-stream architecture for cross-domain feature extraction based on two symmetric spatio-temporal convolutional backbones, e.g., 3D ResNet-50 [21,22].

guidance of forgery-sensitive regions. The ultimate cross-domain features are obtained with **Feature Ensemble** to integrate features of three different levels of abstraction.

FSLR-Guided Feature Enhancement. Many studies [31,53] showed that local textural artifacts represent the high frequency component of shallow features, which is essential for the face forgery detection task. These artifacts are especially salient nearby critical facial regions that are sensitive to forgeries. As aforementioned, we exploit four forgery-sensitive local regions to enhance subtle artifacts and guide the model to localize more possible anomalies in these regions. The module structure is shown in Fig. 4.

The module takes as input low-level RGB (or frequency) features $X \in \mathbb{R}^{c \times d \times h \times w}$ (of c channels, depth d, height h, width w) and FSLR boxes $r \in \mathbb{Z}^{d \times 4 \times 4}$ and outputs the enhanced features of the same shape. First, the region coordinates are scaled down (i.e., region projection) according to the size difference between the RGB image (or frequency map) and low-level features. Then, FSLR features $R \in \mathbb{R}^{4 \times c \times d \times H \times W}$ are obtained with region pooling, which refers to ROI pooling [17] in object detection. Specifically, we crop four sub-features with region coordinates and generate four FSLR features of fixed size

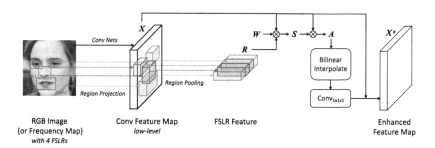

Fig. 4. Structure of FSLR-Guided Feature Enhancement. This module is designed to guide the model to enhance subtle artifacts of shallow features and localize more anomalous regions.

$(H \times W)$ using adaptive max-pooling (Eq. 3). FSLR features condense the irregular semantics of local textural patterns in these four regions, which serve as an attention for global features. Next, transformed features $\boldsymbol{X'} \in \mathbb{R}^{c \times hw}$ are calculated by temporally averaging the RGB (or frequency) features $\boldsymbol{X}$ and flattening spatial dimensions. And features $\boldsymbol{R'} \in \mathbb{R}^{4c \times dHW}$ are also obtained by flattening the FSLR features $\boldsymbol{R}$. Later, the similarity matrix $\boldsymbol{S} \in \mathbb{R}^{hw \times dHW}$ between $\boldsymbol{X'}$ and $\boldsymbol{R'}$ (Eq. 4) is computed, where $\boldsymbol{W} \in \mathbb{R}^{c \times 4c}$ is a learnable weight matrix. Each value in $\boldsymbol{S}$ represents the similarity between each row in $\boldsymbol{X'}^{T}$ and each column in $\boldsymbol{R'}$. By the similarity matrix, we model the internal relevance between those local regions for cross-region forgery mining. And then the attention matrix $\boldsymbol{A} \in \mathbb{R}^{c \times dHW}$ is calculated to enhance the original features (Eq. 5). Moreover, the upsampled $\boldsymbol{A'} \in \mathbb{R}^{c \times d \times h \times w}$ is obtained by reshaping, bilinear interpolation, and $1 \times 1 \times 1$ convolution (Eq. 6, 7). Finally, the enhanced features $\boldsymbol{X}^{e} \in \mathbb{R}^{c \times d \times h \times w}$ are attained by element-wise product and residual addition (Eq. 8). We apply this module to the low-level features of both streams, which enables the model to pay more attention to the regularity and consistency of local semantic regions.

$$\boldsymbol{R} = \mathrm{AdaMaxPool}(\boldsymbol{X}, \mathrm{Proj}(\boldsymbol{r})) \quad (3) \qquad \boldsymbol{A'} = \mathrm{BilinearInterpolate}(\boldsymbol{A}) \quad (6)$$

$$\boldsymbol{S} = \boldsymbol{X'}^{T}\boldsymbol{W}\boldsymbol{R'} \quad (4) \qquad \boldsymbol{A'} = \mathrm{ReLU}(\mathrm{BN}(\mathrm{Conv}_1(\boldsymbol{A'}))) \quad (7)$$

$$\boldsymbol{A} = \boldsymbol{X'}\boldsymbol{S} \quad (5) \qquad \boldsymbol{X}^{e} = \boldsymbol{X} + \boldsymbol{X} \odot \boldsymbol{A'} \quad (8)$$

Cross Attention. In this module, RGB and frequency features are cross-referenced at low, middle, and high levels of the backbone, which enables the model to learn a more comprehensive cross-domain representation. The module takes as input RGB features $\boldsymbol{X}$ and frequency features $\boldsymbol{X}_f$. First, the two features are concatenated on the channel axis and then applied $1 \times 1 \times 1$ convolution (Eq. 9, 10). Next, $3 \times 3 \times 3$ convolution with output channel 2 and sigmoid activation is used to obtain two attention maps (Eq. 11). Finally, the original features are enhanced with attention maps by element-wise product (Eq. 12).

$$\boldsymbol{U} = \mathrm{Concat}(\boldsymbol{X}, \boldsymbol{X}_f) \quad (9) \qquad \boldsymbol{A}, \boldsymbol{A}_f = \mathrm{Sigmoid}(\mathrm{Conv}_3(\boldsymbol{U'})) \quad (11)$$

$$\boldsymbol{U'} = \mathrm{ReLU}(\mathrm{BN}(\mathrm{Conv}_1(\boldsymbol{U}))) \quad (10) \qquad \boldsymbol{X}^{c} = \boldsymbol{X} \odot \boldsymbol{A}, \; \boldsymbol{X}_f^{c} = \boldsymbol{X}_f \odot \boldsymbol{A}_f \quad (12)$$

Feature Fusion. In this module, RGB and frequency features are fused in a complementary way based on Squeeze-and-Excitation (SE) [23]. SE block improves the quality of cross-domain features by explicitly modeling the interdependence between the channels of RGB and frequency features. The module structure is shown in Fig. 5.

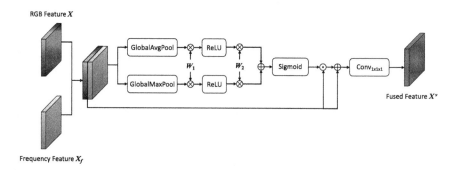

Fig. 5. Structure of Feature Fusion. This module is designed to model the interdependence between RGB and frequency features for improved cross-domain fusion.

This module also takes as input RGB features $X \in \mathbb{R}^{C \times D \times H \times W}$ and frequency features $X_f \in \mathbb{R}^{C \times D \times H \times W}$. The two features are first concatenated to obtain $U \in \mathbb{R}^{2C \times D \times H \times W}$ (Eq. 13). Then, the spatial information is squeezed into a value by global pooling to get channel descriptor $V \in \mathbb{R}^{2C}$ (Eq. 14, 15). Next, we enable channel descriptor V to capture the interdependency between channels and obtain channel attention $A_c \in \mathbb{R}^{2C}$ (Eq. 16, 17, 18), where $W_1 \in \mathbb{R}^{2C \times \frac{2C}{r}}$ and $W_2 \in \mathbb{R}^{\frac{2C}{r} \times 2C}$ are learnable weight matrices, r is the reduction ratio. Finally, the fused features X^v are computed as Eq. 19.

$$U = \text{Concat}(X, X_f) \quad (13) \qquad V'_{\text{avg}} = W_2 \text{ReLU}(W_1 V_{\text{avg}}) \quad (16)$$

$$V_{\text{avg}} = \text{GlobalAvgPool}(U) \quad (14) \qquad V'_{\text{max}} = W_2 \text{ReLU}(W_1 V_{\text{max}}) \quad (17)$$

$$V_{\text{max}} = \text{GlobalMaxPool}(U) \quad (15) \qquad A_c = \text{Sigmoid}(V'_{\text{avg}} + V'_{\text{max}}) \quad (18)$$

$$X^v = \text{ReLU}(\text{BN}(\text{Conv}_{1 \times 1 \times 1}(U + U \odot A_c))) \quad (19)$$

Feature Ensemble. This module aggregates low, middle, and high-level features through adaptive average pooling and concatenation (Eq. 20, 21, 22, 23).

$$\widetilde{X}^{\text{low}} = \lambda_{\text{low}} \text{AdaAvgPool}(X^{\text{low}}, (d_{\text{high}}, h_{\text{high}}, w_{\text{high}})) \quad (20)$$

$$\widetilde{X}^{\text{mid}} = \lambda_{\text{mid}} \text{AdaAvgPool}(X^{\text{mid}}, (d_{\text{high}}, h_{\text{high}}, w_{\text{high}})) \quad (21)$$

$$\widetilde{X}^{\text{high}} = \lambda_{\text{high}} X^{\text{high}} \quad (22)$$

$$\widetilde{X} = \text{Concat}(\widetilde{X}^{\text{low}}, \widetilde{X}^{\text{mid}}, \widetilde{X}^{\text{high}}) \quad (23)$$

where $X^i \in \mathbb{R}^{c_i \times d_i \times h_i \times w_i}$, $i \in \{\text{low}, \text{mid}, \text{high}\}$ are fused features of three abstraction levels, and $\lambda_i \in (0, 1)$, $i \in \{\text{low}, \text{mid}, \text{high}\}$ are three learnable parameters for adaptive feature combination.

4 Experiment and Discussion

4.1 Experiment Setup

Datasets. We used **FaceForensics++** (FF++) [43] for training and validation, and evaluated the cross-dataset generalization on **Celeb-DF** (CDF) [29] and **DeepFake Detection Challenge** (DFDC) [13]. (1) FF++ is the most commonly used benchmark dataset containing 1,000 real videos and 4,000 fake videos. Each real video is manipulated by four face forgery techniques, i.e., DeepFakes (DF) [11], FaceSwap (FS) [15], Face2Face (F2F) [48], and Neural-Textures (NT) [47]. We adopted the slightly-compressed (HQ/c23) and heavily-compressed (LQ/c40) versions for our experiments. (2) CDF is a challenging dataset that includes 590 real videos and 5,639 fake videos synthesized by an improved algorithm. (3) DFDC is a large-scale dataset with extreme filming conditions and various perturbations, which is also very challenging for current deepfake detectors. We used the preview version [14] that includes 1,131 real videos and 4,113 fake videos for our evaluation.

Evaluation Metrics. Following most existing works [20,26,36], Accuracy (ACC) and Area Under the Receiver Operating Characteristic Curve (AUC) were used as the metrics to evaluate our method. As in [20], we reported video-level metrics for fair comparison with image-based methods. Specifically, all frame/clip predictions were averaged across the video and hence all models predicted based on an equal number of frames.

Implementation Details. For each video, we sampled non-overlapping frame clips with a length of 16, and oversampled the minority class (e.g., real in FF++) to tackle label imbalance. We used the state-of-the-art face detector RetinaFace [12] to crop facial images with a size of 224×224 and FSLR box matrices with a size of 4×4. The preset FSLR size is 40×40 for the mouth and 30×30 for the other three. For data augmentations, we applied several traditional image augmentations such as random horizontal flipping. Moreover, as in [41], we conducted Mixup [51] augmentation on aligned real-fake pairs to reduce overfitting. For XDLF, we adopted 3D ResNet-50 [21,22] as the backbone which is pre-trained on large-scale action recognition datasets to accelerate the model convergence. For FSLR-Guided feature enhancement, we set FSLR feature size $H = W = 7$. For feature fusion, we set reduction ratio $r = 16$. For training, we used a batch size of 4 and AdamW [33] optimizer with initial learning rate 1×10^{-4} and weight decay 1×10^{-4}. The learning rate decayed with a cosine annealing [32] strategy with $T_{\max} = 32$.

4.2 In-dataset Evaluation

We evaluated our method in the in-dataset scenario where the training and test sets have identical distributions. Following [20], we compared our method with current state-of-the-art approaches in FF++ under different quality settings (HQ/LQ). As shown in Table 1, we achieve great improvements over most current methods, especially under the challenging low-quality (LQ) setting where frequency statistics are partly destroyed. However, our method still maintains good performance when exploiting frequency spectrum, which we attribute to our two-stream architecture that learns to be biased towards RGB features. Note that we gain comparable results with LipForensics [20], which leverages dynamic lip features from pre-trained lipreading models. Unlike LipForensics, our method does not need any external pre-training data and can be more efficiently trained.

Table 1. In-dataset performance comparisons. We report video-level ACC/AUC (%) when trained and tested on FF++ slightly-compressed (HQ) and heavily-compressed (LQ) videos. The results of other methods are quoted from [20]. The best results are in **bold**, and the second-best results are in underlined.

Method	FF++(HQ)		FF++(LQ)	
	ACC (%)	AUC (%)	ACC (%)	AUC (%)
Xception [43]	97.0	99.3	89.0	92.0
CNN-aug [49]	96.9	99.1	81.9	86.9
Patch-based [5]	92.6	97.2	79.1	78.3
Two-branch [36]	–	99.1	–	91.1
Face X-ray [26]	78.4	97.8	34.2	77.3
CNN-GRU [44]	97.0	99.3	90.1	92.2
LipForensics [20]	**98.8**	**99.7**	94.2	**98.1**
XDLF (ours)	98.1	**99.7**	**94.5**	96.7

Moreover, we show the t-SNE [35] visualization of features extracted from classifiers of LipForensics and our method on FF++ high-quality (HQ) test set in Fig. 6. We observe that although both methods can well distinguish real and fake data, they learn different feature distributions. For LipForensics, the separation distances between real and fake data are smaller than our method, which can easily lead to classification ambiguity in those in-between videos, especially for some real and NeuralTextures-based fake samples. On the other hand, our method learns a more mixed and gathered feature representation of FF++ fake data without obviously separating different forgery types. It proves that our method can learn a generalized feature to detect novel forgeries.

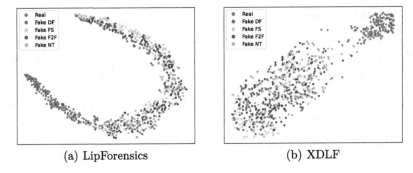

(a) LipForensics   (b) XDLF

Fig. 6. The t-SNE feature visualization of the baseline LipForensics [20] (a) and our proposed XDLF (b) on FF++(HQ) test set. Each dot represents the feature of a video clip. Red dots are real clips, while the rest are fake ones with different forgery types.

4.3   Cross-Dataset Evaluation

In real-world scenarios, a deployed detector is expected to identify videos crafted by unseen manipulations with unknown source videos. Therefore, we conducted cross-dataset evaluation as in [20] to verify the generalization capability of our method. Specifically, we trained the models on FF++(HQ) and tested them on CDF and DFDC. As shown in Table 2, our method outperforms all listed methods on both unseen datasets, surpassing the recent state-of-the-art LipForensics [20] by 0.2% and 0.3% in terms of AUC on CDF and DFDC, respectively.

Table 2. Cross-dataset performance comparisons. We report video-level AUC (%) on CDF and DFDC when trained on FF++(HQ). The results of other methods are quoted from [20]. The best results are in **bold**.

Method	CDF AUC (%)	DFDC AUC (%)
Xception [43]	73.7	70.9
CNN-aug [49]	75.6	72.1
Patch-based [5]	69.6	65.6
Face X-ray [26]	79.5	65.5
CNN-GRU [44]	69.8	68.9
Multi-task [39]	75.7	68.1
DSP-FWA [28]	69.5	67.3
Two-branch [36]	76.7	–
LipForensics [20]	82.4	73.5
XDLF (ours)	**82.6**	**73.8**

4.4   Ablation Study

Evaluations on Core Modules in XDLF. To understand the components responsible for our method's performance, we ablated three core modules in XDLF and examined its in-dataset and cross-dataset generalization performance. The modules are FSLR-Guided Feature Enhancement (**FGFE**), Cross Attention, and Feature Fusion. For the first two, we removed them directly as their inputs and outputs have the same shapes. For Feature Fusion, we replaced it with a simple channel-axis concatenation of RGB and frequency features. We trained all the models on FF++(HQ) and tested them on FF++(HQ), CDF, and DFDC.

Table 3. Evaluations on core modules in XDLF. We report video-level ACC/AUC (%) on FF++(HQ), CDF, and DFDC when trained on FF++(HQ). The highlighted row is our original setting. We ablated core modules in our feature extraction framework to verify their effects. The best results are in **bold**.

Method	FF++(HQ)		CDF		DFDC	
	ACC	AUC	ACC	AUC	ACC	AUC
XDLF (ours)	**98.1**	**99.7**	**74.2**	**82.6**	**66.2**	**73.8**
w/o FGFE	97.9	99.3	71.7	79.2	65.3	69.8
w/o cross sttention	97.8	99.4	72.2	79.9	65.9	71.0
w/o feature fusion	98.0	99.4	73.8	81.5	66.1	72.3

The results are shown in Table 3. We have the following observations: (1) Training our model without FGFE leads to a performance drop on all datasets. In cross-dataset evaluation, the model decreases by 3.4% and 4.0% in terms of AUC on CDF and DFDC, respectively. This suggests that the model learns more generalized features by enhancing subtle artifacts in those forgery-sensitive local regions. (2) Both Cross Attention and Feature Fusion play an essential role in performance improvements. Although they have the same goal to complementarily exploit forgery patterns from different domains, they work differently and enhance the model's performance mutually.

To further understand the effect of FGFE, we show the Grad-CAM [45] visualization of the model without/with FGFE in Fig. 7. It visually explains that FGFE serves as external guidance to help the model focus on four forgery-sensitive local regions. As can be seen, these regions are abundant in motions that contain more subtle artifacts. The model can localize more potential anomalies to detect forgeries with the help of FGFE, which is consistent with our motivation.

Evaluations on Different Information Domains. We altered our feature extraction framework XDLF to prove the necessity to mine forgery clues from three different information domains, i.e., space, frequency, and time domains. Specifically, we trained three variants with each dropping one of the three domains: (1) **Freq-Freq-3D**: The inputs of both streams are the same frequency maps, and the network structure is unchanged. (2) **RGB-RGB-3D**: The inputs of both streams are RGB images, and the network structure is unchanged. (3) **RGB-Freq-2D**: The inputs are still RGB images and frequency maps, but temporal dimension is merged into batch dimension. We replaced the 3D ResNet-50 backbone with 2D ResNet-50 backbone, and replaced all 3D convolutional layers and 3D batch normalization layers with 2D counterparts.

The results are shown in Table 4. We have the following observations: (1) By using 3D spatio-temporal CNN instead of 2D CNN, the in-dataset and cross-dataset generalization performance are all considerably improved. It indicates that our method can leverage 3D CNN to effectively capture temporal defects for forgery detection. (2) Compared to RGB-RGB-3D, Freq-Freq-3D achieves better cross-dataset generalization. It suggests that frequency statistics are more generalizable features than color textures. However, RGB-RGB-3D gains better in-dataset results which may benefit from manipulation-specific artifacts. (3) We note that forgery clues from these three domains work in a complementary way and contribute to the overall performance.

Table 4. Evaluations on different information domains. We report video-level ACC/AUC (%) on FF++(HQ), CDF, and DFDC when trained on FF++(HQ). The highlighted row is our original setting. We developed three variants of feature extraction framework with each dropping one of the three domains, i.e., space, frequency, and time domains. The best results are in **bold**.

Method	Information domains			FF++(HQ)		CDF		DFDC	
	Space	Frequency	Time	ACC	AUC	ACC	AUC	ACC	AUC
RGB-Freq-3D (ours)	✓	✓	✓	**98.1**	**99.7**	**74.2**	**82.6**	**66.2**	**73.8**
Freq-Freq-3D	×	✓	✓	97.6	99.0	73.9	**82.6**	64.5	72.5
RGB-RGB-3D	✓	×	✓	**98.1**	99.5	72.7	81.2	63.6	71.9
RGB-Freq-2D	✓	✓	×	96.4	98.5	68.4	76.1	61.3	69.1

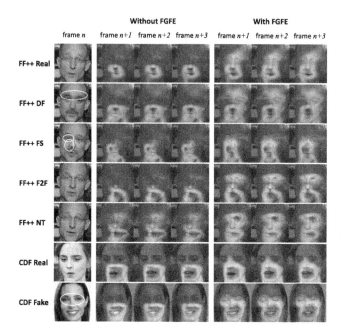

Fig. 7. The Grad-CAM visualization of localized defect regions of the model without/with FSLR-Guided Feature Enhancement (FGFE). We show several examples including four forgery types in FF++ and another dataset CDF. For each example, red circles mark visually noticeable artifacts, and consecutive frames in a video clip are provided to understand temporal defects. The warmer region suggests a higher probability of cross-domain defects the model believes. (Color figure online)

5   Conclusion

In this paper, we propose Cross-Domain Local Forensics (XDLF), a specially designed pipeline for general deepfake video detection. Our approach aims at exploiting forgery patterns from space, frequency, and time domains simultaneously to learn a generalized cross-domain features. We also leverage four forgery-sensitive local regions to guide the model to capture subtle forgery defects. Experiments show that our method achieves impressive performance, especially strong cross-dataset generalization. We hope our work encourages future research on cross-domain forensics for more general deepfake detection.

Acknowledgements. This work was supported by the National Natural Science Foundation of China (62271307, 61771310) and Key Lab of Information Network Security of Ministry of Public Security (The Third Research Institute of Ministry of Public Security).

References

1. Afchar, D., Nozick, V., Yamagishi, J., Echizen, I.: Mesonet: a compact facial video forgery detection network. In: International Workshop on Information Forensics and Security (WIFS), pp. 1–7. IEEE (2018)
2. Agarwal, S., Farid, H., Fried, O., Agrawala, M.: Detecting deep-fake videos from phoneme-viseme mismatches. In: Conference on Computer Vision and Pattern Recognition Workshops (CVPRW), pp. 2814–2822. IEEE/CVF (2020)
3. Agarwal, S., Farid, H., Gu, Y., He, M., Nagano, K., Li, H.: Protecting world leaders against deep fakes. In: Conference on Computer Vision and Pattern Recognition Workshops (CVPRW), pp. 38–45. IEEE/CVF (2019)
4. Amerini, I., Galteri, L., Caldelli, R., Bimbo, A.D.: Deepfake video detection through optical flow based cnn. In: International Conference on Computer Vision Workshops (ICCVW), pp. 1205–1207. IEEE (2019)
5. Chai, L., Bau, D., Lim, S.-N., Isola, P.: What makes fake images detectable? understanding properties that generalize. In: Vedaldi, A., Bischof, H., Brox, T., Frahm, J.-M. (eds.) ECCV 2020. LNCS, vol. 12371, pp. 103–120. Springer, Cham (2020). https://doi.org/10.1007/978-3-030-58574-7_7
6. Chen, S., Yao, T., Chen, Y., Ding, S., Li, J., Ji, R.: Local relation learning for face forgery detection. In: AAAI Conference on Artificial Intelligence (AAAI), pp. 1081–1088. AAAI Press (2021)
7. Chollet, F.: Xception: deep learning with depthwise separable convolutions. In: Conference on Computer Vision and Pattern Recognition (CVPR), pp. 1800–1807. IEEE/CVF (2017)
8. Chugh, K., Gupta, P., Dhall, A., Subramanian, R.: Not made for each other-audiovisual dissonance-based deepfake detection and localization. In: ACM International Conference on Multimedia (ACM MM), pp. 439–447. ACM (2020)
9. Cozzolino, D., Rössler, A., Thies, J., Nießner, M., Verdoliva, L.: Id-reveal: identity-aware deepfake video detection. In: International Conference on Computer Vision (ICCV), pp. 15088–15097. IEEE/CVF (2021)
10. Dang, H., Liu, F., Stehouwer, J., Liu, X., Jain, A.K.: On the detection of digital face manipulation. In: Conference on Computer Vision and Pattern Recognition (CVPR), pp. 5780–5789. IEEE/CVF (2020)
11. DeepFakes: https://github.com/deepfakes/faceswap
12. Deng, J., Guo, J., Ververas, E., Kotsia, I., Zafeiriou, S.: Retinaface: single-shot multi-level face localisation in the wild. In: Conference on Computer Vision and Pattern Recognition (CVPR), pp. 5202–5211. IEEE/CVF (2020)
13. Dolhansky, B., et al.: The deepfake detection challenge dataset. CoRR abs/2006.07397 (2020)
14. Dolhansky, B., Howes, R., Pflaum, B., Baram, N., Canton-Ferrer, C.: The deepfake detection challenge (dfdc) preview dataset. CoRR abs/1910.08854 (2019)
15. FaceSwap. https://github.com/MarekKowalski/FaceSwap
16. Frank, J., Eisenhofer, T., Schönherr, L., Fischer, A., Kolossa, D., Holz, T.: Leveraging frequency analysis for deep fake image recognition. In: International Conference on Machine Learning (ICML), Proceedings of Machine Learning Research, vol. 119, pp. 3247–3258. PMLR (2020)
17. Girshick, R.B.: Fast r-cnn. In: International Conference on Computer Vision (ICCV), pp. 1440–1448. IEEE (2015)

18. Gu, Z., Chen, Y., Yao, T., Ding, S., Li, J., Huang, F., Ma, L.: Spatiotemporal inconsistency learning for deepfake video detection. In: ACM International Conference on Multimedia (ACM MM), pp. 3473–3481. ACM (2021)

19. Guera, D., Delp, E.J.: Deepfake video detection using recurrent neural networks. In: International Conference on Advanced Video and Signal Based Surveillance (AVSS), pp. 1–6. IEEE (2018)

20. Haliassos, A., Vougioukas, K., Petridis, S., Pantic, M.: Lips don't lie: a generalisable and robust approach to face forgery detection. In: Conference on Computer Vision and Pattern Recognition (CVPR), pp. 5039–5049. IEEE/CVF (2021)

21. Hara, K., Kataoka, H., Satoh, Y.: Can spatiotemporal 3d cnns retrace the history of 2d cnns and imagenet? In: Conference on Computer Vision and Pattern Recognition (CVPR), pp. 6546–6555. IEEE/CVF (2018)

22. He, K., Zhang, X., Ren, S., Sun, J.: Deep residual learning for image recognition. In: Conference on Computer Vision and Pattern Recognition (CVPR), pp. 770–778. IEEE/CVF (2016)

23. Hu, J., Shen, L., Sun, G.: Squeeze-and-excitation networks. In: Conference on Computer Vision and Pattern Recognition (CVPR), pp. 7132–7141. IEEE/CVF (2018)

24. Hu, Z., Xie, H., Wang, Y., Li, J., Wang, Z., Zhang, Y.: Dynamic inconsistency-aware deepfake video detection. In: Proceedings of the International Joint Conference on Artificial Intelligence (IJCAI), pp. 736–742. ijcai.org (2021)

25. Li, L., Bao, J., Yang, H., Chen, D., Wen, F.: Advancing high fidelity identity swapping for forgery detection. In: Conference on Computer Vision and Pattern Recognition (CVPR), pp. 5073–5082. IEEE/CVF (2020)

26. Li, L., et al.: Face x-ray for more general face forgery detection. In: Conference on Computer Vision and Pattern Recognition (CVPR), pp. 5000–5009. IEEE/CVF (2020)

27. Li, Y., Chang, M.C., Lyu, S.: In ictu oculi: exposing AI created fake videos by detecting eye blinking. In: International Workshop on Information Forensics and Security (WIFS), pp. 1–7. IEEE (2018)

28. Li, Y., Lyu, S.: Exposing deepfake videos by detecting face warping artifacts. In: Conference on Computer Vision and Pattern Recognition Workshops (CVPRW), pp. 46–52. IEEE/CVF (2019)

29. Li, Y., Yang, X., Sun, P., Qi, H., Lyu, S.: Celeb-df: a large-scale challenging dataset for deepfake forensics. In: Conference on Computer Vision and Pattern Recognition (CVPR), pp. 3204–3213. IEEE/CVF (2020)

30. de Lima, O., Franklin, S., Basu, S., Karwoski, B., George, A.: Deepfake detection using spatiotemporal convolutional networks. CoRR abs/2006.14749 (2020)

31. Liu, H., et al.: Spatial-phase shallow learning: rethinking face forgery detection in frequency domain. In: Conference on Computer Vision and Pattern Recognition (CVPR), pp. 772–781. IEEE/CVF (2021)

32. Loshchilov, I., Hutter, F.: SGDR: stochastic gradient descent with warm restarts. In: International Conference on Learning Representations (ICLR). OpenReview.net (2017)

33. Loshchilov, I., Hutter, F.: Decoupled weight decay regularization. In: International Conference on Learning Representations (ICLR). OpenReview.net (2019)

34. Luo, Y., Zhang, Y., Yan, J., Liu, W.: Generalizing face forgery detection with high-frequency features. In: Conference on Computer Vision and Pattern Recognition (CVPR), pp. 16317–16326. IEEE/CVF (2021)

35. Van der Maaten, L., Hinton, G.: Visualizing data using t-sne. J. Mach. Learn. Res. 9(11), 2579–2605 (2008)

36. Masi, I., Killekar, A., Mascarenhas, R.M., Gurudatt, S.P., AbdAlmageed, W.: Two-branch recurrent network for isolating deepfakes in videos. In: Vedaldi, A., Bischof, H., Brox, T., Frahm, J.-M. (eds.) ECCV 2020. LNCS, vol. 12352, pp. 667–684. Springer, Cham (2020). https://doi.org/10.1007/978-3-030-58571-6_39

37. Matern, F., Riess, C., Stamminger, M.: Exploiting visual artifacts to expose deepfakes and face manipulations. In: Winter Applications of Computer Vision Workshops (WACVW), pp. 83–92. IEEE (2019)

38. Mittal, T., Bhattacharya, U., Chandra, R., Bera, A., Manocha, D.: Emotions don't lie: an audio-visual deepfake detection method using affective cues. In: ACM International Conference on Multimedia (ACM MM), pp. 2823–2832. ACM (2020)

39. Nguyen, H.H., Fang, F., Yamagishi, J., Echizen, I.: Multi-task learning for detecting and segmenting manipulated facial images and videos. In: International Conference on Biometrics Theory, Applications and Systems (BTAS), pp. 1–8. IEEE (2019)

40. Nirkin, Y., Keller, Y., Hassner, T.: FSGAN: subject agnostic face swapping and reenactment. In: International Conference on Computer Vision (ICCV), pp. 7183–7192. IEEE/CVF (2019)

41. NTech-Lab. https://github.com/NTech-Lab/deepfake-detection-challenge

42. Qian, Y., Yin, G., Sheng, L., Chen, Z., Shao, J.: Thinking in frequency: face forgery detection by mining frequency-aware clues. In: Vedaldi, A., Bischof, H., Brox, T., Frahm, J.-M. (eds.) ECCV 2020. LNCS, vol. 12357, pp. 86–103. Springer, Cham (2020). https://doi.org/10.1007/978-3-030-58610-2_6

43. Rössler, A., Cozzolino, D., Verdoliva, L., Riess, C., Thies, J., Nießner, M.: Faceforensics++: learning to detect manipulated facial images. In: International Conference on Computer Vision (ICCV), pp. 1–11. IEEE/CVF (2019)

44. Sabir, E., Cheng, J., Jaiswal, A., AbdAlmageed, W., Masi, I., Natarajan, P.: Recurrent convolutional strategies for face manipulation detection in videos. In: Conference on Computer Vision and Pattern Recognition Workshops (CVPRW), pp. 80–87. IEEE/CVF (2019)

45. Selvaraju, R.R., Cogswell, M., Das, A., Vedantam, R., Parikh, D., Batra, D.: Gradcam: visual explanations from deep networks via gradient-based localization. In: International Conference on Computer Vision (ICCV), pp. 618–626. IEEE (2017)

46. Sun, K., Yao, T., Chen, S., Ding, S., L, J., Ji, R.: Dual contrastive learning for general face forgery detection. CoRR abs/2112.13522 (2021)

47. Thies, J., Zollhöfer, M., Nießner, M.: Deferred neural rendering: image synthesis using neural textures. ACM Trans. Graph. 38(4), 66:1–66:12 (2019)

48. Thies, J., Zollhöfer, M., Stamminger, M., Theobalt, C., Nießner, M.: Face2face: real-time face capture and reenactment of rgb videos. In: Conference on Computer Vision and Pattern Recognition (CVPR), pp. 2387–2395. IEEE/CVF (2016)

49. Wang, S.Y., Wang, O., Zhang, R., Owens, A., Efros, A.A.: CNN-generated images are surprisingly easy to spot... for now. In: Conference on Computer Vision and Pattern Recognition (CVPR), pp. 8692–8701. IEEE/CVF (2020)

50. Yang, X., Li, Y., Lyu, S.: Exposing deep fakes using inconsistent head poses. In: International Conference on Acoustics, Speech and Signal Processing (ICASSP), pp. 8261–8265. IEEE (2019)

51. Zhang, H., Cissé, M., Dauphin, Y.N., Lopez-Paz, D.: Mixup: beyond empirical risk minimization. In: International Conference on Learning Representations (ICLR). OpenReview.net (2018)

52. Zhang, X., Karaman, S., Chang, S.F.: Detecting and simulating artifacts in GAN fake images. In: International Workshop on Information Forensics and Security (WIFS), pp. 1–6. IEEE (2019)

53. Zhao, H., Zhou, W., Chen, D., Wei, T., Zhang, W., Yu, N.: Multi-attentional deepfake detection. In: Conference on Computer Vision and Pattern Recognition (CVPR), pp. 2185–2194. IEEE/CVF (2021)
54. Zhao, T., Xu, X., Xu, M., Ding, H., Xiong, Y., Xia, W.: Learning self-consistency for deepfake detection. In: International Conference on Computer Vision (ICCV), pp. 15003–15013. IEEE/CVF (2021)
55. Zhou, P., Han, X., Morariu, V.I., Davis, L.S.: Two-stream neural networks for tampered face detection. In: Conference on Computer Vision and Pattern Recognition Workshops (CVPRW), pp. 1831–1839. IEEE/CVF (2017)
56. Zhou, Y., Lim, S.N.: Joint audio-visual deepfake detection. In: International Conference on Computer Vision (ICCV), pp. 14780–14789. IEEE/CVF (2021)

Three-Stage Bidirectional Interaction Network for Efficient RGB-D Salient Object Detection

Yang Wang and Yanqing Zhang[✉]

South China University of Technology, Guangzhou, China
202021045142@mail.scut.edu.cn, zyqcs@scut.edu.cn

Abstract. The addition of depth maps improves the performance of salient object detection (SOD). However, most existing RGB-D SOD methods are inefficient. We observe that existing models take into account the respective advantages of the two modalities but do not fully explore the roles of cross-modality features of various levels. To this end, we remodel the relationship between RGB features and depth features from a new perspective of the feature encoding stage and propose a three-stage bidirectional interaction network (TBINet). Specifically, to obtain robust feature representations, we propose three interaction strategies: bidirectional attention guidance (BAG), bidirectional feature supplement (BFS), and shared network, and use them for the three stages of feature encoder, respectively. In addition, we propose a cross-modality feature aggregation (CFA) module for feature aggregation and refinement. Our model is lightweight (3.7 M parameters) and fast (329 ms on CPU). Experiments on six benchmark datasets show that TBINet outperforms other SOTA methods. Our model achieves the best performance and efficiency trade-off.

1 Introduction

Salient object detection (SOD) aims to locate the object(s) most concerned by human eyes from a given scene. It is the pre-task of many computer vision tasks, such as semantic segmentation [1,2], tracking [3,4], image/video compression [5, 6], and image retrieval [7]. Although significant progress has been made in SOD in recent years, it is still challenged to accurately locate objects in complex scenes, such as complex textures, cluttered backgrounds, and low contrast.

With the wide use of depth sensors in smartphones and other devices, RGB-D SOD has attracted the attention of researchers [8–13]. The depth map has illumination invariance and internal consistency, which can provide complementary spatial information for RGB images and improve saliency detection performance.

Supplementary Information The online version contains supplementary material available at https://doi.org/10.1007/978-3-031-26348-4_13.

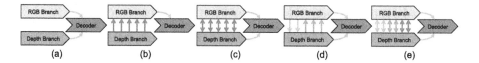

Fig. 1. Comparison of network interaction strategies between existing models and our model. (a) No interaction. (b) Unidirectional interaction. (c) Bidirectional interaction. (d) Cross-modality discrepant interaction. (e) Proposed three-stage bidirectional interaction.

As we all know, RGB and depth are two different modalities. An effective interaction strategy for a two-stream feature encoder can obtain more robust saliency-related features and thereby help the subsequent decoder generate more accurate saliency maps. The existing interaction strategies can be roughly divided into four categories: (i) No interaction mode [9,13,14] shown in Fig. 1(a), which uses two independent branches to learn features of the two modalities separately, and then feds the features into subsequent feature fusion modules or the decoder. (ii) Unidirectional interaction mode [8,10,15] shown in Fig. 1(b), which integrates depth cues into RGB branch, and then feds the integrated features into decoder. (iii) Bidirectional interaction mode [16] shown in Fig. 1(c), which performs the same bidirectional operation on the hierarchical features of the two modalities. (iv) Cross-modality discrepant interaction mode [17] shown in Fig. 1(d), which gives full play to the respective advantages of the two modalities. Most of these interaction strategies are designed based on the modality perspective, while we try to explore the relationship between RGB features and depth features from the perspective of feature encoding stage. The basic observation of hierarchical cross-modality features is that high-level features contain rich global context information, which is conducive to locating salient regions, low-level features contain detailed information that can contribute significantly to refining the boundaries of salient regions [8].

To this end, we propose a novel three-stage bidirectional interaction network (TBINet) for RGB-D SOD. Specifically, the interaction of feature encoding process is divided into three stages (as shown in Fig. 1(e)): the interaction of low-level features in first stage, the interaction of middle-level features in second stage, and the interaction of high-level features in thrid stage. Low-level cross-modality features have specific boundary details, such as RGB image will be difficult to distinguish between salient objects and background in the case of complex texture and low contrast, and depth map will contain misleading information when salient objects and non-salient objects have the same spatial depth. Therefore, in first stage, we propose a bidirectional attention guidance (BAG) module, which can guide the two branches to focus on the important regions of each other while maintaining the modality-specific low-level features. The quality of depth maps tends to be uneven. Decreasing the influence of misleading information from low-quality depth maps is a key and hot issue in RGB-D SOD. We noticed that middle-level features contain approximate location information and rough boundary information. Therefore, in second stage, we propose a bidirectional feature supplement (BFS) module, which extracts cross-modality fusion features and transfers them to two branches separately. The BFS module effectively suppresses the low-quality features of deep branches and helps purify saliency-oriented feature representations. High-level features have the lowest resolution and can locate salient objects. After the abstraction of cross-modality information by the previous layers, the high-level cross-modality features have similar global context information, and the features of the two modalities have strong commonality. Inspired by JL-DCF [18], in thrid stage, we use shared network based on shared CNN layers, which can extract high-level cross-modality features with fewer parameters. The three-stage bidirectional interaction strategy effectively utilizes the characteristics of the three encoding stages. It helps the encoder finally generate multi-level cross-modality feature representations with specificity, purity, and commonality.

In addition, to integrate multi-level cross-modality features, we implement a three-stage refinement decoder. The three stages correspond one-to-one with the three stages of the encoder. Each decoder stage contains a cross-modality feature aggregation (CFA) module. The CFA module performs alternate feature fusion and refinement through two steps to effectively fuse and refine cross-modality features. The decoder generates final accurate saliency maps through feature fusion and refinement of the three CFA modules. Inspired by the channel split and channel shuffle operations in ShuffleNet-v2 [19], we redesign an efficient receptive field block (ERFB) module for the CFA module to expand the receptive field and extract multi-scale features.

Our network adopts the lightweight MobileNet-v3 [20] as the backbone network, and all modules adopt a lightweight design. Our model is lightweight (15.1 MB model size and 3.7 M parameters) and fast (329 ms inference time on CPU and 93 FPS inference speed on GPU). Our main contributions are as follows:

1. We propose a novel three-stage bidirectional interaction network (TBINet) for RGB-D SOD. TBINet adopts different interaction strategies in different stages of the feature encoding process so that the cross-modality features of various levels can give full play to their advantages.
2. We propose a three-stage refinement decoder and a cross-modality feature aggregation (CFA) module. Each decoding stage utilizes a CFA module for feature aggregation. The decoder continuously refines the saliency-oriented feature representation through three-stage feature aggregation and finally generates accurate saliency maps.
3. Our model is based on a lightweight design with fewer parameters and faster speed than cumbersome models. Experiments on six public datasets show that our model outperforms 15 state-of-the-art models and achieves a good balance between efficiency and performance.

2   Related Work

2.1   RGB-D SOD

In some complex scenes, salient objects in RGB images are indistinguishable from the background. Adding depth information may help overcome this challenge. Traditional RGB-D SOD models extract handcrafted features from RGB images and depth maps and fuse them for saliency detection [21–24]. However, due to the limited expressive power of handcrafted features, the performance of traditional methods is not satisfactory.

With the rapid development of deep convolutional neural networks (CNNs), researchers have begun to focus on CNNs-based RGB-D SOD work and push the performance to new peaks [8,12,13,18,25–28]. Two key challenges facing current RGB-D SOD research are dealing with low-quality depth maps and effectively aggregating cross-modality multi-level features. For low-quality depth maps, for example, Fan *et al.* [29] proposed a depth depurator unit to filter low-quality depth maps. Jin *et al.* [9] proposed a complementary depth network, which estimates a depth map from the RGB image, and fuses the estimated depth map with the original depth map. Ji *et al.* [26] proposed a depth calibration and fusion framework capable of calibrating the depth image and correcting the latent bias in the original depth maps. Zhang *et al.* [15] proposed a depth feature manipulation network that can control depth features and avoid feeding misleading depth features. For cross-modality multi-level feature aggregation, for example, Fu *et al.* [18] developed a densely cooperative fusion strategy that uses dense connections to facilitate the fusion of depth and RGB features at different scales. Li *et al.* [30] proposed an adaptive feature selection module that emphasizes the importance of channel features in self-modality and cross-modality while fusing multi-modality spatial features. For more inspiring related works, refer to the recent survey [31,32].

2.2   Efficient RGB-D SOD

Efficiency is also important for models besides performance. Recently, researchers have started to propose some efficient models for RGB-D SOD with lighter size and faster speed. Zhao *et al.* [33] proposed an early fusion single-stream network to make the network lighter. Chen *et al.* [34] constructed a lightweight deep stream to make the network more compact and efficient. More and more computer vision applications are adapting to mobile devices. To this end, many lightweight networks for image classification have been proposed, such as MobileNets [20,35,36] and ShuffleNets [19,37]. Unlike classic cumbersome networks, such as VGG [38] and ResNet [39], lightweight networks can be well adapted to mobile devices due to their extremely high efficiency. Some RGB-D SOD models attempt to use a lightweight network as the backbone network. Wu *et al.* [40] proposed a network named MobileSal, which uses MobileNet-v2 [36] as the backbone network and fuses RGB features with depth features only on the coarsest layers. Zhang *et al.* [15] proposed an efficient model DFMNet based on

MobileNet-v2 [36] and a tailored depth backbone. The current efficient RGB-D SOD models still lacks performance compared with cumbersome models. In this paper, we propose an efficient model that uses MobileNet-v3 [20] as the model backbone network and achieves a good balance between accuracy and efficiency.

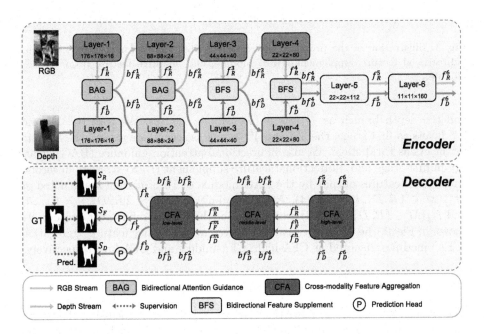

Fig. 2. Overview of our network architecture. Three-stage bidirectional interaction is shown in the upper part of this illustration, and the three stages use the BAG module, BFS module and shared network as the interaction strategy respectively. Three-stage refinement decoder is shown in the lower part of this illustration, and it consists of three CFA modules.

3 Proposed Method

3.1 Overview

Figure 2 shows the framework of the proposed three-stage bidirectional interaction network for RGB-D SOD. Our network consists of encoder and decoder. The encoder generates saliency-related features through a three-stage bidirectional interaction strategy, and the decoder aggregates these features and generates the final saliency map. MobileNet-v3 large [20] is used to build the feature encoder. we divide the encoder into six layers, the output stride is 2 for each layer except 1 for the 5th layer, this means that the feature resolution does not change in the 5th layer, so the 5th layer has the same output resolution as the fourth layer. We denote the features output by the i-th layer of the RGB branch

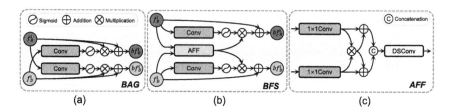

Fig. 3. Illustration of the proposed bidirectional attention guidance (BAG) module, bidirectional feature supplement (BFS) module, and adaptive feature fusion (AFF) module.

and the depth branch as $f_M^i(M \in \{R, D\}, i = 1, ..., 6)$. We take the 1st and 2nd layers as first stage, the 3rd and 4th layers as second stage, and the 5th and 6th layers as thrid stage. We use bidirectional attention guidance (BAG) strategy in first stage and bidirectional feature supplement (BFS) strategy in second stage. The features output by the BAG module or BFS module are denoted as $bf_M^i(M \in \{R, D\}, i = 1, ..., 4)$. After encoding, $bf_M^i(M \in \{R, D\}, i = 1, ..., 4)$ and $f_M^i(M \in \{R, D\}, i = 5, 6)$ are fed into the three-stage refinement decoder. As shown in Fig. 2, the decoder consists of three cross-modality feature aggregation (CFA) modules, denoted as CFA-high, CFA-middle and CFA-low, respectively.

3.2   Three-Stage Bidirectional Interaction (TBI)

The encoder part of Fig. 2 shows the TBI strategy. For the processing of features output by two encoders at different levels, previous works such as SPNet [13], CMWNet [41] and DCFNet [26] tend to fuse the cross-modality features and feed them directly to the decoder. Unlike these works, we process the cross-modality features at each layer and then fed them to the next layer, which enables the use of cross-modality information to improve the network in the feature encoding stage.

First Stage: Bidirectional Attention Guidance (BAG). The detailed structure of the BAG module is shown in Fig. 3(a). The BAG module is based on the spatial attention mechanism. Given the features $f_R^i(i = 1, 2)$ or $f_D^i(i = 1, 2)$, we use a 3×3 convolutional (output channel number is 1) with *Sigmoid* activation function to generate the spatial attention map. To guide one modality to focus on important areas of the other modality. The attention map from one modality is used to enhance another modality. Then, a residual connection is used to combine the enhanced features with their original features. Take the case that depth information enhances RGB information as an example. The process can be described as:

$$bf_R^i = f_R^i + \sigma(Conv_{3\times3}(f_D^i)) \otimes f_R^i, \tag{1}$$

where $Conv_{3\times3}(\cdot)$ denotes a 3×3 convolution, $\sigma(\cdot)$ is the *Sigmoid* activation function, and $\otimes$ represents element-wise multiplication. The features $bf^i_R(i = 1, 2)$ and $bf^i_D(i = 1, 2)$ will be fed into the decoder and the next layer of encoder.

The low-level features of the two modalities have complementary boundary details, so modality specificity should be maintained. The BAG module ensures that the modality specificity is not destroyed while mining more modality correlations.

Second Stage: Bidirectional Feature Supplement (BFS). The detailed structure of the BFS module is shown in Fig. 3(b), an adaptive feature fusion (AFF) module is included in a BFS module. As shown in Fig. 3(c), AFF module is simple and effective, it can adaptively fuse cross-modality features. Given the features $f^i_R(i = 3, 4)$ and $f^i_D(i = 3, 4)$, they are first fed into a 1×1 convolution layer with BatchNorm and ReLU activation to adjust their channel number and obtain their smooth feature representations (*i.e.*, $\hat{f}^i_R = Conv_{1\times1}(f^i_R)$ and $\hat{f}^i_D = Conv_{1\times1}(f^i_D)$, where $Conv_{1\times1}(\cdot)$ denotes a 1×1 convolution with BatchNorm and ReLU activation). Then, we use element-wise multiplication to emphasize the shared feature representation, which can be described as $\hat{f}^i_F = \hat{f}^i_R \otimes \hat{f}^i_D$, where $\otimes$ represents element-wise multiplication. We add $\hat{f}^i_F$ with $\hat{f}^i_R$ and $\hat{f}^i_D$ respectively to get the enhanced features. Finally, the enhanced features are concatenated and fed into a depth-wise separable convolution layer to obtain the final fused features, the process can be described as:

$$f^i_F = DSConv_{3\times3}([\hat{f}^i_F + \hat{f}^i_R, \hat{f}^i_F + \hat{f}^i_D]), \tag{2}$$

where $DSConv_{3\times3}(\cdot)$ denotes a 3×3 depth-wise separable convolution with BatchNorm and ReLU activation, and $[\cdot]$ donates feature concatenation. After these operations, the AFF module adaptively fuses cross-modality features. After obtaining the fused features f^i_F, we use the spatial attention mechanism to enhance f^i_F and then combine the enhanced features with the original features of the two modalities. The entire process can be described as:

$$\begin{cases} bf^i_R = f^i_R + \sigma(Conv_{3\times3}(f^i_R)) \otimes f^i_F, \\ bf^i_D = f^i_D + \sigma(Conv_{3\times3}(f^i_D)) \otimes f^i_F, \end{cases} \tag{3}$$

the features $bf^i_R(i = 3, 4)$ and $bf^i_D(i = 3, 4)$ will be fed into the decoder and the next layer of encoder.

In a word, we first adaptively fuse cross-modality features to obtain pure fused features, then use spatial attention mechanism to enhance the fused features, and finally transfer them to the two modality branches as supplements. The BFS module can effectively suppress low-quality depth information and transfer high-quality cross-modality shared information between branches.

Third Stage: Shared Network. High-level features have rich global contextual information, which is beneficial for localizing salient objects. The saliency-related high-level features of the two modalities have strong commonality.

Inspired by JL-DCF [18], we adopt shared network in the third stage as shown in Fig. 2. Unlike JL-DCF, which uses the strategy of the shared network on the entire feature encoding network, we only share parameters in the most appropriate third stage. Following [18], we concatenate RGB features and depth features in the 4th dimension. The features generated by the 5th and 6th layers of the encoder will be split in the 4th dimension for the decoder. By employing shared network, the two branches share the same parameters in the final stage of the encoder, so the parameters are greatly reduced. Shared network can exploit cross-modality commonality and complementarity, which match the properties of high-level cross-modality features.

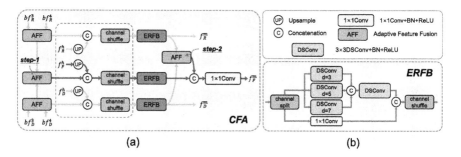

(a)   (b)

Fig. 4. Illustration of the proposed cross-modality feature aggregation (CFA) module and efficient receptive field block (ERFB) module. The CFA-high module does not contain the part circled by the red dotted line in (a). The two AFF modules pointed to by the red arrows are the two-step cross-modality fusion in the CFA module, AFF modules labelled "step-1" and "step-2" denote "fusion before refinement" and "fusion after refinement", respectively. (Color figure online)

3.3   Three-Stage Refinement Decoder

Figure 2 shows the three-stage refinement decoder, whose three stages correspond one-to-one with the three stages of the encoder. The CFA-high module aggregates high-level cross-modality features, CFA-middle and CFA-low are the same.

Cross-Modality Feature Aggregation (CFA). The detailed structure of the CFA module is shown in Fig. 4(a), which consists of AFF modules and efficient receptive field block (ERFB) modules. There are three branches in the CFA module, namely RGB branch, depth branch, and fusion branch. Take CFA-middle as an example, we first fuse the features bf_M^3 with $bf_M^4 (M \in R, D)$, and obtain the features:

$$\begin{cases} bf_R^m = AFF(bf_R^3, bf_R^4), \\ bf_D^m = AFF(bf_D^3, bf_D^4), \end{cases} \tag{4}$$

where $AFF(\cdot)$ is the AFF module. Note that when fusing features of different layers of the same modality, we added squeeze-and-excitation (SE) modules [42] after the 1×1 convolution layer of the AFF module.

After fusing the features from two levels, we conduct the first-step cross-modality feature fusion ("fusion before refinement"):

$$bf_F^m = AFF(bf_R^m, bf_D^m), \tag{5}$$

we concatenate $bf_M^m (M \in R, D, F)$ with the features $f_M^h (M \in R, D, F)$ generated by the CFA-high of the previous stage. Then, we do channel shuffle [37] operations and finally feed them into the ERFB modules of the three branches. The outputs of the three ERFB modules in the CFA-middle are defined by:

$$\begin{cases} f_R^m = ERFB([bf_R^m, up(f_R^h)]), \\ f_D^m = ERFB([bf_D^m, up(f_D^h)]), \\ f_F^{m1} = ERFB([bf_F^m, up(f_F^h)]), \end{cases} \tag{6}$$

where $ERFB(\cdot)$ is the ERFB module, and $up(\cdot)$ represents upsample operation. f_R^m and f_D^m are the final refined features of the RGB branch and the depth branch, respectively.

ERFB is a variant of receptive field block (RFB) module [43] as shown in Fig. 4(b). It has the basic function of the RFB module and has a lower computational cost. Inspired by ShuffleNet-v2 [19], we use channel split and channel shuffle operations on the ERFB module. Features are divided into two parts in the channel dimension. One half is fed into 1×1 convolution as residuals, and the other half is fed into a dilated convolution block with multiple branches to extract multi-scale features. Finally, we concatenate these two parts and use the channel shuffle operation to ensure information communication between different groups of channels.

We conduct the second-step cross-modality feature fusion ("fusion after refinement"), fusing features f_R^m and f_D^m, and obtain the fused features:

$$f_F^{m2} = AFF(f_R^m, f_D^m), \tag{7}$$

then, the features f_F^{m1} and f_F^{m2} are concatenated, and we use a 1×1 convolution to generate the final fused features of the fusion branch:

$$f_F^m = Conv_{1\times1}([f_F^{m1}, f_F^{m2}]), \tag{8}$$

where $Conv_{1\times1}(\cdot)$ denotes a 1×1 convolution with BatchNorm and ReLU activation. Finally, take the fusion branch as an example, the model generates the final saliency maps:

$$S_F = up(\sigma(Conv_{1\times1}(f_F^l))), \tag{9}$$

where $Conv_{1\times1}(\cdot)$ denotes a 1×1 convolution layer.

Two cross-modality feature fusion steps are included in the CFA module, these two steps serve different purposes. The first step fuses the original cross-modality features and uses them to refine the coarse features generated by the previous stage's fusion branch ("fusion before refinement"), the second step fuses the refined features of the current stage's RGB branch and depth branch ("fusion after refinement"). Finally, cross-modality features are effectively fused and refined.

3.4 Loss Function

We employ the pixel position aware loss $\mathcal{L}_{ppa}$ [44] to implement supervision on the three prediction maps S_F, S_R and S_D.

$$\mathcal{L}_{total} = \mathcal{L}_{ppa}(S_F, G) + \mathcal{L}_{ppa}(S_R, G) + \mathcal{L}_{ppa}(S_D, G), \qquad (10)$$

where $\mathcal{L}_{total}$ is the overall loss and G is the ground truth.

Table 1. Quantitative results on seven widely-used datasets. Red, **blue** and **bold** indicate the best, second best, and third best performances respectively. ↑/↓ for a metric denotes that a larger/smaller value is better

Model		Non-efficient model											Efficient model				
		D3Net	UCNet	S2MA	BBSNet	JL-DCF	HAINet	CDINet	DCFNet	DSA2F	RD3D	SPNet	DANet	PGAR	MobileSal	DFMNet	TBINet
		2020	2020	2020	2020	2021	2021	2021	2021	2021	2021	2021	2020	2020	2021	2021	Ours
Params (M) ↓		45.2	31.3	86.6	49.8	70.7	59.8	54.4	108.5	36.5	46.9	175.3	26.7	16.2	**6.5**	2.2	**3.7**
CPU (ms) ↓		862	471	2097	633	3136	3019	1585	1069	2288	701	1217	1139	709	**115**	87	**329**
GPU (FPS) ↑		52	**99**	25	54	6	9	37	36	21	28	31	46	69	**227**	299	93
STERE	F_β^{max} ↑	.891	.899	.882	.903	**.904**	**.906**	.901	.901	.900	**.906**	**.906**	.881	.898	.892	.892	.910
	E_ξ^{max} ↑	.938	.944	.932	.942	**.947**	.944	.942	.945	.942	**.947**	**.949**	.930	.939	.939	.941	.952
	S_α ↑	.899	.903	.890	**.908**	.903	.907	.905	.902	.898	.911	.907	.892	.907	.901	.898	.911
	MAE ↓	.046	.039	.051	.041	.040	.040	.040	**.039**	**.039**	.037	.037	.048	.041	.042	.045	.034
NJU2K	F_β^{max} ↑	.900	.895	.889	.920	.904	.915	**.921**	.915	.907	.914	.928	.893	.907	.895	.910	.928
	E_ξ^{max} ↑	.938	.936	.929	.949	.943	.944	**.951**	.939	.947	.947	**.957**	.893	.940	.937	.947	.958
	S_α ↑	.900	.897	.894	**.921**	.902	.912	**.919**	.912	.904	.916	.925	.897	.909	.896	.906	**.924**
	MAE ↓	.047	.043	.054	**.035**	.041	.038	**.035**	.036	.039	.036	.028	.047	.042	.045	.042	**.029**
NLPR	F_β^{max} ↑	.897	.903	.902	**.918**	.918	.915	.916	.912	.906	**.919**	**.919**	.893	.916	.907	.908	.932
	E_ξ^{max} ↑	.953	.956	.953	.961	**.963**	.960	.960	**.963**	.952	**.965**	.962	.949	.961	.957	.957	.970
	S_α ↑	.912	.920	.916	.930	.925	.924	**.927**	.924	.919	**.930**	**.927**	.909	.930	.919	.923	.937
	MAE ↓	.030	.025	.030	.023	**.022**	.024	.024	.024	**.022**	**.022**	.021	.031	.024	.025	.026	.018
SIP	F_β^{max} ↑	.861	.879	.877	.883	.889	**.892**	.884	.884	.875	.889	**.904**	.884	.876	.872	.887	**.905**
	E_ξ^{max} ↑	.909	.919	.918	.922	.924	.922	.915	.922	.912	.924	**.933**	.920	.915	.911	**.926**	.937
	S_α ↑	.860	.875	.872	.879	.880	.880	.875	.876	.862	**.885**	**.894**	.878	.876	.866	.883	**.894**
	MAE ↓	.063	.051	.057	.055	.049	.053	.054	.052	.057	**.048**	**.043**	.054	.055	.058	.051	**.041**
SSD	F_β^{max} ↑	.834	.854	.847	**.859**	.832	.838	.846	.851	**.863**	.772	**.863**	.849	.798	.835	.851	**.872**
	E_ξ^{max} ↑	.911	.907	.909	**.919**	.902	.903	.899	.913	.859	**.920**	.905	.872	.905	.918	.921	.921
	S_α ↑	.857	.865	.868	**.882**	.860	.857	.853	.864	**.877**	.803	.871	.868	.832	.861	.865	**.874**
	MAE ↓	.059	.049	.053	**.044**	.053	.052	.056	.050	**.048**	.082	**.044**	.050	.068	.053	.051	**.042**
DES	F_β^{max} ↑	.884	.930	**.934**	.927	.923	**.936**	**.934**	.884	.915	.929	.946	.894	.902	.899	.922	**.934**
	E_ξ^{max} ↑	.945	**.976**	.973	.966	.968	.973	.970	.951	.954	.972	**.983**	.957	.945	.945	.972	**.974**
	S_α ↑	.897	.934	**.940**	.934	.931	.935	**.937**	.905	.917	.935	**.945**	.904	.913	.909	.931	.935
	MAE ↓	.031	**.018**	.021	.021	.020	**.018**	.020	.024	.023	**.019**	.014	.029	.026	.025	.021	**.018**

4 Experiments

4.1 Experimental Setup

Datasets and Evaluation Metrics. We evaluate the proposed model on six widely-used datasets to validate its effectiveness, including STERE [45],

NJU2K [46], NLPR [47], SIP [29], SSD [48] and DES [49]. Following previous works [8,13,29], we use 1,485 samples of NJU2K [46] and 700 samples of NLPR [47] for training, and the remaining samples of NJU2K (500) and NLPR (300) for testing. The datasets of STERE (1,000), SIP (929), SSD (80), and DES (135) are used for testing.

We employ four metrics to evaluate various methods, including maximum F-measure (F_β^{max}) [50], maximum E-measure (E_ξ^{max}) [51], S-measure (S_α) [52], and mean absolute error (MAE) [53]. Model parameters, CPU inference time (ms, millisecond) and GPU inference FPS (frame-per-second) are used to evaluate the efficiency of the model.

Implementation Details. We implement our model in PyTorch [54]. Parameters of the backbone network (MobileNet-v3 large [20]) are initialized from the model pre-trained on ImageNet [55]. RGB and depth images are both resized to 352×352 for input. We use a single Nvidia Tesla P100-16GB for training and testing and Intel Xeon (4) @2.199GHz for CPU inference speed test. The training images are augmented using various strategies, including random flipping, rotating, colour enhancement, and border clipping. The initial learning rate is set to 1e-4 and is divided by 5 every 60 epochs. The Adam optimizer is used, and the batch size is 10. It takes about 5 h to train our model for 160 epochs.

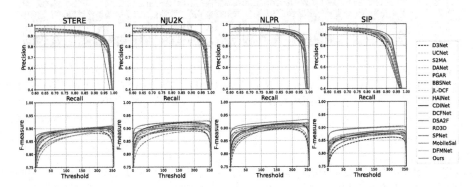

Fig. 5. PR curves [56] and F-measure curves on STERE [45], NJU2K [46], NLPR [47], and SIP [29].

4.2 Comparisons with SOTA Methods

Quantitative Evaluation. We compare the proposed method with 15 RGB-D SOD methods, including 11 non-efficient models (*i.e.*, D3Net [29], UCNet [57], S2MA [58], BBSNet [8], JL-DCF [18], HAINet [14], CDINet [17], DCFNet [26], DSA2F [27], RD3D [25], and SPNet [13]), and 4 efficient models (*i.e.*, DANet [33], PGAR [34], MobileSal [40], and DFMNet [15]). As shown in Table 1, our method outperforms all of the comparison state-of-the-art methods. On the STERE, NLPR, and SIP datasets, our method achieves the best performance on all four evaluation metrics. Our model outperforms most compared RGB-D SOD methods on the NJU2K and SSD datasets except SPNet and BBSNet. As shown in Fig. 5, we plot the PR curves [56] and F-measure curves. For readability, We chose the larger four datasets of the six datasets. The comparison method is still the 15 methods mentioned earlier. In terms of efficiency, among all the compared methods, our method ranks 2nd, 3rd and 4th in model parameters, CPU inference speed, and GPU inference speed, respectively, and is more efficient than most of the compared methods. Overall, our RGB-D SOD method (TBINet) achieves promising performance and efficiency.

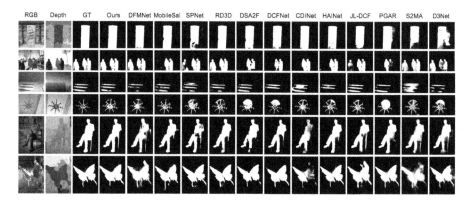

Fig. 6. Visual comparisons of our method (TBINet) with SOTA methods including DFMNet [15],MobileSal [40], SPNet [13], RD3D [25], DSA2F [27], DCFNet [26], CDINet [17], HAINet [14], JL-DCF [18], PGAR [34], S2MA [58], D3Net [29].

Qualitative Evaluation. Figure 6 shows the saliency maps predicted by the proposed method and several state-of-the-art methods on six representative examples. The first row shows a simple example with a single salient object but some misleading information in the depth map. The salient objects predicted by our method, MobileSal, RD3D, and CDINet, have complete boundary details, while results of other methods appear smeared and incomplete to varying degrees. The 2nd and 3rd rows show multiple salient objects, and it is not easy to detect all salient objects accurately. Only our method, DFMNet, and S2MA can completely detect three salient objects in the second row. At the same time, other compared models have missing objects or incomplete segmentation, and the third row is similar. The 4th row shows a salient object with a complex structure. Thanks to the clear depth map, most methods achieve good results. However, some compared methods make poor use of depth information and confuse the background as a salient object. The 5th row shows the low-contrast scene, and it can be observed that our model segments salient objects sharply. The 6th row shows a scene with complex textures. In this example, the depth map is ambiguous. Our model is not misled by low-quality depth information, and accurately locates salient objects.

4.3 Ablation Studies

To verify the effectiveness of the modules and strategies we use in the model, we conduct ablation studies by removing or replacing relevant modules from the full model and reformulating the strategies. We conduct experiments on NJU2K dataset and NLPR dataset.

Table 2. The effectiveness analyses of TBI strategy.

Strategy		Ours	A1	A2	A3	A4	B1	B2	B3	B4	C1	C2	D1	D2	E1	E2	E3	E4
First stage		BAG		UAG-r	UAG-d	BFS	BAG	BAG	BAG	BAG	BAG	BAG	UAG-r	UAG-d		BAG	BFS	Shared
Second stage		BFS	BFS	BFS	BFS	BFS		UFS-r	UFS-d	BAG	BFS	BFS	UFS-r	UFS-d		BAG	BFS	Shared
Third stage		Shared	Shared	Shared	Shared	Shared	Shared	Shared	Shared	Shared		BFS	Shared	Shared		BAG	BFS	Shared
Param (M)		3.7	3.7	3.7	3.7	3.8	3.7	3.7	3.7	3.7	6.4	6.5	3.7	3.7	6.3	6.3	6.5	3.5
NJU2K	F_β^{max} ↑	**.928**	.926	.925	.925	.927	.920	.921	.926	.925	.926	.924	.921	.926	.919	.923	.924	.914
	MAE ↓	**.029**	**.029**	.031	.030	.030	.032	.033	.031	.032	**.029**	.031	.033	.031	.033	.032	.030	.034
NLPR	F_β^{max} ↑	**.932**	.929	.931	.929	.926	.923	.923	.929	.927	**.932**	.928	.923	.929	.929	**.932**	.928	.924
	MAE ↓	**.018**	.020	.019	.019	.021	.022	.021	.019	.021	.019	.019	.021	.019	.020	.019	.019	.021

Effectiveness of TBI Strategy. Our three-stage interaction strategy uses BAG, BFS, and shared network strategies in the first, second and third stages of the encoding process, respectively. For first stage, we first remove the BAG module, this evaluation is denoted as 'A1' in Table 2. Then, we replace the BAG modules with unidirectional attention guidance (UAG) modules. The RGB-enhanced UAG module is abbreviated as UAG-r, and the depth-enhanced UAG module is abbreviated as UAG-d. We denote the RGB-enhanced and depth-enhanced strategies as 'A2' and 'A3', respectively. Finally, the replacement of the BAG module with the BFS module is denoted as 'A4'. Table 2 shows that the BAG module is effective in guiding the network to learn cross-modality correlations. second stage is similar to first stage, we compare the BFS module with four baselines: removing the BFS module (denoted as 'B1'), replacing the BFS module with a unidirectional feature supplement (UFS) module (denoted as 'B2' and 'B3'), and replacing the BFS module with a BAG module (denoted as 'B4'). Table 2 shows the effectiveness of the BFS module. It is worth noting that the performance of 'B3' is significantly better than that of 'B2' and 'B1', which shows that feature supplement to the deep branch can improve the performance very well. The deep reason may be that the BFS module reduces the interference of low-quality depth information. For thrid stage, we do not use shared network strategy (denoted as 'C1') or change to use the BFS module (denoted as 'C2'), the performance of 'C1' is not much different from our strategy, but the parameters are much more. We also changed the BAG module and BFS module to unidirectional interaction at the same time (denoted as 'D1' and 'D2'). These strategies have gaps compared with our strategy. Finally, we evaluate the cases of using the same interaction strategy in all stages (denoted as 'E1', 'E2', 'E3', and 'E4'). As shown in the Table 2, our three-stage bidirectional interaction strategy outperforms the ordinary bidirectional interaction strategy.

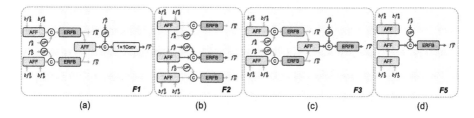

Fig. 7. Illustration of other feature aggregation strategies compared with the CFA module. Channel shuffle operations are hidden for a clear view.

Effectiveness of CFA Module. The CFA module is proposed to aggregate and refine cross-modality features, which adopts a two-step feature fusion and refinement. To verify the effectiveness of the CFA module, we evaluate some different cross-modality feature aggregation strategies, as shown in Fig. 7. We first remove the step of "fusion before refinement" (denoted as 'F1') as shown in Fig. 7(a) or remove the step of "fusion after refinement" (denoted as 'F2') as shown in Fig. 7(b). Table 3 shows that the performance of 'F1' and 'F2' is reduced to varying degrees. Our proposed two-step cross-modality fusion strategy can better fuse and refine cross-modality features. We formulate a strategy denoted as 'F3' as shown in Fig. 7(c): we fuse the refined features of the RGB branch with the refined features of the depth branch and then use the fused features to refine the features generated by the previous stage's fusion branch. The result shows that our strategy outperforms this "refinement-by-refinement" strategy. We remove the supervision of the saliency maps generated by the RGB branch and the depth branch (denoted as 'F4'), the result shows that the supervision of

Table 3. The effectiveness analyses of CFA module.

Strategy		**Ours**	F1	F2	F3	F4	F5	F6
Param(M)		3.7	3.5	3.7	3.7	3.7	3.7	3.6
NJU2K	$F_\beta^{\max}$ ↑	**.928**	.927	.925	.924	.917	.919	.921
	MAE ↓	**.029**	.030	.031	.031	.052	.032	.032
NLPR	$F_\beta^{\max}$ ↑	**.932**	.928	.931	.930	.914	.923	.923
	MAE ↓	**.018**	.020	.020	.019	.037	.021	.021

the RGB branch and the depth branch is effective. The direct removal of the RGB branch and the depth branch in the CFA module is denoted as 'F5' as shown in Fig. 7(d), and Table 3 shows the effectiveness of the three branches in the CFA module. The above evaluation of different cross-modality feature aggregation strategies can conclude that our CFA module can effectively aggregate and refine cross-modality features and generate more accurate saliency maps.

5   Conclusion

We propose a three-stage bidirectional interaction network for RGB-D SOD. Existing works have not explored the relationship between cross-modality features of various levels. Our model employs appropriate interaction strategies at different stages of the encoding process to generate more robust feature representations. In addition, our proposed cross-modality feature aggregation module can effectively aggregate and refine saliency-oriented features to generate accurate saliency maps. Evaluations on six benchmark datasets show promising performance of our TBINet. Our model is lightweight and efficient, which may help the application of RGB-D SOD on mobile devices.

References

1. Shimoda, W., Yanai, K.: Distinct class-specific saliency maps for weakly supervised semantic segmentation. In: Leibe, B., Matas, J., Sebe, N., Welling, M. (eds.) ECCV 2016. LNCS, vol. 9908, pp. 218–234. Springer, Cham (2016). https://doi.org/10.1007/978-3-319-46493-0_14
2. Zeng, Y., Zhuge, Y., Lu, H., Zhang, L.: Joint learning of saliency detection and weakly supervised semantic segmentation. In: Proceedings of the IEEE/CVF International Conference on Computer Vision, pp. 7223–7233 (2019)
3. Hong, S., You, T., Kwak, S., Han, B.: Online tracking by learning discriminative saliency map with convolutional neural network. In: International Conference on Machine Learning, pp. 597–606. PMLR (2015)
4. Mahadevan, V., Vasconcelos, N.: Saliency-based discriminant tracking. In: 2009 IEEE Conference on Computer Vision and Pattern Recognition, pp. 1007–1013. IEEE (2009)
5. Guo, C., Zhang, L.: A novel multiresolution spatiotemporal saliency detection model and its applications in image and video compression. IEEE Trans. Image Process. **19**(1), 185–198 (2009)
6. Ji, Q.G., Fang, Z.D., Xie, Z.H., Lu, Z.M.: Video abstraction based on the visual attention model and online clustering. Signal Process. Image Commun. **28**(3), 241–253 (2013)
7. Cheng, M.M., Hou, Q.B., Zhang, S.H., Rosin, P.L.: Intelligent visual media processing: when graphics meets vision. J. Comput. Sci. Technol. **32**(1), 110–121 (2017)
8. Fan, D.-P., Zhai, Y., Borji, A., Yang, J., Shao, L.: BBS-Net: RGB-D salient object detection with a bifurcated backbone strategy network. In: Vedaldi, A., Bischof, H., Brox, T., Frahm, J.-M. (eds.) ECCV 2020. LNCS, vol. 12357, pp. 275–292. Springer, Cham (2020). https://doi.org/10.1007/978-3-030-58610-2_17

9. Jin, W.D., Xu, J., Han, Q., Zhang, Y., Cheng, M.M.: CDNet: complementary depth network for RGB-D salient object detection. IEEE Trans. Image Process. **30**, 3376–3390 (2021)

10. Liu, Z., Wang, Y., Tu, Z., Xiao, Y., Tang, B.: TritransNet: RGB-D salient object detection with a triplet transformer embedding network. In: Proceedings of the 29th ACM International Conference on Multimedia, pp. 4481–4490 (2021)

11. Luo, A., Li, X., Yang, F., Jiao, Z., Cheng, H., Lyu, S.: Cascade graph neural networks for RGB-D salient object detection. In: Vedaldi, A., Bischof, H., Brox, T., Frahm, J.-M. (eds.) ECCV 2020. LNCS, vol. 12357, pp. 346–364. Springer, Cham (2020). https://doi.org/10.1007/978-3-030-58610-2_21

12. Piao, Y., Rong, Z., Zhang, M., Ren, W., Lu, H.: A2Dele: adaptive and attentive depth distiller for efficient RGB-D salient object detection. In: Proceedings of the IEEE/CVF Conference on Computer Vision and Pattern Recognition, pp. 9060–9069 (2020)

13. Zhou, T., Fu, H., Chen, G., Zhou, Y., Fan, D.P., Shao, L.: Specificity-preserving RGB-D saliency detection. In: Proceedings of the IEEE/CVF International Conference on Computer Vision, pp. 4681–4691 (2021)

14. Li, G., Liu, Z., Chen, M., Bai, Z., Lin, W., Ling, H.: Hierarchical alternate interaction network for RGB-D salient object detection. IEEE Trans. Image Process. **30**, 3528–3542 (2021)

15. Zhang, W., Ji, G.P., Wang, Z., Fu, K., Zhao, Q.: Depth quality-inspired feature manipulation for efficient RGB-D salient object detection. In: Proceedings of the 29th ACM International Conference on Multimedia, pp. 731–740 (2021)

16. Zhang, W., Jiang, Y., Fu, K., Zhao, Q.: BTS-Net: bi-directional transfer-and-selection network for RGB-D salient object detection. In: 2021 IEEE International Conference on Multimedia and Expo (ICME), pp. 1–6. IEEE (2021)

17. Zhang, C., et al.: Cross-modality discrepant interaction network for RGB-D salient object detection. In: Proceedings of the 29th ACM International Conference on Multimedia, pp. 2094–2102 (2021)

18. Fu, K., Fan, D.P., Ji, G.P., Zhao, Q., Shen, J., Zhu, C.: Siamese network for RGB-D salient object detection and beyond. IEEE Trans. Pattern Anal. Mach. Intell. **44**, 5541–5559 (2021)

19. Ma, N., Zhang, X., Zheng, H.-T., Sun, J.: ShuffleNet V2: practical guidelines for efficient CNN architecture design. In: Ferrari, V., Hebert, M., Sminchisescu, C., Weiss, Y. (eds.) Computer Vision – ECCV 2018. LNCS, vol. 11218, pp. 122–138. Springer, Cham (2018). https://doi.org/10.1007/978-3-030-01264-9_8

20. Howard, A., et al.: Searching for mobilenetv3. In: Proceedings of the IEEE/CVF International Conference on Computer Vision, pp. 1314–1324 (2019)

21. Desingh, K., Krishna, K.M., Rajan, D., Jawahar, C.: Depth really matters: Improving visual salient region detection with depth. In: BMVC, pp. 1–11 (2013)

22. Feng, D., Barnes, N., You, S., McCarthy, C.: Local background enclosure for RGB-D salient object detection. In: Proceedings of the IEEE Conference on Computer Vision and Pattern Recognition, pp. 2343–2350 (2016)

23. Lang, C., Nguyen, T.V., Katti, H., Yadati, K., Kankanhalli, M., Yan, S.: Depth matters: influence of depth cues on visual saliency. In: Fitzgibbon, A., Lazebnik, S., Perona, P., Sato, Y., Schmid, C. (eds.) ECCV 2012. LNCS, vol. 7573, pp. 101–115. Springer, Heidelberg (2012). https://doi.org/10.1007/978-3-642-33709-3_8

24. Ren, J., Gong, X., Yu, L., Zhou, W., Ying Yang, M.: Exploiting global priors for RGB-D saliency detection. In: Proceedings of the IEEE Conference on Computer Vision and Pattern Recognition Workshops, pp. 25–32 (2015)

25. Chen, Q., Liu, Z., Zhang, Y., Fu, K., Zhao, Q., Du, H.: RGB-D salient object detection via 3d convolutional neural networks. In: Proceedings of the AAAI Conference on Artificial Intelligence, pp. 1063–1071 (2021)
26. Ji, W., et al.: Calibrated RGB-D salient object detection. In: Proceedings of the IEEE/CVF Conference on Computer Vision and Pattern Recognition, pp. 9471–9481 (2021)
27. Sun, P., Zhang, W., Wang, H., Li, S., Li, X.: Deep RGB-D saliency detection with depth-sensitive attention and automatic multi-modal fusion. In: Proceedings of the IEEE/CVF Conference on Computer Vision and Pattern Recognition, pp. 1407–1417 (2021)
28. Zhang, M., Fei, S.X., Liu, J., Xu, S., Piao, Y., Lu, H.: Asymmetric two-stream architecture for accurate RGB-D saliency detection. In: Vedaldi, A., Bischof, H., Brox, T., Frahm, J.-M. (eds.) ECCV 2020. LNCS, vol. 12373, pp. 374–390. Springer, Cham (2020). https://doi.org/10.1007/978-3-030-58604-1_23
29. Fan, D.P., Lin, Z., Zhang, Z., Zhu, M., Cheng, M.M.: Rethinking RGB-D salient object detection: models, data sets, and large-scale benchmarks. IEEE Trans. Neural Netw. Learn. Syst. **32**(5), 2075–2089 (2020)
30. Li, C., Cong, R., Piao, Y., Xu, Q., Loy, C.C.: RGB-D salient object detection with cross-modality modulation and selection. In: Vedaldi, A., Bischof, H., Brox, T., Frahm, J.-M. (eds.) ECCV 2020. LNCS, vol. 12353, pp. 225–241. Springer, Cham (2020). https://doi.org/10.1007/978-3-030-58598-3_14
31. Zhou, T., Fan, D.P., Cheng, M.M., Shen, J., Shao, L.: RGB-D salient object detection: a survey. Comput. Visual Med. **7**(1), 37–69 (2021)
32. Wang, W., Lai, Q., Fu, H., Shen, J., Ling, H., Yang, R.: Salient object detection in the deep learning era: an in-depth survey. IEEE Trans. Pattern Anal. Mach. Intell. **44**, 3239–3259 (2021)
33. Zhao, X., Zhang, L., Pang, Y., Lu, H., Zhang, L.: A single stream network for robust and real-time RGB-D salient object detection. In: Vedaldi, A., Bischof, H., Brox, T., Frahm, J.-M. (eds.) ECCV 2020. LNCS, vol. 12367, pp. 646–662. Springer, Cham (2020). https://doi.org/10.1007/978-3-030-58542-6_39
34. Chen, S., Fu, Y.: Progressively guided alternate refinement network for RGB-D salient object detection. In: Vedaldi, A., Bischof, H., Brox, T., Frahm, J.-M. (eds.) ECCV 2020. LNCS, vol. 12353, pp. 520–538. Springer, Cham (2020). https://doi.org/10.1007/978-3-030-58598-3_31
35. Howard, A.G., et al.: MobileNets: efficient convolutional neural networks for mobile vision applications. arXiv preprint arXiv:1704.04861 (2017)
36. Sandler, M., Howard, A., Zhu, M., Zhmoginov, A., Chen, L.C.: Mobilenetv 2: Inverted residuals and linear bottlenecks. In: Proceedings of the IEEE Conference on Computer Vision and Pattern Recognition, pp. 4510–4520 (2018)
37. Zhang, X., Zhou, X., Lin, M., Sun, J.: ShuffleNet: an extremely efficient convolutional neural network for mobile devices. In: Proceedings of the IEEE Conference on Computer Vision and Pattern Recognition, pp. 6848–6856 (2018)
38. Simonyan, K., Zisserman, A.: Very deep convolutional networks for large-scale image recognition. arXiv preprint arXiv:1409.1556 (2014)
39. He, K., Zhang, X., Ren, S., Sun, J.: Deep residual learning for image recognition. In: Proceedings of the IEEE Conference on Computer Vision and Pattern Recognition, pp. 770–778 (2016)
40. Wu, Y.H., Liu, Y., Xu, J., Bian, J.W., Gu, Y.C., Cheng, M.M.: Mobile-Sal: extremely efficient RGB-D salient object detection. IEEE Trans. Pattern Anal. Mach. Intell. **44**, 10261–10269 (2021). https://doi.org/10.1109/TPAMI.2021.3134684

41. Li, G., Liu, Z., Ye, L., Wang, Y., Ling, H.: Cross-modal weighting network for RGB-D salient object detection. In: Vedaldi, A., Bischof, H., Brox, T., Frahm, J.-M. (eds.) ECCV 2020. LNCS, vol. 12362, pp. 665–681. Springer, Cham (2020). https://doi.org/10.1007/978-3-030-58520-4_39

42. Hu, J., Shen, L., Sun, G.: Squeeze-and-excitation networks. In: Proceedings of the IEEE Conference on Computer Vision and Pattern Recognition, pp. 7132–7141 (2018)

43. Liu, S., Huang, D., Wang, Y.: Receptive field block net for accurate and fast object detection. In: Ferrari, V., Hebert, M., Sminchisescu, C., Weiss, Y. (eds.) ECCV 2018. LNCS, vol. 11215, pp. 404–419. Springer, Cham (2018). https://doi.org/10.1007/978-3-030-01252-6_24

44. Wei, J., Wang, S., Huang, Q.: F^{3}net: fusion, feedback and focus for salient object detection. In: Proceedings of the AAAI Conference on Artificial Intelligence, pp. 12321–12328 (2020)

45. Niu, Y., Geng, Y., Li, X., Liu, F.: Leveraging stereopsis for saliency analysis. In: 2012 IEEE Conference on Computer Vision and Pattern Recognition, pp. 454–461. IEEE (2012)

46. Ju, R., Ge, L., Geng, W., Ren, T., Wu, G.: Depth saliency based on anisotropic center-surround difference. In: 2014 IEEE International Conference on Image Processing (ICIP), pp. 1115–1119. IEEE (2014)

47. Peng, H., Li, B., Xiong, W., Hu, W., Ji, R.: RGBD salient object detection: a benchmark and algorithms. In: Fleet, D., Pajdla, T., Schiele, B., Tuytelaars, T. (eds.) ECCV 2014. LNCS, vol. 8691, pp. 92–109. Springer, Cham (2014). https://doi.org/10.1007/978-3-319-10578-9_7

48. Zhu, C., Li, G.: A three-pathway psychobiological framework of salient object detection using stereoscopic technology. In: Proceedings of the IEEE International Conference on Computer Vision Workshops, pp. 3008–3014 (2017)

49. Cheng, Y., Fu, H., Wei, X., Xiao, J., Cao, X.: Depth enhanced saliency detection method. In: Proceedings of International Conference on Internet Multimedia Computing and Service, pp. 23–27 (2014)

50. Achanta, R., Hemami, S., Estrada, F., Susstrunk, S.: Frequency-tuned salient region detection. In: 2009 IEEE Conference on Computer Vision and Pattern Recognition, pp. 1597–1604. IEEE (2009)

51. Fan, D.P., Gong, C., Cao, Y., Ren, B., Cheng, M.M., Borji, A.: Enhanced-alignment measure for binary foreground map evaluation. In: Proceedings of the 27th International Joint Conference on Artificial Intelligence, pp. 698–704 (2018)

52. Fan, D.P., Cheng, M.M., Liu, Y., Li, T., Borji, A.: Structure-measure: a new way to evaluate foreground maps. In: Proceedings of the IEEE International Conference on Computer Vision, pp. 4548–4557 (2017)

53. Perazzi, F., Krähenbühl, P., Pritch, Y., Hornung, A.: Saliency filters: contrast based filtering for salient region detection. In: 2012 IEEE Conference on Computer Vision and Pattern Recognition, pp. 733–740. IEEE (2012)

54. Paszke, A., et al.: PyTorch: an imperative style, high-performance deep learning library. In: Advances in Neural Information Processing Systems, vol. 32 (2019)

55. Krizhevsky, A., Sutskever, I., Hinton, G.E.: ImageNet classification with deep convolutional neural networks. In: Advances In Neural Information Processing Systems, vol. 25 (2012)

56. Borji, A., Cheng, M.M., Jiang, H., Li, J.: Salient object detection: a benchmark. IEEE Trans. Image Process. **24**(12), 5706–5722 (2015)

57. Zhang, J., et al.: UC-Net: uncertainty inspired RGB-D saliency detection via conditional variational autoencoders. In: Proceedings of the IEEE/CVF Conference on Computer Vision and Pattern Recognition, pp. 8582–8591 (2020)
58. Liu, N., Zhang, N., Han, J.: Learning selective self-mutual attention for RGB-D saliency detection. In: Proceedings of the IEEE/CVF Conference on Computer Vision and Pattern Recognition, pp. 13756–13765 (2020)

PS-ARM: An End-to-End Attention-Aware Relation Mixer Network for Person Search

Mustansar Fiaz[1], Hisham Cholakkal[1(✉)], Sanath Narayan[2], Rao Muhammad Anwer[1], and Fahad Shahbaz Khan[1]

[1] Department of computer Vision, Mohamed bin Zayed University of Artificial Intelligence, Abu Dhabi, UAE
{mustansar.fiaz,hisham.cholakkal,rao.anwer,fahad.khan}@mbzuai.ac.ae
[2] Inception Institute of Artificial Intelligence, Abu Dhabi, UAE

Abstract. Person search is a challenging problem with various real-world applications, that aims at joint person detection and re-identification of a query person from uncropped gallery images. Although, previous study focuses on rich feature information learning, it's still hard to retrieve the query person due to the occurrence of appearance deformations and background distractors. In this paper, we propose a novel attention-aware relation mixer (ARM) module for person search, which exploits the global relation between different local regions within RoI of a person and make it robust against various appearance deformations and occlusion. The proposed ARM is composed of a relation mixer block and a spatio-channel attention layer. The relation mixer block introduces a spatially attended spatial mixing and a channel-wise attended channel mixing for effectively capturing discriminative relation features within an RoI. These discriminative relation features are further enriched by introducing a spatio-channel attention where the foreground and background discriminability is empowered in a joint spatio-channel space. Our ARM module is generic and it does not rely on fine-grained supervisions or topological assumptions, hence being easily integrated into any Faster R-CNN based person search methods. Comprehensive experiments are performed on two challenging benchmark datasets: CUHK-SYSU and PRW. Our PS-ARM achieves state-of-the-art performance on both datasets. On the challenging PRW dataset, our PS-ARM achieves an absolute gain of 5% in the mAP score over SeqNet, while operating at a comparable speed. The source code and pre-trained models are available at https://github.com/mustansarfiaz/PS-ARM.

Keywords: Person search · Transformer · Spatial attention · Channel attention

L. Wang et al. (Eds.): ACCV 2022, LNCS 13845, pp. 234–250, 2023.
https://doi.org/10.1007/978-3-031-26348-4_14

1 Introduction

Person search is a challenging computer vision problem where the task is to find a target query person in a gallery of whole scene images. The person search methods need to perform pedestrian detection [26,28,41] on the uncropped gallery images and do re-identification (re-id) [7,24,43] of the detected pedestrians. In addition to addressing the challenges associated with these individual sub-tasks, both these tasks need to be simultaneously optimized within person search. Despite numerous real-world applications, person search is highly challenging due to the diverse nature of person detection and re-id sub-tasks within the person search problem.

Person search approaches can be broadly grouped into two-step [4,16,43] and one-step methods [5,36,38]. In two-step approaches, person detection and re-id are performed separately using two different steps. In the first step a detection network such as Faster R-CNN is employed to detect pedestrians. In the second step detected persons are first cropped and re-sized from the *input image*, then utilized in another independent network for the re-identification of the cropped pedestrians. Although two-step methods provide promising results, they are computationally expensive. Different to two-step methods, one step methods employ a unified framework where the backbone networks are shared for the detection and identifications of persons. For a given uncropped image, one-step methods predict the box coordinates and re-id features for all persons in that image. One-step person search approaches such as [5,23] generally extend Faster R-CNN object detection frameworks by introducing an additional branch to produce re-id feature embedding, and the whole network is jointly trained end-to-end. Such methods often struggle while the target person in the galley images has large appearance deformations such as pose variation, occlusion, and overlapping background distractions within the region of interest (RoI) of a target person (see Fig. 1).

1.1 Motivation

To motivate our approach, we first distinguish two desirable characteristics to be considered when designing a Faster R-CNN based person search framework that is robust to appearance deformations (e.g. pose variations, occlusions) and background distractions occurring in the query person (see Fig. 1).

Discriminative Relation Features through Local Information Mixing: The position of different local person regions within an RoI can vary in case of appearance deformations such as pose variations and occlusions. This is likely to deteriorate the quality of re-id features, leading to inaccurate person matching. Therefore, a dedicated mechanism is desired that generates discriminative relation features by globally mixing relevant information from different local regions within an RoI. To ensure a straightforward integration into existing person search pipelines, such a mechanism is further expected to learn discriminative relation features without requiring fine-level region supervision or topological body approximations.

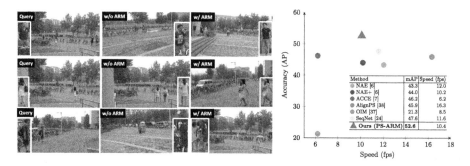

Fig. 1. On the left: Qualitative comparison showing different query examples and their corresponding top-1 matching results obtained with *and* without our ARM module in the same base framework. Here, true and false matching results are marked in green and red, respectively. These examples depict appearance deformations and distracting backgrounds in the gallery images for the query person. Our ARM module that explicitly captures discriminative relation features better handle the appearance deformations in these examples. **On the right:** Accuracy (AP) vs. speed (frames per second) comparison with state-of-the-art person search methods on PRW test set. All methods are reported with a Resnet50 backbone and speed is computed over V100 GPU. Our approach (PS-ARM) achieves an absolute mAP gain of 5% over SeqNet while operating at a comparable speed.

Foreground-Background Discriminability for Accurate Local Information Mixing: The quality of the aforementioned relation features rely on the assumption that the RoI region only contains foreground (person) information. However, in real-world scenarios the RoI regions are likely to contain unwanted background information due to less accurate bounding-box locations. Therefore, discriminability of the foreground from the background is essential for accurate local information mixing to obtain discriminative relation features. Further, such a FG/BG discrimination is expected to also improve the detection performance.

1.2 Contribution

We propose a novel end-to-end one-step person search method with the following novel contributions. We propose a novel attention-aware relation mixer (ARM) module that strives to capture global relation between different local person regions through global mixing of local information while simultaneously suppressing background distractions within an RoI. Our ARM module comprises a relation mixer block and a spatio-channel attention layer. The relation mixer block captures discriminative relation features through a spatially-attended spatial mixing and a channel-wise attended channel mixing. These discriminative relation features are further enriched by the spatio-channel attention layer performing foreground/background discrimination in a joint spatio-channel space. Comprehensive experiments are performed on two challenging benchmark datasets: CUHK-SYSU [36] and PRW [43]. On both datasets, our PS-ARM performs favourably against state-of-the-art approaches. On the challenging PRW benchmark, our PS-ARM achieves a mAP score of 52.6%. Our ARM module is generic and can be easily integrated to any Faster R-CNN based person search methods. Our PS-ARM provides an absolute gain of 5% mAP score over SeqNet, while operating at a comparable speed (see Fig. 1), resulting in a mAP score of 52.6% on the challenging PRW dataset.

2   Related Work

Person search is a challenging computer vision problem with numerous real-world applications. As mentioned earlier, existing person methods can be broadly classified into two-step and one-step methods. Most existing two-step person search approaches address this problem by first detecting the pedestrians, followed by cropping and resizing into a fixed resolution before passing to the re-id network that identifies the cropped pedestrian [4,10,17,22,43]. These methods generally employ two different backbone networks for the detection and re-identification.

On the other hand, several one-step person search methods employ feature pooling strategies such as, RoIPooling or RoIAlign pooling to obtain a scale-invariant representation for the re-id sub-task. [4] proposed a two-step method to learn robust person features by exploiting person foreground maps using pre-trained segmentation network. Han et al. [17] introduced a bounding box refinement mechanism for person localization. Dong et al. [10] utilized the similarity between the query and query-like features to reduce the number of proposals for re-identification. Zhang et al. [43] introduced the challenging PRW dataset. A multi-scale feature pyramid was introduced in [22] for improving person search under scale variations. Wang et al. [34] proposed a method to address the inconsistency between the detection and re-id sub-tasks.

Most one-step person search methods [2,5,9,15,23,25,27,35,36,38] are developed based on Faster R-CNN object detector [30]. These methods generally introduce an additional branch to Faster R-CNN and jointly address the detection and Re-ID subtasks. One of the earliest Faster R-CNN based one-step person approach is [36], which proposed an online instance matching (OIM) loss. Xiao et al. [35] introduced a center loss to explore intra-class compactness. For generating person proposals, Liu et al. [25] introduced a mechanism to iteratively shrink the search area based on query guidance. Similarly, Chang et al. [2] used reinforcement learning to address the person search problem. Chang et al. [38] exploited complementary cues based on graph learning framework. Dont et al. [9] proposed Siamese based Bi-directional Interaction Network (BINet) to mitigate redundant context information outside the BBoxes. On the contrary, Chen et al. [5] proposed Norm Aware Embedding (NAE) to alleviate the conflict between person localization and re-identification by computing magnitude and angle of the embedded features respectively.

Chen at al. [3] developed a Hierarchical Online Instance Matching loss to guide the feature learning by exploiting the hierarchical relationship between detection and re-identification. A query-guided proposal network (QGPN) is proposed by Munjal et al. [27] to learn query guided re-identification score. H

Li et al. [23] proposed a Sequential End-to-end Network (SeqNet) to refine the proposals by introducing Faster R-CNN as a proposal generator into the NAE pipeline to get refined features for detection and re-identification. The Faster R-CNN based one-step person search approaches often struggle while the target undergoes large appearance deformations or come across with distracting background objects within RoI. To address this, we propose a novel person search method, PS-ARM, where a novel ARM module is introduced to capture global relation between different local regions within an RoI. Our PS-ARM enables accurate detection and re-identification of person instances under under challenging scenarios such as pose variation and distracting backgrounds (See Fig. 1).

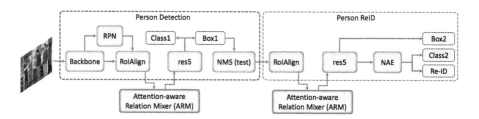

Fig. 2. The overall architecture of the proposed PS-ARM framework. It comprises a person detection branch (shown in green) and person re-ID branch (shown in blue). The person detection branch predicts the initial box locations whereas the person re-id branch refines the box locations and perform a norm-aware embedding (NAE) to disentangle the detection and re-id. The focus of our design is the introduction of a novel *attention-aware relation Mixer* (ARM) module (shown in grey) to the detection and re-id branches. Our ARM module enriches standard RoI Align pooled features by capturing discriminative relation features between different local regions within an RoI. The resulting enriched features are used for box regression and classification in the detection branch, whereas these features are used to refine the box locations, along with generating a norm-aware embedding for box classification (person vs background) and re-id feature prediction in the re-id branch. (Color figure online)

3   Method

3.1   Overall Architecture

Figure 2 shows overall architecture of the proposed framework. It comprises a person detection branch (shown in green) followed by a person re-ID branch (shown in blue). The person detection branch follows the structure of standard Faster R-CNN, which comprises a ResNet backbone (*res1-res4*), a region proposal network (RPN), RoIAlign pooling, and a prediction head for box regression and classification. The person re-id branch takes the boxes predicted by the person detection branch as input and performs RoIAlign pooling on these predicted box locations. The resulting RoI Align pooled features are utilized to perform re-identification. We adopt norm-aware embedding (NAE) that is designed to separate the detection and re-identification using shared feature representation. During inference, the person re-id branch takes only unique boxes (obtained by non-maximum suppression algorithm) from the person detection branch and performs a context bipartite graph matching for the re-id similar to [23]. The above-mentioned standard detection and re-id branches serve as a base network to which we introduce our novel attention-aware relation mixer (ARM) module that enriches the RoI features for accurate person search.

The focus of our design is the introduction of a novel ARM module (shown in grey). Specifically, we integrate our ARM module between the RoIAlign and convolution blocks (res5) in both the person detection and re-id branches of the base framework, without sharing the parameters between both branches. Our proposed ARM module strives to enrich standard RoI Align pooled features by capturing discriminative relation features between different local regions within an RoI through global mixing of local information. To ensure effective enrichment of RoI Align pooled features, we further introduce a foreground/background discrimination mechanism in our ARM module. Our ARM module strives to simultaneously improve both detection and re-id sub-tasks. Therefore, the output is passed to norm-aware embedding to decouple the features for the contradictory detection and re-id tasks. Furthermore, our ARM module is generic and can be easily integrated to other Faster R-CNN based person search methods. Next, we present the details of the proposed ARM module.

3.2 Attention-aware Relation Mixer (ARM) Module

Our ARM module is shown in Fig. 3. It comprises a relation mixer block and a spatio-channel attention layer. Our relation mixer block captures relation between different sub-regions (local regions) within an RoI. The resulting features are further enriched by our spatio-channel attention that attend to relevant input features in a joint spatio-channel space. Our ARM module takes RoIAlign pooled feature $\mathcal{F} \in \mathbb{R}^{C \times H \times W}$ as input. Here, H, W, C are the height, width and number of channels of the RoI feature. For computational efficiency, the number of channels are reduced to $c = C/4$ through a point (1×1) convolution layer before passing to relation mixer and spatio-channel attention blocks.

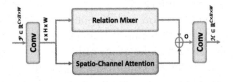

Fig. 3. The network structure of our ARM module. The module takes RoI Align pooled features as input and captures inter-dependency between different local regions, while simultaneously suppressing background distractions for the person search problem. To achieve this objective, ARM module comprises a relation mixer block and a joint spatio-channel attention layer.

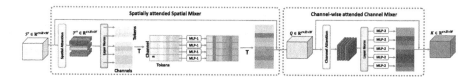

Fig. 4. Structure of relation mixer block within our ARM module. It comprises a spatially attended spatial mixing operation where important local spatial regions will be emphasized using a spatial attention before globally mixing them across all spatial regions (tokens) within each channel using MLP-1 shared across all channels. Following this spatial mixing, we perform a channel attention to emphasize informative channels before globally mixing the channels for each local spatial region (token) using MLP-2 shared across all spatial regions.

Relation Mixer Block: As mentioned earlier, our relation mixer block is introduced to capture the relation between different sub-regions (local regions) within an RoI. This is motivated by the fact that the local regions of a person share certain 'standard' prior relationships among local regions, across RoIs of different person and it is desirable to explicitly learn these inter-dependencies without any supervision. One such module that can learn/encode such inter-dependencies, is the MLP-mixer [32] that performs spatial 'token' mixing followed by 'point-wise' feature refinement. Compared to other context aggregators [19,21,33], the mlp-mixer is more static, dense, and does not share parameters [12]. The core operation of the MLP-Mixer is transposed affinity matrix on a single feature group, which computes the affinity matrix with non-sharing W_{MLP-1} parameters as: $\mathcal{A} = (W_{MLP-1})^T$. To this end, MLP mixer contains a spatial mixer and a channel mixer. The spatial mixer comprise of a layer norm, skip connection and a token-mixing MLP with two fully-connected layers and a GELU nonlinearity. Similarly, the channel mixer employs a channel-mixing MLP, layer norm, skip connection and dropout. The MLP mixer conceptually acts as a persistent relationship memory that can learn and encode the prior relationships among the local regions of an object at a global level. To this end, we introduce our relation mixer comprising a spatially attended spatial mixer and a channel-wise attended channel mixer. our ARM module with residual connection not only enabled using MLP mixer for the first time in the problem of person search, but also provided impressive performance gain over the base framework.

Spatially attended Spatial Mixer: While learning the inter-dependencies of local RoI sub-regions using standard MLP mixer, the background regions are likely to get entangled with the foreground regions, thereby adversely affecting the resulting feature embedding used for the re-id and box predictions. In order to discriminate the irrelevant background information at token level, we introduce a spatial attention before performing token (spatial) mixing within our MLP mixer for emphasizing the foreground regions. In our spatial attention, we employ pooling operations along the channel axis, followed by convolution and sigmoid layers to generate a 2D spatial attention weights $M_s \in \mathbb{R}^{1 \times H \times W}$.

These attention weights are broadcasted along the channel dimension to generate the spatial attention $M_s' \in \mathbb{R}^{c \times H \times W}$. For a given feature $\mathcal{F}' \in \mathbb{R}^{c \times H \times W}$, we obtain the spatially attended feature map $\mathcal{F}'' = \mathcal{F}' \odot M_s'$. Here $\odot$ denotes element-wise multiplication. These spatially attended features ($\mathcal{F}''$) are expected to discriminate irrelevant (background) spatial regions from the foreground. These features are ($\mathcal{F}''$) input to a shared multi-layer perceptron (MLP-1) for globally mixing local features (within $\mathcal{F}''$) across all spatial regions (tokens). Our spatially attended spatial mixing strives to achieve accurate spatial mixing and outputs the feature map Q (see Fig. 4).

Channel-Wise attended Channel Mixer: To further prioritize the feature channels of Q that are relevant for detection and re-id of person instances, we introduce a channel attention before channel mixing. Our channel attention weights $M_c \in \mathbb{R}^{c \times 1}$ are generated through spatial pooling, fully connected (fc) and sigmoid layers, which are broadcasted along the spatial dimension to generate the channel attention weights $M_c' \in \mathbb{R}^{c \times H \times W}$. Similar to spatial attention, these channel weights are element-wise multiplied with the feature map to obtain channel-wise attended feature map. The resulting features are expected to emphasize only the channels that are relevant for effective channel-mixing within our relation mixing block. Our channel mixing employs another shared MLP (MLP-2) for global mixing of channel information. The final output of our relation mixer block results is feature maps $K \in \mathbb{R}^{c \times H \times W}$.

Spatio-Channel Attention Layer: Our relation mixer block performs the mixing operations by treating the spatial and channel information in a disjoint manner. But, in many scenarios, all spatial regions within a channel and all channels at a given spatial location are not equally informative. Hence, it is desired to treat the entire spatio-channel information as a joint space. With this objective, we introduce a joint spatio-channel attention layer within our ARM module to further improve the foreground/background discriminability of RoIAlign pooled features. Our spatio-channel attention layer utilizes parameter-free 3D attention weights obtained based on [39] to modulate the 3D spatio-channel RoI pooled features. These spatio-channel attended features are aggregated with the relation mixer output to produce enriched features O for the person search task. These enriched features projected back to C channels ($\mathcal{H} \in \mathbb{R}^{C \times H \times W}$) and taken as input to the *res*5 block.

In summary, within our ARM module, the relation mixer targets the global relation between different local regions within RoI and captures the discriminative relation features in disjoint spatial and channel spaces. The resulting features are further enriched by a spatio-channel attention that performs foreground/background discrimination in a joint spatio-channel space.

3.3   Training and Inference

For training and inference, we follow a strategy similar to [5, 23]. Our PS-ARM is trained end-to-end with a loss formulation similar to [23]. That is, in the person detection branch, similar to Faster R-CNN, we employ Smooth-L1 and cross entropy losses for box regression and classifications. For the person re-id branch, we employ three additional loss terms similar to [5] for regression, classification and re-ID. Both these branches are trained by utilizing an IoU threshold of 0.5 for selecting positive and negative samples.

During inference, we first obtain the re-id feature for a given query by using the provided bounding box. Then, for the gallery images, the predicted boxes and their re-id features are obtained from the re-id branch. Finally, employ cosine similarity between the re-id features to match a query person with an arbitrarily detected person in the galley.

4   Experiments

We perform the experiments on two person search datasets (*i.e.*, CUHK-SYSU [36]) and PRW [43] to demonstrate the effectiveness of our PS-ARM and compare it with the state-of-the-art methods.

4.1   Dataset and Evaluation Protocols

CUHK-SYSU [36]: is a large scale person search dataset with 96,143 person bounding boxes from a total of 18,184 images. The training and testing sets contains 11,206 images, 55,272 pedestrians, and 5,532 identities and test set includes 6,978 images, 40,871 pedestrians, and 2,900 identities. Instead of using full gallery during inference, different gallery sizes are used for each query from 50 to 4000. The default gallery size is set to 100.

PRW [43]: is composed of video frames recorded by six cameras that are being installed at different location in Tsinghua University. The dataset has a total 11,816 frames containing 43,110 person bounding boxes. In training set, 5,704 images are annotated with 482 identities. The test set has 2,057 frames are labelled as query persons while gallery set has 6,112 images. Hence, the gallery size of PRW dataset is notably larger compared to CUHK-SYSU gallery set.

Evaluation Protocol: We follow two standard protocol for person search performance evaluation of mean Average Precision (mAP) and top-1 accuracy. The mAP is computed by averaging over all queries with an intersection-over-union (IoU) threshold of 0.5. The top-1 accuracy is measured according to the IoU overlaps between the top-1 prediction and ground-truth with the threshold value set to 0.5.

Implementation Details: We used ResNet-50 as our backbone network. We followed [23] and utilized Stochastic Gradient Descent (SGD), set momentum and decay to 0.9 and 5×10^{-4}, respectively. We trained the model for 12 epochs over CUHK-SYSU dataset PRW dataset. During training, we used the batch-size of 3 with input size 900×1500 and set initial learning rate to 0.003 which is warmed up at first epoch and decayed by 0.1 at 8th epoch. During inference, the NMS threshold value is set to 0.4. The code is implemented in PyTorch [29]. The code and trained model will be publicly released.

Table 1. State-of-the-art comparison on CUHK and PRW test sets in terms of mAP and top-1 accuracy. On both datasets, our PS-ARM performs favourably against existing approaches. All the methods here utilize the same ResNet50 backbone. When compared with recently introduced SeqNet, our PS-ARM provides an absolute mAP gain of 5% on the challenging PRW dataset. Also, introducing our novel ARM module to a popular Faster R-CNN based approach (NAE [5]), provides an absolute mAP gain of 3.6%.

Method		CUHK-SYSU		PRW	
		mAP	top-1	mAP	top-1
Two-step	CLSA [22]	87.2	88.5	38.7	65.0
	IGPN [10]	90.3	91.4	42.9	70.2
	RDLR [17]	93.0	94.2	42.9	70.2
	MGTS [4]	83.0	83.7	32.6	72.1
	MGN+OR [40]	93.2	93.8	52.3	71.5
	TCTS [34]	93.9	95.1	46.8	87.5
End-to-end	OIM [36]	75.5	78.7	21.3	49.9
	RCAA [2]	79.3	81.3	–	–
	NPSM [25]	77.9	81.2	24.2	53.1
	IAN [35]	76.3	80.1	23.0	61.9
	QEEPS [27]	88.9	89.1	37.1	76.7
	CTXGraph [38]	84.1	86.5	33.4	73.6
	HOIM [3]	89.7	90.8	39.8	80.4
	BINet [9]	90.0	90.7	45.3	81.7
	AlignPS [37]	93.1	96.4	45.9	81.9
	PGSFL [20]	92.3	94.7	44.2	85.2
	DKD [42]	93.1	94.2	50.5	87.1
	NAE+ [5]	92.1	94.7	44.0	81.1
	PBNet [31]	90.5	88.4	48.5	87.9
	DIOIM [8]	88.7	89.6	36.0	76.1
	APNet [44]	88.9	89.3	41.2	81.4
	DMRN [18]	93.2	94.2	46.9	83.3
	CAUCPS [14]	81.1	83.2	41.7	86.0
	ACCE [6]	93.9	94.7	46.2	86.1
	NAE [5]	91.5	92.4	43.3	80.9
	SeqNet [23]	94.8	95.7	47.6	87.6
	Ours (NAE + ARM)	93.4	94.2	46.9	81.4
	Ours (PS-ARM)	**95.2**	**96.1**	52.6	**88.1**
	Ours (Cascaded PS-ARM)	–	–	**53.1**	**88.3**

4.2 Comparison with State-of-the-Art Methods

Here, we compare our approach with state-of-the-art one-step and two-step person search methods in literature on two datasets: CUSK-SYSU and PRW.

CUHK-SYSU Comparison: Table 1 shows the comparison of our PS-ARM with state-of-the-art two-step and single-step end-to-end methods with the gallery size of 100. Among existing two-step methods, MGN+OR [40] and TCTS [34] achieves mAP of 93.2 and 93.9, respectively. Among existing single-step end-to-end methods, SeqNet [23] and AlignPS [37] obtains mAP of 94.8%, 93.1% respectively.

Fig. 5. Qualitative comparison between the top-1 results obtained from SeqNet (row 2) and our PS-ARM (row3) for the same query input (row 1). Here, true and false matching results are marked in green and red, respectively. SeqNet provides inaccurate predictions due to the appearance deformations in these examples whereas our PS-ARM provides accurate predictions by explicitly capturing discriminative relation features within RoI. (Color figure online)

To further analyse the benefits of our ARM module, we introduced the proposed ARM module in to a Faster R-CNN based method (NAE [5] method) after RoIAlign pooling. We observed that our ARM module can provide an absolute gains of 1.9% and 1.8% to the mAP and top-1 accuracies over NAE (see Table 1). Our PS-ARM outperforms all existing methods, and achieves a mAP score of 95.2. In terms of top-1 accuracy our method sets a state-of-the-art accuracy of 96.1%.

CUHK-SYSU dataset has different range of gallery sizes such as 50, 100, 500, 1000, 2000, and 4000. To further analyze our proposed method, we performed an experiment by varying the gallery size. Our mAP scores across different gallery size are com-

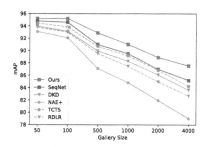

Fig. 6. State-of-the-art comparison of existing methods over CUHK-SYSU dataset with varying gallery sizes. Dotted lines represent two-stage methods whereas solid lines represent one-stage methods. Our PS-ARM shows consistent improvement compared to other methods as the size of gallery increases.

pared with recent one-stage and two-stage methods as shown in Fig. 6.

Fig. 7. Qualitative results of our PS-ARM on challenging PRW dataset. The top-2 matching results for each query image is shown. Our PS-ARM accurately detect and re-identify the query person in both images.

The results shows that our PS-ARM provides consistent performance gain over other approaches across all gallery sizes.

PRW Comparison: Table 1 shows the state-of-the-art comparison on PRW dataset. Among the existing two stage methods, MGN+OR [40] achieves the best mAP score 52.3, but with a very low top-1 accuracy. While comparing the top-1 accuracy, TCTS [34] provides the best performance, but with a very low mAP score. To summarize, the performance of most two-step methods [4, 10, 13, 17, 22, 40] are inferior either in mAP score or top-1 accuracy.

Among one-stage methods, NAE [5] and AlignPS [37], achieved mAP scores of 43.3% and 45.9%. These methods achieved top-1 accuracies of 80.9% and 81.9%. Among the other one-step methods SeqNet [23], PBNet [31], DMRN [18], and DKD [42] also performed well and obtain more than 46% mAP and have more than 86% top-1 accuracy.

To further analyze the effectiveness of our ARM module, we integrate our ARM module to NAE and achieved absolute mAP gain of 3.6% mAP, leading to an mAP score of 46.9%. We observe a similar performance gain over top-1 accuracy, resulting in top-1 score of 81.4%. We also introduced proposed ARM mod-

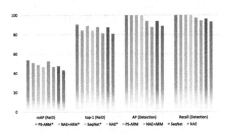

Fig. 8. Person search and detection scores on PRW dataset with and without provided ground-truth detection boxes. The * indicates the results using ground-truth boxes.

ule in Han's [15] method. Compared to the existing methods, [15] utilize a different approach, such as an RoI pooling of 24 × 8 size, instead of 14 × 14. To this end, we modified our PS-ARM to adapt the setting of [15], resulting in an absolute gains of 2% and 1.3% improvement on PRW dataset and obtained 55.3%

mAP and 89.0% top-1 scores, respectively. Our PS-ARM achieve state-of-the-art performance compared the existing one-step and two-step methods. We achieve an mAP score of 52.6% and top-1 score of 88.1%.

Besides similar to cascade RCNN [1], we extend our person search network by introducing an other person re-id branch, called Cascaded PS-ARM. This newly introduced branch takes refined bounding boxes from the Box2 as an input to perform RoIAlign pooling. This strategy further refines the detection and re-identification, producing improved mAP 53.1% and top-1 88.3 % scores.

Qualitative Comparison: Figure 5 shows qualitative comparison between the SeqNet [23] (row 2) and our PS-ARM for the same query input (row 1). Here, true and false matching results are marked in green and red, respectively. The figure shows top-1 results obtained from both methods. It can be observed that SeqNet provides inaccurate top-1 predictions due to the appearance deformations. Our PS-ARM provides accurate predictions on these challenging examples by explicitly capturing discriminative relation features within RoI. Figure 7 shows the qualitative results from our PS-ARM. Here we show the top-2 matching results for each query image. It can be seen that our PS-ARM can accurately detect and re-identify the query person in both gallery images.

Table 2. Ablation study over the PRW dataset by incrementally adding our novel contributions to the baseline. While introducing a MLP mixer to a baseline, both the detection and re-id performance increases over the baseline except top-1. The spatially attended spatial mixing and channel-wise attended channel mixing within our relation mixer captures discriminative relation features within RoI while suppressing distracting background features, hence provides superior re-id performance. Finally, our joint spatio-channel attention removes distracting backgrounds in a joint spatio-channel space, leading to improved detection and re-id performance.

Method	ReID		Detection	
	mAP	top-1	Recall	AP
Baseline	47.6	87.6	96.3	93.1
Baseline + MLP-Mixer	49.1	86.8	96.3	93.3
Baseline + Transformer	47.9	85.8	96.1	93.5
Baseline + Spatio-channel attention layer	48.1	86.2	95.2	93.0
Baseline + Spatial mixing + channel-wise attended channel mixing	49.4	86.7	96.2	93.2
Baseline + Spatially attended spatial mixing + channel mixing	49.5	86.9	96.5	93.4
Baseline + Relation mixer	51.8	87.9	96.6	93.8
PS-ARM (Baseline + ARM)	52.6	88.1	97.1	93.9

4.3   Ablation Study

Here, we perform the ablation study on the PRW dataset. Table 2 shows the performance gain obtained by progressively integrating our novel contributions to the baseline.

First we verify the effectiveness of the context aggregators including MLP-Mixer [32] and Transformer [11] within the proposed framework. The experiment shows that choice of MLP-mixer is better. Moreover, we apply joint spatio-channel attention on the RoI feature maps which results in improved performance compared to baseline. Further, we investigate the introduction of spatially-attended spatial mixing and channel-wise attended channel mixing within our relation mixer which captures discriminative relation features within RoI while suppressing distracting background features. This resulted in superior re-id performance. Introducing our relation mixer comprising of a spatially attended spatial mixing and channel-wise attended channel mixing leads to an overall AP of 93.8 for detection and 51.8 mAP for re-id. To further complement the relation mixer that performs information mixing in the disjoint spatial and channel spaces, we introduce a joint spatio-channel attention. Our joint spatio-channel attention removes distracting backgrounds in a joint spatio-channel space, leading to improved detection and re-id performance by achieving 94.1 and 52.6, respectively.

Relation Between Detection and ReID. In Fig. 8, we validate the effectiveness of the proposed PS-ARM to deal with the contradictory detection and ReID objectives. We compared our PS-ARM with the SOTA SeqNet [23] and NAE [5]. We notice that PS-ARM* and NAE+ARM* outperforms their counter parts provided the ground-truth boxes.

5   Conclusions

We propose a novel person search method named PS-ARM, that strives to capture global relation between different local regions within RoI of a person. The focus of our design is introduction of a novel ARM module, which effectively capturing the global relation within an RoI and make robust against occlusion. The relation mixer block introduces a spatially attended spatial mixing, a channel-wise attended channel mixing, and an input-output feature re-using for capturing discriminative relation features within an RoI. An additional spatio-channel attention layer is introduced within the ARM module to further enrich the discriminability between the foreground/background features in a joint spatio-channel space. Our ARM module is generic and it can be easily integrated to any Faster R-CNN based person search methods. Comprehensive experiments are performed on two benchmark datasets. We achieve state-of-the-art performance on both datasets, demonstrating the merits of our novel contributions.

References

1. Cai, Z., Vasconcelos, N.: Cascade R-CNN: delving into high quality object detection. In: Proceedings of the IEEE Conference on Computer Vision and Pattern Recognition, pp. 6154–6162 (2018)

2. Chang, X., Huang, P.-Y., Shen, Y.-D., Liang, X., Yang, Y., Hauptmann, A.G.: RCAA: relational context-aware agents for person search. In: Ferrari, V., Hebert, M., Sminchisescu, C., Weiss, Y. (eds.) ECCV 2018. LNCS, vol. 11213, pp. 86–102. Springer, Cham (2018). https://doi.org/10.1007/978-3-030-01240-3_6

3. Chen, D., Zhang, S., Ouyang, W., Yang, J., Schiele, B.: Hierarchical online instance matching for person search. In: Proceedings of the AAAI Conference on Artificial Intelligence, vol. 34, pp. 10518–10525 (2020)

4. Chen, D., Zhang, S., Ouyang, W., Yang, J., Tai, Y.: Person search via a mask-guided two-stream CNN model. In: Ferrari, V., Hebert, M., Sminchisescu, C., Weiss, Y. (eds.) ECCV 2018. LNCS, vol. 11211, pp. 764–781. Springer, Cham (2018). https://doi.org/10.1007/978-3-030-01234-2_45

5. Chen, D., Zhang, S., Yang, J., Schiele, B.: Norm-aware embedding for efficient person search. In: CVPR, pp. 12615–12624 (2020)

6. Chen, S., Zhuang, Y., Li, B.: Learning context-aware embedding for person search. arXiv preprint arXiv:2111.14316 (2021)

7. Chen, W., Chen, X., Zhang, J., Huang, K.: Beyond triplet loss: a deep quadruplet network for person re-identification. In: Proceedings of IEEE Conference on Computer Vision and Pattern Recognition (2017)

8. Dai, J., Zhang, P., Lu, H., Wang, H.: Dynamic imposter based online instance matching for person search. Pattern Recogn. **100**, 107120 (2020)

9. Dong, W., Zhang, Z., Song, C., Tan, T.: Bi-directional interaction network for person search. In: Proceedings of the IEEE/CVF Conference on Computer Vision and Pattern Recognition, pp. 2839–2848 (2020)

10. Dong, W., Zhang, Z., Song, C., Tan, T.: Instance guided proposal network for person search. In: CVPR, pp. 2585–2594 (2020)

11. Dosovitskiy, A., et al.: An image is worth 16×16 words: Transformers for image recognition at scale. In: Proceedings of International Conference on Learning Representations (2020)

12. Gao, P., Lu, J., Li, H., Mottaghi, R., Kembhavi, A.: Container: Context aggregation network. arXiv preprint arXiv:2106.01401 (2021)

13. Girshick, R., Iandola, F., Darrell, T., Malik, J.: Deformable part models are convolutional neural networks. In: Proceedings of the IEEE Conference on Computer Vision and Pattern Recognition, pp. 437–446 (2015)

14. Han, B.J., Ko, K., Sim, J.Y.: Context-aware unsupervised clustering for person search. arXiv preprint arXiv:2110.01341 (2021)

15. Han, B.J., Ko, K., Sim, J.Y.: End-to-end trainable trident person search network using adaptive gradient propagation. In: Proceedings of the IEEE/CVF International Conference on Computer Vision, pp. 925–933 (2021)

16. Han, C., et al.: Re-id driven localization refinement for person search. In: Proceedings of IEEE International Conference on Computer Vision (2019)

17. Han, C., et al.: Re-id driven localization refinement for person search. In: Proceedings of the IEEE/CVF International Conference on Computer Vision, pp. 9814–9823 (2019)

18. Han, C., Zheng, Z., Gao, C., Sang, N., Yang, Y.: Decoupled and memory-reinforced networks: Towards effective feature learning for one-step person search. arXiv preprint arXiv:2102.10795 (2021)

19. Kaiser, L., Gomez, A.N., Chollet, F.: DepthWise separable convolutions for neural machine translation. arXiv preprint arXiv:1706.03059 (2017)

20. Kim, H., Joung, S., Kim, I.J., Sohn, K.: Prototype-guided saliency feature learning for person search. In: Proceedings of the IEEE/CVF Conference on Computer Vision and Pattern Recognition, pp. 4865–4874 (2021)

21. Kipf, T.N., Welling, M.: Semi-supervised classification with graph convolutional networks. arXiv preprint arXiv:1609.02907 (2016)

22. Lan, X., Zhu, X., Gong, S.: Person search by multi-scale matching. In: Ferrari, V., Hebert, M., Sminchisescu, C., Weiss, Y. (eds.) ECCV 2018. LNCS, vol. 11205, pp. 553–569. Springer, Cham (2018). https://doi.org/10.1007/978-3-030-01246-5_33

23. Li, Z., Miao, D.: Sequential end-to-end network for efficient person search. In: Proceedings of the AAAI Conference on Artificial Intelligence, vol. 35, pp. 2011–2019 (2021)

24. Liao, S., Hu, Y., Zhu, X., Li, S.Z.: Person re-identification by local maximal occurrence representation and metric learning. In: Proceedings of IEEE Conference on Computer Vision and Pattern Recognition (2015)

25. Liu, H., et al.: Neural person search machines. In: Proceedings of the IEEE International Conference on Computer Vision, pp. 493–501 (2017)

26. Liu, W., Liao, S., Ren, W., Hu, W., Yu, Y.: High-level semantic feature detection: a new perspective for pedestrian detection. In: Proceedings of the IEEE/CVF Conference on Computer Vision and Pattern Recognition, pp. 5187–5196 (2019)

27. Munjal, B., Amin, S., Tombari, F., Galasso, F.: Query-guided end-to-end person search. In: Proceedings of the IEEE Conference on Computer Vision and Pattern Recognition, pp. 811–820 (2019)

28. Pang, Y., Xie, J., Khan, M.H., Anwer, R.M., Khan, F.S., Shao, L.: Mask-guided attention network for occluded pedestrian detection. In: Proceedings of the IEEE/CVF International Conference on Computer Vision, pp. 4967–4975 (2019)

29. Paszke, A., et al.: Pytorch: An imperative style, high-performance deep learning library. In: Advances in Neural Information Processing Systems, vol. 32 (2019)

30. Ren, S., He, K., Girshick, R., Sun, J.: Faster R-CNN: towards real-time object detection with region proposal networks. In: Advances in Neural Information Processing Systems, vol. 28 (2015)

31. Tian, K., Huang, H., Ye, Y., Li, S., Lin, J., Huang, G.: End-to-end thorough body perception for person search. In: Proceedings of the AAAI Conference on Artificial Intelligence, vol. 34, pp. 12079–12086 (2020)

32. Tolstikhin, I.O., Houlsby, N., et al.: MLP-mixer: An all-MLP architecture for vision. In: Advances in Neural Information Processing Systems, vol. 34 (2021)

33. Vaswani, A., et al.: Attention is all you need. In: Proceedings of Advances in Neural Information Processing Systems (2017)

34. Wang, C., Ma, B., Chang, H., Shan, S., Chen, X.: TCTS: a task-consistent two-stage framework for person search. In: Proceedings of the IEEE/CVF Conference on Computer Vision and Pattern Recognition, pp. 11952–11961 (2020)

35. Xiao, J., Xie, Y., Tillo, T., Huang, K., Wei, Y., Feng, J.: IAN: the individual aggregation network for person search. Pattern Recogn. **87**, 332–340 (2019)

36. Xiao, T., Li, S., Wang, B., Lin, L., Wang, X.: Joint detection and identification feature learning for person search. In: Proceedings of the IEEE Conference on Computer Vision and Pattern Recognition, pp. 3415–3424 (2017)

37. Yan, Y., et al.: Anchor-free person search. In: Proceedings of the IEEE/CVF Conference on Computer Vision and Pattern Recognition, pp. 7690–7699 (2021)

38. Yan, Y., Zhang, Q., Ni, B., Zhang, W., Xu, M., Yang, X.: Learning context graph for person search. In: Proceedings of the IEEE/CVF Conference on Computer Vision and Pattern Recognition, pp. 2158–2167 (2019)

39. Yang, L., Zhang, R.Y., Li, L., Xie, X.: Simam: a simple, parameter-free attention module for convolutional neural networks. In: International Conference on Machine Learning, pp. 11863–11874. PMLR (2021)

40. Yao, H., Xu, C.: Joint person objectness and repulsion for person search. IEEE Trans. Image Process. **30**, 685–696 (2020)
41. Zhang, S., Yang, J., Schiele, B.: Occluded pedestrian detection through guided attention in CNNs. In: Proceedings of the IEEE Conference on Computer Vision and Pattern Recognition, pp. 6995–7003 (2018)
42. Zhang, X., Wang, X., Bian, J.W., Shen, C., You, M.: Diverse knowledge distillation for end-to-end person search. In: Proceedings of the AAAI Conference on Artificial Intelligence, vol. 35, pp. 3412–3420 (2021)
43. Zheng, L., Zhang, H., Sun, S., Chandraker, M., Yang, Y., Tian, Q.: Person re-identification in the wild. In: Proceedings of IEEE Conference on Computer Vision and Pattern Recognition, pp. 1367–1376 (2017)
44. Zhong, Y., Wang, X., Zhang, S.: Robust partial matching for person search in the wild. In: Proceedings of the IEEE/CVF Conference on Computer Vision and Pattern Recognition, pp. 6827–6835 (2020)

Weighted Contrastive Hashing

Jiaguo Yu, Huming Qiu, Dubing Chen, and Haofeng Zhang[✉]

School of Computer Science and Engineering, Nanjing University of Science
and Technology, Nanjing 210094, China
{yujiaguo,120106222682,db.chen,zhanghf}@njust.edu.cn

Abstract. The development of unsupervised hashing is advanced by the
recent popular contrastive learning paradigm. However, previous contrastive learning-based works have been hampered by (1) insufficient
data similarity mining based on global-only image representations, and
(2) the hash code semantic loss caused by the data augmentation. In this
paper, we propose a novel method, namely Weighted Contrative Hashing
(WCH), to take a step towards solving these two problems. We introduce
a novel mutual attention module to alleviate the problem of information asymmetry in network features caused by the missing image structure during contrative augmentation. Furthermore, we explore the fine-
grained semantic relations between images, *i.e.*, we divide the images into
multiple patches and calculate similarities between patches. The aggregated weighted similarities, which reflect the deep image relations, are
distilled to facilitate the hash codes learning with a distillation loss, so as
to obtain better retrieval performance. Extensive experiments show that
the proposed WCH significantly outperforms existing unsupervised hashing methods on three benchmark datasets. Code is available at: http://
github.com/RosieYuu/WCH.

Keywords: Unsupervised image retrieval · Deep hashing · Contrastive
learning · Mutual attention · Weighted similarities

1 Introduction

With the advancement of deep neural networks, deep hash has become one of the
most studied approaches for Approximate Nearest Neighbors (ANN) in large-
scale image retrieval. Earlier studies rely heavily on artificial annotations, which
makes it difficult to apply in real-world scenarios due to the high labor costs.
As a result, unsupervised deep hashing [22, 23, 27, 36] has gradually become the
major research direction in this field, with the recent boom in unsupervised
learning [2–4, 12, 13, 26]. The key difficulty with unsupervised hash is that the
ad-hoc encoding process does not extract the key information for hashing, precisely because of the lack of supervised information. Hence, numerous methods
have been proposed to learn better discrete representations for hashing in unsupervised setting.

A large family of recent unsupervised hash learning tasks is based on contrastive learning [3, 4, 13]. These methods build upon instance discrimination,

L. Wang et al. (Eds.): ACCV 2022, LNCS 13845, pp. 251–266, 2023.
https://doi.org/10.1007/978-3-031-26348-4_15

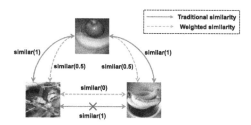

Fig. 1. An example of the conflict of the traditional similarity calculation approach. A typical unsupervised method will treat the top image with both the bottom-left and bottom-right images as similar pairs, because they have a common label. In this case, the two images in the bottom left and bottom right corners should be considered as a similar sample pair. However, the fact is that these two images do not have a common label and they should be considered as a dissimilar pair.

which constructs similar and dissimilar instances and learns the discrete representations by prompting the model to pull in the similar instances and push away the dissimilar instances. With simply the most fundamental concepts for contrastive learning, existing methods [23,27,36] based on contrastive learning have achieved significant success.

Despite their success, most of the current methods mainly focus on adjusting the contrastive loss to fit the hash learning criterion [23,27]. However, directly combining contrastive loss and unsupervised hashing tasks like this leads to two problems. On the one hand, an instance discrimination-based approach leads to the fact that even if the samples are very similar, they still need to be forced apart, *i.e.*, the sample similarity obtained in this way is unreliable. On the other hand, calculating the similarity with the feature vector or hash code of a whole image may lead to the following problem: The top image in Fig. 1 is associated with the labels of apple and banana; the bottom left image is associated with the labels of apple and bird; and the bottom right image is associated with the labels of banana and bowl. In the traditional method, the similarity of both the bottom left image and the bottom right image according to the top image is considered full similar. Therefore, we can say that the bottom left and bottom right images are also very similar. However, the labels of the bottom left and bottom right images do not overlap, *i.e.*, they are actually dissimilar. Based on this, we raise a question: *how to define or even use the similarity between samples to learn high-quality hash codes?*

Curiously, most existing approaches do not focus on this problem. To the best of our knowledge, NSH [36] uses Neural Sorting Operators to obtain the permutation of a vector of similarity scores, and it employs the sorted similarity results to pick the top m positive samples of the anchor, *i.e.*, it improves the comparison by increasing the number of positive samples in the learning framework. However, in the experiment, the optimal number of positive samples is fixed at 3, and all of them are considered fully similar. There are two drawbacks. First, the anchor image and the augmented similar image, especially processed with random crop, are not always similar, *e.g.*, the cropped image only contains the background, which will prompt the network to learn the background rather than the object representation in the image. Second, for multi-labeled images, the positions and sizes of objects vary greatly, and it is difficult to learn a single depth representation that fits all objects. The quality of the hash codes obtained in this way is not high, which will have an impact on the final retrieval results. This can also explain why NSH [36] boosts highly on single-label datasets, but the boost of MAP on multi-label datasets is not very obvious.

In order to solve the above problem, we propose a novel method called Weighted Contrastive Hashing (WCH) to re-weight the similarity of the anchor image and the others. Concretely, we develop a novel metric rule that is more reasonable and efficient for measuring similar samples, and finally apply this rule to the learning of hash codes for better retrieval performance. We divide each image into a number of patches, and exploit the Vision Transformer (ViT) [7] as the encoder to adapt the patches as the input to the model. To obtain the similar samples of an anchor, we use the aggregated vector of similarity between the patches of different samples as weights. Unlike NSH [36], we do not selectively pick the most relevant samples as the positive samples for contrastive learning, but assign trainable weights to all candidate samples, which represent the degree of similarity between samples. That is, we can consider an image pair as less or more similar, rather than stating them as fully similar or dissimilar in absolute terms. Notably, we demonstrate in the experimental section that our method works better than NSH [36]. In addition, to solve the problem of insufficient similarity between augmented images and anchor images, we propose a Mutual Attention (MA) module to reset the weights of each patch of them by calculating their similarity, which can guarantee sufficient similarity of them to make them the most similar pair, so as to facilitate the hash code learning towards the correct direction. In a nutshell, our main contributions are summarized as follows:

- To the best of our knowledge, this is the first time that weighted contrastive learning has been introduced to image retrieval tasks. It alleviates the problem that certain anchor images and negative samples, which are similar enough for the hashing task, are treated as dissimilar pairs.
- We propose a Mutual Attention module to achieve information complementation between the augmented anchor image and the positive sample, avoiding the lack of key information for hashing.
- The excellent performance of our WCH model is extensively demonstrated by comparing it to 18 state-of-the-art hashing frameworks on three benchmark datasets, *i.e.*, CIFAR-10, NUS-Wide, and MS COCO.

2   Related Works

In this section, we will briefly introduce some unsupervised hashing methods here.

Unsupervised Hashing. Early unsupervised hashing methods mainly focus on projecting images to compact representations by constraining the learned hash codes to fit several principles, *e.g.*, quantization [11], balancing [18]. Several recent works using deep learning pay attention to how to generate high quality hash codes [10,17,23,32]. Some others try to preserve the semantic similarity in hash codes [28,30,35], while the majority of methods adopt image pseudo labels with pre-trained networks to convert the unsupervised hashing to fully a supervised learning [37]. Their performance is usually evaluated against ranked candidates. However, they did not try to sort them during training to mine their similarity.

Hashing with Contrastive Learning. Contrastive learning has been a very successful approach for unsupervised hashing tasks. Typical examples include CIMON [23], CIB [27], NSH [36]. All of these methods utilize the contrastive learning framework. As we mentioned before, these methods do not well combine the contrastive learning framework with the hash retrieval task. For example, both CIMON and CIB define the data-enhanced version of an image as a positive sample, and a negative sample is formed by sampling the views of different images. It leads to the possibility that images considered as negative samples may contain positive samples, which will have an impact on the retrieval results. On the other hand, although NSH considers this problem, they simply rank the similarity between anchor samples and select quantitative positive candidates, which does not take into account the similarity degree between the anchor images and augmented images.

Mining Similarity for Unsupervised Hashing. Some methods based on mining similarity aim at solving unsupervised hashing tasks using pairwise methods, *e.g.*, SSDH [34] is a representative method studied in this area. It sets two thresholds at pairwise distances and constructs a similarity structure, and then image features are extracted and hash code learning is performed. However, using two rough thresholds to determine whether they are similar or not is usually unreliable. DistillHash [35] extracts similarity signals using similarity signals from local structures, and further constructs an efficient and adaptive semantic graph, which is updated by decoding it in the context of an autoencoder for hash code learning. MLS^{3}RDUH [33] reconstructs a local semantic similarity structure by exploiting the intrinsic flow structure and cosine similarity in the feature space. DATE [22] improves the commonly used cosine distance by proposing a distribution-based metric. In contrast to these methods, WCH guides the learning of hash codes based on the weighted similarity between patches assigned to each anchor and the rest of the samples.

3   Weighted Contrastive Learning

3.1   Preliminaries

To better explain our method in the next section, we first introduce some concepts and preliminary knowledge here.

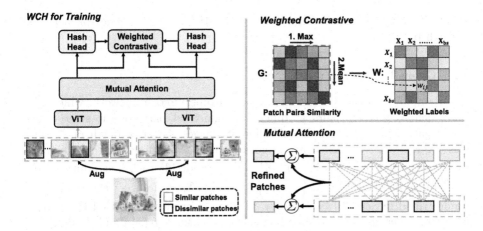

Fig. 2. Overall architecture of the proposed weighted contrastive hashing.

Patch Generation. Following ViT [7], we divide an image $\mathbf{X} \in \mathbb{R}^{s \times s \times c}$ into non-overlapping patches $\mathbf{x}_i \in \mathbb{R}^{p \times p \times c}$, where $i = 1, 2, \cdots, n$. It is obvious that $s^2 = n \times p^2$ and c is the number of image channels.

WCH Encoder. Recent work proposes the use of the ViT model as a universal feature extractor [8]. Inspired by these works, we also use ViT as an encoder for our model. We first flatten the patches $\mathbf{x}_i$ into a vector $\mathbf{p}_i \in \mathbb{R}^{1 \times d}$, where d is the dimension of the vector, and then use a trainable linear projector $\mathbf{LP}$ to map the vector to embedding. The output of this projection is referred to the patch embedding as follows:

$$\mathbf{E}_i = \mathbf{LP}(\mathbf{p}_i), \tag{1}$$

where $\mathbf{E}_i$ is the patch embedding associated with the i-th patch. Unlike the standard ViT, our model does not use the class token. We add the position embedding into the patch embeddings, and the final embedding for ViT Input is:

$$\mathbf{PE}_i = \mathbf{E}_i + \mathbf{PoE}, \tag{2}$$

where, $\mathbf{PoE}$ stands for the position embedding, and $\mathbf{PE}$ is the final projecting embedding, which will be fed into the transformer encoder $f_\theta(\cdot)$.

Binarization. In WCH, we revisit the problem of how to evaluate whether a candidate sample is comparable to an anchor in the contrastive learning framework and obtain higher quality hash codes in the image retrieval task. As for an image retrieval task, the goal is to learn a binary vector $\mathbf{b}_i \in \{-1, 1\}^l$ by mapping the data $\mathbf{x}_i$ into the encoder, where l is the length of the hash code. In general, the hash code is obtained by the sign function:

$$\mathbf{b}_i = \text{sign}(h(\mathbf{x}_i)) \in \{-1, 1\}^l, \tag{3}$$

where $h(\cdot)$ is the encoding function, which mainly consists of the WCH encoder and a one-layer projector. Since the $\text{sign}(\cdot)$ function is non-differentiable, we adopt a straight-through estimator (STE) [1] that allows back-propagation through $\mathbf{b}_i$.

3.2 Overall Architecture

WCH employs contrastive learning as an unsupervised framework, which typically defines positive pairs as different augmented parts of the same image and negative pairs as samples of different images. Given a batch of N samples $\left\{\mathbf{X}^1, \cdots, \mathbf{X}^i, \cdots, \mathbf{X}^N\right\}$, it first goes through two different data augmenters to get two different views $\tilde{\mathbf{X}}^i$ and $\hat{\mathbf{X}}^i$. Then, we divide each image into n non-overlapping patches, $\tilde{\mathbf{X}}^i = \left[\tilde{\boldsymbol{x}}_1^i; \cdots; \tilde{\boldsymbol{x}}_k^i; \cdots; \tilde{\boldsymbol{x}}_n^i\right], \hat{\mathbf{X}}^i = \left[\hat{\boldsymbol{x}}_1^i; \cdots; \hat{\boldsymbol{x}}_k^i; \cdots; \hat{\boldsymbol{x}}_n^i\right]$. As we described in Sect. 3.1, we employ ViT as the encoder, and feed the patches into it to generate the corresponding encoded features $f_\theta(\tilde{\boldsymbol{x}}_k^i)$ and $f_\theta(\hat{\boldsymbol{x}}_k^i)$.

During traditional augmentation, the augmented two images are usually taken as a similar pair to guide the training direction. However, some cropped images containing background only are totally different from others, which might damage the training process. Therefore, we employ the Mutual Attention (MA) module to re-weight the image patches to guarantee similarity between them. After that, the weighted image similarity is calculated by computing the patch similarity between different images, and it is subsequently used to construct the final weighted contrastive loss function. The overall architecture is illustrated in Fig. 2.

3.3 Mutual Attention

Given any two encoded patches $f_\theta(\tilde{\boldsymbol{x}}_k^i)$ and $f_\theta(\hat{\boldsymbol{x}}_t^i)$ in the corresponding augmented pair, the similarity of them can be calculated as

$$\mathbf{s}_{k,t} = f_\theta(\tilde{\boldsymbol{x}}_k^i)^T f_\theta(\hat{\boldsymbol{x}}_t^i). \tag{4}$$

Therefore, we can construct a similarity matrix $\mathbf{S} \in \mathbb{R}^{n \times n}$, which measures the similarity between each patch of the augmented pair. Then, we normalize the row vectors and column vectors respectively with softmax function:

$$\begin{cases} \mathbf{S}^1 = softmax(\mathbf{s}_{1*}, \cdots, \mathbf{s}_{i*}, \cdots, \mathbf{s}_{n*}) \\ \mathbf{S}^2 = softmax(\mathbf{s}_{*1}, \cdots, \mathbf{s}_{*j}, \cdots, \mathbf{s}_{*n}) \end{cases}, \tag{5}$$

where $\mathbf{s}_{i*}$ stands for the i-th row of $\mathbf{S}$, and $\mathbf{s}_{*j}$ means the j-th column of $\mathbf{S}$. Then, the refined patch vector is reconstructed with the following calculation:

$$\begin{cases} \tilde{\boldsymbol{f}}_k^i = \sum_{j=1}^n s_{j,k}^2 f_\theta(\tilde{\boldsymbol{x}}_j^i) \\ \hat{\boldsymbol{f}}_k^i = \sum_{j=1}^n s_{k,j}^1 f_\theta(\hat{\boldsymbol{x}}_j^i) \end{cases}, \tag{6}$$

where $s_{j,k}^1$ is the j-th row and k-column element of $\mathbf{S}^1$, and $s_{k,j}^2$ is analogously defined. After this operation, the refined augmented pair can be guaranteed to be similar, and thence be undoubtedly treated as a positive pair.

3.4   Weighted Similarities Calculation

Previous works often use embedding vectors to explore the relationship between different images. Specifically, most existing unsupervised hashing methods assume binary similarity between two images, *i.e.*, two images can be similar (positive sample) or dissimilar (negative sample). For example, NSH [36] uses hash codes to calculate the degree of similarity between images, and then ranks them according to the magnitude of similarity, and selects the top m positive samples according to the result of the ranking, *i.e.*, determines that these m samples and anchor are similar. However, selecting positive samples based on similarity like this will cause two problems. First, there may be noise in the positive samples. Since the number of positive samples is set to a fixed value, forcing a fixed number of positive samples based on the ranking results will slow down the convergence of the model. Second, the results of the first m closest samples may not be equivalent. For an anchor image $\mathbf{X}_i$, $\mathbf{X}_j$ is one of the selected m positive samples that are similar to $\mathbf{X}_i$ in one iteration. However, it is possible that in another iteration, $\mathbf{X}_j$ is not one of the m closest samples since there are more similar images in this training batch. In this situation, $\mathbf{X}_j$ will be treated as a negative sample of $\mathbf{X}_i$, which results in inconsistency with the former one, and this conflict will damage the training process and cause the training to fail to converge.

Weighted Labels Processing. To tackle these problems, instead of adopting a strategy such as selecting positive samples, we reformulate the rules for computing the similarity between images and use the obtained similarity to re-weight the contrastive loss to capture the semantic information that may overlap between the anchor and negative samples. In WCH, we exploit the fine-grained interaction results between patches to explore the relationship between different images.

Specifically, suppose there are two patch features $\boldsymbol{f}_k^i$ and $\boldsymbol{f}_t^j$ extracted from two different images $\mathbf{X}^i$ and $\mathbf{X}^j$, respectively. The similarity between them can be defined as

$$g_{kt}^{ij} = (\boldsymbol{f}_k^i)^T \boldsymbol{f}_t^j. \tag{7}$$

Therefore, the similarity matrix of $\mathbf{X}^i$ and $\mathbf{X}^j$ can be constructed as $\mathbf{G}^{ij} \in \mathbb{R}^{n \times n}$. For each row in $\mathbf{G}^{ij}$, the max value represents the most similar batches in $\mathbf{X}^i$ and all patches in $\mathbf{X}^j$, and the mean of the max values of each row is the similarity of $\mathbf{X}^i$ and $\mathbf{X}^j$:

$$w_{ij} = \text{mean}(\max_{row}(\mathbf{G}^{ij})), \tag{8}$$

where $max(\cdot)$ means to take the maximum value according to the row direction, and $mean(\cdot)$ stands for calculating the mean value of the vector. For a mini-batch containing bs images, including one augmented image and $bs - 1$ other images, the similarity matrix $\mathbf{W} \in \mathbb{R}^{bs \times bs}$ can be constructed with Eq. 8. To fit the value of $\mathbf{W}$ within a proper range, we conduct a temperature weighted row softmax as $\mathbf{W}_{i*} = softmax(\mathbf{W}_{i*}/\tau_w)$, where τ_w [14] is the temperature coefficient. Furthermore, to guarantee that the augmented images are equivalent to the anchor images, we divide each row with the element w_{ii}:

$$\mathbf{W} = \text{diag}(\text{diag}(\mathbf{W}))^{-1}\mathbf{W}, \tag{9}$$

where $\text{diag}(\cdot)$ means extracting the diagonal vector from a matrix or constructing a diagonal matrix with a vector.

3.5   Training and Inference

For training, we use the maximum similarity between patches to guide the contrastive objective [25]:

$$\mathcal{L}_{\text{WCE}} = -\sum_{i=1}^{bs}\sum_{j=1}^{bs} w_{ij} \log \frac{\exp\left(\tilde{\mathbf{b}}_i \hat{\mathbf{b}}_j^{\mathsf{T}}/l/\tau\right)}{\sum_{k=1}^{bs} \exp\left(\tilde{\mathbf{b}}_i \hat{\mathbf{b}}_k^{\mathsf{T}}/l/\tau\right)}, \tag{10}$$

where τ is the temperature scale. Finally, the loss function is formulated as

$$\mathcal{L}_{\text{WCH}} = \mathcal{L}_{\text{WCE}} + \mathcal{L}_R, \tag{11}$$

where $\mathcal{L}_R$ refers to the quantization loss and bit balancing loss [9]. The whole learning procedure is shown in Algorithm 1.

Inference Process. In the inference process, WCH abandons the MA and weighted labeling modules dedicated to training and keeps only the encoder and hash head for generating hash codes characterizing the semantic information of the images. The Hamming distance between the hash codes of the images is then computed to accomplish the retrieval task.

4   Discussion

Remark 1: Why Do We Choose the ViT Encoder? In WCH, our key idea is to use the patch-level semantic information captured by ViT [7] as a benchmark to measure the degree of similarity between arbitrary images and assign corresponding weights to each pair of images by aggregating the similarity between patches to measure the degree of similarity. Unlike the recent self-supervised visual representation learning-based approaches [27,36], they only determine similar samples by the global feature similarity of the whole image. Instead, we introduce a novel inter-patch-based fine-grained interaction module using the ViT model, enabling fine-grained interactions between patches and each pair of images to mine more detailed semantic alignment.

Furthermore, we use the ViT model to address the problems posed by contrastive learning methods that rely on instance discrimination tasks. As mentioned before, positive native pairs are defined as different views of the same image, while negative pairs are formed by sampling views of different images. This common approach ignores their semantic content. Our approach, on the

---

Algorithm 1: The Training Procedure of WCH.

Input: Dataset $\mathcal{D} = \{\mathbf{x}_i\}_{i=1}^{N}$ and batch size n.
Output: Network parameters θ.
for batch in $\mathcal{D}.repeat()$ **do**
    batch$_1$, batch$_2$ = aug(batch), aug(batch)
    f$_1$, f$_2$ = M$_\theta$(batch$_1$), M$_\theta$(batch$_2$)
    # mutual attention
    sim = einsum($'$nid, njd $\rightarrow$ nij$'$, f$_1$, f$_2$)
    f$_1$ = einsum($'$nid, ndj $\rightarrow$ nij$'$, softmax(sim.T), f$_1$)
    f$_2$ = einsum($'$nid, ndj $\rightarrow$ nij$'$, softmax(sim), f$_2$)
    # weighted similarities calculation
    sim = einsum($'$nid, mjd $\rightarrow$ nmij$'$, f$_1$, f$_2$)
    sim = softmax(sim.max(-1).mean(-1)/τ_w)
    weighted = matmul(diag(diag(sim))$^{-1}$, sim)
    # hashing
    b$_1$, b$_2$ = hash(f$_1$.mean(1)), hash(f$_2$.mean(1))
    logits = softmax(matmul(b1, b2.T)/l/τ)
    # weighted corss entropy
    loss= cross_entropy(logits, weighted)
    loss.backward()
end

---

other hand, fully exploits the semantic content of the images and makes reasonable use of fine-grained interaction results as a measure of similarity between images, as detailed in Sect. 3.4.

Remark 2: Why Mutual Attention Helps? First, note the phenomenon that most models construct positive and negative samples by treating the same images produced by different augmenters as positive pairs, while the rest of the samples are considered as negative pairs. However, this manually designed approach involves many manual choices, and inappropriate data augmentation schemes may severely alter the image structure, resulting in data-enhanced images that do not possess label-preserving properties, *i.e.*, images undergo transformations that may lose high-level semantic information. For example, a common data augmentation scheme is random cropping, which may randomly crop out the sample information that contains label-related information for single-labeled images. Similarly, for multi-labeled images, where the position and size of objects vary greatly, the random cropping method will most likely crop out some objects in multi-labeled images, making the sample information contained in multi-labeled images reduced. This operation will lead to asymmetric semantic information between the anchor and the positive samples.

The mutual attention module in Fig. 2 reconstructs the feature vectors associated with the pictures based on the similarity between positive sample pairs of patches. Therefore, it can be seen as a specific attention mechanism. Intuitively, it focuses our attention on the degree of similarity of patch pairs. The attention fraction is used so that the feature vectors of each patch carry information about

Table 1. Performance comparison (mAP) of WCH and the state-of-the-art **unsupervised** hashing methods. *Note that we use a more common setting on NUS-WIDE with the 21 most frequent classes, while some papers report results on 10 classes.

Method	Reference	CIFAR-10			NUS-WIDE			MS COCO		
		16 bits	32 bits	64 bits	16 bits	32 bits	64 bits	16 bits	32 bits	64 bits
AGH [21]	ICML11	0.333	0.357	0.358	0.592	0.615	0.616	0.596	0.625	0.631
ITQ [11]	PAMI13	0.305	0.325	0.349	0.627	0.645	0.664	0.598	0.624	0.648
DGH [20]	NeurIPS14	0.335	0.353	0.361	0.572	0.607	0.627	0.613	0.631	0.638
SGH [6]	ICML17	0.435	0.437	0.433	0.593	0.590	0.607	0.594	0.610	0.618
BGAN [31]	AAAI18	0.525	0.531	0.562	0.684	0.714	0.730	0.645	0.682	0.707
BinGAN [38]	NeurIPS18	0.476	0.512	0.520	0.654	0.709	0.713	0.651	0.673	0.696
GreedyHash [32]	NeurIPS18	0.448	0.473	0.501	0.633	0.691	0.731	0.582	0.668	0.710
HashGAN [10]	CVPR18	0.447	0.463	0.481	–	–	–	–	–	–
DVB [29]	IJCV19	0.403	0.422	0.446	0.604	0.632	0.665	0.570	0.629	0.623
DistillHash [35]	CVPR19	0.284	0.285	0.288	0.667	0.675	0.677	–	–	–
TBH [30]	CVPR20	0.532	0.573	0.578	0.717	0.725	0.735	0.706	0.735	0.722
MLS^{3}RDUH [33]	IJCAI20	0.369	0.394	0.412	0.713	0.727	0.750	0.607	0.622	0.641
DATE [22]	MM21	0.577	0.629	0.647	0.793	0.809	0.815	–	–	–
MBE [17]	AAAI21	0.561	0.576	0.595	0.651	0.663	0.673	–	–	–
CIMON [23] *	IJCAI21	0.451	0.472	0.494	–	–	–	–	–	–
CIBHash [27]	IJCAI21	0.590	0.622	0.641	0.790	0.807	0.815	0.737	0.760	0.775
SPQ [15]	ICCV21	0.768	0.793	0.812	0.766	0.774	0.785	–	–	–
NSH [36]	IJCAI22	0.706	0.733	0.756	0.758	0.811	0.824	0.746	0.774	0.783
WCH	Proposed	**0.897**	**0.910**	**0.932**	**0.799**	**0.823**	**0.838**	**0.776**	**0.808**	**0.834**

other patches to different degrees. More specifically, this attention mechanism is very useful for multiple patches, especially when there are many classes of objects and the positions are highly variable.

Remark 3: Why Do We Gather W in This Way? The purpose of Weighted Labels is to use the maximum similarity between patches to guide the contrastive objective. Using the maximum similarity between patches in Eq. 8, we can get the most similar patch pair among all patches in the two images. Then the sum of the maximum similarity is averaged. The model learns the fine-grained semantic alignment between patches by applying the weighted label to the contrastive loss.

5   Experiments

In this section, we conduct experiments on three datasets, including one single-labeled dataset and two multi-labeled datasets, to evaluate our method.

5.1   Datasets and Evaluation Metrics

Three benchmark datasets are used in our experiments. **CIFAR-10** [16] consists of 60,000 images from 10 classes. We follow the common setting [10] and select 10,000 images (1000 per class) as the query set. The remaining 50,000 images are regarded as the database. **NUS-WIDE** [5] has of 81 categories of images. We adopt the 21-class subset following [36]. 100 images of each class are utilized as a query set, with the remaining being the gallery. **MS COCO** [19] is a benchmark for multiple tasks. We use the conventional set with 12,2218 images. We randomly select 5,000 images as queries with the remaining ones the database.

Evaluation Metric. To compare the proposed method with the baselines, we adopt several widely-used evaluation metrics, including the mean average precision (mAP), top-K precision (P@K), precision-recall (PR) curves [37].

5.2   Implementation Details

For all three datasets, the images were resized to $224 \times 224 \times 3$ and we adopt the image augmentation strategies of [3]. The standard ViT-Base [7] was used as the backbone, with patches of size and number 16 and 196, respectively. As in previous work [23,27], we loaded a pre-trained model trained on ImageNet to accelerate the convergence. We used the cosine decay method and trained 50 epochs for all models, with the initial learning rate set to 1×10^{-5}.

5.3   Comparison with the SotA

Baselines. We compare WCH against 18 state-of-the-art baselines, including 3 traditional unsupervised hashing methods and 15 recent unsupervised hashing methods. For fair comparisons, all the methods are reported with identical training and test sets. Additionally, the shallow methods are evaluated with the same deep features as the ones we are using.

Results. Table 1 shows the retrieval performance in mAP and Table 2 demonstrates the precision of the first 1000 returned images. It can be clearly observed that WCH obtains the best results on all three datasets for the two metrics. Another interesting observation is that WCH significantly outperforms the previous works CIBHash and NSH on different hash bits and datasets. Note that all three methods use contrastive learning. In addition, the P-R curves of WCH and several baselines on CIFAR-10 and MS COCO are reported in Fig. 3, from which it can also be discovered that the curves of our method are highly above those of other methods for all three different code lengths.

5.4   Ablation Studies

In this subsection, we considered the following ablation experiments to verify the effectiveness and contribution of each component of WCH, and the specific results are shown in Table 3.

(i) **ViT Baseline.** We first investigate the enhancements that the ViT backbone brings to the unsupervised hashing domain. In this baseline, the class token covering global features is applied directly to the hash head to generate a hash code characterizing the image. Subsequent contrastive loss is used to update the network parameters, which form a design close to the CIB [27] except that the network backbone differs. Regrettably, the application of the ViT backbone alone is not sufficient to improve the performance of the unsupervised hashing.

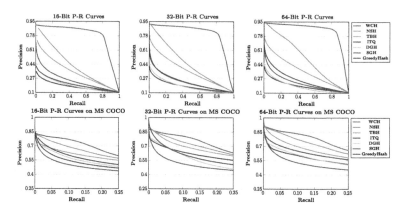

Fig. 3. P-R curves comparison with other methods on CIFAR-10 and MS COCO.

Table 2. P@1000 results on CIFAR-10 and MS COCO.

Method	CIFAR-10			MS COCO		
	16 bits	32 bits	64 bits	16 bits	32 bits	64 bits
AGH	0.306	0.321	0.317	0.602	0.635	0.644
DGH	0.315	0.323	0.324	0.623	0.642	0.650
HashGAN	0.418	0.436	0.455	–	–	–
GreedyHash	0.322	0.403	0.444	0.603	0.624	0.675
TBH	0.497	0.524	0.529	0.646	0.698	0.701
CIBHash	0.526	0.570	0.583	0.734	0.767	0.785
NSH	0.691	0.716	0.744	0.733	0.770	0.805
WCH	**0.889**	**0.902**	**0.923**	**0.795**	**0.830**	**0.855**

(ii) **Without $\mathcal{L}_R$.** We also reveal the impact of traditional quantization loss and bit balance loss [9] on WCH. It can be seen that these conventional regularizers have no significant improvement in the encoding quality. As a result, we can attribute the good performance entirely to our design.

(iii) **MA → mean.** We use this baseline to demonstrate the validity of our MA module. Here we remove the mutual attention mechanism of anchor and positive samples in Eq. 6 and replace it with the averaging operation. Although it also achieves trivially good results, there is still a noticeable margin of difference with the performance of WCH, which indicates that the motivation of mutual attention can play a positive role.

(iv) **Weighted → hard.** This baseline does not use weighted labels, but rather the most fundamental hard labels, which means that the weighted contrastive learning degrades to the standard contrastive learning. The non-negligible performance degradation in Table 3 precisely illustrates the shortcoming of standard contrastive learning, which cannot close the distance between the anchor and similar negative samples in the feature space. This also highlights the crucial role played by our core motivation from the side.

Table 3. Ablation study results of mAP@1000 on MS COCO. The baselines are constructed by replacing some key modules of WCH.

	Baseline	16 bits	32 bits	64 bits
(i)	ViT Baseline	0.573	0.595	0.622
(ii)	Without $\mathcal{L}_R$	0.773	0.810	0.828
(iii)	MA → mean	0.742	0.782	0.805
(iv)	weighted → hard	0.738	0.777	0.799
(v)	Without scale	0.461	0.479	0.491
	WCH	**0.776**	**0.808**	**0.834**

Fig. 4. Examples of top-10 retrieved results of 32-bit on CIFAR-10.

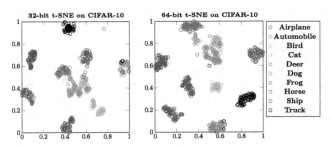

Fig. 5. 32-bit and 64-bit t-SNE visualization results on CIFAR-10.

(v) **Without scale.** In this baseline, we remove the operation defined in Eq. 9 and simply use the similarity matrix **W** in Eq. 8 as a weighted label during the calculation of the loss. We can strikingly see an unexpectedly dramatic performance slippage. Hence, affine mapping based on positive sample similarity is a key factor to guarantee the effectiveness of weighted comparison learning.

Results. Baseline (i) contradicts our intuition that directly replacing the backbone network with ViT can not bring meaningful performance improvement. Baseline (iii) shows that MA is an effective solution to deal with the information asymmetry problem for positive samples. We use baseline (iv) to validate our core motivation that weighted contrast learning can substantially alleviate the class collision problem of negative samples and thus further improve the retrieval performance.

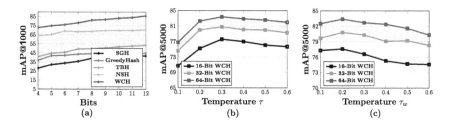

Fig. 6. (a) mAP@1000 results with extremely short code lengths on CIFAR-10. (b)&(c)Effects of different temperatures τ and τ_w on MS COCO.

5.5   Visualization and Hyper-parameters

Visualization. To more intuitively demonstrate the performance of our method, we show the retrieved top-10 images on CIFAR-10 in Fig. 4, where a high semantic accuracy can be observed from the results. In addition, to show whether the embedded hash codes are discriminative enough for retrieval, the t-SNE plots [24] of hash codes for both 32-bit and 64-bit on CIFAR-10 are also illustrated in Fig. 5, where the plotted dots of different classes show obvious boundaries between them, which means that the generated codes are separable and shows the consistency with other results.

Hyper-parameters. In Fig. 6(a), we show the results for very short hash code lengths on CIFAR10. Although the performance varies slightly depending on the hyperparameter settings, it is generally stable and state-of-the-art. We also evaluated the impact of the temperature coefficient τ of the WCE loss and the temperature coefficient τ_w of computing the weighted labels on the final performance of MSCOCO, and we depict these trends in Fig. 6(b) and (c).

6   Conclusion

In this paper, we propose a weighted contrastive hashing model to explore semantic information based on fine-grained information interactions between patches for image retrieval. The proposed mutual attention module can well solve the inconsistency of the anchor image and the augmented images. A weighted coefficient is calculated to weigh the similarities of the images in a training batch, and it can better improve the hash code learning. Extensive experiments show that the proposed method improves the state-of-the-art unsupervised hashing scheme in image retrieval.

Acknowledgements. This work was supported in part by the National Natural Science Foundation of China (NSFC) under Grants No. 61872187, and No. 62072246, in part by the Natural Science Foundation of Jiangsu Province under Grant No. BK20201306, and in part by the "111" Program under Grant No. B13022.

References

1. Bengio, Y., Léonard, N., Courville, A.: Estimating or propagating gradients through stochastic neurons for conditional computation. arXiv preprint arXiv:1308.3432 (2013)
2. Caron, M., Misra, I., Mairal, J., Goyal, P., Bojanowski, P., Joulin, A.: Unsupervised learning of visual features by contrasting cluster assignments. In: NeurIPS (2020)
3. Chen, T., Kornblith, S., Norouzi, M., Hinton, G.: A simple framework for contrastive learning of visual representations. In: ICML (2020)
4. Chen, X., Fan, H., Girshick, R., He, K.: Improved baselines with momentum contrastive learning. arXiv preprint arXiv:2003.04297 (2020)
5. Chua, T.S., Tang, J., Hong, R., Li, H., Luo, Z., Zheng, Y.: NUS-WIDE: a real-world web image database from national university of Singapore. In: CIVR (2009)
6. Dai, B., Guo, R., Kumar, S., He, N., Song, L.: Stochastic generative hashing. In: ICML (2017)
7. Dosovitskiy, A., et al.: An image is worth 16x16 words: transformers for image recognition at scale. ArXiv abs/2010.11929 (2021)
8. Dubey, S.R., Singh, S.K., Chu, W.: Vision transformer hashing for image retrieval. ArXiv abs/2109.12564 (2021)
9. Liong, V.E., Lu, J., Wang, G., Moulin, P., Zhou, J.: Deep hashing for compact binary codes learning. In: CVPR (2015)
10. Dizaji, K.G., Zheng, F., Sadoughi, N., Yang, Y., Deng, C., Huang, H.: Unsupervised deep generative adversarial hashing network. In: CVPR (2018)
11. Gong, Y., Lazebnik, S., Gordo, A., Perronnin, F.: Iterative quantization: a procrustean approach to learning binary codes for large-scale image retrieval. IEEE Trans. Pattern Anal. Mach. Intell. **35**(12), 2916–2929 (2013)
12. Grill, J.B., et al.: Bootstrap your own latent: a new approach to self-supervised learning. ArXiv abs/2006.07733 (2020)
13. He, K., Fan, H., Wu, Y., Xie, S., Girshick, R.: Momentum contrast for unsupervised visual representation learning. In: CVPR (2020)
14. Hinton, G., Vinyals, O., Dean, J.: Distilling the knowledge in a neural network. arXiv preprint arXiv:1503.02531 (2015)
15. Jang, Y.K., Cho, N.I.: Self-supervised product quantization for deep unsupervised image retrieval. In: 2021 IEEE/CVF International Conference on Computer Vision (ICCV), pp. 12065–12074 (2021)
16. Krizhevsky, A., Hinton, G.: Learning multiple layers of features from tiny images. Technical Report, University of Toronto (2009)
17. Li, Y.Q., van Gemert, J.C.: Deep unsupervised image hashing by maximizing bit entropy. In: AAAI (2021)
18. Lin, K., Lu, J., Chen, C.S., Zhou, J.: Learning compact binary descriptors with unsupervised deep neural networks. In: CVPR (2016)
19. Lin, T.-Y., et al.: Microsoft COCO: common objects in context. In: Fleet, D., Pajdla, T., Schiele, B., Tuytelaars, T. (eds.) ECCV 2014. LNCS, vol. 8693, pp. 740–755. Springer, Cham (2014). https://doi.org/10.1007/978-3-319-10602-1_48
20. Liu, W., Mu, C., Kumar, S., Chang, S.F.: Discrete graph hashing. In: NeurIPS (2014)
21. Liu, W., Wang, J., Kumar, S., Chang, S.F.: Hashing with graphs. In: ICML (2011)
22. Luo, X., et al.: A statistical approach to mining semantic similarity for deep unsupervised hashing. In: MM (2021)
23. Luo, X., et al.: CIMON: towards high-quality hash codes. In: IJCAI (2021)

24. van der Maaten, L., Hinton, G.: Visualizing data using t-SNE. J. Mach. Learn. Res. **9**, 2579–2605 (2008)
25. van der Oord, A., Li, Y., Vinyals, O.: Representation learning with contrastive predictive coding. arXiv preprint arXiv:1807.03748 (2018)
26. Qi, Y., Song, Y.Z., Zhang, H., Liu, J.: Sketch-based image retrieval via Siamese convolutional neural network. In: IEEE International Conference on Image Processing (ICIP) (2016)
27. Qiu, Z., Su, Q., Ou, Z., Yu, J., Chen, C.: Unsupervised hashing with contrastive information bottleneck. In: IJCAI (2021)
28. Shen, F., Xu, Y., Liu, L., Yang, Y., Huang, Z., Shen, H.T.: Unsupervised deep hashing with similarity-adaptive and discrete optimization. IEEE Trans. Pattern Anal. Mach. Intell. **40**(12), 3034–3044 (2018)
29. Shen, Y., Liu, L., Shao, L.: Unsupervised binary representation learning with deep variational networks. Int. J. Comput. Vision **127**(11–12), 1614–1628 (2019)
30. Shen, Y., et al.: Auto-encoding twin-bottleneck hashing. In: CVPR (2020)
31. Song, J., He, T., Gao, L., Xu, X., Hanjalic, A., Shen, H.T.: Binary generative adversarial networks for image retrieval. In: AAAI (2018)
32. Su, S., Zhang, C., Han, K., Tian, Y.: Greedy hash: towards fast optimization for accurate hash coding in CNN. In: NeurIPS (2018)
33. Tu, R.C., Mao, X.L., Wei, W.: MLS3RDUH: deep unsupervised hashing via manifold based local semantic similarity structure reconstructing. In: IJCAI (2020)
34. Yang, E., Deng, C., Liu, T., Liu, W., Tao, D.: Semantic structure-based unsupervised deep hashing. In: IJCAI (2018)
35. Yang, E., Liu, T., Deng, C., Liu, W., Tao, D.: DistillHash: unsupervised deep hashing by distilling data pairs. In: CVPR (2019)
36. Yu, J., Shen, Y., Zhang, H., Torr, P.H.S., Wang, M.: Learning to hash naturally sorts. In: IJCAI (2022)
37. Zhang, H., Liu, L., Long, Y., Shao, L.: Unsupervised deep hashing with pseudo labels for scalable image retrieval. IEEE Trans. Image Process. **27**(4), 1626–1638 (2018)
38. Zieba, M., Semberecki, P., El-Gaaly, T., Trzcinski, T.: BinGAN: learning compact binary descriptors with a regularized GAN. In: NeurIPS (2018)

Content-Aware Hierarchical Representation Selection for Cross-View Geo-Localization

Zeng Lu[1,3], Tao Pu[2], Tianshui Chen[1(✉)], and Liang Lin[2]

[1] Guangdong University of Technology, Guangzhou, China
tianshuichen@gmail.com
[2] Sun Yat-Sen University, Guangzhou, China
[3] Guangzhou Quwan Network Technology Co., Ltd., Guangzhou, China

Abstract. Cross-view geo-localization (CVGL) aims to retrieve the images that contain the same geographic target content and are from different views. However, the target content usually scatters over the whole image, and they are indiscernible from the background. Thus, it is difficult to learn feature representation that focuses on these contents, rendering CVGL a challenging and unsolved task. In this work, we design a Content-Aware Hierarchical Representation Selection (CA-HRS) module, which can be seamlessly integrated into current deep networks to facilitate CVGL. This module can help focus more on the target content while ignoring the background region, thus as to learn more discriminative feature representation. Specifically, this module learns hierarchical important factors to each location of the feature maps according to their importance and enhances the feature representation based on the learned factors. We conduct experiments on several large-scale datasets (i.e., University-1652, CVUSA and CVACT), and the experiment results show the proposed module can obtain obvious performance improvement over current competing algorithms. Codes are available at https://github.com/Allen-lz/CA-HRS.

Keywords: Geo localization · Feature selection · Image retrieval

1 Introduction

As a practical and challenging sub-task of image retrieval [14,27,30], cross-view geo-localization (CVGL) aims to find the target images in one view among large-scale candidates (gallery) that have the same contents with the input query image in another view. Formally, there are three views of images, i.e., satellite-view,

This work is supported in part by National Natural Science Foundation of China (NSFC) under Grant No. 62206060 and in part by Science and Technology Project of Guangdong Province under Grant No. 2021A1515011341.

L. Wang et al. (Eds.): ACCV 2022, LNCS 13845, pp. 267–280, 2023.
https://doi.org/10.1007/978-3-031-26348-4_16

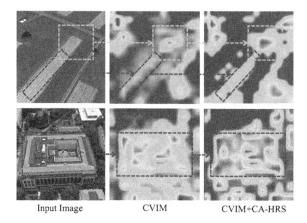

Input Image   CVIM   CVIM+CA-HRS

Fig. 1. Two examples of the input image (left), the learned feature maps by the baseline CVIM (middle), and the learned feature maps by integrating the CA-HRS module (right).

drone-view, and ground-view images. It contains three types of tasks according to different views of the input and target images: **Drone → Satellite** with the input image of drone-view and the target images of satellite-view; **Satellite → Drone** with the input image of satellite-view and the target images of drone-view; **Ground → Satellite** with the input image of ground-view and the target images of satellite-view.

Recently, CVGL receives increasing attention as it benefits variant applications such as agriculture, aerial photography, event detection, and accurate delivery [10,18,27,32]. Current works for this task combine metric learning [2,14,15] or domain adaptation [12,23] with deep neural networks to learn view-invariant feature representation. More recent works further introduce manually annotated orientation information to regularize training to improve CVGL performance. However, this works either increase the model complexity and inference time or incur additional annotation overhead, making them impractical and unscalable. On the other hand, the target contents usually scatter over the whole image. Current algorithms roughly find the content regions but can not well highlight these regions to learn more discriminative feature representation. As shown in Fig. 1, the learned feature representation is slightly obvious but can not be distinguished from the surrounding background regions.

To address these issues, we design a novel yet effective Content-Aware Hierarchical Representation Selection (CA-HRS) module that helps to better focus on the target content meanwhile suppress the background regions. We experimentally find that it has higher activation values on the content regions and slightly lower activation values on the background regions. Thus, it is expected to set the regions with higher activation values the higher value and set the regions with lower activation values with lower values, and thus make the contents distinguished from the background regions. To achieve this end, the CA-HRS module computes an average representation as a threshold. Then, it sets the locations with the activation values higher than this threshold as positive while those values lower than this threshold as positive to obtain an enhancement coefficient map. Moreover, the content regions usually have different scales for different images, and we propose to compute multi-scale enhancement coefficient maps, and combine them to obtain the hierarchical enhancement coefficient maps. Finally, we design an adaptive residual fusion mechanism to seamlessly and flexibly integrate the CA-HRS into current CVGL algorithms for feature enhancement to facilitate the performance of CVGL. As shown in Fig. 1, by integrating the CA-HRS module into the current cross-view image matching (CVIM) algorithm [30], it can learn feature maps that obviously focus on the content regions while ignoring most of the background regions. Moreover, the CA-HRS incur no additional parameters and very limited computational overhead (i.e., about 1.0%), and thus it can be integrated into variant CVGL algorithms to boost their performance.

The contributions of this work can be summarized in the following. First, we design a novel yet effective content-aware hierarchical feature selection (CA-HRS) module that can help feature enhancement by focusing more on content regions while ignoring the background regions. Second, we introduce an adaptive residual fusion mechanism that can integrate the CA-HRS into current algorithms flexibly and seamlessly. Finally, we conduct extensive experiments on the large-scale University-1652, CVUSA, and CVACT datasets, and the experiment results show that the proposed module can obviously improve the performance of current state-of-the-art algorithms.

2 Relate Work

With the advancement of deep learning in images [4,16,17], cross-view geo-localization based on deep learning has achieved significant development. Siamese network [6] and metric learning are often used in image retrieval. The contrastive loss can reduce the distance between two matched positive samples and increase the distance between two unmatched negative samples [15]. The triple loss can simultaneously reduce and increase the distance between positive and negative samples [2,14]. There is still a lot of works that use metric learning to train deep neural networks to learn discriminative features [7,8,11,22].

In order to reduce the distance between two different domains, the most direct method is to transform the image features in one domain to another domain, namely cross-domain adaptation task [5,28]. Lin et al. introduce a cross-view feature translation approach to greatly extend the reach of image geo-localization methods [12]. Shi et al. applied a regular polar transform to warp a satellite image such that its domain is closer to that of a ground-view panorama [21]. Shi et al. proposed a novel Cross-View Feature Transport (CVFT) technique to explicitly establish cross-view domain transfer that facilitates feature alignment between ground and satellite images [23].

The orientation information is also integrated into the neural network for learning. Liu et al. integrated the orientation information of each pixel into the convolution neural network for cross-view geo-localization, which improved the geo-localization accuracy [14]. Vo et al. proposed a new loss function, which combined rotation invariance and orientation regression in the training process, so that the network learned orientation and got a better feature representation [24]. Rodrigues et al. proposed a semantic-driven data enhancement technology that enables Siamese Network to discover objects that are difficult to capture [20]. Then, the enhanced samples are input to a multi-scale attention embedding network to perform the matching task. Zhu et al. [33] propose to estimate the orientation and align a pair of cross-view images with unknown alignment information.

In order to enable the network to focus on the feature extraction of images in different domains. CBMA [26] simply combines convolutional layers with sigmoid to extract key features. Zheng et al. applied Instance loss [31] to cross-view geo-localization [30]. Satellite images, ground images, and drone images were extracted by using corresponding backbone network to extract features. These features share the same classifier. They continued to use this network structure in their subsequent work LPN [25], LPN used a fixed division method to extract local features using context, so that the features were discriminative. Arandjelovic et al. proposed NetVLAD [1], which is a scene recognition method. It can extract local features and aggregate them to enhance the expressive ability of features. the method can also be applied to CVGL. Hu et al. introduced a CVM-Net [11], in which NetVLAD is embedded in Siamese network [6]. CVM-Net extracts the local features and then integrate them for image retrieval and geo-localization. Experiments show that the network with local features is better than that with only global features.

3   Hierarchical Enhancement Coefficient Map

In this section, we present the computing process of the hierarchical enhancement coefficient map (HECM) which helps to pay more attention to the important content regions while ignoring the unimportant background regions. In the context of the CVGL task, we observe the activation values of the content regions are usually slightly higher than those of the background regions. Thus, it is expected to increase the higher activation values even higher to emphasize the content regions and meanwhile to decrease the smaller activation values to even smaller to suppress the background. On the other hand, the different content regions usually share different scales. To achieve the above end, we propose to compute HECM that has higher important factors for regions with higher activation values and has smaller important factors otherwise.

Specifically, given the input feature maps of layer l, denoted as $\mathbf{f}^l \in \mathcal{R}^{W^l \times H^l \times C^l}$ in which W^l, H^l and C^l are the width, height, and channel number, we first compute an mean activation value for each location, formulated as

$$a^l = \frac{1}{C^l} \sum_{c=1}^{C^l} \mathbf{f}^{l,c}, \tag{1}$$

where $\mathbf{f}^{l,c}$ is the c-th feature map of $\mathbf{f}^l$. Then, average the activation values over all locations to obtain the mean representation, formulated as

$$thr_0^l = \frac{1}{W^l H^l} \sum_{w=1}^{W^l} \sum_{h=1}^{H^l} a_{wh}^l. \tag{2}$$

As discussed above, we consider the regions with activation values higher than the mean representation as important content region while those with activation values smaller than the mean representation as unimportant background regions. Intuitively, we can compute the ECM $\mathbf{m}_0^l \in \mathcal{R}^{W^l \times H^l}$, in which the value $\mathbf{m}_{0,wh}^l$ denote the important of location (w, h) and it can be computed by

$$\mathbf{m}_{0,wh}^l = \mathbf{1}(a_{wh}^l \geq thr_0^l). \tag{3}$$

In this way, we can obtain an ECM $\mathbf{m}_0^l$ to indicate the importance of each location. Considering different scales of content regions, we further introduce the average pooling with different kernel sizes that operates on the mean activation map a^l to obtain the thresholds and ECMs for different scales. For scale i, we first perform an average pooling with a kernel size of $k_i^l \times k_i^l$ on a^l to obtain an new mean activation map $a_i^l \in \mathcal{R}^{W_i^l \times H_i^l}$. Then, the threshold can be computed by

$$thr_i^l = \frac{1}{W_i^l H_i^l} \sum_{w=1}^{W_i^l} \sum_{h=1}^{H_i^l} a_{i,wh}^l. \tag{4}$$

Similarly, we compare the activation value of each location of a with the threshold to obtain the corresponding ECM $\mathbf{m}_i^l$, in which $\mathbf{m}_{i,wh}^l$ can be computed by

$$\mathbf{m}_{i,wh}^l = \mathbf{1}(a_{wh}^l \geq thr_i^l). \tag{5}$$

Finally, we combine all the ECMs to obtain the HECM $\mathbf{m}^l$. For each location (w, h), the value can be compute by

$$\mathbf{m}_{wh}^l = 1 + \log_{10}(1 + \sum_{i=0}^{K} \mathbf{m}_{i,wh}^l), \tag{6}$$

where K is the number of scales, and the log functions are used to normalize the important values for more stable training.

Selection of the Kernel Sizes. To ensure seamless and flexible integration with current CVGL algorithms, the kernel sizes of the pooling operations should be automatically adjusted according to the size of the input feature maps. Concretely, it is expected that the kernel size of the largest kernel can not large than $min(W^l, H^l)/2$ and the kernel sizes of different pooling have great variance. Suppose there are K scales of pooling operation, we can first obtain the maximal kernel size and base kernel variation stride:

$$k_m^l = min(h, w)/s, \tag{7}$$

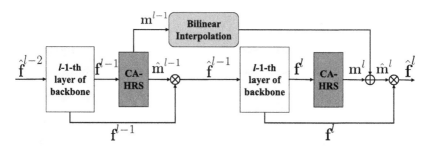

Fig. 2. A illustration of adaptive residual fusion (ARF). In particular, $\hat{f}_l$ is the final enhanced feature map of l-th layer, $\hat{m}_l$ is the final hierarchical enhancement coefficient map of CA-HRS of l-th layer.

$$s_b^l = max(1, min(W_l, H^l)/K - 1). \tag{8}$$

Then, we can compute the kernel size for the i-th by:

$$k_i^l = k_m^l - i \times s_b^l. \tag{9}$$

4   Adaptive Residual Fusion Mechanism

As suggested in previous works, local information may be lost if the network goes deeper. This may lead to fuzzy boundaries of the target content, and thus resulting in degration of the CVGL performance. Inspired by previous work [9], we design an adaptive residual fusion (ARF) mechanism that takes the HECM for enhancement to avoid losing the local information. Figure 2 presents an overall computing process of the ARF mechanism. It first uses the bilinear interpolation to re-sample the previous HECM $\mathbf{m}^{l-1}$ to the same size with $\mathbf{m}^l$, and adds them to obtain the final HECM for layer l, formulated as

$$\hat{\mathbf{m}}^l = \mathbf{m}^l + \phi_{bi}(\mathbf{m}^{l-1}, W^l, H^l), \tag{10}$$

where ϕ_{bi} is the bilinear interpolation operation that re-samples the $\mathbf{m}^{l-1}$ from the size of $W^{l-1} \times H^{l-1}$ to the size of $W^l \times H^l$. Once we obtain $\hat{\mathbf{m}}^l$, we perform dot product of the final HECM $\hat{\mathbf{m}}^l$ and each channel of the feature maps:

$$\hat{\mathbf{f}}_c^l = \mathbf{f}_c^l \cdot \hat{\mathbf{m}}^l. \tag{11}$$

We perform the operation for all channels and obtain the final enhanced feature representation $\hat{\mathbf{f}}^l$.

5   Experiments

In this section, we present in-depth ablative studies to analyze the effect of each component of the proposed CA-HRS module. We also combine the CA-HRS module with current leading algorithms and compare it with state-of-the-art algorithms to show its superiority.

5.1 Experimental Settings

Datasets. For a fair comparison, we follow previous works [25] to conduct experiments on the CVUSA [27], CVACT [14], and University-1652 [30] datasets. CVUSA and CVACT are two most-used datasets that cover the ground-view and satellite-view images and are used to evaluate the subtask of Ground → Satellite. Therein, CVUSA contains a training set of 35,532 ground-and-satellite image pairs and a validation set of 8884 image pairs. CVACT contains 35,532 ground-and-satellite image pairs for training, 8884 image pairs for validation, and 92,802 image pairs for testing. There exists merely one true-matched image for each query image on the CVUSA test set and exist several true-matched images for each query image on the CVACT test set. Different from the above two datasets, University-1652 covers the satellite-view and drone-view images, which are used to evaluate both two subtasks of Satellite → drone and Drone → Satellite. Specifically, in the Satellite → Drone task, it provides 37,855 drone-view images in the query set and 701 true-matched satellite-view images and 250 satellite-view distractors in the gallery. There is only one true-matched satellite-view image under this setting. In the Drone → Satellite task, it provides 701 satellite-view query images, and 37,855 true-matched drone-view images and 13,500 drone-view distractors in the gallery. There are multiple true-matched drone-view images under this setting.

Implementation Details. We only perform simple data enhancement with random cropping at a certain size and a 0.5 probability of horizontal flipping for all images that are input to the network. Since there are few aerial images, we also perform a 90° random rotation operation on images. None of the above data enhancements are used in the testing stage. The height and width of the input image are set to 256. n and s are set to 3 and 2 respectively, the corresponding scale number is 3. For the first 5 epochs of training, a warmup strategy is utilized to slowly increase the learning rate to its initial value. And, after every 80 epochs, the learning rate change to 1/10 of its original value. The Stochastic Gradient Descent (SGD) is used as the optimizer. We adopt the structure of D2 for backbone ResNet-50, and adopt CA-HRS module after the last convolutional layer for backbone VGG-16 since VGG-16 does not divide the layers like ResNet-16.

Evaluation Protocol. In the evaluation phase, the feature map output by backbone is transformed into a vector through the shape change of the tensor. Then the vectors belonging to the query image and gallery image will be normalized. Finally, the cosine similarity between them is calculated to measure the similarity between images, and the retrieval result is generated according to the similarity. The images ranked in the top-10 of similarity will be used as the retrieved results.

Table 1. Comparison of R@1 and AP of different integration strategies. The best results are highlighted in bold.

Structure	Drone → Satellite		Satellite → Drone	
	R@1	AP	R@1	AP
S1	62.85	66.54	75.46	62.28
S2	63.01	67.32	77.03	62.89
S3	63.85	68.36	78.07	64.82
S4/D4	63.16	67.74	77.19	63.05
D3	63.54	68.42	77.51	63.62
D2	**64.87**	**69.28**	**80.03**	**64.73**
D1	64.02	68.76	78.73	64.17

Table 2. Comparison of R@1 and AP of different scale numbers. The best results are highlighted in bold.

Scale number	Drone → Satellite		Satellite → Drone	
	R@1	AP	R@1	AP
1	63.13	67.52	76.89	63.01
2	63.93	68.01	77.85	63.28
3	**64.87**	**69.28**	**80.03**	**64.73**
4	64.16	68.53	79.19	63.92
5	64.09	68.34	78.75	63.64
6	63.82	68.17	78.46	63.33

Table 3. Comparison of R@1 and AP of the CVIM+CA-HRS with and without the ARF mechanism. The best results are highlighted in bold.

Method	Drone → Satellite		Satellite → Drone	
	R@1	AP	R@1	AP
CVIM+CA-HRS w/o ARF	63.12	67.48	78.32	64.11
CVIM+CA-HRS w/ ARF	**64.87**	**69.28**	**80.03**	**64.73**

5.2   Analyses the CA-HRS Module

To analyze the effect of the CA-HRS module, and integrate it into two baselines, namely cross-view image matching (CVIM) [30] and local pattern network (LPN) [25]. Both two algorithms use the ResNet-50 and VGG-16 that have four layer blocks as the backbone.

Analysis of Integration Strategy. As a plug-and-play module, CA-HRS can be integrated into any layer of the deep neural network. However, it may lead

Table 4. Comparison of R@1, R@Top1% and AP of the LPN and CVIM with and without the CA-HRS module foar the Satellite → Drone, Drone → Satellite and Ground → Satellite subtasks on the University-1652, CVUSA and CVACT datasets. The best results are highlighted in bold.

Dataset	Task	Methods	Backbone	R@1	R@Top1%	AP
University-1652	Satellite → Drone	CVIM [30]	ResNet-50	74.47	97.15	59.45
		CVIM [30] + CA-HRS	ResNet-50	80.03	98.29	64.27
		LPN [25]	ResNet-50	86.45	–	74.79
		LPN [25] + CA-HRS	ResNet-50	86.88	98.72	74.83
	Drone → Satellite	CVIM [30]	ResNet-50	58.23	86.00	62.91
		CVIM [30] + CA-HRS	ResNet-50	64.87	90.45	69.28
		LPN [25]	ResNet-50	75.93	–	79.14
		LPN [25] + CA-HRS	ResNet-50	76.67	93.76	79.77
CVUSA	Ground → Satellite	CVIM [30]	VGG-16	43.91	91.78	–
		CVIM [30] + CA-HRS	VGG-16	48.83	93.96	53.82
		LPN [25]	VGG-16	79.69	98.50	–
		LPN [25] + CA-HRS	VGG-16	84.89	99.39	87.18
		LPN [25]	ResNet-50	85.79	99.41	–
		LPN [25] + CA-HRS	ResNet-50	87.16	99.49	89.15
CVACT	Ground → Satellite	LPN [25]	VGG-16	73.83	95.87	–
		LPN [25] + CA-HRS	VGG-16	77.15	96.96	80.11
		LPN [25]	ResNet-50	79.99	97.03	–
		LPN [25] + CA-HRS	ResNet-50	80.91	97.07	83.20

to different effects if integrating this module into different layers. In this part, we analyze the effect of this choice. Here, we conduct experiments using the CVIM baseline with ResNet-50 backbone on the University-1652 dataset. We design two categories of integration strategies: shallow layer integration that mainly integrates the CA-HRS in shallow layers and deep layer integration that mainly integrates the CA-HRS in deep layers. As shown in Table 1, the backbone of ResNet-50 contains 4 layers, Sx indicates that CA-HRS is preferentially embedded in the shallow layers of ResNet-50, Dx indicates that CA-HRS is preferentially embedded in the deep layers of ResNet-50, and x represents the number of CA-HRS. We find the performance increasingly becomes better from strategy S1 to S3, as stacking more CA-HRS may better enhance feature representation. However, the performance degrades when adding more CA-HRS, i.e., S4. One possible reason for this phenomenon is that may over-emphasize the content regions and lose some less-obvious but equally-important regions. Thus, the performance inversely increases from setting D4 to D2. As shown D2 achieves the best performance for both the Drone → Satellite and Satellite → Drone subtasks. Thus, we select the D2 strategy.

Analysis of Scale Number. The number of scales in the CA-HRS module controls the richness of the scale information and it also plays key roles in the CVGL tasks. To analyze its effect, we further conduct experiments that vary the scale number from 1 to 6, and present the performance comparisons on the University-1652 dataset. As shown in Table 2, the R@1 and AP both the Drone

$\rightarrow$ Satellite and Satellite $\rightarrow$ Drone subtasks increases obviously when increasing the scale number from 1 to 3, as it enhances feature representation from more scale and thus focus more and better on the content regions. However, the R@1 and AP become saturate or even worse when further increasing it from 3 to 6. Obviously, the scale information is saturated, and thus adding more scale can not help capture more information and have the risk to be over-fitting. Based on these analyses, we set the scale number as 3 in the experiments.

Analysis of the ARF Mechanism. In this work, we introduce the ARF mechanism to better update the HECM. Here, we further conduct an experiments to analyze its contribution by comparing the results that removes this mechanism. As shown Table 3, we find the R@1 and AP suffer from evident drop on both Drone $\rightarrow$ Satellite and Satellite $\rightarrow$ Drone subtasks.

Analysis of Complexity and Efficiency. As we introduce an additional CA-HRS module, we also analyze the model complexity and efficiency. As discussed above, the CA-HRS module does not contain any learnable parameters, and thus the model size is the same as the baselines without integrating the CA-HRS module. Here, we main analyze the number of multiply-accumulate operations (MAC) and inference time with and without the CA-HRS module. We find the number of MAC are nearly the same for both the CVIM and LPN baselines with and without the CA-HRS modules. In addition, the inference time increases from 6.80 ms to 6.87 ms and from 6.95 ms to 7.01 ms, with the relative increases of 1.03% and 0.86%, respectively. These comparisons suggest the CA-HRS does not incur additional computation overhead and is practical for real-world applications.

Table 5. The comparison results of LPN with CA-HRS module and current state-of-the-art competitors for the Drone $\rightarrow$ Satellite and Satellite $\rightarrow$ Drone subtasks on the University-1652 dataset. The best results are highlighted in bold.

Methods	Dataset	Backbone	Drone $\rightarrow$ Satellite		Satellite $\rightarrow$ Drone	
			R@1	AP	R@1	AP
CVIM [30]	University-1652	ResNet-50	58.49	63.31	71.18	58.74
Contrastive Loss [13]	University-1652	ResNet-50	52.39	57.44	63.91	52.24
Triplet Loss (M = 0.3) [3]	University-1652	ResNet-50	55.18	59.97	63.62	53.85
Triplet Loss (M = 0.5) [3]	University-1652	ResNet-50	55.58	58.60	64.48	53.15
Soft Margin Triplet Loss [11]	University-1652	ResNet-50	53.21	58.03	65.62	54.47
LPN [25]	University-1652	ResNet-50	75.93	79.14	86.45	74.79
LPN+CA-HRS	University-1652	ResNet-50	**76.67**	**79.77**	**86.88**	**74.83**

Contribution of CA-HRS Module. As the above-mentioned description, we use the CVIM and LPR algorithms as baselines. Here, we emphasize the comparison with these two baselines to show the contribution of the CA-HRS module.

Table 6. The comparison results of LPN with CA-HRS module and current state-of-the-art competitors for Ground → Satellite subtask on the CVUSA and CVACT datasets. The best results are highlighted in bold. - indicates the corresponding results are not provided.

Methods	Backbone	CVUSA				CVACT			
		R@1	R@5	R@10	R@Top1%	R@1	R@5	R@10	R@Top1%
MCVPlaces [27]	AlexNet	–	–	–	34.40	–	–	–	–
Regmi [19]	X-Fork	48.75	–	81.27	95.98	–	–	–	–
Siam-FCANet [2]	ResNet-34	–	–	–	98.30	–	–	–	–
CVM-Net [11]	VGG-16	18.80	44.42	57.47	91.54	20.15	45.00	56.87	87.57
Zhai [29]	VGG-16	–	–	–	43.20	–	–	–	–
Orientation [14]	VGG-16	27.15	54.66	67.54	93.91	46.96	68.28	75.48	92.04
CVIM [30]	VGG-16	43.91	66.38	74.58	91.78	31.20	53.64	63.00	85.27
CVFT [23]	VGG-16	61.43	84.69	90.94	99.02	61.05	81.33	86.52	95.93
LPN [25]	VGG-16	79.69	91.70	94.55	98.50	73.85	87.54	90.66	95.87
LPN+CA-HRS	VGG-16	**84.89**	**95.18**	**97.04**	**99.39**	**77.15**	**90.11**	**92.50**	**96.96**
LPN [25]	ResNet-50	85.79	95.38	96.98	99.41	79.99	90.63	92.56	97.03
LPN+CA-HRS	ResNet-50	**87.16**	**95.98**	**97.55**	**99.49**	**80.91**	**90.95**	**92.93**	**97.07**

(i) Comparisons with the CVIM baseline. To ensure fair comparisons, we conduct experiments to compare the results in paper [30]. Here, we perform the comparison with the ResNet-50 as the baseline on the University-1652 dataset and with the VGG-16 as the baseline on the CVUSA dataset. The results are presented in Table 4. On the University-1652 dataset, integrating the CA-HRS module obviously improves all metrics on both Satellite → Drone, Drone → Satellite subtasks. For example, it outperforming the baseline CVIM by 5.56%, 1.14%, 4.82% in R@1, R@Top1 and AP for the Satellite → Drone task, and 6.64%, 4.45%, and 6.37% for the Drone → Satellite subtask, respectively. On the CVUSA dataset, it also obtains evident improvement by integrating the CA-HRS module. Specifically, the R@1 and R@Top1 improvements are 4.92% and 2.18%. These comparisons well demonstrate the effectiveness of the CA-HRS module.

(ii) Comparison with the LPN baseline. LPR is a more recent-proposed algorithm and it achieves better overall performance. Here, we also compare with the results that are reported in the original paper [25] for fair comparisons. Here, we conduct experiments with ResNet-50 as the backbone on the University-1652 dataset and with both ResNet-50 and VGG-16 on the CVUSA and CVACT datasets. As shown in Table 4, integrating the CA-HRS with the LPN algorithm also leads to performance improvement over all settings. On the University-1652 dataset, integrating the CA-HRS module improves the R@1 and AP from 86.45% and 74.79% to 86.88% and 74.83% for the Satellite → Drone and from 75.93% and 79.14% to 76.67% to 79.77%, respectively. On the CVUSA and CVACT datasets with VGG-16 as the backbone, the R@1 and R@Top1 improvements are 5.20% and 0.89% on the CVUSA dataset and 3.32% and 1.09% on the CVACT dataset. These comparisons further suggest that the CA-HRS can generalize to different baseline algorithms to facilitate the CVGL task.

5.3 Comparison with State of the Arts

In this part, we present the comparisons with current state-of-the-art algorithms to show the superiority of the proposed CA-HRS. Here, we present the results of LPN+CA-HRS as it achieves the overall best performance.

Performance on University-1652. As all of the current algorithms that have reported their results on University-1652 use the ResNet-50 as the backbone, we also present our results with the same backbone for fair comparisons. As shown in Table 5, LPN is the previous best-performing algorithm for both the Drone → Satellite and Satellite → Drone tasks, which obtains very obvious improvement compared with early works. By integrating the CA-HRS module, it can further improve the performance. Specifically, it leads to 0.74% and 0.43% R@1 improvement on both two subtasks, respectively.

Performance on CVUSA and CVACT. On the CVUSA and CVACT datasets, current algorithms use ResNet-50, VGG-16, and some other networks as backbones. For fair comparisons, we divide them into three groups according to the used backbone networks for fair comparisons, i.e., ResNet-50-based, VGG-16-based, and other-net-based. Besides, current algorithms [25] [23] mainly present the R@K (K=1,5,10) and R@Top1% and do not report the AP, and thus we also present these metrics for comparisons. The comparison results are presented in Table 6. When using the VGG-16 as the backbone, the current best-performing algorithm is also LPN that achieves the R@1, R@5, R@10, R@Top1% of 79.69%, 91.70%, 94.55%, 98.50% on the CVUSA dataset and 73.85%, 87.54%, 90.66%, 95.87% on the CVACT dataset. By integrating the CA-HRS module into the LPN, it boosts these metrics by 5.20%, 3.48%, 2.49%, 0.89% on the CVUSA dataset and 3.30%, 2.57%, 1.84%, 1.09% on the CVACT dataset. It is noteworthy that the improvement is more obvious for the more strict metric. When using the ResNet-50 as the backbone, the LPN can achieve even better performance compared with those using VGG-16. Expectedly, it can still improve the performance when integrating the CA-HRS module.

6 Conclusion

In this work, we design a novel yet effective content-aware hierarchical representation selection module that can be seamlessly integrated into current CVGL algorithms to facilitate the performance of CVGL. The proposed module helps to locate the content regions while ignoring the background regions to learn discriminative feature representation. We conduct extensive experiments on multiple CVGL datasets (e.g., University-1652, CVUSA and CVACT) to demonstrate the superiority of our proposed module.

References

1. Arandjelovic, R., Gronat, P., Torii, A., Pajdla, T., Sivic, J.: NetVLAD: CNN architecture for weakly supervised place recognition. In: Proceedings of the IEEE Conference on Computer Vision and Pattern Recognition, pp. 5297–5307 (2016)
2. Cai, S., Guo, Y., Khan, S., Hu, J., Wen, G.: Ground-to-aerial image geo-localization with a hard exemplar reweighting triplet loss. In: Proceedings of the IEEE/CVF International Conference on Computer Vision, pp. 8391–8400 (2019)
3. Chechik, G., Sharma, V., Shalit, U., Bengio, S.: Large scale online learning of image similarity through ranking (2010)
4. Chen, T., Pu, T., Wu, H., Xie, Y., Lin, L.: Structured semantic transfer for multi-label recognition with partial labels. In: Proceedings of the AAAI Conference on Artificial Intelligence, vol. 36, pp. 339–346 (2022)
5. Chen, T., Pu, T., Wu, H., Xie, Y., Liu, L., Lin, L.: Cross-domain facial expression recognition: a unified evaluation benchmark and adversarial graph learning. IEEE Trans. Pattern Anal. Mach. Intell. 44, 9887–9903 (2021)
6. Chopra, S., Hadsell, R., LeCun, Y.: Learning a similarity metric discriminatively, with application to face verification. In: 2005 IEEE Computer Society Conference on Computer Vision and Pattern Recognition (CVPR2005), vol. 1, pp. 539–546. IEEE (2005)
7. Deng, W., Zheng, L., Ye, Q., Kang, G., Yang, Y., Jiao, J.: Image-image domain adaptation with preserved self-similarity and domain-dissimilarity for person re-identification. In: Proceedings of the IEEE Conference on Computer Vision and Pattern Recognition, pp. 994–1003 (2018)
8. Hadsell, R., Chopra, S., LeCun, Y.: Dimensionality reduction by learning an invariant mapping. In: 2006 IEEE Computer Society Conference on Computer Vision and Pattern Recognition (CVPR2006), vol. 2, pp. 1735–1742. IEEE (2006)
9. He, K., Zhang, X., Ren, S., Sun, J.: Identity Mappings in Deep Residual Networks. In: Leibe, B., Matas, J., Sebe, N., Welling, M. (eds.) ECCV 2016. LNCS, vol. 9908, pp. 630–645. Springer, Cham (2016). https://doi.org/10.1007/978-3-319-46493-0_38
10. Hsieh, M.R., Lin, Y.L., Hsu, H.W.: Drone-based object counting by spatially regularized regional proposal network. In: ICCV, pp. 4165–4173 (2017)
11. Hu, S., Feng, M., Nguyen, R.M., Lee, G.H.: CVM-Net: cross-view matching network for image-based ground-to-aerial geo-localization. In: Proceedings of the IEEE Conference on Computer Vision and Pattern Recognition, pp. 7258–7267 (2018)
12. Lin, T.Y., Belongie, S., Hays, J.: Cross-view image geolocalization. In: Proceedings of the IEEE Conference on Computer Vision and Pattern Recognition, pp. 891–898 (2013)
13. Lin, T.Y., Cui, Y., Belongie, S., Hays, J.: Learning deep representations for ground-to-aerial geolocalization. In: Proceedings of the IEEE Conference on Computer Vision and Pattern Recognition, pp. 5007–5015 (2015)
14. Liu, L., Li, H.: Lending orientation to neural networks for cross-view geo-localization. In: Proceedings of the IEEE/CVF Conference on Computer Vision and Pattern Recognition, pp. 5624–5633 (2019)
15. Melekhov, I., Kannala, J., Rahtu, E.: Siamese network features for image matching. In: 2016 23rd International Conference on Pattern Recognition (ICPR), pp. 378–383. IEEE (2016)
16. Pu, T., Chen, T., Wu, H., Lin, L.: Semantic-aware representation blending for multi-label image recognition with partial labels. arXiv preprint arXiv:2203.02172 (2022)

17. Pu, T., Chen, T., Xie, Y., Wu, H., Lin, L.: Au-expression knowledge constrained representation learning for facial expression recognition. In: 2021 IEEE International Conference on Robotics and Automation (ICRA), pp. 11154–11161. IEEE (2021)

18. Qian, Y., et al.: Building information modeling and classification by visual learning at a city scale (2019)

19. Regmi, K., Shah, M.: Bridging the domain gap for ground-to-aerial image matching. In: Proceedings of the IEEE/CVF International Conference on Computer Vision, pp. 470–479 (2019)

20. Rodrigues, R., Tani, M.: Are these from the same place? seeing the unseen in cross-view image geo-localization. In: Proceedings of the IEEE/CVF Winter Conference on Applications of Computer Vision, pp. 3753–3761 (2021)

21. Shi, Y., Liu, L., Yu, X., Li, H.: Spatial-aware feature aggregation for image based cross-view geo-localization. Adv. Neural. Inf. Process. Syst. **32**, 10090–10100 (2019)

22. Shi, Y., Yu, X., Campbell, D., Li, H.: Where am i looking at? joint location and orientation estimation by cross-view matching. In: Proceedings of the IEEE/CVF Conference on Computer Vision and Pattern Recognition, pp. 4064–4072 (2020)

23. Shi, Y., Yu, X., Liu, L., Zhang, T., Li, H.: Optimal feature transport for cross-view image geo-localization. In: Proceedings of the AAAI Conference on Artificial Intelligence, vol. 34, pp. 11990–11997 (2020)

24. Vo, N.N., Hays, J.: Localizing and orienting street views using overhead imagery. In: Leibe, B., Matas, J., Sebe, N., Welling, M. (eds.) ECCV 2016. LNCS, vol. 9905, pp. 494–509. Springer, Cham (2016). https://doi.org/10.1007/978-3-319-46448-0_30

25. Wang, T., et al.: Each part matters: local patterns facilitate cross-view geo-localization. IEEE Transactions on Circuits and Systems for Video Technology (2021)

26. Woo, S., Park, J., Lee, J.-Y., Kweon, I.S.: CBAM: Convolutional Block Attention Module. In: Ferrari, V., Hebert, M., Sminchisescu, C., Weiss, Y. (eds.) ECCV 2018. LNCS, vol. 11211, pp. 3–19. Springer, Cham (2018). https://doi.org/10.1007/978-3-030-01234-2_1

27. Workman, S., Souvenir, R., Jacobs, N.: Wide-area image geolocalization with aerial reference imagery. In: Proceedings of the IEEE International Conference on Computer Vision, pp. 3961–3969 (2015)

28. Xie, Y., Chen, T., Pu, T., Wu, H., Lin, L.: Adversarial graph representation adaptation for cross-domain facial expression recognition. In: Proceedings of the 28th ACM International Conference on Multimedia, pp. 1255–1264 (2020)

29. Zhai, M., Bessinger, Z., Workman, S., Jacobs, N.: Predicting ground-level scene layout from aerial imagery. In: Proceedings of the IEEE Conference on Computer Vision and Pattern Recognition, pp. 867–875 (2017)

30. Zheng, Z., Wei, Y., Yang, Y.: University-1652: a multi-view multi-source benchmark for drone-based geo-localization. In: Proceedings of the 28th ACM International Conference on Multimedia, pp. 1395–1403 (2020)

31. Zheng, Z., Zheng, L., Garrett, M., Yang, Y., Xu, M., Shen, Y.D.: Dual-path convolutional image-text embeddings with instance loss. ACM Trans. Multimedia Comput. Commun. Appl. (TOMM) **16**(2), 1–23 (2020)

32. Zhu, P., Wen, L., Bian, X., Ling, H., Hu, Q.: Vision meets drones: a challenge. arXiv: Computer Vision and Pattern Recognition (2018)

33. Zhu, S., Yang, T., Chen, C.: Revisiting street-to-aerial view image geo-localization and orientation estimation. In: Proceedings of the IEEE/CVF Winter Conference on Applications of Computer Vision, pp. 756–765 (2021)

CLUE: Consolidating Learned and Undergoing Experience in Domain-Incremental Classification

Chengyi Cai[1], Jiaxin Liu[1], Wendi Yu[1], and Yuchen Guo[2]([⊠])

[1] Tsinghua-Berkeley Shenzhen Institute, Shenzhen, China
{ccy20,liujiaxi20,ywd20}@mails.tsinghua.edu.cn
[2] Beijing National Research Center for Information Science and Technology,
Tsinghua University, Beijing 100084, China
yuchen.w.guo@gmail.com

Abstract. Deep neural networks tend to be vulnerable to catastrophic forgetting when learning new tasks. To address it, continual learning has become a promising and popular research field in recent years. It is noticed that plentiful research predominantly focuses on class-incremental (CI) settings. However, another practical setting, domain-incremental (DI) learning, where the domain distribution shifts in new tasks, also suffers from deteriorating rigidity and should be emphasized. Concentrating on the DI setting, in which the learned model is overwritten by new domains and is no longer valid for former tasks, a novel method named Consolidating Learned and Undergoing Experience (CLUE) is proposed in this paper. In particular, CLUE consolidates former and current experiences by setting penalties on feature extractor distortion and sample outputs alteration. CLUE is highly applicable to classification models as neither extra parameters nor processing steps are introduced. It is observed through extensive experiments that CLUE achieves significant performance improvement compared with other baselines in the three benchmarks. In addition, CLUE is robust even with fewer replay samples. Moreover, its feasibility is supported by both theoretical derivation and model interpretability visualization. The code is available at: https://github.com/Multiplied-by-1/CLUE.

1 Introduction

Humans are born with the ability to learn continuously, adapting to new scenarios without forgetting previous knowledge when facing the ever-changing world. However, when deep neural networks are applied in learning new tasks, sharp declines will be observed in the performance of previous ones, which is called catastrophic forgetting [12]. Since multiple new training samples, unpredictable scenario changes, and novel requirements in new tasks keep emerging as time

Supplementary Information The online version contains supplementary material available at https://doi.org/10.1007/978-3-031-26348-4_17.

L. Wang et al. (Eds.): ACCV 2022, LNCS 13845, pp. 281–296, 2023.
https://doi.org/10.1007/978-3-031-26348-4_17

passes by, making it impractical to seek a once-for-all training scheme [1, 23, 35], alleviating forgetting in continual learning [23, 35] is of great significance.

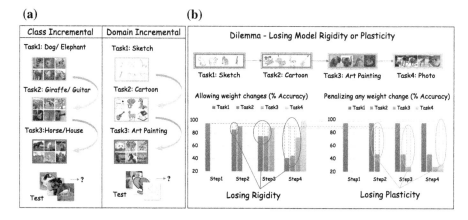

Fig. 1. (a) Difference between CI and DI learning. CI learning shares all the domains and divides the classes into tasks, while DI learning shares all the classes and possess different domains in each task. (b) DI learning also faces a tradeoff between rigidity and plasticity - allowing weights to be accessible to all changes leads to forgetting while penalizing and forbidding any weight change results in failure of learning new tasks.

Though abundant works have focused on continual learning [6, 18, 21, 23, 35], the majority only consider class-incremental (CI) scenario [18, 21], where different classes are distributed into sequentially-appearing tasks. Nevertheless, in real cases, sometimes it is the domain distribution that changes in the incoming tasks instead of the classes. For example, regarding autonomous driving, though the target objects to classify, such as cars, cyclists, and pedestrians, are always the same, the domain may shift due to different weather, location, and time. Thus, though being ignored currently, domain-incremental (DI) [35] learning is also an important and urgent research topic. The difference between CI continual learning and DI continual learning is depicted in Fig. 1(a).

Besides, the same as its CI counterpart, DI classification also faces the rigidity-and-plasticity delimma [6]. Namely, for continual learning methods, allowing enough plasticity might result in catastrophic forgetting, while allowing no degradation in previous tasks may lead to the incapability of transferring to new tasks, which is illustrated in Fig. 1(b).

Since the up-to-date research in domain-incremental continual learning [32, 33, 36, 39] does not have a unified problem setting, we first clarify the problem setting following the principles in [35]. Then, to find the main culprit of catastrophic forgetting in DI settings, we conduct preliminary experiments to observe the performance of deep learning models facing sequential tasks. As is shown in Fig. 2, when learning novel tasks, the changes in data distribution lead to an alteration of the original feature extractor, which further creates a distorted attention map and invalid logits (output before Softmax) of previous data, contributing to the misclassification.

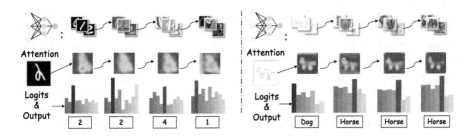

Fig. 2. Results of preliminary experiments. Facing novel tasks, the changes in domain distribution lead to distorted attention maps and invalid logits outputs, contributing to the misclassification. *Left* shows an example in the Digits benchmark while *right* shows an example in the Pictures benchmark.

In order to alleviate forgetting, the feature extractor updated with data in novel distribution should remain valid for previous data. Therefore, a Consolidating Learned and Undergoing Experience (CLUE) method is proposed to maintain the performance of old tasks when training on novel ones. CLUE can be applied to all deep classification models and ensure an unchanged network architecture without additional calculation overhead. In CLUE, a network knowledge distillation loss and a stored data regularization loss are respectively utilized to increase the rigidity of the feature extractor and maintain the logits output distribution. CLUE greatly reduces forgetting and achieves good performance compared to baselines on three domain-incremental benchmarks - Digits, Pictures, Processing - we arrange in this paper. Besides, it shows remarkable robustness towards changes in the buffer size (numbers of stored replay samples), which greatly lessen storage overhead. Moreover, the feasibility of CLUE is supported by mathematical derivation and neural network interpretability visualization. Our main contributions are as follows:

- We clarify the problem setting of domain-incremental (DI) classification, define the metrics, and find the primary cause of forgetting in DI settings.
- For DI settings, we propose a Consolidating Learned and Undergoing Experience (CLUE) method, which consolidates knowledge by two additional losses from network and data aspects without introducing extra parameters or processing steps, and can be applied to mainstream deep classifiers.
- CLUE achieves remarkable performance compared with other baselines on three benchmarks - Digits, Pictures, and Processing - from different domain-incremental aspects, which are arranged using existing open-sourced datasets.
- Further experiments show that CLUE also possesses strong robustness when buffer size changes. Besides, the feasibility of CLUE is also supported by theoretical derivation and model interpretability visualization.

2   Related Work

Continual Learning

An increasing amount of research [18,21] tries to alleviate forgetting in continual learning settings. Regarding image classification, the most common mainstream method is the replay-based method [6,18,21]. The simple-yet-efficient method store samples from previous tasks and merge them with data of novel tasks [6]. Recent years has witnessed a great evolution of replay-based methods, from storing raw samples [3,4,17,25] to synthetic image generators [31], intermediate features [38] or hidden representations [34], compressed embeddings [9] and prototypes [43]. Other non-replay methods may set some regularization. By attaching a term to the loss function to prevent weights from unwanted changes [12,42], orthogonally modifying weights [41] or applying knowledge distillation [15] to maintain the performance, those methods help to avoid forgetting. Another typical method is to redesign the architecture into a growable network [29,40] or apply a masked network [7,19,20,30] to allow training of novel classes. However, those architecture-based methods tend to be efficient only when the task label is available in reference [6,18,35].

Nevertheless, the above research mainly focuses on class-incremental classification tasks, where novel classes following the same domain distribution will appear in new tasks. In this paper, we concentrate on the domain-incremental setting and propose our method to avoid catastrophic forgetting depending on why performance drops facing newly appeared alien domains.

Continual Domain Adaptation

Domain adaptation [24,37], where training data and target test data are from different domains, has been another popular research topic. Nevertheless, it only prioritizes performance in the target domain without caring about catastrophic forgetting, which is partially different from continual learning settings. Some recent work begins to handle forgetting in continual domain adaptation. However, the up-to-date research [32,33,36,39] does not have a unified problem setting. [32] focuses on class-incremental learning when cross-domain training sets are available. In [39], both domains and classes will increment in future tasks. Instead of sequentially appearing tasks, the source and target domains' performance is focused in [33]. Besides, most of them are based on heavy image preprocessing [36] or expandable network [32,39] structure, which introduces additional computational overhead and may be invalid for some specific deep learning classifiers and practical application.

Therefore, this paper simplifies and clarifies the definition of DI image classification as stated in [35]. Besides, we propose a continual learning method without additional training overhead, such as image pre-processing or meta-learning calculations, and ensure that the network architecture remains unchanged facing novel tasks, which is applicable for all deep learning classifiers.

3   Method

3.1   Problem Statement

Assuming $\mathbb{D}_{train} = \{\mathbf{D}_t\}_{t=1}^{T}$ is the domain-incremental classfication training dataset with T tasks, each task satisfies $\mathbf{D}_t = \{(I_{t,i}, c_{t,i})\}_{i=1}^{|\mathbf{D}_t|}$ with $|\mathbf{D}_t|$ samples. In a training image sample of task t, $(I_{t,i}, c_{t,i})$, $I_{t,i}$ represents for the ith image, and $c_{t,i}$ is its class label. Categories are shared in all tasks. Namely, $c_{t,i} \in \mathbb{C}, t = 1, 2, ..., T$, where $\mathbb{C}$ is the set of all classes. Besides, each task t belongs to one specific domain d without overlapping with others.

While training on task t, only current data $\mathbf{D}_t$ is available. Besides, a small buffer $\mathbf{B} = \{(I_b, info)\}_{b=1}^{|\mathbf{B}|}$ is also allowed to store former samples and corresponding information, where the buffer size $|\mathbf{B}|$ should be far less than $|\mathbf{D}_t|, t = 1, 2, ..., T$.

The Goal of DI Classification

The optimizing goal for learning task t is to minimize the loss of classifying all classes in learned and undergoing tasks when only $\mathbf{D_t} \cap \mathbf{B}$ is available, which can be presented as:

$$\min[\sum_{j=1}^{|\mathbf{D}_i|} L_{CE}(F(\mathbf{\Theta}_t; I_{i,j}), c_{i,j})]_{i=1}^{t} \qquad (1)$$

where L_{CE} represents the cross-entropy loss function, F represents forwarding propagation calculation, and $\mathbf{\Theta}_t$ is the current model.

During reference, $\mathbb{D}_{test}$ is used to evaluate the performance, where all the domains $\{d_1, d_2, ..., d_t\}$ are included.

Metrics of DI classification

In the DI continual learning setting, task-average accuracy and Forgetting are the evaluation metrics.

In realistic situations, sample numbers might vary between domains because data in a particular domain (e.g., hand-written digits) might be easier to collect. To avoid overlooking the performance in some scenes with fewer test samples, we calculate the mean classification accuracy of every task, which can be formulated as:

$$\mathbf{ACC}_{avg} = \frac{1}{T} \sum_{t=1}^{T} ACC(t, T) \qquad (2)$$

where $ACC(t, T)$ means the classification accuracy of task t after learning the Tth task, the end of continual learning.

Forgetting is another commonly used continual learning metric reflecting the severity of forgetting [4]. However, only measuring the absolute accuracy drops in DI settings might be unreasonable since learning difficulty varies between domains. For example, training hand-written digits might achieve a higher accuracy than digits in street scenes. Thus, the absolute accuracy decay when catastrophic forgetting happens might be more significant in hand-written digits.

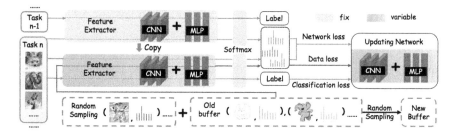

Fig. 3. The framework of CLUE. Replay images and corresponding logits are stored in the fixed memory. Network loss and Data loss are utilized to maintain the rigidity of feature extractor to consolidate the learned and undergoing experiences.

Therefore, ensuring the balance between domains, we also calculate task-average forgetting:

$$\mathbf{F}_{avg} = \frac{1}{T} \sum_{t=1}^{T} \frac{\max(ACC(t,t) - ACC(t,T), 0)}{ACC(t,t)} \tag{3}$$

3.2   Consolidating Learned and Undergoing Experience (CLUE)

Motivation and Framework

As shown in Fig. 2, when a deep neural network faces sequential tasks from alien domain distributions, the weights in the original model adapt themselves to classify samples following the new domain distribution. The overwritten weights then lead to distorted attention maps of former task samples, which results in the invalid logits output and incorrect classification. Figure 2 shows how the handwritten digit '2' is misclassified into '4' and '1' after learning street scene and synthetic numbers because of no-more appropriate feature extractor and how the sketch image of 'Dog' is classified into 'Horse' after training on comic, painting and photos.

Thus, the main reason for forgetting tends to differ from that of the CI setting, where unbalanced, fully connected layers contribute most to forgetting. In this paper, focusing on DI problems, we propose CLUE that maintains the feature extractor's performance on previous tasks.

Figure 3 shows the overall framework of CLUE. Two loss functions concerning the original network and data are used to consolidate the learned and undergoing tasks, apart from the common cross-entropy classification loss. The training for each task requires two steps - updating the model and refreshing the buffer. In the first step, a network loss is calculated as the knowledge distillation between former and updated networks using current data. On the other hand, a data loss controls the logits alteration of stored images in the buffer. With both of the loss functions, the model is updated. In the second step, the buffer is renewed with random sampling, where the allowed account of stored images is equally arranged for each task. Old samples are randomly dropped from the buffer if it is full, while new ones are selected from the current task to fill the vacancy. Both the image and its logits are stored.

Loss of Learned Network

Since $(I_{i,j}, c_{i,j}) \sim \mathbf{D}_i, i = 1, 2, ..., t - 1$ in equation (1) is not available when learning task t, we need to estimate the optimization goal in (1) using available data. Knowledge distillation loss [11] can be used to measure the changes when training the model. Thus, the feature extractor can be prevented from becoming invalid in former tasks. Utilizing data in current tasks, we can calculate the model's output before updating and compare it with the output after training the current task. The final network loss applying knowledge distillation can be written as:

$$L_{net} = -\sum_{j=1}^{|\mathbf{D}_t|}(F(\mathbf{\Theta}_{t-1}; I_{t,j})/Tem)\log(F(\mathbf{\Theta}_t; I_{t,j})/Tem) \qquad (4)$$

where $|\mathbf{D}_t|$ represents the number of samples in current task t, $\mathbf{\Theta}_{t-1}$ and $\mathbf{\Theta}_t$ respectively represents the model before and after training task t, and $Tem \geq 1$ is the temperate that rescale the output and is set to be 2 following [15] in this paper.

The supplement will deliver a detailed formula derivation about why L_{net} estimate equation (1) and can be used in optimization.

Loss of Learned Samples

Another method to estimate equation (1) is to take advantage of the stored samples in the replay buffer. The replay samples directly guarantee the model performance on old tasks. To keep the output logits the same as when the replay sample is trained, we minimize the distance between logits of the current model and stored logits. The data loss L_{data} can be represented as:

$$L_{data} = \sum_{b=1}^{|\mathbf{B}|} ||F(\mathbf{\Theta}_t; I_b) - z_b||_2^2 \qquad (5)$$

where $|\mathbf{B}|$ is the buffer size, I_b represents the stored image, and z_b represents its corresponding stored logits.

$L2$ norm is used instead of the cross-entropy loss, whose effectiveness is further discussed in the ablation study. Besides, please refer to the supplement for a detailed theoretical derivation of how L_{data} relates to equation (1).

Overall Loss Function

Attaching L_{net} and L_{data} to the training loss of current classification task L_{CE}, we will get the overall loss function to update training models under the CLUE framework:

$$L_{all} = L_{CE} + \lambda_1 L_{net} + \lambda_2 L_{data} \qquad (6)$$

where λ_1 and λ_2 are both hyper-parameters balancing different losses.

4   Results

Benchmarks

The following three benchmarks are used in this paper:

Digits. We organize commonly-used digit datasets as the Digits benchmark, sorting them from simple to complex. The four tasks appear sequentially in the order of MNIST [13], MNIST-M [8], SVHN [22], Synthesis [28] to test the continual learning ability of methods.

Pictures. PACS dataset [14] contains object images of various types, from line sketches to real pictures, where the object categories are shared between domains. We also arrange four tasks orderly from simple to complex. Sketch, cartoons, art paintings, and authentic photo images appear sequentially.

Processing. STL10 [5] contains ten categories of animal images. Here, we perform image processing on STL10 to divide four different types of domains. We use PyTorch's official transform tool [26] for image processing and set four sequential tasks for brightness, grayscale, sharpness, and contrast changes.

Through benchmarks from the three different aspects, we investigate the performance of methods for continual learning in the DI settings. Please refer to the supplement for more details concerning benchmarks.

Compared Baselines

We first implement two basic methods as upper and lower bounds for performance comparison in continual learning. *Naive*, which means that all weights in the original classifier are variable for finetuning facing novel tasks, without adding any continual learning approach, is considered the theoretical lower bound. *Joint training*, where it is assumed that data of all tasks are available simultaneously to train the model together, is considered the theoretical upper bound. It is worth noting that they are not actual experimental upper and lower bounds. In some cases, continual learning methods may exceed the range. For other comparison methods, we implement *LwF* [15], *EWC* [12], and *SI* [42] three regularization-based method, where knowledge distillation or weight update methods are utilized to control network changes. With regard to replay-based methods, *Replay* [27], *DER* [2], *DER++* [2] are implemented where replaying raw samples and feature vectors are both included.

Implement Details

All experiments are conducted on RTX 3090 GPUs. We use the same classifier backbone to ensure fair comparisons between methods and implement all methods in the Avalanche [16] continual learning framework with PyTorch [26].

ResNet-18 [10] is used for all tasks. Besides, we use an SGD optimizer with a learning rate of 0.01 and a momentum of 0.9 for all experiments. The batch size for the Digits benchmark is 128, while the Pictures and Processing benchmark have a batch size of 64. Training epochs are set to be 20 for Digits and Pictures and 40 for Processing.

Table 1. Performance comparision on different benchmarks (%)

	Benchmark 1: Digits	Benchmark 2: Pictures	Benchmark 3: Processing
Matrics	$\mathbf{ACC_{avg}}$ ↑	$\mathbf{ACC_{avg}}$ ↑	$\mathbf{ACC_{avg}}$ ↑
Naïve	64.52 ± 0.25	66.91 ± 1.82	68.88 ± 0.54
LwF	71.89 ± 1.26	84.03 ± 0.75	73.98 ± 1.35
EWC	62.28 ± 3.06	69.80 ± 2.78	69.98 ± 1.20
SI	64.60 ± 2.22	69.65 ± 2.96	66.47 ± 1.82
Replay	78.73 ± 0.54	87.32 ± 0.64	77.52 ± 0.73
DER	79.37 ± 0.53	87.84 ± 0.99	75.58 ± 1.02
DER++	80.27 ± 0.93	88.68 ± 0.28	75.28 ± 1.01
Ours	**84.81 ± 0.31**	**90.16 ± 0.82**	**79.13 ± 0.98**
Joint	90.67 ± 0.14	88.30 ± 1.70	88.12 ± 0.41
Matrics	$\mathbf{F_{avg}}$ ↓	$\mathbf{F_{avg}}$ ↓	$\mathbf{F_{avg}}$ ↓
Naïve	42.24 ± 0.24	38.24 ± 2.45	23.29 ± 0.57
LwF	31.86 ± 1.88	13.56 ± 1.58	18.18 ± 1.49
EWC	44.97 ± 4.56	34.61 ± 4.15	21.85 ± 1.83
SI	42.32 ± 3.45	34.21 ± 4.33	27.38 ± 2.85
Replay	22.91 ± 0.72	9.36 ± 0.70	6.80 ± 2.04
DER	20.52 ± 0.71	3.77 ± 0.55	6.61 ± 2.22
DER++	19.23 ± 1.57	3.95 ± 1.05	8.28 ± 1.93
Ours	**7.23 ± 0.44**	**1.34 ± 0.18**	**5.03 ± 1.21**
Joint	-	-	-

To determine the hyperparameters λ_1 and λ_2 in our method, we use the grid parameter tuning method. λ_1 is determined to be 0.8, 0.2, and 0.5 for the Digits, Pictures, Processing benchmark. λ_2 is decided to be 0.1 for all benchmarks.

4.1 Performance Comparison

Results are indicated in Table 1 and Fig. 4. Table 1 shows the metric results of different methods on the three benchmarks. Besides, the dynamic change plot of overall accuracy, which is different from the task-average accuracy ACC_{avg}, is depicted in the left part of Fig. 4. The eventual accuracy of all tasks after learning the last one is plotted in the right part of Fig. 4.

Table 1 shows that CLUE exceeds other methods by a large margin, with 84.81%, 90.16%, and 79.13% of ACC_{avg} in each benchmark. It is remarkable that in the Picture benchmark, CLUE even surpasses the upper bound Joint. It is probably because the obvious difference between different domains confuses the neural network when training them jointly. However, the neural network can be trained on single novel tasks without confusion carrying former knowledge by applying CLUE. CLUE is also superb in solving catastrophic forgetting, with only 7.23% forgetting compared with 19.23% of the second-best method in the Digits benchmark. In the Picture and Processing benchmarks, CLUE is also excellent in alleviating forgetting, only having 1.34% and 5.03% F_{avg} respectively.

Besides, CLUE also maintains a high overall classification accuracy throughout learning steps, shown in Fig. 4 *left*. While other methods suffer from various degrees of forgetting, CLUE maintains comparatively stable even when the domain distribution in the current task changes. Figure 4 *right* shows the eventual accuracy of all learned tasks when complete training is finished. In each benchmark, the histograms of the first three tasks reflect the rigidity of the method and the last histogram of task 4 mirrors plasticity. Through this graph, CLUE maintains high accuracy in former tasks while preserving plasticity for future tasks, almost always being the closest to the upper bound accuracy.

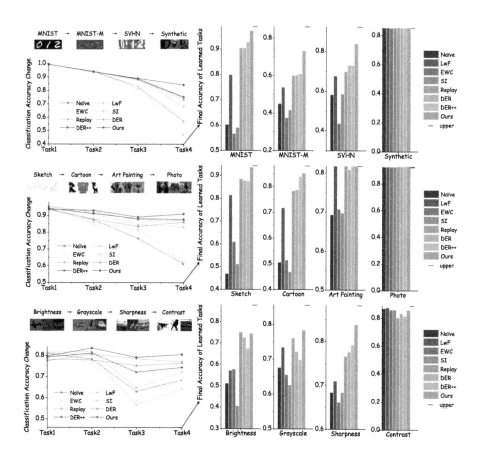

Fig. 4. Performance comparisons of different methods. The dynamic changes of overall accuracy are depicted in the *left* while the final accuracy distribution of all tasks after completing all tasks is in the *right*. Our method maintains high accuracy in former tasks while preserving plasticity for future tasks.

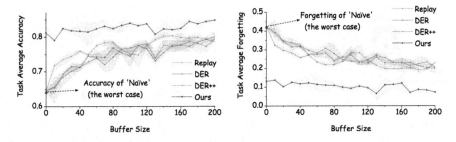

Fig. 5. The dynamic changes of ACC_{avg} and F_{avg} on the Digits benchmark using different continual learning methods when the buffer size changes from 0 to 200. The results of method 'Naive' is marked in red as a comparison. Our method is observed to be robust, facing reduced buffer sizes. (Color figure online)

4.2 Buffer Size Analysis

In continual learning settings, buffer sizes (the number of replay images allowed to be stored) are fixed. Therefore, as the task number grows, stored image numbers of each task will gradually decrease, significantly impacting the performance. Superior continual learning methods must maintain high performance even when buffer sizes are limited. Thus, it is fundamental to study the changes in metrics when the buffer size becomes smaller.

In Fig. 5, the dynamic changes of ACC_{avg} and F_{avg} is plotted for all the replay-based methods. It is observed that when the buffer size is reduced to 0, other replay-based methods degenerate into 'Naive', while CLUE still maintains decent performance. When the buffer size is gradually reduced to one-tenth, in CLUE, both ACC_{avg} and F_{avg} have a significantly smaller change than other methods. ACC_{avg} is maintained at around 80%, while F_{avg} is approximately 10%.

Therefore, we can safely conclude that CLUE is a robust method facing reduced buffer size. On the one hand, it helps maintain continual learning performance as the task number increases. On the other hand, when the task number remains unchanged, it significantly reduces the number of required replay samples, saving storage overhead.

4.3 Model Interpretability Analysis

As is illustrated above in Fig. 2, the changes in domain distribution may lead to a distorted attention map and invalid logits outputs. In this part, we conduct experiments to interpret the effectiveness of CLUE in correcting the attention map and preserving the logits distribution.

Figure 6 shows the output logits distribution of nine typical samples from the first three tasks of the three benchmarks. Before forgetting shows the results right after learning the task to which the samples belong, while after forgetting

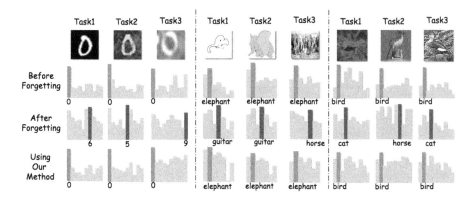

Fig. 6. Logits distributions of typical samples. *Left* shows the Digits benchmark, *middle* depicts the Pictures benchmark, and *right* corresponds to the Processing benchmark. CLUE works in maintaining the original logits distributions and classifying correctly.

shows the performance of the eventual model. The last row in Fig. 6 shows the final results when applying CLUE.

It is observed that without our model, '0' might lose the logits distribution of the original domain, leading to the misclassification as '6', '5', or '9'. The sample 'elephant' in the Picture benchmark and 'bird' in the Processing benchmark might also be misclassified without an effective continual learning method. With CLUE, accurate classification can be achieved by maintaining the original logits distribution as much as possible to prevent catastrophic forgetting.

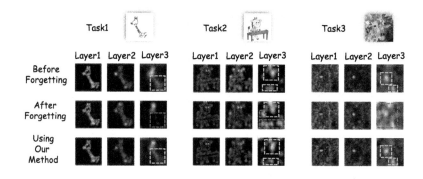

Fig. 7. Attention Maps of typical samples from the Picture benchmark without or with CLUE. The mean output of channels in layer3 (ResNet-18) is used as the attention map, where bright color corresponds to the vital area. CLUE works in maintaining proper attention maps and classifying correctly. (Color figure online)

The attention maps of typical samples are shown in Fig. 7. The Pictures benchmark is used as an example as its 224 * 224-sized images are large enough to be observed and analyzed.

As is shown, in the sketch image, for correct classification, the neck of the giraffe should be focused. By applying CLUE, the attribute can be maintained. The cartoon giraffe can be classified successfully by its head and legs. However, after forgetting, the neural network tends to pay more attention to table legs and the giraffe's tail. Our method can preserve important areas without being disturbed by noisy information. The eye and noise in the art painting of giraffes are the prime areas for classification. CLUE keeps them well. Therefore, applying CLUE guarantees a more accurate attention map.

4.4   Ablation Studies

Table 2 shows the ablation studies, where the significance of L_{net} and L_{data} is tested. Metrics of the classifier when removing L_{net} or L_{data} or both are calculated and recorded. Additionally, we test the performance where L_{data} is replaced by commonly used cross-aentropy L_{CE}.

As is observed in Table 2, when both L_{net} and L_{data} are omitted, the method degrades into naive finetuning. Attaching either L_{net} or L_{data} will promote ACC_{avg} by a large margin in the Digits ($\approx 15\%$) and Pictures ($\approx 10\%$) benchmark. In the Processing benchmark, L_{net} is more effective than L_{data}. Regarding F_{avg}, L_{net} works better in alliviating forgetting in all benchmarks. Obviously, utilizing both loss functions will further improve the performance.

Table 2. Ablation studies

	$L_{network}$	L_{data}	$ACC_{avg}(\%) \uparrow$	$F_{avg}(\%) \downarrow$
Benchmark 1: Digits	✗	✗	64.51	42.23
	✓	✗	80.15	14.38
	✗	✓	79.00	20.96
	✓	L_{CE}	82.41	11.64
	✓	✓	**84.81**	**7.23**
Benchmark 2: Pictures	✗	✗	66.91	38.24
	✓	✗	87.99	3.78
	✗	✓	88.27	6.03
	✓	L_{CE}	89.12	2.45
	✓	✓	**90.16**	**1.34**
Benchmark 3: Processing	✗	✗	68.88	23.29
	✓	✗	76.55	8.27
	✗	✓	72.60	14.03
	✓	L_{CE}	74.25	7.67
	✓	✓	**79.13**	**5.03**

Besides, when changing L_{net} into L_{CE}, drops are witnessed in the performance, with approximately 2%, 1% and 5% decrease in the ACC_{avg} on the three benchmarks. It is probably because L_{CE} focuses more on the between-class difference of samples, while L_{net} focuses more on the class logits distributions facing new domains.

As a result, both L_{net} and L_{data} are effective and contributing parts for our CLUE.

5   Conclusion

In this paper, focused on the DI setting, we clarify the problem setting and define the metrics. Firstly, the main culprit of forgetting when training on shifted domains is found through preliminary experiments. It is because the feature extractor has adjusted to the new domain distribution, being invalid for former tasks. Next, to deal with this issue, a Consolidating Learned and Undergoing Experience (CLUE) method is proposed to mitigate forgetting in DI classification. It consists of two more loss functions to control network changes and can be applied in any mainstream classification model without introducing extra parameters or processing steps. Comprehensive experiments show the superior performance of CLUE compared with other baselines on three benchmarks - Digits, Pictures, and Processing, with higher task-average accuracy and less forgetting. Besides, extensive experiments show that CLUE remains robust when there are limited replay samples, maintaining higher performance and saving storage overhead. Moreover, both theoretical derivation and model interpretability visualization justify the feasibility of CLUE.

Acknowledgements. This work was supported by the National Key R&D Program of China (No. 2020AAA0105500) the National Natural Science Foundation of China (No. U21B2013, No. 61971260)

References

1. Biesialska, M., Biesialska, K., Costa-Jussa, M.R.: Continual lifelong learning in natural language processing: a survey. arXiv preprint arXiv:2012.09823 (2020)
2. Buzzega, P., Boschini, M., Porrello, A., Abati, D., Calderara, S.: Dark experience for general continual learning: a strong, simple baseline. Adv. Neural. Inf. Process. Syst. **33**, 15920–15930 (2020)
3. Castro, F.M., Marín-Jiménez, M.J., Guil, N., Schmid, C., Alahari, K.: End-to-end incremental learning. In: Ferrari, V., Hebert, M., Sminchisescu, C., Weiss, Y. (eds.) ECCV 2018. LNCS, vol. 11216, pp. 241–257. Springer, Cham (2018). https://doi.org/10.1007/978-3-030-01258-8_15
4. Chaudhry, A., Dokania, P.K., Ajanthan, T., Torr, P.H.S.: Riemannian walk for incremental learning: understanding forgetting and intransigence. In: Ferrari, V., Hebert, M., Sminchisescu, C., Weiss, Y. (eds.) ECCV 2018. LNCS, vol. 11215, pp. 556–572. Springer, Cham (2018). https://doi.org/10.1007/978-3-030-01252-6_33
5. Coates, A., Ng, A., Lee, H.: An analysis of single-layer networks in unsupervised feature learning. In: Proceedings of the Fourteenth International Conference on Artificial Intelligence and Statistics, pp. 215–223. JMLR Workshop and Conference Proceedings (2011)

6. Delange, M., et al.: A continual learning survey: defying forgetting in classification tasks. IEEE Trans. Pattern Anal. Mach. Intell. **44**, 3366–3385 (2021)
7. Fernando, C., et al.: PathNet: evolution channels gradient descent in super neural networks. arXiv preprint arXiv:1701.08734 (2017)
8. Ganin, Y., et al.: Domain-adversarial training of neural networks. J. Mach. Learn. Res. **17**(1), 2030–2096 (2016)
9. Hayes, T.L., Kafle, K., Shrestha, R., Acharya, M., Kanan, C.: REMIND your neural network to prevent catastrophic forgetting. In: Vedaldi, A., Bischof, H., Brox, T., Frahm, J.-M. (eds.) ECCV 2020. LNCS, vol. 12353, pp. 466–483. Springer, Cham (2020). https://doi.org/10.1007/978-3-030-58598-3_28
10. He, K., Zhang, X., Ren, S., Sun, J.: Deep residual learning for image recognition. In: Computer Science (2015)
11. Hinton, G., Vinyals, O., Dean, J., et al.: Distilling the knowledge in a neural network. arXiv preprint arXiv:1503.02531 2(7) (2015)
12. Kirkpatrick, J., et al.: Overcoming catastrophic forgetting in neural networks. Proc. Natl. Acad. Sci. **114**(13), 3521–3526 (2017)
13. LeCun, Y.: The MNIST database of handwritten digits (1998). http://yann.lecun.com/exdb/mnist/
14. Li, D., Yang, Y., Song, Y.Z., Hospedales, T.M.: Deeper, broader and artier domain generalization. In: Proceedings of the IEEE International Conference on Computer Vision, pp. 5542–5550 (2017)
15. Li, Z., Hoiem, D.: Learning without forgetting. IEEE Trans. Pattern Anal. Mach. Intell. **40**(12), 2935–2947 (2017)
16. Lomonaco, V., et al.: Avalanche: an end-to-end library for continual learning. In: Proceedings of the IEEE/CVF Conference on Computer Vision and Pattern Recognition, pp. 3600–3610 (2021)
17. Lopez-Paz, D., Ranzato, M.: Gradient episodic memory for continual learning. In: Advances in Neural Information Processing Systems, vol. 30 (2017)
18. Mai, Z., Li, R., Jeong, J., Quispe, D., Kim, H., Sanner, S.: Online continual learning in image classification: an empirical survey. Neurocomputing **469**, 28–51 (2022)
19. Mallya, A., Davis, D., Lazebnik, S.: Piggyback: adapting a single network to multiple tasks by learning to mask weights. In: Ferrari, V., Hebert, M., Sminchisescu, C., Weiss, Y. (eds.) ECCV 2018. LNCS, vol. 11208, pp. 72–88. Springer, Cham (2018). https://doi.org/10.1007/978-3-030-01225-0_5
20. Mallya, A., Lazebnik, S.: PackNet: adding multiple tasks to a single network by iterative pruning. In: Proceedings of the IEEE Conference on Computer Vision and Pattern Recognition, pp. 7765–7773 (2018)
21. Masana, M., Liu, X., Twardowski, B., Menta, M., Bagdanov, A.D., van de Weijer, J.: Class-incremental learning: survey and performance evaluation on image classification. arXiv preprint arXiv:2010.15277 (2020)
22. Netzer, Y., Wang, T., Coates, A., Bissacco, A., Wu, B., Ng, A.Y.: Reading digits in natural images with unsupervised feature learning (2011)
23. Parisi, G.I., Kemker, R., Part, J.L., Kanan, C., Wermter, S.: Continual lifelong learning with neural networks: a review. Neural Netw. **113**, 54–71 (2019)
24. Patel, V.M., Gopalan, R., Li, R., Chellappa, R.: Visual domain adaptation: a survey of recent advances. IEEE Signal Process. Mag. **32**(3), 53–69 (2015)
25. Prabhu, A., Torr, P.H.S., Dokania, P.K.: GDumb: a simple approach that questions our progress in continual learning. In: Vedaldi, A., Bischof, H., Brox, T., Frahm, J.-M. (eds.) ECCV 2020. LNCS, vol. 12347, pp. 524–540. Springer, Cham (2020). https://doi.org/10.1007/978-3-030-58536-5_31

26. Pytorch, A.D.I.: Pytorch (2018)
27. Riemer, M., et al.: Learning to learn without forgetting by maximizing transfer and minimizing interference. arXiv preprint arXiv:1810.11910 (2018)
28. Roy, P., Ghosh, S., Bhattacharya, S., Pal, U.: Effects of degradations on deep neural network architectures. arXiv preprint arXiv:1807.10108 (2018)
29. Rusu, A.A., et al.: Progressive neural networks. arXiv preprint arXiv:1606.04671 (2016)
30. Serra, J., Suris, D., Miron, M., Karatzoglou, A.: Overcoming catastrophic forgetting with hard attention to the task. In: International Conference on Machine Learning, pp. 4548–4557. PMLR (2018)
31. Shin, H., Lee, J.K., Kim, J., Kim, J.: Continual learning with deep generative replay. In: Advances in Neural Information Processing Systems, vol. 30 (2017)
32. Simon, C., et al.: On generalizing beyond domains in cross-domain continual learning. In: Proceedings of the IEEE/CVF Conference on Computer Vision and Pattern Recognition, pp. 9265–9274 (2022)
33. Tang, S., Su, P., Chen, D., Ouyang, W.: Gradient regularized contrastive learning for continual domain adaptation (2021)
34. van de Ven, G.M., Siegelmann, H.T., Tolias, A.S.: Brain-inspired replay for continual learning with artificial neural networks. Nat. Commun. 11(1), 1–14 (2020)
35. Van de Ven, G.M., Tolias, A.S.: Three scenarios for continual learning. arXiv preprint arXiv:1904.07734 (2019)
36. Volpi, R., Larlus, D., Rogez, G.: Continual adaptation of visual representations via domain randomization and meta-learning. In: Proceedings of the IEEE/CVF Conference on Computer Vision and Pattern Recognition. pp. 4443–4453 (2021)
37. Wang, M., Deng, W.: Deep visual domain adaptation: a survey. Neurocomputing 312, 135–153 (2018)
38. Xiang, Y., Fu, Y., Ji, P., Huang, H.: Incremental learning using conditional adversarial networks. In: Proceedings of the IEEE/CVF International Conference on Computer Vision, pp. 6619–6628 (2019)
39. Xie, J., Yan, S., He, X.: General incremental learning with domain-aware categorical representations. In: Proceedings of the IEEE/CVF Conference on Computer Vision and Pattern Recognition, pp. 14351–14360 (2022)
40. Yoon, J., Yang, E., Lee, J., Hwang, S.J.: Lifelong learning with dynamically expandable networks. arXiv preprint arXiv:1708.01547 (2017)
41. Zeng, G., Chen, Y., Cui, B., Yu, S.: Continual learning of context-dependent processing in neural networks. Nat. Mach. Intell. 1(8), 364–372 (2019)
42. Zenke, F., Poole, B., Ganguli, S.: Continual learning through synaptic intelligence. In: International Conference on Machine Learning, pp. 3987–3995. PMLR (2017)
43. Zhu, F., Zhang, X.Y., Wang, C., Yin, F., Liu, C.L.: Prototype augmentation and self-supervision for incremental learning. In: Proceedings of the IEEE/CVF Conference on Computer Vision and Pattern Recognition, pp. 5871–5880 (2021)

Continuous Self-study: Scene Graph Generation with Self-knowledge Distillation and Spatial Augmentation

Yuan Lv, Yajing Xu$^{(\boxtimes)}$, Shusen Wang, Yingjian Ma, and Dengke Wang

Beijing University of Posts and Telecommunications, Beijing, China
{lvyuan,xyj,ShusenW,wangdk}@bupt.edu.cn

Abstract. As an extension of visual detection tasks, scene graph generation (SGG) has drawn increasing attention with the achievement of complex image understanding. However, it still faces two challenges: one is the distinguishing of objects with high visual similarity, the other is the discriminating of relationships with long-tailed bias. In this paper, we propose a Continuous Self-Study model (CSS) with self-knowledge distillation and spatial augmentation to refine the detection of hard samples. We design a long-term memory structure for CSS to learn its own behavior with the context feature, which can perceive the hard sample of itself and focus more on similar targets in different scenes. Meanwhile, a fine-grained relative position encoding method is adopted to augment spatial features and supplement relationship information. On the Visual Genome benchmark, experiments show that the proposed CSS achieves obvious improvements over the previous state-of-the-art methods. Our code is available at https://github.com/LINYE1998/Continuous_Self_Study.

1 Introduction

Scene graph [1,2] structure is a medium bridging the image and the text [3,4]. It is comprised by the detection of a list of ⟨*subject-predicate-object*⟩ triplets [5] to describe the objects and their relationships in an image. With feature augmentation by the extraction of context information [1,6,7] and the introduction of external semantic knowledge [8–10], it can not only improve the accuracy of classification in upstream tasks, such as object detection [11] and visual relationship detection [10,12,13], but also provide a more comprehensive and specific structure for its downstream visual understanding tasks [14,15], including image retrieval [16], visual question answering [17,18] and image captioning [19], thus has been drawing increasing attention.

Supported by National Natural Science Foundation of China.

Supplementary Information The online version contains supplementary material available at https://doi.org/10.1007/978-3-031-26348-4_18.

L. Wang et al. (Eds.): ACCV 2022, LNCS 13845, pp. 297–315, 2023.
https://doi.org/10.1007/978-3-031-26348-4_18

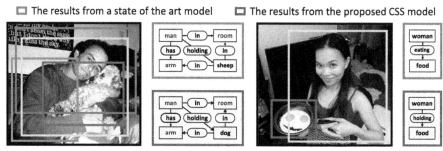

Fig. 1. Examples of the two challenges for SGG task. (1) The distinguish of the unclear target, e.g., the object *dog* is predicted as *sheep* wrongly in (a). (2) The discriminating of the long-tailed relationship, e.g., the *holding* is misclassified as *eating* in (b), due to the common sense bias [2] from ⟨*woman-eating-food*⟩. (Color figure online)

To generate high-quality scene graphs, multifarious scene graph generation (SGG) methods [1,2,6,20,21] have been proposed to optimize the prediction of objects and relations. It can be mainly classified as the traditional SGG types [1,6] and the unbiased SGG types [20,22]. Both approaches refine the targets by passing visual or semantic messages with the extraction of context information [1,7]. Differently, the first types focus on a better feature extraction network [2,23,24] to mine useful information from more perspectives, while the second types concentrate on the debiasing work [20,21,25] to recall more semantic relationships [2] and obtain a more balancing result for the application of downstream tasks [20]. Although the previous methods have promising improvement in performance, most of them suffer from the limitations of existing SGG datasets [26]: the inadequate training data with hard sample, and the unbalanced distribution of the long-tailed relation.

The atypical objects with high visual similarity are always hard to be distinguished. For example, in the red box in Fig. 1(a), the *dog* is identified as *sheep* by a state-of-the-art model. While the *dog* is ambiguous and difficult to be distinguished, it's easy to recognize it by inference with the context information. To extract the scene information from the image, numerous researchers [2,6,7,10,27] struggle for better feature extraction networks [20]. However, it's still difficult to understand the scene and focus on the hard samples under dozens of predicted objects and biased relation of square growth. For human beings, how do we think when observing objects that are difficult to distinguish? Focusing on the unclear targets, we usually realize that we are confused and list several alternative possibilities, and then make the judgment in combination with the scene information. Inspired by the recent knowledge distillation work [28–31] , we want to enable the network to perceive the hard samples of itself and distinguish them with the supplement of scene context information.

Meanwhile, the recent research focuses on the debiasing work to balance the results from long-tailed bias. However, with the increase of the mean recall

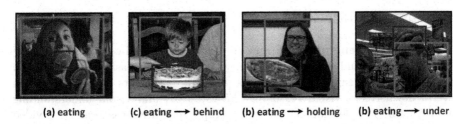

(a) eating   **(c) eating ⟶ behind**   **(b) eating ⟶ holding**   **(b) eating ⟶ under**

Fig. 2. Four typical cases of the long-tailed bias. With the subject *human* and the object *food*, different relations are more likely to be predicted as eating. By supplementing the relative position information, the relationship can be predicted correctly, e.g., from eating to *behind* in (b), *holding* in (c), and *under* in (d).

(mR@K) [23] among each predicate, most of the debiasing methods cause an unacceptable decline in Recall (R@K). As illustrated in Fig. 2, the unbiased SGG approaches choose a preference for the relationships with similar semantics, but finitely to predict them more accurately. It means that the upper limit of the predicates' prediction is limited by the information the neural networks extract. Hence, the unbiased SGG methods still need to optimize the feature extraction networks for more useful information. A large proportion of predicates have a high correlation with the spatial relations between their objects. Therefore, Both SGG types [2,12,20,22,25] and much more visual detection tasks [32,33] adopt position encoding to extract spatial information. The common position encoding cuts the image averagely into a set resolution and encodes each object separately. Nevertheless, it cannot extract the relative position information explicitly, while the relative position can provide more details between the object pairs. Besides, for small objects, it's hard to extract accurate spatial information. This motivates us to augment the spatial feature with fine-grained encoding.

Hence, for the distinguishing of objects with high visual similarity, we proposed a novel self-distillation method: Continuous Self-Study (CSS) for SGG model to learn from its own behavior with a real-time updated long-term memory structure. Focusing on the hard sample, CSS transfers the detection task from the prediction of objects to the distinguishing of similar targets. Moreover, for the discriminating of relationships with long-tailed bias, we propose a spatial augmentation (SA) of the relative position to improve the spatial information from ambiguous and directionless to accurate and directional.

Our contributions can be summarized as follows:

- A Continuous Self-Study method for SGG models is proposed to learn self-behavior, so as to obtain better visual understanding and distinguish similar targets in complex scenes.
- A spatial augmentation method is designed for visual relation detection to effectively improve the recall (R@K) and mean recall (mR@K) among each predicate in unbiased SGG field.
- Experiments on the benchmark dataset show that our approach can improve on the state-of-the-art baseline.

2   Related Work

Scene Graph Generation. Scene graph [1,2] is a mid-connection [3,4,16] of visual domain and semantic domain, which drawn increasing attention with its refinement of visual detection tasks [5,11–13,34] and its potential value in several downstream visual reasoning [17–19,35–37] and visual understanding tasks [14,15]. The development of scene graph generation task can be divided into two stages. In the first stage, various methods [1,6,8,13,23,27,38–40] are proposed to explore multiple ways of the extraction of the feature. With the supplement of context feature [1,2,6,7] and the introduction of external language information [3,8–10], these methods access promising improvement of function and performance. However, This scene graph generation is far from practical, due to the biased SGG problem [3,20] with the long-tailed dataset.

In the second stage, multiple approaches are proposed to generate unbiased scene graph. Zellers *et al.* [2] firstly pointed out the bias problem of SGG and the followers [9,23,26] proposed the unbiased metric to evaluate SG with increased attention on tail relationships. Tang *et al.* [20] draw the counterfactual causality from the trained graph to infer the effect from the bias. Yu *et al.* [22] proposed a cognition tree loss to make the tail classes receive more attention in a coarse-to-fine mode. Guo *et al.* [21] tackled the bias problem with semantic adjustment and balanced predicate learning. Chiou *et al.* [25] used a dynamic label frequency estimation to balance the head and the tail data. However, the recent approaches struggle for the identification of tail predicates and focus on the promotion of the mean recall. With the improvement of mR@K, the recall of the head data got a severe drop, which made the SGG still far from practical.

Knowledge Distillation. Knowledge distillation [41–45] is a method of extraction, generalization, and transmission of knowledge. By transferring the knowledge [42–44,46,47] of a complex pre-trained teacher network [42,43,46], a simple student network [29,31,41,48,49] can be trained effectively with the pseudo labels. To address the problem of confirmation bias [50] in pseudo-labeling, Pham *et al.* [51] trained the teacher along with the student and corrected the bias with the feedback of the student's performance. However, these traditional methods depended on a well-trained teacher network [28]. Several self-knowledge distillation methods [30,52–55] are proposed to reduce the necessity of training a large network. Nevertheless, because of the square growth relation [7] with the targets, it's still hard for SGG model to overcome the limited computing [28,41,56,57] which inversely optimizes the detection of hard samples. To this end, we distill the knowledge from the network with a memory structure that enables the network to study from its own behavior.

3   Methodology

As illustrated in Fig. 3, the CSS model consists of two parts: (1) the Self-Study module (SS) for object refinement. It retains its behavior information in a real-

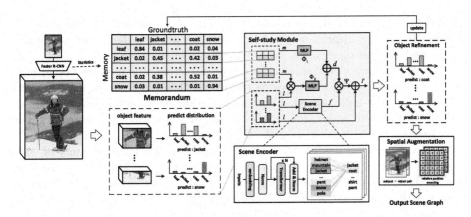

Fig. 3. Overview of the proposed Continuous Self-Study model. For the input image, the proposals are generated by a detector. With the object refinement by the self-study module, the hard sample **jacket** which is predicted wrongly by the detector can be refined to **coat** correctly. Then the relationship prediction is optimized with the spatial augmentation module. The output scene graph is generated with the combination of the predicted pair-wise objects and their detected relationships.

time updated memory *Memorandum*. The hard samples are perceived by distilling the knowledge from *Memorandum* and combining it with the detected results. Then the hard samples are focused on to distinguished and refined with the supplement of the scene context. (2) the spatial augmentation module for relationship optimization. It embeds the spatial feature of the relative position through a fine-grained encoding with an explicit spatial constraint, which distinguishes the bidirectional relationships.

The input of the CSS are proposals generated by a detector. In order to describe better, the following definitions are given. For an input image I, We use a pre-trained Faster RCNN [58] as an underlying detector [2] to predict a set of region proposals $B = \{b_i\}$ and their corresponding detected object $O = \{o_i\}$. The proposal $b_i \in \mathbb{R}^4$ is represented by a bounding box $b_i = [x_i, y_i, w_i, h_i]$, where (x_i, y_i) are the coordinates of the box's top left corner, w_i and h_i are the width and height of the bounding box respectively. Meanwhile, the detector extracts a set of visual feature vector $V = \{v_i\}$ for each proposal b_i. With the feature vector, the Roi Box Head outputs a set of predicted vector $L = \{L_i\}$ which represents the per-class confidence distribution. In addition, the $C = \{c_i\}, i \in \{1, ..., R_c\}$ is the category set of the object, where R_c is the dimension.

3.1   Self-study Module

Memorandum. We design a long-term memory structure, Memorandum, for CSS to retain its behavior information. The Memorandum $M \in \mathbb{R}^{R_c \times R_c}$ is a square matrix represented by a set of memory vectors $M = \{m_i.\}$, where $m_i.$ is the memory of the object class $c_i \in R_c$ which records the CSS's historical predicted behavior. More specifically, the scalar m_{ij} represents the conditional probability $P(gt = c_j | pred = c_i)$ that the detector predicts the object as c_i while its ground truth is c_j, as illustrated in Fig. 3.

Intuitively, for a well-trained class c_w of CSS model, the memory vector $m_{w\cdot}$ will just activate at node m_{ww} with the rest of inactive nodes of $m_{w\cdot}$. On the contrary, for the indistinguishable classes, the memory vector will activate at pairwise even more nodes. Hence, the Memorandum structure can be regarded as a summary note organized by the CSS itself.

The CSS is trained with a two-stage strategy to avoid the difficulty of convergence caused by error accumulation at the beginning of training. In the first stage, the SGG network is trained without Memorandum until the network achieves the performance of the baseline. In the second stage, we initialize a heatmap $H \in \mathbb{R}^{R_C \times R_C}$ with the statistical matrix $S_0 = \{s_{ij}\} \in \mathbb{R}^{R_C \times R_C}$, where s_{ij} is the statistical quantity of the c_i predicted by CSS with the ground truth of c_j. Then, the heatmap is updated with the new statistics S_i each iteration, where S_i is the i-th statistical matrix of its iteration similar to S_0. Considering that the relevance between the current and historical state of the heatmap on CSS decreases over time, the historical data of each iteration is attenuated during the training process with the variant formula of Newton's law of cooling [59]:

$$T(t) = T(0)e^{-\alpha t} \tag{1}$$

where $T(0)$ and $T(t)$ are the temperature of time 0 and t respectively, and α is the attenuation factor. Equation 1 can be regarded as a cooling process for an impulse response, which is widely used to calculate the heat of events in today's social network [60]. By summing the impulse responses after each iteration of training, the heatmap H at moment t can be calculated with Eq. 2:

$$H(t) = \sum_{i=0}^{t} S_i e^{-\alpha(t-i)} \tag{2}$$

With the increasing decay over time, the $H(t)$ can reflect the behavior of CSS with an appropriate cycle, which addresses the accumulation of too much historical behavior. To simplify the calculating process of $H(t)$, Eq. 2 can be further converted by making a difference between $H(t+1)$ and $H(t)$:

$$H(t+1) - H(t) = \sum_{i=0}^{t+1} S_i e^{-\alpha(t+1-i)} - \sum_{i=0}^{t} S_i e^{-\alpha(t-i)} \tag{3}$$

which can be simplified as:

$$H(t+1) = e^{-\alpha}H(t) + S_{t+1} \tag{4}$$

Equation 4 will be derived in detail in the supplementary materials. The heatmap can be updated iteratively only through the statement of the last moment and the statistical matrix of current moment by Eq. 4. Then the Memorandum can be calculated with H:

$$M_i = LineNorm(H(i)) \tag{5}$$

where $LineNorm$ is a function that normalizes each row of H separately. By repeatedly distilling and updating knowledge from Memorandum, CSS can continuously study from its own behavior and finally get a well-trained network with an ideal Memorandum, which is a dynamic balanced diagonal matrix.

Knowledge Distillation. This subsection is to perceive and refine the hard samples with the behavior information distilling from the Memorandum and the supplement of the scene information. As illustrated in Fig. 3, each object O_i can be predicted with a pre-labeled c_i by getting the maximum value of the confidence vector l_i. The memory m_{c_i} is drawn out from the Memorandum with the pre-predicted c_i. Then we use a perceiving layer to get the focus feature d_i which focus on the hard sample:

$$d_i = \Phi_1(l_i \odot m_{c_i}) + \alpha\Phi_2(m_{c_i}) \tag{6}$$

where Φ_1 and Φ_2 are multi-layer perceptron, α is a balance hyperparameter, and $\odot$ denotes the element-wise product. The focus feature can be regarded as a confusion vector with the confusion degree of each class. For the confidence vector l_i and the drawn-out memory m_{c_i} with only one activate node, the confusion degree will be very low for all nodes of d_i so that the CSS can decrease the correction with the scene feature for o_i. On the contrary, if l_i or m_{c_i} has two or more activate nodes, the confusion degree will be high between the class relative to the activate nodes, which will increase the refinement with the scene attention feature $F = \{f_i\}$.

As shown in the bottom of Fig. 3, the scene attention feature is extracted with the scene encoder network. In this work, we embed and normalize the l_i first, and then we use N transformer-based Encoder which connected end to end to adaptively gather contextual information for a certain object. The f_i can be regarded as an inference that predicts the probability distribution of c_i type object under a certain scenario.

Object Refinement. The objects are refined with the combination and fusion of the focus feature and the scene attention feature. To avoid the deviation from the image, the refinement needs to be constraint with the original visual information. Through the supplement of the original confidence distribution l, the predicted label of the object is refined by:

$$l_i' = \text{softmax}(l_i + \beta\Psi(f_i \odot d_i)) \tag{7}$$

where l_i' is the confidence distribution of the object after the refinement, β is a balance hyperparameter, and Ψ is a projection function. We denote $\beta\Psi(f_i \odot d_i)$

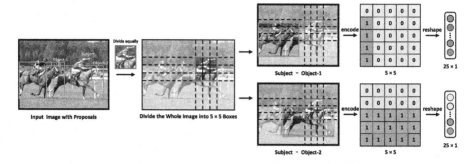

Fig. 4. The fine-grained encoding of the relative position. For the pairwise targets, the subject box is divided equally into nine parts. The whole image can be divided into twenty-five little boxes with the split lines obtained in the previous step. Then these boxes are encoded into a matrix, and the boxes which intersect with the object box are encoded as one. Finally, The matrixes are reshaped as a vector.

as $\tilde{l}$ with the statistical of the refined predicts, This Self-Study structure refine the distribution of classification probability by:

$$P(L') = P(L|V) + P(\tilde{L}|D, O_1, ..., O_n) \cdot P(D|L, M) \tag{8}$$

where $\tilde{L} = \tilde{l}_i$ is the refinement of the object, $L' = \{l'_i\}$ is the final output prediction distribution, and $D = d_i$ is the focus feature set. Equation 8 embodies the essence of the self-study method. $P(D|L, M)$ is the knowledge distilled from the CSS. For hard samples with high $P(D|L, M)$, CSS will focus more on scene context between the confusion classes with less attention from visual features. It is a positive feedback process to continuous self-study because the network will be refined by learning of the Memorandum, while the Memorandum will transform to a better distribution with the better performance of CSS.

3.2   Spatial Augmentation Module

To enhance the spatial constraint for the prediction of relations, we augment the relation feature with a fine-grained relative position spatial encoding. For better description, we use b_s and b_o to distinguish the bounding box of subject and object. As illustrated Fig. 4, for each triplet $\langle subject\ \text{-}predicate\text{-}object\rangle$, $b_s = [x, y, w, h]$ is divided equally into nine little boxes. Then, b_s is expanded into a larger box b_u, which obtained the whole view on the image I. With the nine boxes inside b_s and sixteen boxes outside b_s, the image I can be divided into twenty-five region $b_u = \{z_{ij}\}$, where z_{ij} can be represented by:

$$z_{ij} = [x_{ij}, y_{ij}, w_{ij}, h_{ij}], \quad i, j = 0, ..., 4 \tag{9}$$

where x_{ij}, w_{ij} can be further represented as:

$$x_{ij} = \begin{cases} 0, & j = 0 \\ x - (1 - j)\frac{1}{3}w, & j = 1, ..., 4 \end{cases} \tag{10}$$

$$w_{ij} = \begin{cases} x, & i = 0 \\ \frac{1}{3}w, & i = 1,2,3 \\ w_I - x - w, & i = 4 \end{cases} \tag{11}$$

where w_I is the width of the image. Meanwhile, y_{ij}, h_{ij} can be calculated in the same way with Eqs. 10 and 11 respectively. The set Q_b is defined as the region within the bounding box b, and b_o is defined as the bounding box of the triplet's object. The spatial embedding of $\langle b_s, bo \rangle$ is encoded by a boolean matrix $H = \{h_{ij}\}$:

$$h_{ij} = \begin{cases} 0, & Q_{b_o} \cap Q_{z_{ij}} = \varnothing \\ 1, & Q_{b_o} \cap Q_{z_{ij}} \neq \varnothing \end{cases} , \quad i, j = 0, ..., 4 \tag{12}$$

where h_{ij} represents the existence of intersection of b_o with z_{ij}. This encoding method can not only describe all possible spatial relationships in a consistent way but also make a distinction between $\langle subject, object \rangle$ and $\langle object, subject \rangle$. Then, H is reshaped to a vector $s \in \mathbb{R}^{25}$. The spatial feature is extracted from s with a fully connected layer and then fuse with the conventional encoding feature.

3.3   Scene Graph Generation

A scene graph consists of the class labels with the locations of individual objects and the relationship between each pairwise object [9], which can be defined as :

$$G = \{B, O, R\} \tag{13}$$

where $B = \{b_1, b_2, ..., b_n\}$ is a set of bounding boxes, $O = \{O_1, O_2, ..., O_n\}$ is the set of class labels corresponding to B, which is refined in Sect. 3.1 with the self-study method, $R = \{r_{O_0 \to O_1}, r_{O_0 \to O_2}, ..., r_{O_n \to O_{n-1}}\}$ is the set of relation between O_i and O_j with $n(n-1)$ elements. The relationships R is predicted with a Roi Relation Head. In this paper, we use MOTIFS [2], as the Roi Relation Head, and debias the predict of the relation R with TDE [20]. Finally, the triplet list is ranked with the comprehensive confidence score of the object and the predicate. The scene graph is generated with the combination of the detected pairwise objects and their predicted relationships, and finally ordered by its joint probability $P(O_i)P(R_{O_i \to O_j})P(O_j)$.

4   Experiment

4.1   Experimental Settings

Datasets. Following the recent works [2,9,20] in SGG, we trained and evaluated our model on the Visual Genome (VG) [26] dataset. It consists of 108k images with 75k object categories and 37k predicate classes. Since 92% predicate classes have no more than 10 samples, we followed previous works [1] and adopted a widely used VG split, containing the 150 most frequent object categories with 50 predicate classes. Meanwhile, the VG dataset is split into a training set (70%) and a test set (30%) with a validation set (5k) sampled from the training set for parameter tuning.

Table 1. The SGG performances of Relationship Retrieval on mean Recall@K [9, 23], and the CSS is our proposed model.

Model	Method	Predicate classification			Scene graph classification			Scene graph detection		
		mR@20	mR@50	mR@100	mR@20	mR@50	mR@100	mR@20	mR@50	mR@100
IMP+ [1,9]	–	–	9.8	10.5	–	5.8	6.0	–	3.8	4.8
FREQ [2,23]	–	8.3	13.0	16.0	5.1	7.2	8.5	4.5	6.1	7.1
KERN [9]	–	–	17.7	19.2	–	9.4	10.0	–	6.4	7.3
PA [61]	–	15.2	19.2	20.9	8.7	10.9	11.6	5.7	7.7	8.8
GPS-Net [62]	–	17.4	21.3	22.8	10.0	11.8	12.6	6.9	8.7	9.8
GB-Net-β [63]	–	–	22.1	24.0	–	12.7	13.4	–	7.1	8.5
VTranseE [24]	baseline	11.6	14.7	15.8	6.7	8.2	8.7	3.7	5.0	6.0
	TDE [24]	18.9	25.3	28.4	9.8	13.1	14.7	6.0	8.2	10.2
MOTIFS [2]	baseline	10.8	14.0	15.3	6.3	7.7	8.2	4.2	5.7	6.6
	Focal	10.9	13.9	15.0	6.3	7.7	8.3	3.9	5.3	6.6
	Reweight	16.0	20.0	21.9	8.4	10.1	10.9	6.5	8.4	9.8
	Resample	14.7	18.5	20.0	9.1	11.0	11.8	5.9	8.2	9.7
	Lu+cKD [64]	14.4	18.5	20.2	8.7	10.7	11.4	5.8	8.1	9.6
	CogTree [22]	20.9	26.4	29.0	12.1	14.9	16.1	7.9	10.4	11.8
	TDE [20]	18.5	24.9	28.3	11.1	13.9	15.2	5.8	8.2	9.8
	TDE-CSS	**20.0**	**26.1**	**28.5**	**11.8**	**14.8**	**16.2**	**6.7**	**8.9**	**10.8**
	DLFE [25]	22.1	26.9	28.8	12.8	15.2	15.9	8.6	11.7	13.8
	DLFE-CSS	**23.9**	**28.8**	**30.7**	**13.6**	**16.0**	**16.9**	**8.7**	**12.0**	**14.1**
VCTree [23]	baseline	14.0	17.9	19.4	8.2	10.1	10.8	5.2	6.9	8.0
	Reweight	16.3	19.4	20.4	10.6	12.5	13.1	6.6	8.7	10.1
	Lu+cKD [64]	14.4	18.4	20.0	9.7	12.4	13.1	5.7	7.7	9.1
	CogTree [22]	22.0	27.6	29.7	15.4	18.8	19.9	7.8	10.4	12.1
	EBM [65]	14.2	18.2	19.8	10.4	12.5	13.5	5.7	7.7	9.1
	EBM-CSS	**17.1**	**20.8**	**22.3**	**10.9**	**13.0**	**14.1**	**6.0**	**7.1**	**9.7**
	TDE [20]	18.4	25.4	28.7	8.9	12.2	14.0	6.9	9.3	11.1
	TDE-CSS	**19.4**	**25.9**	**29.4**	**9.2**	**12.9**	**14.9**	**7.1**	**9.6**	**11.8**
	DLFE [25]	20.8	25.3	27.1	15.8	18.9	20.0	8.6	11.8	13.8
	DLFE-CSS	**23.7**	**28.6**	**30.5**	**16.0**	**18.9**	**20.4**	**8.7**	**11.9**	**14.0**

Task and Evaluation. We followed the previous work [2] to divide the SGG task into three sub-tasks: (1) Predicate Classification (**PredCls**) which takes the ground truth bounding boxes with its object labels for relation prediction; (2) Scene Graph Classification (**SGCls**) which takes ground-truth bounding boxes to predict the object label and the relation between the pairwise objects. (3) Scene Graph Detection (**SGDet**) which detects scene graph from scratch. The metric of the traditional SGG task is **Recall@K(R@K)**, which is the fraction of ground-truth targets that are recalled correctly in top K predictions [12]. Due to the long-tailed bias, the good performance on R@K caters to "head" predicates, e.g. *on* [20]. The metric of the recent unbiased SGG task is **mean Recall@K(mR@K)** [9,23], which retrieves each class of relation separately and averages R@K for each relation. The good performance on mR@K achieves more balanced results among different predicates.

Model Configuration. In this paper, we evaluated our method with the roi relation head based on two classic baselines: MotifNet [2] and VCTree [23]. The fusion function for the relation head is set to sum in PredCls and SGDet, and gate in SGCls. Other hyperparameters can be viewed in **Model Zoo** [66]. All models share the same pre-trained detector and the same settings as well.

4.2   Implementation Details

Following the previous work [20], we used a pre-trained Faster R-CNN [58] with a ResNeXt-101-FPN [67,68] and freeze the weights during the training process. For SGCls and SGDet tasks, we first train the original SGG models with the source domain recommended from the configuration for all tasks, including the learning rate. The batch size is set to 12. Then we initialized the Memorandum with the statistics of the results for its counterparts, respectively. The attenuation factor α is set to 0.998 in this paper.

Table 2. The results of Relationship Retrieval on Recall@K, , and the CSS is our proposed model. The Motifs-TDE and VCTree-TDE are traditional SGG approaches, and the others are the unbiased SGG approaches.

Model	Predicate classification			Scene graph classification			Scene graph detection		
	R@20	R@50	R@100	R@20	R@50	R@100	R@20	R@50	R@100
IMP+ [1,9]	52.7	59.3	61.3	31.7	34.6	35.4	14.6	20.7	24.5
FREQ [2,23]	53.6	60.6	62.2	29.3	32.3	32.9	20.1	26.2	30.1
KERN [9]	–	65.8	67.6	–	36.7	37.4	–	27.1	29.8
VTransE [24]	59.0	65.7	67.6	35.4	38.6	39.4	23.0	29.7	34.3
Motifs-TDE [2,20]	38.7	50.8	55.8	21.8	27.2	29.5	12.4	16.9	20.3
VCTree-TDE [20,23]	39.1	49.9	54.5	22.8	28.8	31.2	14.3	19.6	23.3
MOTIFS [2]	58.5	65.2	67.1	32.9	35.8	36.5	21.4	27.2	30.3
MOTIFS-CSS	**59.5**	**66.1**	**67.9**	**35.9**	**39.1**	**39.9**	**25.2**	**32.3**	**37.2**
VCTree [23]	60.1	66.4	68.1	35.2	38.1	38.8	22.0	27.9	31.3
VCTree-CSS	**61.6**	**66.9**	**68.5**	**41.6**	**45.6**	**46.6**	**24.5**	**31.4**	**36.0**

Table 3. Ablation studies of individual components of our method. The baseline model mentioned below is Motifs-TDE unless otherwise indicated.

SS	SA	Predicate classification		Scene graph classification		Scene graph detection	
		mR@20/R@20	mR@50/R@50	mR@20/R@20	mR@50/R@50	mR@20/R@20	mR@50/R@50
–	–	18.5 / 38.7	24.9 / 50.8	11.1 / 22.1	13.9 / 27.2	5.8 / 12.4	8.2 / 16.9
✓	–	– / –	– / –	11.4 / 21.2	14.6 / 27.9	6.7 / 12.9	8.9 / 17.5
–	✓	**20.0 / 42.0**	**26.1 / 53.1**	11.2 / 24.8	14.5 / 30.4	6.4 / 13.3	8.7 / 18.4
✓	✓	– / –	– / –	**11.8 / 26.2**	**14.8 / 31.7**	6.4 / 12.9	**8.9 / 18.6**

4.3   Experiment Results

We evaluated the CSS with the comparison of the conventional unbiased SGG approaches and traditional SGG approaches. As illustrated in Tabs.1 and 2, we compared our performance with several state-of-the-art unbiased SGG methods: TDE [20], EBM [65] and DLFE [25] with mR@K, and the classic traditional SGG model: Motifs [2] and Vctree [23] with R@K.

Table 4. The average precision (left) and the average recall (right) of the object detection for the bounding boxes with different sizes.

	Predicate classification				Scene graph detection			
SS		✓		✓		✓		✓
SA			✓	✓			✓	✓
Small	46.2/53.6	**47.4/54.7**	46.4/53.7	47.2/**54.7**	5.7/21.7	5.7/**21.8**	**5.8**/21.7	**5.8/21.8**
Medium	54.8/62.0	**55.6**/62.8	55.1/62.0	55.6/**62.9**	11.9/32.3	11.9/32.3	11.9/32.3	11.9/32.3
Large	53.1/60.7	53.9/61.4	53.2/60.7	**54.0/61.5**	17.9/35.4	17.9/**35.6**	17.9/35.4	17.9/**35.6**
All	56.6/64.2	**57.4/64.9**	56.8/64.2	**57.4/64.9**	12.9/34.4	**13.0/34.5**	12.9/34.4	**13.0/34.5**

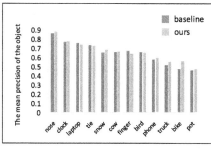
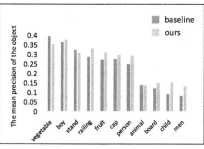

(a) The average precision of the easy sample   (b) The average precision of the hard sample

Fig. 5. The mean precision of the object between the easy and hard sample in the SGCls task. The results of objects with high precision are close between the baseline and CSS, while CSS obtain an obvious improvement to the low-precision objects.

Object Retrieval. We accumulated the average precision and the average recall of objects with different sizes by the COCO-API [69]. As illustrated in Table 4, both the precision and the recall achieve promotion from the baseline with an average of 1.4% and 1.5% relative gain in SGCls sub-task. However, the optimization of both the precision and recall are under 0.5% in SGDet.

Relationship Retrieval (RR). The results are listed in Tables 1 and 2. The CSS model improves on the baseline by an average of 4.6%, 4.5%, 9.3% relative gain of mR@K in each subtask respectively. Meanwhile, it is obvious that the debiasing method causes an unacceptable decline in Recall, as shown with the unbiased SGG model in Table 2. Moreover, the recall of each predicate is applied in the supplementary materials. The CSS improves the RR with an average of 37.8% of the head predicates and 26.72% of the tail predicates, which can be illustrated intuitively in Fig. 6.

Ablation Study. We considered the ablations of each module to investigate the effectiveness of each part of the proposed CSS. The results of SS and CSS are vacant in PredCls task because the label of the object has already been provided as the input of this task, therefore there is no need for object refinement.

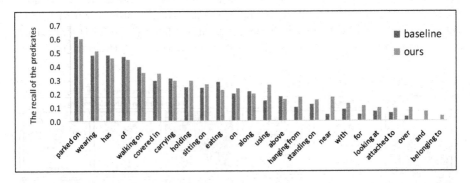

Fig. 6. The recall of the relationships between baseline and CSS. It indicated that the recall of the overall predicates is improved with CSS.

As illustrated in Table 3, each module obtains an obvious promotion: over the baseline, the SS and SA module achieves an average of 8.2% and 5.2% respectively. Besides, we observed that the average precision and recall between SS and CSS are close in Table 4 which obtain an obvious improvement, while the results of SA are almost consistent with the baseline. Moreover, the recall of each predicate shows that the SS module improves the tail predicates with an average of 40% without any optimization of the head predicates, while SA promotes the RR with 19.3% of the head predicates and 32.6% of the hard ones. In addition, we evaluated the influence of the granularity of the relative position encoding in the scene augmentation module. The granularity is set to be 5×5, 8×8 and 13×13. The results are recorded in the supplementary materials, which show limited promotion with the increase of the encoding granularity. However, the increased cost of time and resources is unacceptable.

4.4   Quantitative Studies

Object Detection. As illustrated in Table 4, the SS module achieves promising improvement on object detection in SGCls, especially the targets with small bounding boxes. However, the promotion is limited in the SGDet sub-task. Since the scene information of complex image contains unexpected noise, the SS module is struggle to refine objects with confounding factors. While the labeling process selects the bounding box with human focus, it naturally mitigates the scene noise in SGCls. With the denoising of the bounding box proposals, the SS has great potential to optimize the objects in complex scenes in the future.

Visual Relation Detection. The RR results verify that CSS can refine the SGG effectively with the promising promotion. The recall among each relation shows that SA optimizes the prediction of the overall predicates, while SS focuses more on the tail predicates. Combining the results of SS in Tabs. 3 and 4, it shows that the prediction of the relationships is sensitive to the object, which enables the SS module still work with limited refinement of the object detection. Further, we can infer that the hard sample has a strong constraint on the tail predicates.

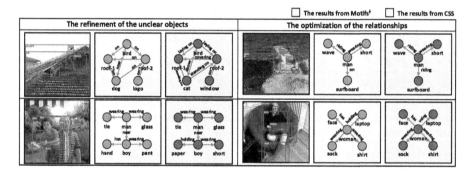

Fig. 7. The visualization results of SG generated from MOTIFS-TDE [20] and CSS.

Scene Graph Generation. The improvement of both the R@K and mR@K illustrated the effectiveness of the CSS model. Moreover, we observed that the CSS achieves the best performance in mR@K and R@K from different aspects. It illustrates that the two modules have different preferences and the CSS balances the results of the head and the tail predicates.

4.5   Visualization Results

For a more intuitive explanation, we generated several SGCls examples from MOTIFS-TDE and Motifs-CSS. As illustrated in Fig. 7, the first row shows the optimization of the unclear objects with small bounding boxes. The top example of the first row also consistent with our analysis in Sec. 4.4 that the relationships are sensitive with the objects. With the refinement of *cat* from *dog*, the misclassified predicate *on* is also optimized with *standing on* correctly. The second row shows examples of the debiasing work. We can observe the refinement of both the head predicate *riding* and the tail predicate *using*, which can provide richer information for the downstream tasks.

5   Conclusion

In this work, we introduced a Continuous Self-Study model (CSS) for scene graph generation. By learning the self-behavior and combining the scene information, the CSS improves the accuracy of identifying the ambiguous targets in complex images. Meanwhile, with the fine-grained relative position encoding, the CSS is able to discriminate the visual relationships with long-tailed bias effectively. Since our proposed method achieves improvements of two basic tasks: object detection and visual relationship detection, it will be helpful to improve the performance in much more visual understanding tasks in the future.

Acknowledgement. This work was supported by the National Natural Science Foundation of China (NSFC No.62076031).

References

1. Xu, D., Zhu, Y., Choy, C.B., Fei-Fei, L.: Scene graph generation by iterative message passing. In: Proceedings of the IEEE Conference on Computer Vision and Pattern Recognition, pp. 5410–5419 (2017)
2. Zellers, R., Yatskar, M., Thomson, S., Choi, Y.: Neural motifs: scene graph parsing with global context. In: Proceedings of the IEEE Conference on Computer Vision and Pattern Recognition, pp. 5831–5840 (2018)
3. Ye, K., Kovashka, A.: Linguistic structures as weak supervision for visual scene graph generation. In: Proceedings of the IEEE/CVF Conference on Computer Vision and Pattern Recognition, pp. 8289–8299 (2021)
4. Wang, Y.S., Liu, C., Zeng, X., Yuille, A.: Scene graph parsing as dependency parsing. arXiv preprint arXiv:1803.09189 (2018)
5. Marino, K., Salakhutdinov, R., Gupta, A.: The more you know: using knowledge graphs for image classification. arXiv preprint arXiv:1612.04844 (2016)
6. Wang, W., Wang, R., Shan, S., Chen, X.: Exploring context and visual pattern of relationship for scene graph generation. In: Proceedings of the IEEE/CVF Conference on Computer Vision and Pattern Recognition, pp. 8188–8197 (2019)
7. Li, Y., Ouyang, W., Zhou, B., Shi, J., Zhang, C., Wang, X.: Factorizable Net: an efficient subgraph-based framework for scene graph generation. In: Ferrari, V., Hebert, M., Sminchisescu, C., Weiss, Y. (eds.) ECCV 2018. LNCS, vol. 11205, pp. 346–363. Springer, Cham (2018). https://doi.org/10.1007/978-3-030-01246-5_21
8. Gu, J., Zhao, H., Lin, Z., Li, S., Cai, J., Ling, M.: Scene graph generation with external knowledge and image reconstruction. In: Proceedings of the IEEE/CVF Conference on Computer Vision and Pattern Recognition, pp. 1969–1978 (2019)

9. Chen, T., Yu, W., Chen, R., Lin, L.: Knowledge-embedded routing network for scene graph generation. In: Proceedings of the IEEE/CVF Conference on Computer Vision and Pattern Recognition, pp. 6163–6171 (2019)

10. Hung, Z.S., Mallya, A., Lazebnik, S.: Contextual translation embedding for visual relationship detection and scene graph generation. IEEE Trans. Pattern Anal. Mach. Intell. **PP**(99), 1 (2020)

11. Fang, Y., Kuan, K., Lin, J., Tan, C., Chandrasekhar, V.: Object detection meets knowledge graphs. In: International Joint Conferences on Artificial Intelligence (2017)

12. Dai, B., Zhang, Y., Lin, D.: Detecting visual relationships with deep relational networks. In: Proceedings of the IEEE Conference on Computer Vision and Pattern Recognition, pp. 3076–3086 (2017)

13. Li, Y., Ouyang, W., Zhou, B., Wang, K., Wang, X.: Scene graph generation from objects, phrases and region captions. In: Proceedings of the IEEE International Conference on Computer Vision, pp. 1261–1270 (2017)

14. Chiou, M.J., Liu, Z., Yin, Y., Liu, A.A., Zimmermann, R.: Zero-shot multi-view indoor localization via graph location networks. In: Proceedings of the 28th ACM International Conference on Multimedia, pp. 3431–3440 (2020)

15. Armeni, I., et al.: 3D scene graph: a structure for unified semantics, 3d space, and camera. In: Proceedings of the IEEE/CVF International Conference on Computer Vision, pp. 5664–5673 (2019)

16. Schuster, S., Krishna, R., Chang, A., Fei-Fei, L., Manning, C.D.: Generating semantically precise scene graphs from textual descriptions for improved image retrieval. In: Proceedings of the Fourth Workshop on Vision and Language, pp. 70–80 (2015)

17. Norcliffe-Brown, W., Vafeias, E., Parisot, S.: Learning conditioned graph structures for interpretable visual question answering. arXiv preprint arXiv:1806.07243 (2018)

18. Zhu, Z., Yu, J., Wang, Y., Sun, Y., Hu, Y., Wu, Q.: Mucko: multi-layer cross-modal knowledge reasoning for fact-based visual question answering. arXiv preprint arXiv:2006.09073 (2020)

19. Yang, X., Tang, K., Zhang, H., Cai, J.: Auto-encoding scene graphs for image captioning. In: Proceedings of the IEEE/CVF Conference on Computer Vision and Pattern Recognition, pp. 10685–10694 (2019)

20. Tang, K., Niu, Y., Huang, J., Shi, J., Zhang, H.: Unbiased scene graph generation from biased training. In: Proceedings of the IEEE/CVF Conference on Computer Vision and Pattern Recognition, pp. 3716–3725 (2020)

21. Guo, Y., et al.: From general to specific: informative scene graph generation via balance adjustment. In: Proceedings of the IEEE/CVF International Conference on Computer Vision, pp. 16383–16392 (2021)

22. Yu, J., Chai, Y., Wang, Y., Hu, Y., Wu, Q.: CogTree: cognition tree loss for unbiased scene graph generation. arXiv preprint arXiv:2009.07526 (2020)

23. Tang, K., Zhang, H., Wu, B., Luo, W., Liu, W.: Learning to compose dynamic tree structures for visual contexts. In: Proceedings of the IEEE/CVF Conference on Computer Vision and Pattern Recognition, pp. 6619–6628 (2019)

24. Zhang, H., Kyaw, Z., Chang, S.F., Chua, T.S.: Visual translation embedding network for visual relation detection. In: Proceedings of the IEEE Conference on Computer Vision and Pattern Recognition, pp. 5532–5540 (2017)

25. Chiou, M.J., Ding, H., Yan, H., Wang, C., Zimmermann, R., Feng, J.: Recovering the unbiased scene graphs from the biased ones. In: Proceedings of the 29th ACM International Conference on Multimedia, pp. 1581–1590 (2021)

26. Krishna, R., et al.: Visual genome: connecting language and vision using crowd-sourced dense image annotations. arXiv preprint arXiv:1602.07332 (2016)

27. Yang, J., Lu, J., Lee, S., Batra, D., Parikh, D.: Graph R-CNN for scene graph generation. In: Ferrari, V., Hebert, M., Sminchisescu, C., Weiss, Y. (eds.) ECCV 2018. LNCS, vol. 11205, pp. 690–706. Springer, Cham (2018). https://doi.org/10.1007/978-3-030-01246-5_41

28. Ji, M., Shin, S., Hwang, S., Park, G., Moon, I.C.: Refine myself by teaching myself: feature refinement via self-knowledge distillation. In: Proceedings of the IEEE/CVF Conference on Computer Vision and Pattern Recognition, pp. 10664–10673 (2021)

29. Tian, Y., Krishnan, D., Isola, P.: Contrastive representation distillation. arXiv preprint arXiv:1910.10699 (2019)

30. Zhang, L., Song, J., Gao, A., Chen, J., Bao, C., Ma, K.: Be your own teacher: improve the performance of convolutional neural networks via self distillation. In: Proceedings of the IEEE/CVF International Conference on Computer Vision, pp. 3713–3722 (2019)

31. Peng, B., et al.: Correlation congruence for knowledge distillation. In: Proceedings of the IEEE/CVF International Conference on Computer Vision, pp. 5007–5016 (2019)

32. Dosovitskiy, A., et al.: An image is worth 16x16 words: transformers for image recognition at scale. arXiv preprint arXiv:2010.11929 (2020)

33. Radford, A., et al.: Learning transferable visual models from natural language supervision. In: International Conference on Machine Learning, pp. 8748–8763. PMLR (2021)

34. Lu, C., Krishna, R., Bernstein, M., Fei-Fei, L.: Visual relationship detection with language priors. In: Leibe, B., Matas, J., Sebe, N., Welling, M. (eds.) ECCV 2016. LNCS, vol. 9905, pp. 852–869. Springer, Cham (2016). https://doi.org/10.1007/978-3-319-46448-0_51

35. Shi, J., Zhang, H., Li, J.: Explainable and explicit visual reasoning over scene graphs. In: Proceedings of the IEEE/CVF Conference on Computer Vision and Pattern Recognition, pp. 8376–8384 (2019)

36. Krishna, R., Chami, I., Bernstein, M., Fei-Fei, L.: Referring relationships. In: Proceedings of the IEEE Conference on Computer Vision and Pattern Recognition, pp. 6867–6876 (2018)

37. Johnson, J., Gupta, A., Fei-Fei, L.: Image generation from scene graphs. In: Proceedings of the IEEE Conference on Computer Vision and Pattern Recognition, pp. 1219–1228 (2018)

38. Woo, S., Kim, D., Cho, D., Kweon, I.S.: LinkNet: relational embedding for scene graph. arXiv preprint arXiv:1811.06410 (2018)

39. Dunn, G., et al.: Evaluation and validation of social and psychological markers in randomised trials of complex interventions in mental health: a methodological research programme. Health Technol. Assess. (Winchester, England) 19, 1–116 (2015)

40. Qi, M., Li, W., Yang, Z., Wang, Y., Luo, J.: Attentive relational networks for mapping images to scene graphs. In: Proceedings of the IEEE/CVF Conference on Computer Vision and Pattern Recognition, pp. 3957–3966 (2019)

41. Hinton, G., Vinyals, O., Dean, J.: Distilling the knowledge in a neural network. arXiv preprint arXiv:1503.02531 (2015)

42. Romero, A., Ballas, N., Kahou, S.E., Chassang, A., Gatta, C., Bengio, Y.: FitNets: hints for thin deep nets. arXiv preprint arXiv:1412.6550 (2014)

43. Zagoruyko, S., Komodakis, N.: Paying more attention to attention: improving the performance of convolutional neural networks via attention transfer. arXiv preprint arXiv:1612.03928 (2016)

44. Kim, J., Park, S., Kwak, N.: Paraphrasing complex network: network compression via factor transfer. arXiv preprint arXiv:1802.04977 (2018)
45. Ahn, S., Hu, S.X., Damianou, A., Lawrence, N.D., Dai, Z.: Variational information distillation for knowledge transfer. In: Proceedings of the IEEE/CVF Conference on Computer Vision and Pattern Recognition, pp. 9163–9171 (2019)
46. Yim, J., Joo, D., Bae, J., Kim, J.: A gift from knowledge distillation: Fast optimization, network minimization and transfer learning. In: Proceedings of the IEEE Conference on Computer Vision and Pattern Recognition, pp. 4133–4141 (2017)
47. Koratana, A., Kang, D., Bailis, P., Zaharia, M.: LIT: learned intermediate representation training for model compression. In: International Conference on Machine Learning, pp. 3509–3518. PMLR (2019)
48. Liu, Y., Cao, J., Li, B., Yuan, C., Hu, W., Li, Y., Duan, Y.: Knowledge distillation via instance relationship graph. In: Proceedings of the IEEE/CVF Conference on Computer Vision and Pattern Recognition, pp. 7096–7104 (2019)
49. Park, W., Kim, D., Lu, Y., Cho, M.: Relational knowledge distillation. In: Proceedings of the IEEE/CVF Conference on Computer Vision and Pattern Recognition, pp. 3967–3976 (2019)
50. Arazo, E., Ortego, D., Albert, P., O'Connor, N.E., McGuinness, K.: Pseudo-labeling and confirmation bias in deep semi-supervised learning. In: 2020 International Joint Conference on Neural Networks (IJCNN), pp. 1–8. IEEE (2020)
51. Pham, H., Dai, Z., Xie, Q., Le, Q.V.: Meta pseudo labels. In: Proceedings of the IEEE/CVF Conference on Computer Vision and Pattern Recognition, pp. 11557–11568 (2021)
52. Xu, T.B., Liu, C.L.: Data-distortion guided self-distillation for deep neural networks. Proceed. AAAI Conf. Artif. Intell. **33**, 5565–5572 (2019)
53. Yun, S., Park, J., Lee, K., Shin, J.: Regularizing class-wise predictions via self-knowledge distillation. In: Proceedings of the IEEE/CVF Conference on Computer Vision and Pattern Recognition, pp. 13876–13885 (2020)
54. Lee, H., Hwang, S.J., Shin, J.: Self-supervised label augmentation via input transformations. In: International Conference on Machine Learning, pp. 5714–5724. PMLR (2020)
55. Lan, X., Zhu, X., Gong, S.: Knowledge distillation by on-the-fly native ensemble. arXiv preprint arXiv:1806.04606 (2018)
56. Howard, A.G., et al.: MobileNets: efficient convolutional neural networks for mobile vision applications. arXiv preprint arXiv:1704.04861 (2017)
57. Zhang, X., Zhou, X., Lin, M., Sun, J.: ShuffleNet: an extremely efficient convolutional neural network for mobile devices. In: Proceedings of the IEEE Conference on Computer Vision and Pattern Recognition, pp. 6848–6856 (2018)
58. Ren, S., He, K., Girshick, R., Sun, J.: Faster R-CNN: towards real-time object detection with region proposal networks. Adv. Neural. Inf. Process. Syst. **28**, 91–99 (2015)
59. Vollmer, M.: Newton's law of cooling revisited. Eur. J. Phys. **30**, 1063 (2009)
60. Yang, Z., Huang, X., Xiu, J., Liu, C.: SocialRank: social network influence ranking method. In: 2012 IEEE 2nd International Conference on Cloud Computing and Intelligence Systems, vol. 2, pp. 591–595. IEEE (2012)
61. Tian, H., Xu, N., Liu, A.A., Zhang, Y.: Part-aware interactive learning for scene graph generation. In: Proceedings of the 28th ACM International Conference on Multimedia, pp. 3155–3163 (2020)
62. Lin, X., Ding, C., Zeng, J., Tao, D.: GPS-Net: graph property sensing network for scene graph generation. In: Proceedings of the IEEE/CVF Conference on Computer Vision and Pattern Recognition, pp. 3746–3753 (2020)

63. Zareian, A., Karaman, S., Chang, S.-F.: Bridging knowledge graphs to generate scene graphs. In: Vedaldi, A., Bischof, H., Brox, T., Frahm, J.-M. (eds.) ECCV 2020. LNCS, vol. 12368, pp. 606–623. Springer, Cham (2020). https://doi.org/10. 1007/978-3-030-58592-1_36

64. Wang, T.J.J., Pehlivan, S., Laaksonen, J.: Tackling the unannotated: scene graph generation with bias-reduced models. arXiv preprint arXiv:2008.07832 (2020)

65. Suhail, M., et al.: Energy-based learning for scene graph generation. In: Proceedings of the IEEE/CVF Conference on Computer Vision and Pattern Recognition, pp. 13936–13945 (2021)

66. Tang, K.: A scene graph generation codebase in PyTorch (2020). https://github. com/KaihuaTang/Scene-Graph-Benchmark.pytorch

67. Lin, T.Y., Dollár, P., Girshick, R., He, K., Hariharan, B., Belongie, S.: Feature pyramid networks for object detection. In: Proceedings of the IEEE Conference on Computer Vision and Pattern Recognition, pp. 2117–2125 (2017)

68. Xie, S., Girshick, R., Dollár, P., Tu, Z., He, K.: Aggregated residual transformations for deep neural networks. In: Proceedings of the IEEE Conference on Computer Vision and Pattern Recognition, pp. 1492–1500 (2017)

69. Lin, T.-Y., et al.: Microsoft COCO: common objects in context. In: Fleet, D., Pajdla, T., Schiele, B., Tuytelaars, T. (eds.) ECCV 2014. LNCS, vol. 8693, pp. 740–755. Springer, Cham (2014). https://doi.org/10.1007/978-3-319-10602-1_48

Spatial Group-Wise Enhance: Enhancing Semantic Feature Learning in CNN

Yuxuan Li, Xiang Li$^{(\boxtimes)}$, and Jian Yang

Nankai University, 38 Tongyan Road, Jinnan District, Tianjin 300350,
People's Republic of China
`xiang.li.implus@njust.edu.cn`

Abstract. The success of attention modules in CNN has attracted increasing and widespread attention over the past years. However, most existing attention modules fail to consider two important factors: (1) For images, different semantic entities are located in different areas, thus they should be associated with different spatial attention masks; (2) most existing framework exploits individual local or global information to guide the generation of attention masks, which ignores the joint information of local-global similarities that can be more effective. To explore these two ingredients, we propose the Spatial Group-wise Enhance (SGE) module. SGE explicitly distributes different but accurate spatial attention masks for various semantics, through the guidance of local-global similarities inside each individual semantic feature group. Furthermore, SGE is lightweight with *almost no extra parameters and calculations*. Despite being trained with only category supervisions, SGE is effective in highlighting multiple active areas with various high-level semantics (such as the dog's eyes, nose, etc.). When integrated with popular CNN backbones, SGE can significantly boost their performance on image recognition tasks. Specifically, based on ResNet101 backbones, SGE improves the baseline by 0.7% Top-1 accuracy on ImageNet classification and 1.6~1.8% AP on COCO detection tasks. The code and pretrained models are available at https://github.com/implus/PytorchInsight.

Keywords: Computer vision · Backbone · Attention mechanism

1 Introduction

Recently, attention mechanisms have become extremely popular in convolutional neural networks. SENet [1] first proposes feature recalibration using the global information in a channel-wise manner. Subsequently, more works [2,3] extend the recalibration to the spatial dimension, enabling the attention factors to be spatially redistributed. Despite their great success, there are at least two aspects have been ignored by most existing work, which limits the rationality and effectiveness of attention modules:

© The Author(s), under exclusive license to Springer Nature Switzerland AG 2023
L. Wang et al. (Eds.): ACCV 2022, LNCS 13845, pp. 316–332, 2023.
https://doi.org/10.1007/978-3-031-26348-4_19

For Spatial Attention Modeling: The natural image usually contains multiple semantic objects distributed in different image regions. However, almost all the existing spatial attention modules [2–4] only use one single global spatial attention mask, which obviously has no way to reasonably reflect the spatial distribution of different semantic features.

For Attention Mask Generation: Existing attention modules strive to guide the generation of the attention mask by utilizing global [1–7], or local [7,8], or local-local pair [9,10] information, but unfortunately lose the chances of gaining benefits from the joint information of the local-global pairs.

In this paper, we aim to propose a novel attention mechanism by taking into account the two factors:

For the first factor, inspired by the CapsuleNet [11] where the *grouped sub-features* can represent the instantiation parameters of a specific type of entity, we propose a group-wise attention mechanism. To be specific, the feature vector is first divided into groups, which are supposed to be learnt with multiple semantics (similarly as Capsules do). Then different spatial attention masks are designed and generated between different semantic feature groups, in the purpose of achieving a more *reasonable and explainable* spatial attention modeling.

For the second factor, in order to fully utilize both global and local information, and to lighten the complexity of the designed module as much as possible, we propose to use the similarity between the global feature descriptor and the local feature vector to guide the generation of the attention mask, which introduces rich information from local-global pairs.

To combine both factors above, the two solutions are merged naturally and completely into a unified framework by requiring *almost no additional parameters and calculations*, which is termed Spatial Group-wise Enhance (SGE) module.

We show on the ImageNet [12]benchmark that the SGE module performs better or comparable to a series of recently proposed state-of-the-art attention modules, despite its superiority in both model capacity and complexity. Similar trend is also observed on smaller dataset like CIFAR-100 [13]. Meanwhile, based on ResNet101 [14] backbones, SGE improves the baseline by 0.7% Top-1 accuracy on ImageNet classification and 1.6~1.8% AP on COCO detection tasks, which demonstrates its remarkable advantages in accurate spatial modeling.

In the ablation study, we show that both solutions of the two factors play an important role for improving the final performance. We also examine the changes in the distribution of the semantic feature activations for each group after the SGE module. The results show that SGE significantly improves the spatial distribution of different semantic sub-features within its group, which strengthens the feature learning in semantic regions and compresses the possible noise and interference. The visualization of activation maps by Grad-CAM [15] also shows that SGE is able to make better use of accurate spatial features.

2   Related Work

Spatial Attention Modeling: In this part, we mainly focus on spatial attention mechanism, where the exsiting work mainly generates a *single* spatial mask for the entire tensor. BAM [2] and CBAM [3] utilize the convolutional layers or channel-based max/avg pooling layer to produce a unified attention map for spatial refinement. GCNet [4] proposes a context modeling, where a convolutional layer is also utilized to produce one spatial mask. The variants of GENet [7] with local extent ratio can be regarded as that each channel has its own attention spatial mask obtained by local information. However, [11] shows that a single scalar is difficult to characterize a semantic entity well, and local attention is also very limited in terms of semantic enhancement. Conversely, the proposed SGE explicitly assign different spatial attention masks in different semantic feature groups, leading to accurate feature enhancement.

Table 1. Summary of major differences among popular lightweight attention modules. The additional costs comprehensively consider the situation of multiple backbones.

Features	SGE (ours)	SE	SK	SRM	GE	BAM	CBAM	GC	GCT
Multiple Spatial Attention Mask	✓				✓ (local version)				
Spatial Attention on *Feature Vectors*	✓								
Global Feature for Attention Generation	✓	✓	✓	✓	✓ (global version)	✓	✓	✓	✓
Local Feature for Attention Generation	✓				✓ (local version)				
Additional Parameter Cost < ~1%	✓				✓ (GE-θ^-)				
Additional FLOPs Cost < ~1%	✓	✓		✓	✓ (GE-θ^-)		✓	✓	✓

Attention Mask Generation: The existing methods can be mainly attributed into the following three groups:

- **Global Only:** A series of work like SENet [1], SKNet [5], SRM [6], GCT [16], BAM [2], CBAM [3] and ECANet [17] performs feature recalibration via the guidance of global averaged statistics. The gather operator in GENet [7] aggregates neuron responses over a given spatial extent to guide the production of the refined tensor. Among the different parameter-free versions of GENet (GE-θ^-), the one with global extent ratio achieves the best performance. Different from the global average operator, GCNet [4] utilizes the context modeling block to weighted average the global statistics. FcaNet [18] decomposes channel features in the frequency domain and utilizes multi-frequency components with the selected DCT bases to replace global average pooling. Instead of squeezing a 3D feature tensor into a single feature vector, Coordinate Attention Network [19] and Triplet Attention Network [20] utilize global pooling along height and width dimensions separately to capture fine-grained global spatial attention.

- **Local Only:** Residual Attention Network [8] constructs a light encoder-decoder architecture between stages to utilize the local spatial information for generating attention masks. The variants of GENet [7] with local extent ratio aggregate the local spatial neuron responses to produce the refined feature tensors. SCNet [21] utilizes a self-calibration branch to allow local spatial information adaptively interact with its surrounding context.

- **Local-Local Pair Only:** [9] gives a thorough study of spatial attention mechanisms designed for broad application, where four types of attention terms are investigated in different combinations of context/position encodings of dense key-query pairs [22]. Such a key-query pair essentially reflects the property of local-local pairs. Another representative structure based on local-local pairs is Non-Local [10] Network, which aims at strengthening the features of the query position via aggregating information from all other positions. However, the time and space complexity of the Non-Local blocks are both quadratic to the number of positions, which are considerably heavy for lightweight modules.

In contrast, our proposed SGE module explores a novel and rich guidance which is generally ignored by the related work: the local-global similarity. Such operator can not only make good use of both global and local information, but also utilize the advantage of the joint statistics between them. Compared to other attention modules, SGE also has fewer parameters, less computational complexity (Table 2), and a clear interpretable mechanism (Fig. 3). Table 1 summarizes the essential differences between SGE and other existing lightweight attention modules for better reference.

Grouped Features: Learning and distributing features into groups in convolutional networks has been widely studied recently. AlexNet [23] initially presents the group convolution and divides features into two groups on different GPUs to save computing budgets. ResNeXt [24] examines the importance of grouping in feature transfer and suggests that the number of groups should be increased to obtain higher accuracy under similar model complexity. The MobileNet series [25–27] and Xception [28] treat each channel as a group and model only spatial relationships inside these groups. The ShuffleNet [29,30] family rearranges the grouped features to produce efficient feature representation. Res2Net [31] uses a hierarchical mode to transfer grouped sub-features, enabling the network to incorporate multi-scale features in a single bottleneck. CapsuleNet [11] models each of the grouped neurons as a capsule, where the activities of the neurons within an active capsule represent the various properties of a particular entity that is present in the image. The overall length of the vector of instantiation parameters is used to represent the existence of the entity and the orientation of the vector is forced to represent the properties of the entity. In SGE, all enhancements are operated inside groups, which saves computational overhead similarly as in group convolution. Conceptually, the SGE module adopts the basic modeling assumptions of CapsuleNet, and believes that the features of each group are able to actively learn various semantic entity representations. At the same time, in the process of visualization of this paper, we also use the length of the sub-feature to measure as its activation value, analogous to the probability of the existence of entities in CapsuleNet.

3   Method

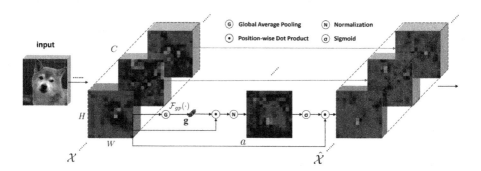

Fig. 1. Illustration of the proposed lightweight SGE module. It processes the sub-features of each group in parallel, and uses the similarity between global statistical feature and local positional features in each group as the attention guidance to enhance the features, thus obtaining well-distributed semantic feature representations in space.

Here we describe the detailed implementation of SGE module, which unifies the above aforementioned two solutions: various semantic spatial attention mask and local-global similarity guidance. We consider a C channel, $H \times W$ convolutional feature map and divide it into G groups along the channel dimension. Without loss of generality, we first examine a certain group separately (see the bottom black box in Fig. 1). Then the group has a vector representation at every position in space, namely $\mathcal{X} = \{\mathbf{x}_{1...m}\}, \mathbf{x}_i \in \mathbb{R}^{\frac{C}{G}}, m = H \times W$. Conceptually inspired by the capsules [11], we further assume that this group gradually captures a specific semantic response (such as the dog's eyes) during the course of network learning. In this group space, ideally we can get features with strong responses at the eye positions (i.e., features with a larger vector length and similar vector directions among multiple eye regions), whilst other positions almost have no activation and become zero vectors. However, due to the unavoidable noise and the existence of similar patterns, it is usually difficult for CNNs to obtain the well-distributed feature responses. We propose to utilize the overall information of the entire group space to further enhance the learning of semantic features in critical regions, given the fact that the features of the entire space are not dominated by noise (otherwise the model learns nothing from this group). Therefore we can use the global statistical feature through spatial averaging function

$\mathcal{F}_{gp}(\cdot)$ to approximate the semantic vector that this group learns to represent:

$$\mathbf{g} = \mathcal{F}_{gp}(\mathcal{X}) = \frac{1}{m}\sum_{i=1}^{m}\mathbf{x}_i. \tag{1}$$

Next, using this global feature, we can generate the corresponding importance coefficient for each feature, which is obtained by simple dot product that measures the similarity between the global semantic feature $\mathbf{g}$ and local feature $\mathbf{x}_i$ to some extent. Thereby for each position, we have the following expression:

$$c_i = \mathbf{g} \cdot \mathbf{x}_i. \tag{2}$$

Note that c_i can also be expanded as $\|\mathbf{g}\|\|\mathbf{x}_i\|\cos(\theta_i)$, where θ_i is the angle between $\mathbf{g}$ and $\mathbf{x}_i$. It indicates that features that have a larger vector length (i.e., $\|\mathbf{x}_i\|$) and a direction (i.e., θ_i) closer to $\mathbf{g}$ are more likely to obtain a larger initial coefficient, which is in line with our assumptions. In order to prevent the biased magnitude of coefficients between various samples, we normalize c over the space, as is widely practiced in [32–34]:

$$\hat{c}_i = \frac{c_i - \mu_c}{\sqrt{\sigma_c^2 + \epsilon}}, \quad \mu_c = \frac{1}{m}\sum_{j}^{m}c_j, \quad \sigma_c^2 = \frac{1}{m}\sum_{j}^{m}(c_j - \mu_c)^2, \tag{3}$$

where ϵ (e.g., 1e-5) is a constant added for numerical stability. To make sure that the normalization inserted in the network can represent the identity transform, we introduce a pair of parameters γ, β for each coefficient $\hat{c}_i$, which scale and shift the normalized value:

$$a_i = \gamma\hat{c}_i + \beta. \tag{4}$$

Note that γ, β here are the only parameters introduced in our module. In a single SGE unit, the number of γ, β is the same as the number of groups G, and the order of their magnitude is about tens (typically, 32 or 64), which is basically *negligible* compared to the millions of parameters of the entire network. Finally, to obtain the enhanced feature vector $\hat{\mathbf{x}}_i$, the original $\mathbf{x}_i$ is scaled by the generated importance coefficients a_i via a sigmoid function gate $\sigma(\cdot)$ over the space:

$$\hat{\mathbf{x}}_i = \mathbf{x}_i \cdot \sigma(a_i), \tag{5}$$

and all the enhanced features form the resulted feature group as

$$\hat{\mathcal{X}} = \{\hat{\mathbf{x}}_{1...m}\}, \hat{\mathbf{x}}_i \in \mathbb{R}^{\frac{C}{G}}, m = H \times W. \tag{6}$$

4 Experiments

4.1 Image Classification

We first compare SGE with a set of SOTA attention modules on ImageNet benchmark. The ImageNet 2012 dataset [12] comprises 1.28 million training images and 50k validation images from 1k classes. We train networks on the training set and report the Top-1 and Top-5 accuracies on the validation set with single 224×224 central crop. For data augmentation, we follow the standard practice [35] and perform the random-size cropping to 224 × 224 and random horizontal flipping. The practical mean channel subtraction is adopted to normalize the input images. All networks are trained with naive softmax cross entropy without label-smoothing regularization [36]. We train all the architectures from scratch by synchronous SGD with weight decay 0.0001 and momentum 0.9 for 100 epochs, starting from learning rate 0.1 and decreasing it by a factor of 10 every 30 epochs. The total batch size is set as 256 and 8 GPUs (32 images per GPU) are utilized for training, using the weight initialization strategy in [37]. Our codes are implemented in the pytorch [38] framework in which all results

Table 2. Comparisons between various guidance for spatial attention mask generation on ImageNet validation set, based on ResNet50. The best and the second best records are marked as red and **blue**, respectively.

Backbone	Param	GFLOPs	Top-1 (%)	Top-5 (%)
ResNet50 [14]	25.56 M	4.122	76.38	92.91
SE-ResNet50 [1]	28.09 M	**4.130**	77.18	**93.67**
SK-ResNet50* [5]	26.15 M	4.185	77.54	93.70
BAM-ResNet50 [2]	25.92 M	4.205	76.90	93.40
CBAM-ResNet50 [3]	28.09 M	4.139	77.63	93.66
SRM-ResNet50 [6]	**25.62 M**	4.139	77.13	93.51
GCT-ResNet50 [16]	25.68 M	4.134	77.30	93.70
GE-ResNet50 [7]	25.56 M	4.127	76.78	93.22
SGE-ResNet50 (ours)	25.56 M	4.127	**77.58**	93.66
ResNet101 [14]	44.55 M	7.849	78.20	93.91
SE-ResNet101 [1]	49.33 M	**7.863**	78.47	94.10
SK-ResNet101* [5]	45.68 M	7.978	**78.79**	**94.27**
BAM-ResNet101 [2]	44.91 M	7.933	78.22	94.02
CBAM-ResNet101 [3]	49.33 M	7.879	78.35	94.06
SRM-ResNet101 [6]	**44.68 M**	7.879	78.47	94.20
GCT-ResNet101 [16]	44.76 M	7.869	78.60	94.10
GE-ResNet101 [7]	44.55 M	7.858	78.42	94.14
SGE-ResNet101 (ours)	44.55 M	7.858	78.90	94.37
ResNeXt50 [24]	25.03 M	4.273	77.15	93.52
SE-ResNeXt50 [1]	27.56 M	**4.281**	78.09	93.96
SK-ResNeXt50 [5]	27.42 M	4.505	**78.21**	**94.07**
BAM-ResNeXt50 [2]	25.39 M	4.356	77.44	93.60
CBAM-ResNeXt50 [3]	27.56 M	4.290	78.08	94.05
GCT-ResNeXt50 [16]	**25.19 M**	4.285	78.20	94.00
GE-ResNeXt50 [7]	25.03 M	4.279	77.48	93.69
SGE-ResNeXt50 (ours)	25.03 M	4.279	78.25	94.09
DenseNet121 [39]	7.98 M	2.883	75.36	92.60
SE-DenseNet121 [1]	**7.99 M**	2.884	**76.21**	93.00
SK-DenseNet121* [5]	8.10 M	2.930	75.83	92.88
BAM-DenseNet121 [2]	8.07 M	2.904	76.20	**93.01**
CBAM-DenseNet121 [3]	**7.99 M**	**2.886**	76.10	92.78
GE-DenseNet121 [7]	7.98 M	2.884	76.18	92.88
SGE-DenseNet121 (ours)	7.98 M	2.884	76.45	93.06

are reproduced. Note that in the following tables, Param. denotes the number of parameter and the definition of FLOPs follows [29], i.e., the number of multiply-adds.

Comparisons with State-of-the-Art Attention Modules. We select a series of state-of-the-art attention modules, which is considered to be relatively lightweight, and demonstrate their performance based on ResNet50, ResNet101 [14,40], ResNeXt50 [24] and DenseNet121 [39]. For a fair comparison, we implement all the attention modules (partially refer to the official codes[1]) with their respective best settings using a unified pytorch framework. Following [1,3], these attention modules are placed after the last BatchNorm [32] layer inside each bottleneck except for BAM and SK. BAM [2] is naturally designed between stages. SK [5] is originally designed on ResNeXt-like bottlenecks with multiple large-kernel group convolutions. To transfer it to the ResNet/DenseNet backbones, we make a slight modification and only append one additional 3×3 group ($G = 32$) convolution upon each original 3×3 convolutions, to prevent the parameters and calculations of the corresponding SKNets from being too large or too small. For GE, we select the best performed parameter-free settings with global extent ratio, namely GE-θ^-, for comparisons (the other variants increase the number of parameters too much). From the results of Table 2, we observe that based on ResNet50, SGE is on par with the best entries from CBAM (Top-1) and SK/SE (Top-5) but has much fewer parameters and slightly less calculations. As for ResNet101, it outperforms most other competing modules. The similar trend is also hold for ResNeXt50 [24] and DenseNet121 [39].

The Effectiveness of Local-Global Similarities. To validate the effectiveness of local-global similarities, we conduct extensive experiments by comparing SGE with global-only and local-only variants of the state-of-the-art SE and GE modules. Specifically in Table 3, to keep the comparisons more fair under the settings of multiple spatial semantics in global-only type, we extend SE with group settings (denoted as SE*), where the fc layers are replaced by group conv1×1 layers with group number G. We also extend GE-θ^- as GE-θ^{-*} with groups. Considering the parameter-free settings of GE-θ^-, we simply average the elements in each group of the global pooled vector to reweight the activations. For the modified group versions of SE* and GE-θ^{-*}, we choose the two settings G=32 and G=64 for experiments. In local-only type, we select the GE modules with spatial extent ratio e=8. Furthermore, we validate the importance of local-global similarities by deleting the similarity part but only using the length of each local sub-feature itself to guide the attention generation in SGE, which is denoted as SGE (- similarity). For the comparisons with local-local pairs, four variants of Non-Local [10] blocks are applied. As the module adds a lot of extra complexity, it is forced to place only one instance on the last stage of ResNet50. From the above results, we notice that the joint information of local-global similarities is considerably efficient and beneficial for achieving the best performance.

---

[1] https://github.com/Jongchan/attention-module.

Table 3. Comparisons between various guidance for spatial attention mask generation on ImageNet validation set, based on ResNet50. The best records are marked as **bold**.

Type	Backbone	Param	GFLOPs	Top-1/5 (%)
Global-only	+ SE [1]	28.09 M	4.130	77.2/93.7
	+ SE* (G=32)	26.20 M	4.128	77.2/93.6
	+ SE* (G=64)	25.89 M	4.128	77.0/93.5
	+ GE-θ^- [7]	25.56 M	4.127	76.8/93.2
	+ GE-θ^{-*} (G=32)	25.56 M	4.127	76.6/93.2
	+ GE-θ^{-*} (G=64)	25.56 M	4.127	76.7/93.4
Local-only	+ GE-θ^- (e=8)	25.56 M	4.127	76.5/93.1
	+ GE-θ^{-*} (G=64, e=8)	25.56 M	4.127	76.5/93.2
	+ SGE (- similarity)	25.56 M	4.125	77.0/93.5
Local-local pair	+ Non-Local [10] (Gaussian)	29.76 M	4.328	75.8/92.7
	+ Non-Local [10] (Embedded Gaussian)	33.95 M	4.534	75.6/92.6
	+ Non-Local [10] (Dot Product)	33.95 M	4.534	76.2/92.8
	+ Non-Local [10] (Concatenation)	33.96 M	4.534	76.2/92.9
Local-global pair	+ SGE (**ours**)	25.56 M	4.127	**77.6/93.7**

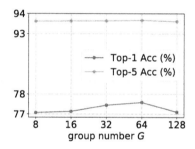

Fig. 2. Performance of SGE-ResNet50 w.r.t. group number G.

Group Number G. In the SGE module, the number of groups G controls the number of semantic sub-features. Too many groups will result in a reduction in the sub-feature dimension within each group, leading to weaker feature representation for each semantic response; On the contrary, too few groups will make the diversity of semantics limited. It is natural to speculate that there is a moderate hyperparameter G that balances semantic diversity and the ability of representing each semantic to optimize network performance. From Fig. 2, we can see that with the increase of G, the performance of the network shows a trend of increasing first and then decreasing (especially in terms of Top-1 accuracy), which is highly consistent with our deduction. Through the experimental results, we recommend the group number G to be 32 or 64. In subsequent experiments, we use $G = 64$ by default.

Initialization of the γ and β. During the experiment, we find that the initialization of the parameter γ and β has a not negligible effect on the result. We use

Table 4. Performance of SGE-ResNet50 as a function of initializations of γ and β.

γ	β	Top-1 (%)	Top-5 (%)
0	0	77.38	**93.71**
0	1	**77.58**	93.66
1	0	77.22	93.58
1	1	77.08	93.70

Table 5. Performance of SGE-ResNet50 with/without the normalization part.

Type	Top-1 (%)	Top-5 (%)
w/ Normalization	**77.58**	**93.66**
w/o Normalization	76.50	93.16

Table 6. Comparisons to the state-of-the-art attention modules on CIFAR-100 test set. The best and the second best records are marked as red and **blue**, respectively.

Backbone	Param	GFLOPs	Top-1 (%)
ResNet50 [14]	23.71 M	1.306	78.06
SE [1]	26.24 M	1.310	78.96
SK* [5]	24.30 M	1.329	79.42
BAM [2]	24.06 M	1.335	79.35
CBAM [3]	26.24 M	1.317	78.44
SRM [6]	23.77 M	1.316	78.62
GCT [16]	23.75 M	1.312	79.10
GE [7]	23.71 M	1.310	78.83
SGE (ours)	23.71 M	1.310	79.47

values 0, 1 for grid search to see the effects of the initialization. From Table 4 we find that initializing γ to 0 tends to get better results. We speculate that when the ordinary patterns of semantic learning has not yet been completely formulated in convolutional feature maps during the initial stage of network training, it may be appropriate to temporarily discard the attention mechanism, but let the network learn a basic semantic representation first. After the initial training period, the attention modules then need to be gradually turned in effect. Therefore, in the early moments of network learning, the attention mechanism of SGE is not suggested to participate heavily in training by setting γ to 0. Such an operation is almost equivalent to simulate the learning process of a network without attention modules during the very early training stage, since each sub-feature of each location is linearly multiplied by the same constant (i.e., $\sigma(\beta)$), whose effect can be cancelled by the following BatchNorm layer.

Normalization. To investigate the importance of normalization in SGE modules, we conduct experiments by eliminating the normalization part from SGE (as shown in Table 5) and find that performance is considerably reduced. The central reason is that the variance of the activation values of different samples in the same group can be statistically very different, indicating that normalization is essential for SGE to work.

Image Classification on CIFAR-100. We also compare SGE with a set of SOTA modules on the 32×32 image dataset CIFAR-100 [13] benchmark. We perform random cropping on images with 4-pixel padding, random horizontal flipping and random rotation with $15°$C. We train networks on the train set and report the Top-1 accuracy on the test set. We adopt a standard training strategy as stated in [41]. Total batch size is set as 128. From the results in Table 6, we observe that based on the ResNet50 backbones, the SGE outperforms all other competing modules in Top-1 classification accuracy, with minimal parameters and relatively lowest computations. SGE's good performance on small image dataset demonstrates its robustness to the scale of the input images.

4.2   Object Detection

We further evaluate the SGE module on object detection on COCO 2017 [42], whose train set is comprised of 118k images, validation set of 5k images. We follow the standard setting [43] of evaluating object detection via the standard mean Average-Precision (AP) scores at different box IoUs or object scales, respectively. The input images are resized with their shorter side being 800 pixels [44]. We train on 8 GPUs with 2 images per each. The backbones of all models are pretrained on ImageNet [12] (directly borrowed from the models listed in Table 2), then all layers except for the first two stages are jointly finetuned with FPN [44] neck and a set of detector heads. Following the conventional finetuning setting [43], the BatchNorm layers are frozen during finetuning. All models are trained for 24 epochs using synchronized SGD with a weight decay of 0.0001 and momentum of 0.9. The learning rate is initialized to 0.02, and decays by a factor of 10 at the 18th and 22nd epochs. The choice of hyper-parameters follows the latest release of the detection benchmark [45].

Table 7. $AP_{50:95}$ (%) scores via embedding SGE on the backbones of state-of-the-art detectors on COCO [42] dataset. The best records are marked as **bold**.

Backbone	Param.	GFLOPs	Retina [46]	Faster [47]	Mask [43]	Cascade [48]
ResNet50	23.51M	88.0	36.4	37.5	38.6	41.1
+ SGE	23.51M	88.1	**37.5**	**38.7**	**39.6**	**42.6**
ResNet101	42.50M	167.9	38.1	39.4	40.4	42.6
+ SGE	42.50M	168.1	**38.9**	**41.0**	**42.1**	**44.4**

Experiments on State-of-the-Art Detectors . We embed the SGE modules into the popular detector framework separately to check if the enhanced feature map helps to detect objects. We select four popular detection frameworks, including RetinaNet [46], Faster RCNN [47], Mask RCNN [43], and Cascade RCNN [48], and choose the widely used FPN [44] as the detection neck. For a fair comparison, we only replace the pretrained backbone model on ImageNet while keeping the other components in the entire detector intact. Table 7 shows the performance of embedding the backbone with the SGE module on these state-of-the-art detectors. We find that although SGE introduces almost no additional parameters and calculations, the gain of detection performance is still very noticeable with basically more than 1% AP point. It is worth noting that SGE can be more prominently advanced on stronger detectors (**+1.5%** AP on ResNet50 and **+1.8%** on ResNet101 in Cascade RCNN).

Table 8. Various AP (%) comparisons based on the state-of-the-art detectors (Faster [47]/Mask [43]/Cascade [48] RCNN) and backbone ResNet101 [14] on COCO [42] dataset. The Parm. and GFLOPs are only with the backbone parts, given that all the remaining structures are kept the same. The numbers in brackets denote the improvements over the baseline backbones. The best records are marked as **bold**.

Backbone	Param	GFLOPs	Detector	$AP_{50:95}$	AP_{50}	AP_{75}	AP_{small}	AP_{media}	AP_{large}
ResNet101 [14]	42.5 M	167.9	Faster	39.4	60.7	43.0	22.1	43.6	52.1
+ SE [1]	47.3 M	168.3	Faster	40.4(+1.0)	61.9	44.2	23.7(+1.6)	44.5	51.9
+ CBAM [3]	47.3 M	168.5	Faster	40.1(+0.7)	61.9	43.6	23.3(+1.2)	44.5	51.2
+ GC(r16) [4]	47.3 M	168.3	Faster	40.3(+0.9)	62.1	43.8	23.4(+1.3)	44.8	51.8
+ GE(-θ^-) [7]	**42.5 M**	**168.1**	Faster	39.5(+0.1)	61.2	43.4	23.2(+1.1)	44.4	50.5
+ SGE	**42.5 M**	**168.1**	Faster	**41.0**(+1.6)	**63.0**	**44.3**	**24.5**(+2.4)	**45.1**	**52.9**
ResNet101 [14]	42.5 M	167.9	Mask	40.4	61.6	44.2	22.3	44.8	52.9
+ SE [1]	47.3 M	168.3	Mask	41.5(+1.1)	63.0	45.3	23.8(+1.5)	45.5	54.7
+ CBAM [3]	47.3 M	168.5	Mask	41.2(+0.8)	62.9	44.8	24.6(+2.3)	45.5	53.1
+ GC(r16) [4]	47.3 M	168.3	Mask	41.6(+1.2)	63.2	45.6	24.7(+2.4)	45.8	53.8
+ GE(-θ^-) [7]	**42.5 M**	**168.1**	Mask	40.6(+0.2)	62.5	44.0	24.0(+1.7)	45.2	52.8
+ SGE	**42.5 M**	**168.1**	Mask	**42.1**(+1.7)	**63.7**	**46.1**	**24.8**(+2.5)	**46.6**	**55.1**
ResNet101 [14]	42.5 M	167.9	Cascade	42.6	60.9	46.4	23.7	46.1	56.9
+ SE [1]	47.3 M	168.3	Cascade	43.4(+0.8)	62.2	47.2	24.1(+0.4)	47.5	57.9
+ CBAM [3]	47.3 M	168.5	Cascade	43.3(+0.7)	62.1	47.1	24.5(+0.8)	47.4	57.7
+ GC(r16) [4]	47.3 M	168.3	Cascade	43.4(+0.8)	62.2	47.4	24.8(+1.1)	47.4	57.9
+ GE(-θ^-) [7]	**42.5 M**	**168.1**	Cascade	42.8(+0.2)	61.8	46.5	24.1(+0.4)	47.0	57.2
+ SGE	**42.5 M**	**168.1**	Cascade	**44.4**(+1.8)	**63.2**	**48.4**	**25.7**(+2.0)	**48.3**	**58.7**

Comparisons with State-of-the-Art Attention Modules . Next, based on backbone ResNet101, we compare SGE with several representative strong attention modules on various competitive state-of-the-art detectors, and report the detailed AP scores including the metrics over three different scales. The original backbones are replaced with the corresponding attention embedded ResNets, which are pretrained on ImageNet. In Table 8, thanks to the enhancement of critical regions, SGE greatly improves the accuracy of detection for small objects ($> 2\%$ absolute AP gain) while its performance of the media and large objects still significantly competitive. This is consistent with our visualization in Fig. 3, which demonstrates that the SGE module is able to retain the feature representation of the spatial region well. Conversely for the others, in each channel, they give the same importance coefficient for every single location, resulting in a loss of the expression of the micro-region to some extent. In the case of general metric $AP_{50:95}$, SGE outperforms the popular SE by a considerably nonnegligible margin, including 0.6% absolute improvement on Faster/Mask RCNN and 1% on Cascade RCNN.

4.3   Visualization and Interpretation

In order to verify that our approach achieves the goal of improving the semantic feature representation, we first demonstrate *several examples* with specific semantic visual clues (in Fig. 3) and show how SGE helps to improve detection accuracy especially in small objects (in Fig. 4).

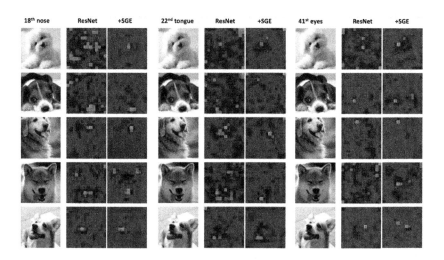

Fig. 3. We select several feature groups with representative semantics to display before and after using SGE on ResNet50. We sample images of different shapes, categories, and angles to verify the robustness of the SGE module.

Visualization of Different Semantic Enhancement. We train a network based on ResNet50 on ImageNet [12] and place the SGE module after the last BatchNorm [32] layer of each bottleneck with reference to SENet [1], by setting G = 64. To better reflect the semantic information while preserving the large spatial

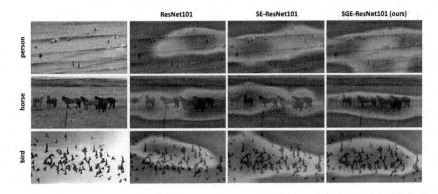

Fig. 4. Grad-CAM [15] visualization results for detection backbone. We compare the visualization results of SE-ResNet101 and SGE-ResNet101 with the ResNet101 baseline. It is clear that our SGE module shows good coverage of target semantic spatial features than other counterparts.

resolution as much as possible, we choose to examine the feature maps of the 4th stage with output size of 14 × 14. For each feature vector of each group, we use its length (i.e., $\|\mathbf{x}_i\|$) to indicate their activation value and linearly normalize it to the interval $[0, 1]$ for a better view. Figure 3 shows three representative groups with semantic responses. As listed in three large columns, they are the 18th, 22nd, and 41st group, which are empirically found to correspond to the concept of the nose, tongue, and eyes. Each large column contains three small columns, where the first small column is the original image, the second small column is the feature map response from the original ResNet50, and the third one is the feature map response enhanced by the SGE module. We select images of dogs of different angles and types to test the robustness of SGE for feature enhancement. Despite its simplicity, the SGE module is very effective in improving the feature representation of specific semantics at corresponding locations while suppressing a large amount of noise. It is worth noting that in the 4th and 7th rows, SGE can strongly emphasize the activation of the eye areas, although their eyes are almost closed. In contrast, the original ResNet fails to capture such patterns.

Activation Map for Detecting Objects. We apply Grad-CAM [15] to several backbones using the images from COCO test set. Grad-CAM can explicitly emphasize the critical regions for semantic feature representations through the gradient guidance. As the regions are considered as important clues for the network to predict correctly, we attempt to judge how the model is making good use of image features. From Fig. 4, thanks to the explicit spatial enhancement mechanism, SGE module is able to cover more critical and accurate locations for semantic expressions, which clearly explains why the detection performance of small or middle objects could be boosted significantly as show in Table 8.

5   Conclusion

To explore the two missing ingredients for attention mechanism in CNN: multiple spatial semantics and local-global similarities, we propose a Spatial Group-wise Enhance (SGE) module that enables each of its feature groups to enhance the learnt semantic representation, guided by its respective local-global similarities. SGE is designed nearly without introducing additional parameters and computational complexity. We visually show that the feature groups have the ability to express different semantics, while the SGE module can significantly enhance this ability. Despite its simplicity, SGE has achieved a steady improvement in both image classification and detection tasks, which demonstrates its compelling effectiveness in practice.

References

1. Hu, J., Shen, L., Sun, G.: Squeeze-and-excitation networks. In: CVPR (2018)
2. Park, J., Woo, S., Lee, J.Y., Kweon, I.S.: BAM: bottleneck attention module. arXiv preprint arXiv:1807.06514 (2018)
3. Woo, S., Park, J., Lee, J.Y., Kweon, I.S.: CBAM: convolutional block attention module. arXiv preprint arXiv:1807.06521 (2018)
4. Cao, Y., Xu, J., Lin, S., Wei, F., Hu, H.: GCNet: non-local networks meet squeeze-excitation networks and beyond. arXiv preprint arXiv:1904.11492 (2019)
5. Li, X., Wang, W., Hu, X., Yang, J.: Selective kernel networks. In: CVPR (2019)
6. Lee, H., Kim, H.E., Nam, H.: SRM: a style-based recalibration module for convolutional neural networks. arXiv preprint arXiv:1903.10829 (2019)
7. Hu, J., Shen, L., Albanie, S., Sun, G., Vedaldi, A.: Gather-excite: exploiting feature context in convolutional neural networks. In: NeurIPs (2018)
8. Wang, F., et al.: Residual attention network for image classification. arXiv preprint arXiv:1704.06904 (2017)
9. Zhu, X., Cheng, D., Zhang, Z., Lin, S., Dai, J.: An empirical study of spatial attention mechanisms in deep networks. arXiv preprint arXiv:1904.05873 (2019)
10. Wang, X., Girshick, R., Gupta, A., He, K.: Non-local neural networks. In: CVPR (2018)
11. Sabour, S., Frosst, N., Hinton, G.E.: Dynamic routing between capsules. In: NeurIPs (2017)
12. Russakovsky, O., et al.: ImageNet large scale visual recognition challenge. In: IJCV (2015)
13. Krizhevsky, A., Hinton, G.: Learning multiple layers of features from tiny images. Handbook of Systemic Autoimmune Diseases (2009)
14. He, K., Zhang, X., Ren, S., Sun, J.: Deep residual learning for image recognition. In: CVPR (2016)
15. Selvaraju, R.R., Cogswell, M., Das, A., Vedantam, R., Parikh, D., Batra, D.: Grad-CAM: visual explanations from deep networks via gradient-based localization. In: CVPR (2017)
16. Yang, Z., Zhu, L., Wu, Y., Yang, Y.: Gated channel transformation for visual recognition. In: CVPR (2020)
17. Wang, Q., Wu, B., Zhu, P., Li, P., Zuo, W., Hu, Q.: ECA-Net: efficient channel attention for deep convolutional neural networks. In: CVPR (2020)

18. Qin, Z., Zhang, P., Wu, F., Li, X.: FcaNet: frequency channel attention networks. In: ICCV (2021)
19. Hou, Q., Zhou, D., Feng, J.: Coordinate attention for efficient mobile network design. In: CVPR (2021)
20. Misra, D., Nalamada, T., Arasanipalai, A.U., Hou, Q.: Rotate to attend: convolutional triplet attention module. In: WACV (2021)
21. Liu, J.-J., Hou, Q., Cheng, M.-M., Wang, C., Feng, J.: Improving convolutional networks with self-calibrated convolutions. In: CVPR (2020)
22. Vaswani, A., et al.: Attention is all you need. In: NeurIPs (2017)
23. Krizhevsky, A., Sutskever, I., Hinton, G.E.: ImageNet classification with deep convolutional neural networks. In: NeurIPs (2012)
24. Xie, S., Girshick, R., Dollár, P., Tu, Z., He, K.: Aggregated residual transformations for deep neural networks. In: CVPR (2017)
25. Howard, A.G., et al.: MobileNets: efficient convolutional neural networks for mobile vision applications. arXiv preprint arXiv:1704.04861 (2017)
26. Sandler, M., Howard, A., Zhu, M., Zhmoginov, A., Chen, L.C.: MobileNetV2: inverted residuals and linear bottlenecks. In: CVPR (2018)
27. Howard, A., et al.: Searching for mobileNetV3. arXiv preprint arXiv:1905.02244 (2019)
28. Chollet, F.: Xception: deep learning with depthwise separable convolutions. In: CVPR (2017)
29. Zhang, X., Zhou, X., Lin, M., Sun, J.: ShuffleNet: an extremely efficient convolutional neural network for mobile devices. arxiv 2017. arXiv preprint arXiv:1707.01083 (2017)
30. Ma, N., Zhang, X., Zheng, H.T., Sun, J.: ShuffleNetV2: practical guidelines for efficient CNN architecture design. arXiv preprint arXiv:1807.11164 (2018)
31. Gao, S.H., Cheng, M.M., Zhao, K., Zhang, X.Y., Yang, M.H., Torr, P.: Res2Net: a new multi-scale backbone architecture. arXiv preprint arXiv:1904.01169 (2019)
32. Ioffe, S., Szegedy, C.: Batch normalization: accelerating deep network training by reducing internal covariate shift. arXiv preprint arXiv:1502.03167 (2015)
33. Wu, Y., He, K.: Group normalization. In: Ferrari, V., Hebert, M., Sminchisescu, C., Weiss, Y. (eds.) ECCV 2018. LNCS, vol. 11217, pp. 3–19. Springer, Cham (2018). https://doi.org/10.1007/978-3-030-01261-8_1
34. Qiao, S., Wang, H., Liu, C., Shen, W., Yuille, A.: Weight standardization. arXiv preprint arXiv:1903.10520 (2019)
35. Szegedy, C., et al.: Going deeper with convolutions. In: CVPR (2015)
36. Szegedy, C., Vanhoucke, V., Ioffe, S., Shlens, J., Wojna, Z.: Rethinking the inception architecture for computer vision. In: CVPR (2016)
37. He, K., Zhang, X., Ren, S., Sun, J.: Delving deep into rectifiers: surpassing human-level performance on imagenet classification. In: ICCV (2015)
38. Paszke, A., Gross, S., Chintala, S., Chanan, G.: PyTorch: tensors and dynamic neural networks in python with strong GPU acceleration (2017)
39. Huang, G., Liu, Z., Van Der Maaten, L., Weinberger, K.Q.: Densely connected convolutional networks. In: CVPR (2017)
40. He, K., Zhang, X., Ren, S., Sun, J.: Identity mappings in deep residual networks. In: Leibe, B., Matas, J., Sebe, N., Welling, M. (eds.) ECCV 2016. LNCS, vol. 9908, pp. 630–645. Springer, Cham (2016). https://doi.org/10.1007/978-3-319-46493-0_38
41. Devries, T., Taylor, G.W.: Improved regularization of convolutional neural networks with cutout. CoRR (2017)

42. Lin, T.-Y., et al.: Microsoft COCO: common objects in context. In: Fleet, D., Pajdla, T., Schiele, B., Tuytelaars, T. (eds.) ECCV 2014. LNCS, vol. 8693, pp. 740–755. Springer, Cham (2014). https://doi.org/10.1007/978-3-319-10602-1_48
43. He, K., Gkioxari, G., Dollár, P., Girshick, R.: Mask R-CNN. In: ICCV (2017)
44. Lin, T.Y., Dollár, P., Girshick, R., He, K., Hariharan, B., Belongie, S.: Feature pyramid networks for object detection. In: CVPR (2017)
45. Chen, K., et al.: MMDetection. https://github.com/open-mmlab/mmdetection (2018)
46. Lin, T.Y., Goyal, P., Girshick, R., He, K., Dollár, P.: Focal loss for dense object detection. In: ICCV (2017)
47. Ren, S., He, K., Girshick, R., Sun, J.: Faster R-CNN: towards real-time object detection with region proposal networks. In: NeurIPS (2015)
48. Cai, Z., Vasconcelos, N.: Cascade R-CNN: delving into high quality object detection. In: CVPR (2018)

SEIC: Semantic Embedding
with Intermediate Classes for Zero-Shot
Domain Generalization

Biswajit Mondal[(✉)] [iD] and Soma Biswas[iD]

Indian Institute of Science, Bangalore, India
mondalb682@gmail.com, somabiswas@iisc.ac.in

Abstract. In this work, we address the Zero-Shot Domain Generalization (ZSDG) task, where the goal is to learn a model from multiple source domains, such that it can generalize well to both unseen classes and unseen domains during testing. Since it combines the tasks of Domain Generalization (DG) and Zero-Shot Learning (ZSL), here we explore whether advances in these fields also translate to improved performance for the ZSDG task. Specifically, we build upon a state-of-the-art approach for domain generalization and appropriately modify it such that it can generalize to unseen classes during the testing stage. Towards this goal, we propose to make the feature embedding space semantically meaningful, by not only making an image feature close to its semantic attributes, but also taking into account its similarity with the other neighbouring classes. In addition, in order to reserve space for the unseen classes in the embedding space, we propose to introduce pseudo intermediate classes in between the semantically similar classes during training. This reduces confusion of the similar classes and thus increases the discriminability of the embedding space. Extensive experiments on two large-scale benchmark datasets, namely DomainNet and DomainNet-LS and comparisons with the state-of-the-art approaches show that the proposed framework outperforms all the other techniques on both the datasets.

1 Introduction

The recent advancement in deep neural networks has achieved enormous success in numerous areas of computer vision, such as classification [13,49], segmentation [40], retrieval [34,38,50], playing Atari games with reinforcement learning [32], etc. In standard supervised training, we assume that the training and the test data belong to the same distribution, and the test data contains only the classes that were seen during training. Such models can fail when they encounter images from classes and domains unseen during the training process, as often encountered in real scenarios. Since it is impractical to collect examples from all possible classes and domains during training, it is important that the learnt models generalize well to these challenging scenarios. This has led to a significant

L. Wang et al. (Eds.): ACCV 2022, LNCS 13845, pp. 333–350, 2023.
https://doi.org/10.1007/978-3-031-26348-4_20

amount of research focused in areas like domain generalization (DG) and zero-shot learning (ZSL). The DG task [25,27,39,48] aims to classify samples from unseen target domains after learning from multiple source domains which have the same classes as the target data. On the other hand, the ZSL task [23,33] aims to classify samples belonging to classes unseen during training, but from the same domain as the training data. It is only recently, that researchers have started addressing the more realistic and challenging zero-shot domain generalization (ZSDG) task [8,28,29], where during testing, the samples can not only belong to unseen classes, but also unseen domains.

ZSDG being a combination of DG and ZSL tasks, an advancement in any of these fields should translate to an advancement in the ZSDG problem. But recent research [8] indicates that naively combining DG and ZSL approaches does not help to improve the performance on the ZSDG task. In this work, we explore whether a state-of-the-art DG approach can be appropriately modified so that it also achieves state-of-the-art performance for the ZSDG task. Specifically, we build upon MixStyle [51], which computes the convex combination of instance-level feature statistics of different samples to generate diverse domains/styles for training, while keeping the semantic information intact for the DG task.

In this work, we propose simple, yet effective modifications which can generalize the MixStyle [51] framework for classifying unseen classes (from unseen domains) during testing. Towards this goal, we want to make the feature embedding space semantically meaningful, so that unseen class images/features can be matched with their semantic attributes, as well as discriminative, so that the classification performance in this space is satisfactory. To account for both these objectives, we propose two modifications to the original DG approach, namely (1) We introduce intermediate (pseudo) classes between semantically similar classes in the embedding space, to reserve space for the unseen classes during testing; (2) Each image feature is encouraged to be not only close to its true attribute vector, but also at semantically meaningful distances from the attributes of its neighbouring classes. The combined framework is termed as **S**emantic **E**mbedding with **I**ntermediate **C**lasses (SEIC). To summarize, our contributions are as follows:

1. We propose a simple, yet effective framework termed SEIC, to address the problem of zero-shot domain generalization.
2. We propose to make the feature embedding space semantically meaningful and discriminative by accounting for the neighbouring class information as well as by introducing intermediate pseudo classes.
3. We show that a state-of-the-art DG method can be appropriately modified to get state-of-the-art result for the related ZSDG problem.
4. Extensive experiments and comparisons on the challenging DomainNet and DomainNet-LS datasets [35] justify the effectiveness of the SEIC framework.

2   Related Work

Here, we briefly describe the related literature on domain generalization (DG), zero-shot learning (ZSL) and finally zero-shot domain generalization (ZSDG).

Domain Generalization: First proposed in [5], domain generalization is a problem gaining rapid attention in the vision community. A broad category of approaches can be summarized by domain-invariant representation learning, i.e., learning representations that eliminate domain-specific variations within the dataset. This approach was first examined in the context of domain adaptation in [4], which was used to construct a domain-adversarial neural network in [11]. Several algorithms have been proposed for domain generalization via adversarial learning [25,27,39,48]. MixStyle [51] is motivated by the observation that visual domain is closely related to image style (or domain). [20] augments the feature-space by identifying the dominant modes of change in the source domain. [14,47] transform images into frequency space to perform domain generalization. Single-source DG methods, tackle a more challenging scenario, where only a single source domain is available during training [14,37,39,42]. Some works also address the DG problem during the testing phase [16].

Zero Shot Learning: ZSL [23,33] aims to transfer the model trained on the seen classes to the unseen ones, usually using a semantic space between seen classes and unseen classes. Early works in ZSL focused on the conventional ZSL [1,2,6,7,22], where the test data only belongs to the unseen classes, and the predicted class is calculated based on the feature similarity with the attributes of the unseen test classes in the embedding space. In generalized ZSL (GZSL), both seen and unseen classes can be present during testing, making it a more challenging task. The works in [3,9,15,46] addresses the overfitting problem that arises due to training on only the seen classes [44]. Many works [19,31,45] employ generative methods for converting the problem into a supervised learning problem using Generative Adversarial Networks (GANs) [12] and Variational Autoencoder (VAEs) [21] to synthesize images of the unseen classes. [18] uses adaptive metric learning to check the compatibility of a sample with the class semantics.

Zero Shot Domain Generalization: In general, the ZSL and DG tasks have been considered separately. But recently, ZSDG is being researched actively because of its more realistic and practical applications. Cumix [28] aims to simulate the test-time domain and semantic shift using images from unseen domains and categories by mixing up the images available in source domains and categories during the training phase. It also uses a curriculum-based mixing policy to generate increasingly complex training samples. SLE-Net [8] uses visual and semantic encoders to learn domain-agnostic structured latent embeddings by projecting images from different domains and their class-specific semantic representations to a common latent space. SnMpNet [34] addresses the problem of image retrieval, where the test data can belong to classes or domains which are unseen during training. Our work is similar in spirit to [29], which effectively exploits semantic information of the classes to adapt the existing DG methods to tackle the ZSDG task. Zero-shot domain adaptation is another research area similar to ZSDG, which aims to transfer the knowledge from a single source domain to a target domain. [26] projects the samples of source and target domains to a common space and then learns unseen class prototypes of the target domain. [17] learns class-agnostic domain feature representations and prevents negative transfer effects using adversarial learning. [41] introduces a new scenario where labelled samples are available for a subset of target domain classes and proposes a method to transform samples from source domain to target domain without loss of class information.

3   Problem Definition

Zero-shot domain generalization (ZSDG) aims to classify unseen classes in unseen domains. Let $\mathcal{X}$ denote the image space, $\mathcal{Y}$ the set of possible classes and $\mathcal{D}$ the set of possible domains. The classes are divided into two sets, one is used for training or the *seen classes* ($\mathcal{Y}^s \in \mathcal{Y}$) and the other for testing or the *unseen classes* ($\mathcal{Y}^u \in \mathcal{Y}$). Similarly, we have *seen domains* ($\mathcal{D}^s \in \mathcal{D}$) and *unseen domains* ($\mathcal{D}^u \in \mathcal{D}$). For training, we are given the set, $\mathcal{M} = \{(\mathbf{x}, y, \mathbf{a}_y, d) | \mathbf{x} \in \mathcal{X}, y \in \mathcal{Y}^s, \mathbf{a}_y \in \mathcal{E}, d \in \mathcal{D}^s\}$, where $\mathbf{x}$ is an image belonging to a seen class and a seen domain, and has a class label y belonging to the seen class set $\mathcal{Y}^s$. $\mathbf{a}_y$ is the semantic embedding in $\mathcal{E}$ for class y in $\mathcal{Y}^s$, where $\mathcal{E}$ is the embedding space. d is $\mathbf{x}$'s domain label from the seen domain set.

During testing, the goal is to classify the test data $\mathcal{N} = \{\mathbf{x}\}$, which belong to an unseen class, i.e. $y \in \mathcal{Y}^u$ and also an unseen domain, i.e. $d \in \mathcal{D}^u$. In standard ZSL, training is done on the set of seen classes and testing on the set of unseen classes which are mutually disjoint, but the domains remain the same, i.e., $\mathcal{Y}^s \cap \mathcal{Y}^u = \phi$ and $\mathcal{D}^s \equiv \mathcal{D}^u$. In DG, training is done on images belonging to a set of domains that is disjoint to the set of domains used for testing, but the set of classes is shared, i.e., $\mathcal{D}^s \cap \mathcal{D}^u = \phi$ and $\mathcal{Y}^s \equiv \mathcal{Y}^u$. Each domain can have different distributions, i.e., $p_{\mathcal{X}}(\mathbf{x}|d_i) \neq p_{\mathcal{X}}(\mathbf{x}|d_j), \forall i \neq j$. Here, we address the more challenging ZSDG problem where testing is done on domains and classes unseen during training, i.e., $\mathcal{D}^s \cap \mathcal{D}^u = \phi$ and $\mathcal{Y}^s \cap \mathcal{Y}^u = \phi$.

4   Proposed Method

Now, we describe the proposed framework, termed SEIC for the ZSDG task. First, we describe the recent state-of-the-art DG technique MixStyle [51] that we use as the backbone for SEIC framework, followed by the proposed modifications.

4.1   Handling Unseen Domains Using Domain Generalization

Here, we briefly describe the MixStyle [51] approach, where given training data from multiple source domains, the goal is to learn a model which can generalize well to unseen target domains. MixStyle regularizes CNN training by perturbing the style information of the samples from the source domains. It mixes the feature statistics of two instances with a random convex weight to simulate new styles. The framework broadly consists of a feature extractor $\mathcal{F}^{DG}$ and a classifier g^{DG}

to get the output, $y^{DG} = g^{DG} \circ \mathcal{F}^{DG}$. The mixing is done using the statistics of features from the output of different CNN layers in the feature extractor. Let $\mathbf{f}_i$ and $\mathbf{f}_j$ be the feature maps corresponding to samples $\mathbf{x}_i$ and $\mathbf{x}_j$ after a particular CNN layer. It computes the mixed style feature statistics for $\mathbf{f}_i$ using $\mathbf{f}_j$ as follows:

$$\mu_{ms}(\mathbf{f}_i; \mathbf{f}_j) = \lambda\mu(\mathbf{f}_i) + (1 - \lambda)\mu(\mathbf{f}_j) \tag{1}$$

$$\sigma_{ms}(\mathbf{f}_i; \mathbf{f}_j) = \lambda\sigma(\mathbf{f}_i) + (1 - \lambda)\sigma(\mathbf{f}_j) \tag{2}$$

where $\lambda \sim \beta(\alpha, \alpha)$ and $\alpha \in (0, \infty)$ is a hyper parameter. $\sigma(.)$ and $\mu(.)$ are standard deviation and mean, respectively, computed along the height and width of each channel. Finally, the style-normalized features $f_{ms}(\mathbf{f}_i; \mathbf{f}_j)$ are computed by using the mixed feature statistics as follows:

$$f_{ms}(\mathbf{f}_i; \mathbf{f}_j) = \sigma_{ms}(\mathbf{f}_i; \mathbf{f}_j) * \mathbf{f}_i' + \mu_{ms}(\mathbf{f}_i; \mathbf{f}_j) \tag{3}$$

$$\text{where, } \mathbf{f}_i' = \frac{\mathbf{f}_i - \mu(\mathbf{f}_i)}{\sigma(\mathbf{f}_i)} \tag{4}$$

The mixing of the statistics does not alter the class information (i.e. class is same as that of $\mathbf{x}_i$) even if the two features being mixed belong to different classes. This module can be easily plugged in after different layers of the CNN to get more diversity in the source domains and achieve better generalizability for unseen domains.

4.2   Handling Unseen Classes Using the Proposed SEIC Framework

We will now describe the proposed modifications, such that the above model also performs well for unseen classes during testing. Specifically, we make the following three modifications: (i) First, to establish the connection between the seen and unseen classes, we replace the classifier weights using the semantic vectors, which are automatically obtained using the class names. (ii) To reserve space for the unseen classes which will be encountered during testing, we introduce intermediate pseudo-classes in the training process; (iii) We utilize the neighbourhood class information to make the feature embedding space semantically meaningful. These modifications (details below) enable the proposed framework (SEIC) to handle unseen classes as well during the testing stage.

(i) *Utilizing Class Attributes to Link the Seen and Unseen Classes:* In ZSDG task, since unseen classes can be encountered during testing, it is important to link the seen and unseen classes. Specifically, the goal is to learn the relation between the feature embeddings and the class semantics, such that the class label of the test data can be predicted by comparing it with the semantic embeddings of all the classes. The semantic embeddings can be obtained using unsupervised Natural Language Processing algorithms like Word2Vec [30], GloVe vectors [36], etc. As discussed earlier, in MixStyle, the model architecture has a feature extractor $\mathcal{F}^{DG}$ followed by a classifier g^{DG}. For handling unseen classes, we replace the weights in the classification layer by the semantic vectors of each

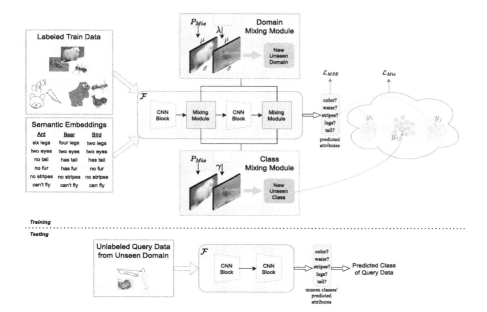

Fig. 1. Depiction of the proposed method. For training, samples from seen domains and seen classes are fed to the feature extractor, which consists of CNN blocks, each followed by a mixing module. The mixing module consists of the domain and class mixing modules to simulate unseen domains and classes. To learn distinctive features, especially between similar classes, the mixing module is given the information of pairwise mixing probabilities of the classes, P_{Mix} and intermediate pseudo classes are inserted at the output layer. During testing, the model has no mixing module and pseudo-class nodes, and predicts the samples from unseen classes and unseen domains.

class, i.e., we want the predicted semantic embeddings extracted from the model to be similar to the semantic embeddings, $g : \mathcal{Y} \rightarrow \mathcal{E}$. The modified feature extractor and embedding function are denoted as $\mathcal{F} : \mathcal{X} \rightarrow \mathcal{E}$ and $g : \mathcal{Y} \rightarrow \mathcal{E}$ respectively. In this work, g is the Word2Vec embedding of the class name. For an image $\mathbf{x}$, the model predicts the class as follows:

$$y^* = \underset{y}{\operatorname{argmax}} \, g(y)^T \mathcal{F}(\mathbf{x}). \tag{5}$$

With this modification, the model is capable of recognizing unseen classes. But currently, since the model is trained using only the seen classes, the feature embedding space may not be discriminative enough to distinguish between the seen and unseen classes, which we address using the intermediate pseudo-classes.

(ii) *Introducing Intermediate Pseudo Classes.* Now, we discuss how we address the challenge of correctly classifying unseen classes during testing. The current model will try to embed the features of the unseen classes to its class attributes, but since these classes were not used for model training, their embeddings are usually confused, specifically with those of the semantically similar seen classes. For example, a new *bee* class feature may be easily confused with features of other insects, but will probably not be confused with features from animals like cat or vehicles like buses, etc. To improve class discrimination and thus reduce this confusion, we propose to introduce additional pseudo classes in between the existing training classes, with more emphasis on semantically similar classes. These intermediate classes act as proxies for the unseen classes, that might be encountered during testing. We propose to generate data for the intermediate classes using the available training data as explained below.

Given two classes $y_i, y_j \in \mathcal{Y}^s$ in the seen class set with their respective semantic embeddings as $\mathbf{a}_{y_i}, \mathbf{a}_{y_j} \in \mathcal{E}$, we form an intermediate pseudo class y_{ij} (or equivalently y_{ji}) and assign it a semantic embedding equal to the average of the semantic embeddings of y_i and y_j, i.e.,

$$\mathbf{a}_{y_{ij}} = \frac{\mathbf{a}_{y_i} + \mathbf{a}_{y_j}}{2} \tag{6}$$

To generate the training data for these intermediate pseudo-classes, we propose to mix pairs of samples belonging to classes y_i and y_j in the feature space after different CNN layers. Given the feature maps $\mathbf{f}_i$ and $\mathbf{f}_j$ of two training instances after some CNN layer, we obtain the intermediate class feature f_{ic} as:

$$f_{ic}(\mathbf{f}_i; \mathbf{f}_j) = \gamma \mathbf{f}_i + (1 - \gamma)\mathbf{f}_j \tag{7}$$

where $\gamma \in \mathbb{R}$ is sampled from a uniform distribution, i.e., $\gamma \sim \mathbf{U}(t_1, t_2)$. Since the pseudo classes are generated using the average of two semantic embeddings, as shown in Eq. (6), we choose t_1 and t_2 such that the feature of one class does not overshadow the feature of the other class.

We have discussed handling of unseen domains by mixing the statistics and handling the unseen classes by mixing the features. Now, we combine them to get our final mixing module:

$$f_{Mix}(\mathbf{f}_i; \mathbf{f}_j) = f_{ic}(\mathbf{f}_i'; \mathbf{f}_j') * \sigma_{ms}(\mathbf{f}_i; \mathbf{f}_j) + \mu_{ms}(\mathbf{f}_i; \mathbf{f}_j) \tag{8}$$

where $\mathbf{f}_i'$ and $\mathbf{f}_j'$ are defined as per Eq. (4). Since we also want to retain the original samples with their class and domain information, the mixing is done only if a generated random number (r) is less than a pre-set threshold (τ), otherwise the original features with their class information are used for training. This random number is generated independently for each mixing module and for every batch in every epoch. Depending on whether mixing has happened or not, we have two separate losses. The final loss for the image $\mathbf{x}_i$ belonging to class y_i is given as:

$$\mathcal{L}_{Mix}(\mathbf{x}_i; \mathbf{x}_j) = \begin{cases} \mathcal{L}_{CE}(g(y)^T \mathcal{F}(\mathbf{x}_i; \mathbf{x}_j), y_{ij}), & r < \tau \\ \mathcal{L}_{CE}(g(y)^T \mathcal{F}(\mathbf{x}_i), y_i), & \text{otherwise} \end{cases} \tag{9}$$

where $\mathcal{L}_{CE}$ is the cross-entropy loss, $g(y)$ is the set of semantic embeddings corresponding to each class node, $\mathcal{F}$ is the feature extractor consisting of the CNN layers and the mixing modules inserted between the layers, y_{ij} is the pseudo class in-between the classes corresponding to $\mathbf{x}_i$ and $\mathbf{x}_j$. When the random number r (generated uniformly in the range 0 to 1) is greater than or equal to the threshold τ, then the mixing module acts as an identity function.

As discussed earlier, to learn distinctive features for closely related classes, we focus more on learning the pseudo classes that are between two semantically close classes. First, we calculate the Euclidean distance between the semantic embeddings of every pair of classes to find the similarity among them. A class is mixed with another class following a probability distribution based on the semantic similarity of the classes. For e.g., if we are given only three classes y_i, y_j and y_k, with their pair-wise distances from $\mathbf{a}_{y_i}$ as, $dist(\mathbf{a}_{y_i}, \mathbf{a}_{y_j}) = d_{ij}$ and $dist(\mathbf{a}_{y_i}, \mathbf{a}_{y_k}) = d_{ik}$. Then, the probability of mixing a sample of class y_i with a sample of class y_j will be:

$$P_{Mix}(y_i, y_j; y_k) = \frac{exp(-d_{ij})}{exp(-d_{ij}) + exp(-d_{ik})} \tag{10}$$

The probability of mixing a sample of class y_i with a sample of class y_k can be calculated in a similar manner. Another reason why we mix semantically close classes is that it has more potential to generate meaningful novel classes. For example, mixing two types of insects may produce another insect, but mixing an insect with a dog might not produce anything realistic.

(iii) *Incorporating Information from Neighbouring Classes:* With the above modifications, the model is now capable of handling both unseen classes and unseen domains during testing. But, the success of the unseen class predictions depend upon how semantically meaningful the feature embeddings are. Here, we propose to guide the feature embeddings not only using its correct ground truth attribute (using the classification loss), but also using the information of its semantically similar neighbouring classes. For e.g., the feature embedding of an insect class *wasp* can be guided by its class attribute, and also by its relative distances from the other insect classes. This is specially important during the feature computation of the unseen classes, where in absence of its ground truth attributes, the embedding has to be solely guided by the seen class attributes.

The standard classification loss encourages the model to predict a score of 1 for the ground truth class and 0 for all other classes. In contrast, we propose to calculate the loss not just with respect to the ground truth attribute, but also with respect to the other classes, appropriately weighted by their similarity with the ground truth class. Given an image $\mathbf{x}_i \in \mathcal{X}$, belonging to class $y_i \in \mathcal{Y}^s$ with semantic vector $\mathbf{a}_{y_i} \in \mathcal{E}$, we propose to use an additional loss term as follows:

$$\mathcal{L}_{MSE}(\mathbf{x}_i) = \sum_{y \in \mathcal{Y}^s} exp\left(-\frac{\|(\mathbf{a}_{y_i} - \mathbf{a}_y)\|^2}{max_{z \in \mathcal{Y}^s} \|(\mathbf{a}_{y_i} - \mathbf{a}_z)\|^2}\right)(\|\mathcal{F}(\mathbf{x}_i) - \mathbf{a}_y\|^2 - \|\mathbf{a}_{y_i} - \mathbf{a}_y\|^2)^2 \tag{11}$$

where $\mathbf{a}_y$ is the semantic vector of an arbitrary class y. $\|\mathcal{F}(\mathbf{x}_i) - \mathbf{a}_y\|^2$ is the distance between the predicted embedding of the sample x_i and the ground truth semantic vector of class y. Similarly, $\|\mathbf{a}_{y_i} - \mathbf{a}_y\|^2$ is the distance between the ground truth semantic vector of x_i and the ground truth semantic vector of class y. The term inside the second parentheses encourages the embedding of the image feature and its ground truth attribute vector with respect to the attributes of the neighbouring classes to be similar. The term inside the first parentheses is an exponential weighting factor so that this strict relative positioning is mainly applied for the semantically similar classes.

Note that when the mixing module is not activated, we directly follow Eq. (11). For the case when mixing is done, we calculate the $\mathcal{L}_{MSE}$ loss by replacing a_{y_i} with the average of the attributes of the mixed classes, i.e., $a_{y_{ij}}$ as given in Eq. (6). We combine the two losses to get the final loss as:

$$\mathcal{L} = \mathcal{L}_{Mix} + \eta \mathcal{L}_{MSE} \tag{12}$$

where η is a hyper-parameter to balance the relative effects of the two losses. Similar idea has been explored in SnMpNet [34] for the retrieval task.

5   Experimental Evaluation

Here, we describe in detail the datasets used, implementation details, results and further analysis of the proposed approach.

Datasets Used: For evaluation of our method, we use two large-scale benchmark datasets, namely DomainNet and DomainNet-LS, as used in the recent works in ZSDG [8,28]. **DomainNet** [35] consists of 345 classes and 6 domains, namely *clipart, infograph, painting, quickdraw, real* and *sketch*, spread across approximately 0.6 million images. We follow the same experimental protocol defined in the literature. Out of 345 classes, we use 300 for training as the seen classes and the remaining 45 for testing as unseen classes. Out of 6 domains, we use 5 domains at a time for training as seen domains and the remaining domain is used for testing as unseen domain. We hold-out each domain (except *real*) one-by-one and repeat the training process using the 300 classes in the remaining 5 domains. The testing is not done on the *real* domain, since the backbone is pre-trained on the ImageNet dataset, and thus the *real* domain can not be considered as an unseen domain. For the **DomainNet-LS** benchmark, only *real* and *painting* domains are used for training and the rest are used for testing, the splitting of the classes remains same. Clearly, it is a more challenging setting, since the domain invariant features have to be learnt only using two domains.

Implementation Details: For fair comparison with the state-of-the-art approaches, we use the ResNet-50 backbone, which has four CNN blocks. We have the mixing modules only after the 1st three blocks, as done in [51]. To learn the features in the semantic space, we use the 300-dimension semantic vectors from the Word2Vec [30] representation. Following [51], we use $\alpha = 0.1$ as the input parameter of the Beta distribution. Inspired from [52], we sample γ from a distribution uniform in $t_1 = 0.4$ to $t_2 = 0.6$, i.e. $\gamma \sim U(0.4, 0.6)$. Based on the analysis shown in Fig. 3(b), we set $\eta = 1$ to give equal weightage to both the losses. For training, we use the Adam optimizer with a learning rate of 10^{-5} and a batch size of 80. We find that setting the probability threshold, with which the mixing module is activated, equal to 0.2, i.e., $\tau = 0.2$ gives the best results, as shown in Fig. 3(a).

5.1   Results on DomainNet and DomainNet-LS Datasets

Table 1. Leave-one-domain-out ZSDG results on DomainNet using average per-class accuracy metric.

Method		Target Domain					Average
DG	ZSL	Clipart	Infograph	Painting	Quickdraw	Sketch	
–	DEVISE [10]	20.1	11.7	17.6	6.1	16.7	14.4
	ALE [1]	22.7	12.7	20.2	6.8	18.5	16.2
	SPNet [43]	26.0	16.9	23.8	8.2	21.8	19.4
DANN	DEVISE [10]	20.5	10.4	16.4	7.1	15.1	13.9
	ALE [1]	21.2	12.5	19.7	7.4	17.9	15.7
	SPNet [43]	25.9	15.8	24.1	8.4	21.3	19.1
Epi-FCR	DEVISE [10]	21.6	13.9	19.3	7.3	17.2	15.9
	ALE [1]	23.2	14.1	21.4	7.8	20.9	17.5
	SPNet [43]	26.4	16.7	24.6	9.2	23.2	20.0
CuMix (img only) [28]		25.2	16.3	24.4	8.7	21.7	19.2
CuMix (two-level) [28]		26.6	17.0	25.3	8.8	21.9	19.9
CuMix [28]		27.6±0.5	17.8±0.2	25.5±0.4	9.9±0.3	22.6±0.3	20.7±0.3
SLE-Net [8]		27.8±0.3	**18.4±0.4**	26.6±0.3	11.5±0.2	25.2±0.3	21.9±0.3
Proposed SEIC		**29.9±0.2**	17.4±0.1	**26.7±0.4**	**12.0±0.4**	**27.3±0.3**	**22.7±0.3**

Table 2. Leave-one-domain-out ZSDG results on DomainNet using standard accuracy metric.

Method	Clipart	Infograph	Painting	Quickdraw	Sketch	Average
CuMix [28]	27.8	16.3	27.6	9.7	25.9	21.5
SLE-Net [8]	29.1	17.6	**28.8**	11.5	26.3	22.7
Proposed SEIC	**32.7**	**18.3**	27.5	**11.9**	**30.4**	**24.2**

Here, we perform extensive experiments to evaluate the effectiveness of the proposed SEIC framework for the ZSDG task.

Results on DomainNet Dataset: For DomainNet dataset, we compare our results using two metrics: average per-class accuracy and standard accuracy. We follow the same experimental protocol as the previous works in the literature, namely Cumix [28] and SLE-Net [8]. Fisrt, we report the results on standalone ZSL methods such as DEVISE [10], ALE [1] and SPNet [43] and combination of the ZSL methods with DG methods, like DANN [11] and Epi-FCR [24].

Along with SLE-Net [8], we report the results of CuMix and its variants: CuMix (img only) where MixUp is applied only at the image level and CuMix (two-level) where MixUp is applied at both image and feature level, as given in [28].

In Table 1, we report the average per-class accuracy for the five test domains using various methods. The results of all the previous approaches have been directly taken from [8]. On using only the standalone ZSL methods DEVISE [10], ALE [1] and SPNet [43], we get 14.4%, 16.2% and 19.4% accuracy, respectively. On integrating the DANN [11] framework with the above ZSL methods, there is a drop in accuracy. Instead of DANN [11], if we combine Epi-FCR [24] with the ZSL methods, the average accuracies improve to 15.9%, 17.5% and 20.0%, respectively. The proposed SEIC framework outperforms the state-of-the-art SLE-Net [8] on four out of the five domains with an average accuracy of 22.7%, which is better than [8] by 0.8%.

Table 3. Leave-one-domain-out ZSDG results on DomainNet-LS using average per-class accuracy metric.

Method	Clipart	Infograph	Quickdraw	Sketch	Average
SPNet [43]	21.5	14.1	4.8	17.3	14.4
Epi-FCR+SPNet [43]	22.5	14.9	5.6	18.7	15.4
CuMix (img only) [28]	21.2	14.0	4.8	17.3	14.3
CuMix (two-level) [28]	22.7	16.5	4.9	19.1	15.8
CuMix (reverse) [28]	22.9	15.8	4.8	18.2	15.4
CuMix [28]	23.7	**17.1**	5.5	19.7	16.5
SLE-Net [8]	24.0	16.0	7.2	20.5	16.9
Proposed SEIC	**25.9**	16.0	**8.5**	**22.9**	**18.3**

In Table 2, we show the standard accuracies of the proposed approach for the DomainNet dataset and compare it with the previous two ZSDG methods. SLE-Net [8] obtains average accuracy of 22.7%. Here also, our method outperforms the other methods with an average accuracy of 24.2%, which is an increase of 1.5% over SLE-Net [8].

Results on DomainNet-LS Dataset: In Table 3, we show the results on the DomainNet-LS dataset, where we train the model only on 2 domains: *real* and *painting*. An average accuracy of 14.4% is attained by SPNet [8]. On combining it with Epi-FCR [24], the accuracy improves by 1.0%. CuMix [28] achieves an average accuracy of 16.5% beating its other variants like CuMix (reverse) [28]. Our method achieves an average accuracy of 18.3% which is better than SLE-Net [8] by 1.4%.

5.2   Additional Analysis

Ablation Study: In Table 4, each of the components is deactivated one-by-one while keeping the others activated. The baseline here is the original backbone with the fixed semantic embeddings as the classifiers. We do this analysis on the DomainNet-LS dataset by taking the following five cases:

(a) Here, domain mixing, which generates mixed style features is deactivated, rest of the components are active. The mixing of two samples is done only on the feature level, the statistics of the features are not altered. Here, the model's ability to generalize to unseen domains would be hampered.

(b) Here, the CE loss corresponding to the intermediate class features f_{ic} is absent, thereby making the model less effective at recognizing unseen test classes.

(c) Our proposed method uses the knowledge of semantically similar classes for generating the pseudo intermediate classes for increasing the class discriminability. Here, we turn off this component making the model inefficient at distinguishing between similar classes. Here, for creating the intermediate classes, two randomly picked samples are used instead of semantically similar classes.

(d) Here, the information provided by the neighbouring classes i.e. $\mathcal{L}_{MSE}$ defined in Eq. (11) is not used.

(e) This is the proposed SEIC framework which uses all the components. We observe that all the proposed modules help towards improving the performance of the SEIC framework for the ZSDG task.

Table 4. Analysis of the contribution of each component in the proposed method using DomainNet-LS dataset.

	f_{ms}	f_{ic}	P_{Mix}	$\mathcal{L}_{MSE}$	Clipart	Infograph	Quickdraw	Sketch	Average
Case (a):	✗	✓	✓	✓	24.9	15.0	6.5	21.9	17.1
Case (b):	✓	✗	✓	✓	25.8	14.2	6.4	21.7	17.0
Case (c):	✓	✓	✗	✓	24.2	15.4	5.4	20.2	16.3
Case (d):	✓	✓	✓	✗	25.2	15.9	8.0	21.2	17.6
Case (e):	✓	✓	✓	✓	**25.9**	**16.0**	**8.5**	**22.9**	**18.3**

Visualization of the Semantic Space: Here, we visualize the feature embedding space which is learnt using the proposed SEIC framework. Figure 2 shows the t-SNE plots of the feature embeddings for 10 randomly chosen unseen test classes for the DomainNet dataset. We observe that the unseen test classes form reasonably nice clusters in the embedding space, even though the model has not been trained using data from these classes or domains. Also, the clusters are semantically meaningful, for example, in Fig. 2(c), we observe that semantically similar classes (living creatures) like *parrot, dolphin* and *octopus* are closer to each other compared to other different classes like *scissors* and *boomerang*.

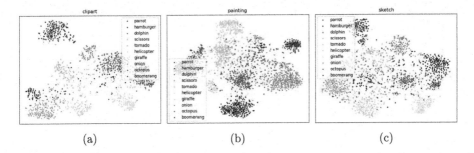

Fig. 2. t-SNE plots of the semantic space of test domains: (a) Clipart, (b) Painting and (c) Sketch in the DomainNet dataset for 10 unseen classes.

Table 5. Model predictions for some test images in 5 domains of the DomainNet dataset. The ground truth is given at the top of each image. The correct (green) and incorrect (red) predictions are shown at the bottom of each image.

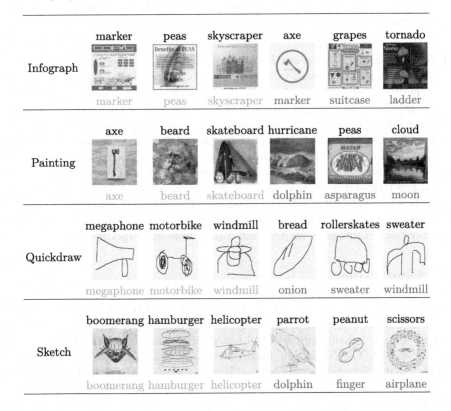

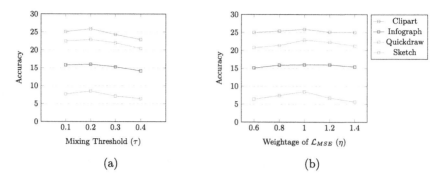

Fig. 3. Effect of variations in hyperparameters.

Visual Examples of Correct and Incorrect Predictions: Table 5 shows few examples which are correctly and wrongly classified by the proposed SEIC framework. These images are from two domains of the DomainNet dataset. We observe that many of the wrong predictions are quite intuitive and may be wrongly classified even by humans. For example, the last image of painting, i.e. *cloud* is wrongly predicted as *moon*, the last image of quickdraw, i.e. *sweater* is wrongly predicted as *windmill*, etc.

Effect of Varying Different Hyperparameters: First, we show the variation in performance for different values of τ, which controls how frequently a mixing module is activated. We analyze the performance for $\tau = \{0.1, 0.2, 0.3, 0.4\}$, on DomainNet-LS. In Fig. 3(a), we observe that the best result for each domain is obtained at $\tau = 0.2$. But the degradation of performance for different values of τ is very gradual, indicating that the model performance is quite stable with respect to this hyperparameter. In Fig. 3(b), we show the accuracy variation for different values of η, which is a hyperparameter weighting the importance of the MSE loss term in Eq. (12). We analyze the results for $\eta = \{0.6, 0.8, 1.0, 1.2, 1.4\}$. Clearly, the best result is achieved when the value of η is set equal to 1. The trend is consistent across each of the four domains. Therefore, both the loss terms are given equal weightage in the final loss equation in our experiments.

6   Conclusion

In this work, we propose a novel framework termed SEIC, to address the ZSDG task. Specifically, we extend a state-of-the-art DG method capable of generalizing across unseen domains into a ZSDG framework which can handle unknown test classes as well. Generalization across unseen domains is achieved by generating intermediate domains by mixing the feature statistics of the different training samples. Similarly, generalization across unseen classes is handled by generating pseudo classes between similar seen classes using mixed features of the training samples. In addition, we also utilize the information of the neighbourhood classes to learn the semantically meaningful feature embeddings. Extensive experiments on two large-scale benchmark datasets and comparison with the state-of-the-art show the effectiveness of the proposed SEIC framework.

Acknowledgements. This work is partly supported through a research grant from SERB, Department of Science and Technology, Govt. of India.

References

1. Akata, Z., Perronnin, F., Harchaoui, Z., Schmid, C.: Label-embedding for attribute-based classification. In: IEEE Conference on Computer Vision and Pattern Recognition, pp. 819–826 (2013). https://doi.org/10.1109/CVPR.2013.111
2. Akata, Z., Reed, S.E., Walter, D., Lee, H., Schiele, B.: Evaluation of output embeddings for fine-grained image classification. In: IEEE Conference on Computer Vision and Pattern Recognition (CVPR), pp. 2927–2936 (2015). https://doi.org/10.1109/CVPR.2015.7298911
3. Annadani, Y., Biswas, S.: Preserving semantic relations for zero-shot learning. In: IEEE Conference on Computer Vision and Pattern Recognition (CVPR), pp. 7603–7612 (2018)
4. Ben-David, S., Blitzer, J., Crammer, K., Kulesza, A., Pereira, F., Vaughan, J.: A theory of learning from different domains. Mach. Learn. **79**, 151–175 (2010)
5. Blanchard, G., Lee, G., Scott, C.: Generalizing from several related classification tasks to a new unlabeled sample. In: Advances in Neural Information Processing Systems (2011)
6. Bucher, M., Herbin, S., Jurie, F.: Improving semantic embedding consistency by metric learning for zero-shot classification. In: Leibe, B., Matas, J., Sebe, N., Welling, M. (eds.) ECCV 2016. LNCS, vol. 9909, pp. 730–746. Springer, Cham (2016). https://doi.org/10.1007/978-3-319-46454-1_44
7. Cacheux, Y.L., Borgne, H.L., Crucianu, M.: Modeling inter and intra-class relations in the triplet loss for zero-shot learning. In: IEEE/CVF International Conference on Computer Vision (ICCV), pp. 10332–10341 (2019). https://doi.org/10.1109/ICCV.2019.01043
8. Chandhok, S., et al.: Structured latent embeddings for recognizing unseen classes in unseen domains. In: British Machine Vision Conference (BMVC) (2021)
9. Chao, W.-L., Changpinyo, S., Gong, B., Sha, F.: An empirical study and analysis of generalized zero-shot learning for object recognition in the wild. In: Leibe, B., Matas, J., Sebe, N., Welling, M. (eds.) ECCV 2016. LNCS, vol. 9906, pp. 52–68. Springer, Cham (2016). https://doi.org/10.1007/978-3-319-46475-6_4
10. Frome, A., et al.: DeViSE: a deep visual-semantic embedding model. In: Advances in Neural Information Processing Systems, pp. 2121–2129 (2013)
11. Ganin, Y., et al.: Domain-adversarial training of neural networks. J. Mach. Learn. Res. **17**, 1–35 (2016)
12. Goodfellow, I.J., et al.: Generative adversarial nets. In: Advances in Neural Information Processing Systems, pp. 2672–2680 (2014)
13. Huang, C., Li, Y., Loy, C.C., Tang, X.: Learning deep representation for imbalanced classification. In: Proceedings of the IEEE Conference on Computer Vision and Pattern Recognition (CVPR) (2016)
14. Huang, J., Guan, D., Xiao, A., Lu, S.: FSDR: frequency space domain randomization for domain generalization. In: IEEE Conference on Computer Vision and Pattern Recognition (CVPR), pp. 6891–6902 (2021)

15. Huynh, D., Elhamifar, E.: Fine-grained generalized zero-shot learning via dense attribute-based attention. In: IEEE/CVF Conference on Computer Vision and Pattern Recognition (CVPR), pp. 4482–4492 (2020). https://doi.org/10.1109/CVPR42600.2020.00454

16. Iwasawa, Y., Matsuo, Y.: Test-time classifier adjustment module for model-agnostic domain generalization. In: Advances in Neural Information Processing Systems (NeurIPS), pp. 2427–2440 (2021)

17. Jhoo, W.Y., Heo, J.: Collaborative learning with disentangled features for zero-shot domain adaptation. In: 2021 IEEE/CVF International Conference on Computer Vision, ICCV 2021, Montreal, QC, Canada, 10–17 October 2021, pp. 8876–8885. IEEE (2021). https://doi.org/10.1109/ICCV48922.2021.00877

18. Jiang, H., Wang, R., Shan, S., Chen, X.: Adaptive metric learning for zero-shot recognition. IEEE Signal Process. Lett. **26**(9), 1270–1274 (2019). https://doi.org/10.1109/LSP.2019.2917148

19. Jurie, F., Bucher, M., Herbin, S.: Generating visual representations for zero-shot classification. In: 2017 IEEE International Conference on Computer Vision Workshops (ICCV), pp. 2666–2673 (2017). https://doi.org/10.1109/ICCVW.2017.308

20. Khan, M.H., Zaidi, T., Khan, S., Khan, F.S.: Mode-guided feature augmentation for domain generalization. In: British Machine Vision Conference (BMVC), p. 176 (2021)

21. Kingma, D.P., Welling, M.: Auto-encoding variational Bayes. In: International Conference on Learning Representations (ICLR) (2014)

22. Kodirov, E., Xiang, T., Gong, S.: Semantic autoencoder for zero-shot learning. In: IEEE Conference on Computer Vision and Pattern Recognition (CVPR), pp. 4447–4456 (2017). https://doi.org/10.1109/CVPR.2017.473

23. Lampert, C.H., Nickisch, H., Harmeling, S.: Learning to detect unseen object classes by between-class attribute transfer. In: IEEE Computer Society Conference on Computer Vision and Pattern Recognition (CVPR), pp. 951–958 (2009). https://doi.org/10.1109/CVPR.2009.5206594

24. Li, D., Zhang, J., Yang, Y., Liu, C., Song, Y., Hospedales, T.M.: Episodic training for domain generalization. In: IEEE/CVF International Conference on Computer Vision (ICCV), pp. 1446–1455 (2019). https://doi.org/10.1109/ICCV.2019.00153

25. Li, H., Pan, S.J., Wang, S., Kot, A.C.: Domain generalization with adversarial feature learning. In: IEEE/CVF Conference on Computer Vision and Pattern Recognition, pp. 5400–5409 (2018). https://doi.org/10.1109/CVPR.2018.00566

26. Li, X., Fang, M., Chen, B.: Generalized zero-shot domain adaptation with target unseen class prototype learning. In: Neural Computing and Applications 34 (2022). https://doi.org/10.1007/s00521-022-07413-z

27. Li, Y., et al.: Deep domain generalization via conditional invariant adversarial networks. In: Ferrari, V., Hebert, M., Sminchisescu, C., Weiss, Y. (eds.) ECCV 2018. LNCS, vol. 11219, pp. 647–663. Springer, Cham (2018). https://doi.org/10.1007/978-3-030-01267-0_38

28. Mancini, M., Akata, Z., Ricci, E., Caputo, B.: Towards recognizing unseen categories in unseen domains. In: Vedaldi, A., Bischof, H., Brox, T., Frahm, J.-M. (eds.) ECCV 2020. LNCS, vol. 12368, pp. 466–483. Springer, Cham (2020). https://doi.org/10.1007/978-3-030-58592-1_28

29. Maniyar, U., Joseph, K.J., Deshmukh, A.A., Dogan, U., Balasubramanian, V.N.: Zero-shot domain generalization. In: British Machine Vision Conference (BMVC) (2020)

30. Mikolov, T., Chen, K., Corrado, G., Dean, J.: Efficient estimation of word representations in vector space. In: International Conference on Learning Representations (ICLR) (2013)
31. Mishra, A., Reddy, M.S.K., Mittal, A., Murthy, H.A.: A generative model for zero shot learning using conditional variational autoencoders. In: Conference on Computer Vision and Pattern Recognition (CVPR), pp. 2188–2196 (2018). https://doi.org/10.1109/CVPRW.2018.00294
32. Mnih, V., et al.: Playing atari with deep reinforcement learning. ArXiv preprint arXiv:1312.5602. (2013)
33. Palatucci, M., Pomerleau, D., Hinton, G.E., Mitchell, T.M.: Zero-shot learning with semantic output codes. In: Advances in Neural Information Processing Systems, pp. 1410–1418 (2009)
34. Paul, S., Dutta, T., Biswas, S.: Universal cross-domain retrieval: Generalizing across classes and domains. In: Proceedings of the IEEE/CVF International Conference on Computer Vision (ICCV), pp. 12056–12064 (2021)
35. Peng, X., Bai, Q., Xia, X., Huang, Z., Saenko, K., Wang, B.: Moment matching for multi-source domain adaptation. In: Proceedings of the IEEE International Conference on Computer Vision, pp. 1406–1415 (2019)
36. Pennington, J., Socher, R., Manning, C.D.: GloVe: global vectors for word representation. In: Proceedings of the Conference on Empirical Methods in Natural Language Processing (EMNLP), pp. 1532–1543 (2014). https://doi.org/10.3115/v1/d14-1162
37. Tobin, J., Fong, R., Ray, A., Schneider, J., Zaremba, W., Abbeel, P.: Domain randomization for transferring deep neural networks from simulation to the real world. In: IEEE/RSJ International Conference on Intelligent Robots and Systems, pp. 23–30 (2017). https://doi.org/10.1109/IROS.2017.8202133
38. Vo, N., et al.: Composing text and image for image retrieval - an empirical odyssey. In: Proceedings of the IEEE/CVF Conference on Computer Vision and Pattern Recognition (CVPR) (2019)
39. Volpi, R., Namkoong, H., Sener, O., Duchi, J.C., Murino, V., Savarese, S.: Generalizing to unseen domains via adversarial data augmentation. In: Advances in Neural Information Processing Systems (NeurIPS), pp. 5339–5349 (2018)
40. Wang, K., Liew, J.H., Zou, Y., Zhou, D., Feng, J.: PANet: few-shot image semantic segmentation with prototype alignment. In: The IEEE International Conference on Computer Vision (ICCV) (2019)
41. Wang, Q., Breckon, T.P.: Generalized zero-shot domain adaptation via coupled conditional variational autoencoders. CoRR abs/2008.01214 (2020)
42. Wang, Z., Luo, Y., Qiu, R., Huang, Z., Baktashmotlagh, M.: Learning to diversify for single domain generalization. In: IEEE/CVF International Conference on Computer Vision (ICCV), pp. 814–823 (2021). https://doi.org/10.1109/ICCV48922.2021.00087
43. Xian, Y., Choudhury, S., He, Y., Schiele, B., Akata, Z.: Semantic projection network for zero- and few-label semantic segmentation. In: IEEE Conference on Computer Vision and Pattern Recognition (CVPR), pp. 8256–8265 (2019). https://doi.org/10.1109/CVPR.2019.00845
44. Xian, Y., Schiele, B., Akata, Z.: Zero-shot learning - the good, the bad and the ugly. In: 2017 IEEE Conference on Computer Vision and Pattern Recognition (CVPR), pp. 3077–3086. IEEE Computer Society (2017). https://doi.org/10.1109/CVPR.2017.328

45. Xian, Y., Sharma, S., Schiele, B., Akata, Z.: F-VAEGAN-D2: a feature generating framework for any-shot learning. In: Conference on Computer Vision and Pattern Recognition (CVPR), pp. 10275–10284 (2019). https://doi.org/10.1109/CVPR.2019.01052

46. Xie, G., et al.: Attentive region embedding network for zero-shot learning. In: IEEE Conference on Computer Vision and Pattern Recognition (CVPR), pp. 9384–9393 (2019). https://doi.org/10.1109/CVPR.2019.00961

47. Xu, Q., Zhang, R., Zhang, Y., Wang, Y., Tian, Q.: A Fourier-based framework for domain generalization. In: IEEE Conference on Computer Vision and Pattern Recognition (CVPR), pp. 14383–14392 (2021)

48. Yang, F., Cheng, Y., Shiau, Z., Wang, Y.F.: Adversarial teacher-student representation learning for domain generalization. In: Advances in Neural Information Processing Systems (NeurIPS), pp. 19448–19460 (2021)

49. Zhang, H., Liu, S., Zhang, C., Ren, W., Wang, R., Cao, X.: SketchNet: sketch classification with web images. In: CVPR (2016). https://doi.org/10.1109/CVPR.2016.125

50. Zhang, J., et al.: Generative domain-migration hashing for sketch-to-image retrieval. In: Ferrari, V., Hebert, M., Sminchisescu, C., Weiss, Y. (eds.) ECCV 2018. LNCS, vol. 11206, pp. 304–321. Springer, Cham (2018). https://doi.org/10.1007/978-3-030-01216-8_19

51. Zhou, K., Yang, Y., Qiao, Y., Xiang, T.: Domain generalization with mixStyle. In: ICLR (2021)

52. Zhu, F., Cheng, Z., Zhang, X., Liu, C.: Class-incremental learning via dual augmentation. In: Advances in Neural Information Processing Systems, pp. 14306–14318 (2021)

Co-attention Aligned Mutual Cross-Attention for Cloth-Changing Person Re-identification

Qizao Wang[1], Xuelin Qian[2(✉)], Yanwei Fu[2], and Xiangyang Xue[1,2]

[1] School of Computer Science, Shanghai Key Lab of Intelligent Information Processing, Fudan University,
Shanghai, China
qzwang22@m.fudan.edu.cn

[2] School of Data Science, and MOE Frontiers Center for Brain Science, Shanghai Key Lab of Intelligent Information Processing, Fudan University, Shanghai, China
{xlqian,yanweifu,xyxue}@fudan.edu.cn

Abstract. Person re-identification (Re-ID) has been widely studied and achieved significant progress. However, traditional person Re-ID methods primarily rely on cloth-related color appearance, which is unreliable under real-world scenarios when people change their clothes. Cloth-changing person Re-ID that takes this problem into account has received increasing attention recently, but it is more challenging to learn discriminative person identity features, since larger intra-class variation and smaller inter-class easily occur in the image feature space with clothing changes. Beyond appearance features, some known identity-related features can be implicitly encoded in images (*e.g.*, body shapes). In this paper, we first design a novel Shape Semantics Embedding (SSE) module to encode body shape semantic information, which is one of the essential clues to distinguish pedestrians when their clothes change. To better complement image features, we further propose a Co-attention Aligned Mutual Cross-attention (CAMC) framework. Different from previous attention-based fusion strategies, it first aligns features from multiple modalities, then effectively interacts and transfers identity-aware but cloth-irrelevant knowledge between the image space and the body shape space, resulting in a more robust feature representation. To the best of our knowledge, this is the first work to adopt Transformer to handle the multi-modal interaction for cloth-changing person Re-ID. Extensive experiments demonstrate the effectiveness of our proposed method and show the superior performance achieved on several cloth-changing person Re-ID benchmarks. Codes will be available at https://github.com/QizaoWang/CAMC-CCReID.

1 Introduction

Person re-identification (Re-ID) aims at identifying and associating the same person across different cameras, which has great potential applications in

---

Supplementary Information The online version contains supplementary material available at https://doi.org/10.1007/978-3-031-26348-4_21.

L. Wang et al. (Eds.): ACCV 2022, LNCS 13845, pp. 351–368, 2023.
https://doi.org/10.1007/978-3-031-26348-4_21

video surveillance, including suspect tracking, activity analysis, human-computer interaction, *etc.*. Depending on application scenarios, existing person re-identification approaches can be broadly grouped into two categories, short-term and long-term. Short-term person Re-ID has been widely studied in the past decades, involved in multiple challenges and research directions, including occlusion [30,37,55] and infrared-visible modalities [53,54], supervised [26,34,68] and unsupervised learning [21,51,58], representation [5,50,56] and metric learning [7,13,45]. However, all of these methods assume that the same person would always wear the same clothes, so the learned features may mostly rely on clothing appearances. On the contrary, long-term person Re-ID focuses more on the application in real-world scenarios, which takes into account practical problems, such as changing clothes and incremental identities. Among them, the cloth-changing problem has attracted more and more attention, which is also known as cloth-changing person Re-ID.

To deal with the challenge of clothing changes, it is important to use robust identity-related features. Many researchers naturally turn their attention to human body shape information, since the body shape of a person usually remains unchanged for a relatively long duration. However, it is extremely difficult to mine it from RGB color images. Consequently, most cloth-changing Re-ID methods draw support from other modalities, such as 2D human posture keypoints [39], contour sketches [57], gaits [22] and 3D shapes [6]. In this paper, we also target cloth-changing person Re-ID, but propose a novel Shape Semantics Embedding (SSE) module. It uses heatmaps of human postures to encode body shape semantic information, which is more lightweight and robust in long-term scenarios.

When integrating useful information from multiple modalities, one of the challenges is how to make them interact effectively. Unfortunately, in previous cloth-changing Re-ID methods, the interaction between appearance features and features extracted from other modalities is relatively simple. Intuitively, there is an inevitable gap in the representations of different modalities even for the same person, so a simple interaction between modalities could not make full of use of abundant multi-modal information. In recent years, Transformer [48] has been widely used, and many researches have shown its effectiveness in computer vision tasks, such as object detection [3,70], and person re-identification [12,30,35]. The attention mechanism in Transformer can effectively capture the contextual semantic information of input sequences. Inspired by it, we propose a Co-Attention Aligned Mutual Cross-attention (CAMC) framework for cloth-changing person Re-ID.

More specifically, an appearance branch and a shape branch are first designed in our framework. The former uses a conventional backbone (*e.g.*, ResNet-50 [10]) to extract appearance features, and the latter is exactly our proposed SSE module for obtaining body shape information. To relieve the gap between both features from different modalities, we adopt element-wise attention for alignment. Subsequently, a symmetrical mutual cross-attention module is applied to effectively distill appearance and body shape information from each other. With our proposed CAMC framework, appearance features are refined with the help of body shape semantic information which is cloth-irrelevant, while body shape semantic features are supplemented with aligned appearance features robust to clothing changes.

In summary, our contributions are listed as follows:

1. We propose a Shape Semantics Embedding module based on the self-attention mechanism to encode body shape semantics irrelevant to clothes, which is essential to identify a person when changing clothes.
2. We propose a novel mutual interaction module based on the cross-attention mechanism to interact appearance and body shape features effectively, resulting in a fused feature more robust to clothing changes. To mitigate the feature gap from different modalities and improve the efficiency and effectiveness of their feature interaction, we additionally introduce an element-wise co-attention alignment module for alignment.
3. To the best of our knowledge, it is the first work to adopt Transformer to handle multi-modal interaction for cloth-changing person Re-ID. Extensive experiments demonstrate the efficacy of our proposed model on several cloth-changing Re-ID benchmarks.

2   Related Work

2.1   Person Re-identification

Person Re-ID task aims at identifying a specific person across different cameras and locations. With the rise of deep learning, person Re-ID technology has made great progress and is widely used in smart cities, intelligent security, human-computer interaction, *etc.*. Many works try to explore fine-grained pedestrian identity features via metric learning, for instance, hard triplet loss [7,13] to encourage a closer feature distance among the same identity, and classification loss [41,64,66] to learn a high-level global feature from the whole input. There are also some other works dealing with spatial misalignment problems, such as occlusion [11,65], variant camera views [33,44], diverse poses [28,38], different resolutions [27], and manifold domains [18,23]. However, these models are well-trained based on the assumption that the same person has the same clothing in a short duration, which seriously hinders their applications in long-term real-world scenarios. In this paper, we focus on the more realistic long-term cloth-changing person Re-ID task and further explore a robust person identity feature extraction model to solve the problem of unreliable appearance information.

2.2   Cloth-Changing Person Re-identication

To further improve the applicability and practicability of person Re-ID models in real-world scenarios, more and more researchers turn to studying the cloth-changing person Re-ID task, which targets to match the same person across different locations over a long duration and inevitably faces the cases of changing clothes. In this situation, appearance/texture information can no longer be used as an accurate representation to distinguish different pedestrians, which makes it difficult for many previous methods to achieve satisfactory results. Thus, nowadays, many studies have tried to solve the problem via learning more stable biological representation, for example, Wan et al. [49] and Yu et al. [60] extract facial features to improve the person Re-ID accuracy; Yang et al. [57] utilize contour sketches to indicate discriminative characteristics; and Qian et al. [39] and Li et al. [29] use shape information to help feature learning. Different from existing works, we not only focus on extracting precise biological body shape semantic information, but also try to better align the two modality features of appearance and body shape, which can further boost the cloth-changing person Re-ID performance.

2.3   Transformer-Based Person Re-identification

Transformer [48] has made great achievements in the field of natural language processing. Inspired by the self-attention mechanism, many researchers apply Transformers to computer vision tasks and find such Transformers can be as effective as CNNs over feature extraction. For example, Dosovitskiy et al. propose ViT [9] which processes images directly as sequences, Touvron et al. introduce a teacher-student strategy specific for Transformers to speed up the ViT training without using any large-scale pretraining data, and Carion et al. design DETR [3] performing cross-attention between the object query and the feature map to transform the detection task into a one-to-one matching problem. Since Transformer can capture long-distance dependency and help models pay attention to different parts of the human body, such as the head, shoulder, waist, and thigh, and obtain rich local relevant semantic information, Li et al. [30] and He et al. [12] adopt Transformer to solve the person partial-observation problem in occlusion person Re-ID task. In this paper, we take the advantage of the Transformer cross-attention mechanism to interact appearance and body shape semantic information, and generate a robust fused pedestrian feature under the cloth-changing scenario. Different from the latest work [2], which uses ViT as backbone only, we explore to effectively use Transformer to interact multiple modalities in cloth-changing person Re-ID.

3   Methodology

3.1   Overview

In this paper, we aim to address the problem of person Re-ID under the long-term cloth-changing setting, where the clothing appearance is unreliable and even would hinder the network to extract discriminative features. Considering that body shape is more robust against clothing changes, we draw support from heatmaps of human postures to encode body shape semantic information. Furthermore, we propose a Co-attention Aligned Mutual Cross-attention framework to effectively align and interact multi-modality features.

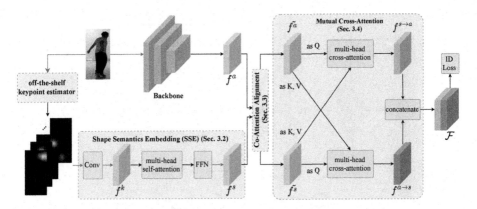

Fig. 1. Overview of our Co-attention Aligned Mutual Cross-attention (CAMC) framework. It consists of an appearance branch, and a body shape branch implemented by our proposed Shape Semantics Embedding (SSE) module. We propose a co-attention alignment module to align the two modalities. Then a symmetrical mutual cross-attention module is applied to effectively interact and fuse appearance and body shape semantic information, outputting a fused feature robust to clothing changes.

The overall framework is shown in Fig. 1, which consists of an appearance branch and a body shape branch. Concretely, the former is designed to extract appearance features from the given person image $x \in \mathbb{R}^{H \times W \times c}$. Following [46], we use ResNet-50 [10] as backbone, and the stride of the first convolution layer in *res4* block is set to 1 to increase the feature resolution. The output feature is flattened in the spatial dimensions to obtain the appearance feature sequence $f^a \in \mathbb{R}^{hw \times d}$, where h and w are the height and width of the feature map, d denotes the feature dimension.

In the rest of this section, we first introduce how the body shape branch extracts rich semantic information about body shapes (see Sect. 3.2). Second, we propose a co-attention alignment module to align multi-modal information (see Sect. 3.3). Then, we elaborate on the mutual cross-attention module, which plays a key role in the multi-modal feature interaction (see Sec. 3.4). Lastly, we briefly describe the procedures of training and inference in Sect. 3.5.

3.2 Shape Semantics Embedding Module

When people change their clothes, although most appearance clues, such as the color of clothes, change significantly, their body shapes are relatively stable even for a long time. Therefore, it is useful to encode and utilize the body shape semantic information, which is more robust to clothing changes. To achieve such a goal, we propose a Shape Semantics Embedding (SSE) module to encode it from heatmaps of human postures. Our proposed SSE module especially leverages the self-attention mechanism to learn relations between different human posture keypoints. The intuition is that biological information is contained in

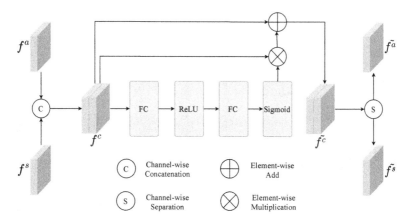

Fig. 2. Illustration of our proposed co-attention alignment module. By encoding the concatenated appearance and body shape semantic features, element-wise attention scores are obtained to effectively align features from different branches.

these relations (*e.g.*, hip to knee). Compared with encoding individual keypoint, such a design is more stable and robust, even if people change their postures.

As shown in Fig. 1, we employ an off-the-shelf estimator HRNet [43] to obtain heatmaps of human postures. Given an input image x, HRNet returns K heatmaps of human postures, where each heatmap represents the location distribution of one posture keypoint. We regard K as the feature dimension and apply a convolution layer to increase the dimension from K to d, outputting a human posture feature $f^k \in \mathbb{R}^{hw \times d}$, where h and w represent the height and width of the feature map. To encode the relations between each pair of human posture parts, we feed f^k into an encoder layer, which consists of one multi-head self-attention layer [48] with a skip connection and layer normalization [1]. A standard feed-forward network [48] is further attached to output the body shape semantic feature $f^s \in \mathbb{R}^{hw \times d}$. More details on the structure design are discussed in the supplemental material.

Benefiting from the ability to effectively capture the long-distance and short-distance semantic information of inputs, the self-attention mechanism can encode the correlations between different parts of human postures, which present the body shape semantics. Meanwhile, the multi-head mechanism enables different heads to effectively focus on all kinds of semantic details.

3.3   Co-attention Alignment Module

Intuitively, the appearance feature f^a and the body shape feature f^s are from two different modalities, so they may not correspond, containing mismatched redundancy and interference information. To effectively match and integrate their useful information, we propose a co-attention alignment module, as shown in Fig. 2, to align both features before further interaction and fusion.

Specifically, the input of the co-attention alignment module is the concatenated two branch features $f^c = [f^a; f^s] \in \mathbb{R}^{hw \times 2d}$, where $[* ; *]$ means the concatenation operation. Then it further goes through two fully-connected (FC) layers. Similar to [16], the first FC layer compresses the feature dimension to a quarter and the second one decodes it back to the original dimension, not only reducing the number of parameters, but also achieving the information bottleneck effect. After that, a *sigmoid* function is attached to produce element-wise attention scores for feature alignment. For training stability, we further introduce a skip connection from the input to the output. The overall formulation can be expressed as,

$$s = \sigma(\phi(f^c W_1 + b_1) W_2 + b_2) \tag{1}$$

$$[\tilde{f}^a; \tilde{f}^s] = \tilde{f}^c ; \quad \tilde{f}^c = s \otimes f^c + f^c \tag{2}$$

where $W_1 \in \mathbb{R}^{2d \times (d/2)}$, $W_2 \in \mathbb{R}^{(d/2) \times 2d}$, $b_1 \in \mathbb{R}^{1 \times (d/2)}$ and $b_2 \in \mathbb{R}^{1 \times 2d}$ are weights and biases of two fully-connected layers; ϕ is the *ReLU* activation function and σ denotes the *sigmoid* function; $\otimes$ means the element-wise multiplication. We divide the aligned features $\tilde{f}^c \in \mathbb{R}^{hw \times 2d}$ in half, to separately get the aligned appearance feature sequence $\tilde{f}^a \in \mathbb{R}^{hw \times d}$ and the aligned body shape semantic feature sequence $\tilde{f}^s \in \mathbb{R}^{hw \times d}$.

3.4 Mutual Cross-Attention Module

After successfully aligning appearance and body shape semantic features, we introduce a cross-attention module to realize the information interaction between the two modalities. As shown in Fig. 1, this module is performed mutually and symmetrically, that is, the body shape features are integrated into the appearance features and vice versa.

Take one side as an example, we first apply the multi-head cross-attention mechanism to do interaction by computing the dot-product similarity between the appearance feature $\tilde{f}^a$ and the body shape feature $\tilde{f}^s$. The similarity is scaled by $\sqrt{d}$ and normalized by a softmax function. Subsequently, the result is regarded as an attention weight to perform a weighted sum of $\tilde{f}^s$. In this way, for each appearance feature $\tilde{f}_i^a \in \mathbb{R}^{1 \times d}$, where $i \in [1, hw]$, we can find body shape semantic information with similar responses in $\tilde{f}^s$, and integrate it effectively. In other words, the appearance features $\tilde{f}^a$ are well refined with the help of the cloth-irrelevant body shape semantic information $\tilde{f}^s$. The formulation is expressed as,

$$Q = \tilde{f}^a ; \quad K = \tilde{f}^s ; \quad V = \tilde{f}^s \tag{3}$$

$$f^{s \to a} = \psi \left(\text{softmax} \left(\frac{Q K^\mathsf{T}}{\sqrt{d}} \right) V + \tilde{f}^a \right) \tag{4}$$

where $f^{s \to a} \in \mathbb{R}^{hw \times d}$ and ψ denotes the layer normalization [1]. It is similar to the standard multi-head cross-attention module in Transformer [48], but there is no necessary for linear projection for query, key, and value space, thanks to our proposed co-attention alignment module.

For the other side of the interaction, we adopt the same formulation but set $Q = \tilde{f}^s$, $K = \tilde{f}^a$, $V = \tilde{f}^a$, to integrate the matched appearance features into the body shape semantic features. Finally, we get $f^{a \to s} \in \mathbb{R}^{hw \times d}$, which represents body shape semantic features that are supplemented with aligned appearance features robust to clothing changes.

3.5 Training and Inference

Thanks to the symmetrical mutual cross-attention interaction between the aligned features from the two branches, we take full use of appearance and body shape semantic information. We concatenate the two features after interaction together to obtain a more robust and discriminative feature, which can be formulated as follows:

$$\mathcal{F} = [f^{s \to a}; f^{a \to s}] \in \mathbb{R}^{hw \times 2d} \tag{5}$$

For training, we adopt a linear classifier with the input of the fused feature $\mathcal{F}$, and optimize the model by cross-entropy loss with label smoothing. For inference, we directly use $\mathcal{F}$ as final features to compute the cosine distance between two person images for retrieval.

4 Experiment

4.1 Experimental Setup

Datasets. We mainly evaluate our method on two widely used long-term cloth-changing person Re-ID datasets: Celeb-reID [20] and LTCC [39]. **Celeb-reID** is acquired from the Internet using street snapshots of celebrities, which contains 34,186 images of 1,052 identities. Specifically, more than 70% images of each person show different clothes on average. **LTCC** is an indoor cloth-changing person Re-ID dataset, which has 17,138 images of 152 identities with 478 different outfits captured from 12 camera views. LTCC is challenging as it contains diverse human poses, large changes of illumination, and large variations of occlusion. To better illustrate our model efficacy on the general person Re-ID task, we also evaluate our method on **Market-1501** [63], which is a benchmark dataset for the standard person Re-ID without clothing changes.

Implementation Details. Our method is implemented on the Pytorch framework. We adopt ResNet-50 [10] initialized by ImageNet [8] as backbone to extract person appearance features. The input images are resized to 256×128. For data augmentation, color jitter, random horizontal flipping, padding, random cropping, and random erasing [67] are used. We use Adam optimizer [24] for 150 epochs, with the warmup strategy that linearly increases the learning rate from 3×10^{-5} to 3×10^{-4} in the first 10 epochs. Then decrease the learning rate by a factor of 10 at epoch 40 and 80. The batch size is set to 64 for Celeb-reID and Market-1501, and 32 for LTCC, with 4 images per ID. To get the heatmaps

Table 1. Comparison of our method with the state-of-the-art methods on Celeb-reID. The best results are shown in bold.

Methods	Rank-1	Rank-5	mAP
ResNet-Mid [59]	43.3	54.6	5.8
Two-Stream [66]	36.3	54.5	7.8
MLFN [4]	41.4	54.7	6.0
HACNN [26]	47.6	63.3	9.5
Part-Bilinear [42]	19.4	40.6	6.4
PCB [46]	37.1	57.0	8.2
MGN [52]	49.0	64.9	10.8
ReIDCaps [20]	51.2	65.4	9.8
CESD [39]	50.9	66.3	9.8
RCSANet [19]	54.9	–	11.0
Baseline (ResNet-50)	52.9	66.2	9.9
Ours	**57.5**	**71.5**	**12.3**

of human postures, we employ HRNet [43] pre-trained on COCO dataset [32], where the number of heatmaps is 17. We merge the 5 heatmaps corresponding to the nose, ears, and eyes as "face", resulting in 13 heatmaps. We simply freeze all weights of HRNet during training.

Evaluation Metrics. For evaluation, we adopt standard metrics as in most person Re-ID literature, namely Cumulative Matching Characteristic (CMC) curves and mean average precision (mAP). To make a fair comparison with the existing research works, for LTCC, we evaluate our method under both the standard setting and the cloth-changing setting. Specifically, for the standard setting, images in the testing set with the same identity and the same camera view are discarded when computing evaluation scores. In other words, there are both cloth-consistent and cloth-changing samples in the testing set. For the cloth-changing setting, images with the same identity, camera view, and clothes are discarded during testing, so there are only cloth-changing samples in the testing set.

4.2   Quantitative Results

Performance on the Celeb-reID Dataset. We evaluate our proposed method on Celeb-reID and compare it with other state-of-the-art competitors. Results are shown in Table 1. Among them, ReIDCaps [20], CESD [39] and RCSANet [19] are specially designed for the cloth-changing person Re-ID problem. For a fair comparison, the results of ReIDCaps [20] and RCSANet [19] are achieved without applying the fine-grained body parts learning strategy. The results of ReIDCaps [20] are copied from the original paper and it uses deeper DenseNet-121 [17] as the backbone. Our method outperforms all compared methods on the challenging cloth-changing dataset Celeb-reID which contains great clothing variations. Our method outperforms the state-of-the-art

Table 2. Comparison of our method with the state-of-the-art methods on LTCC. The best results of the state-of-the-art method and our method are shown in bold. "Standard" and "Cloth-Changing" mean the standard setting and the cloth-changing setting, respectively.

Methods	Cloth-Changing		Standard	
	Rank-1	mAP	Rank-1	mAP
LOMO [31] + KISSME [25]	10.75	5.25	26.57	9.11
LOMO [31] + XQDA [31]	10.95	6.2	25.35	9.54
PCB [46]	23.52	10.03	61.86	27.52
HACNN [26]	21.59	9.25	60.24	26.71
RGA-SC [62]	31.4	14.0	65.0	27.5
ISP [69]	27.8	11.9	66.3	29.6
GI-ReID [22]	23.7	10.4	63.2	29.4
CESD [39]	25.15	12.40	71.39	34.41
Chen *et al.* [6]	31.2	14.8	–	–
FSAM [14]	**38.5**	**16.2**	**73.2**	35.4
Baseline (ResNet-50)	31.89	13.07	69.17	33.16
Ours	**35.97**	**15.43**	**73.23**	**35.31**

method RCSANet [19] by 2.6% in Rank-1 accuracy. The great improvement of our method compared with our baseline model (ResNet-50) also demonstrates that our method can help tackle the cloth-changing challenge of person Re-ID.

Performance on the LTCC Dataset. We also evaluate our proposed method on LTCC and compare it with several competitors. In Table 2, competitors include methods based on hand-crafted feature representations, deep learning baselines, and methods specially designed for cloth-changing person Re-ID. All state-of-the-art standard person Re-ID methods achieve relatively inferior performance, because they do not take the clothing changes into account. To reduce the interference of clothes, some cloth-changing person Re-ID methods use information from different modalities. For example, FSAM [14] integrates three modalities and fine-tunes the parsing network while training, while our method only uses an off-the-shelf human posture keypoints extractor. Results show that our method achieves comparable results with the state-of-the-art cloth-changing person Re-ID methods.

Performance on the Market-1501 Dataset. To further show the feasibility of our method for the cases without clothing changes in the short term, we additionally evaluate our method on the standard benchmark person Re-ID dataset Market-1501. As shown in Table 3, our method is comparable with the state-of-the-art methods on Market-1501. Specifically, our method still achieves improvement compared with the baseline model (ResNet-50), which shows that our method can take advantage of the body shape information to extract more discriminative person identity features.

Table 3. Comparison of our method with state-of-the-art methods on Market-1501.

Methods	Rank-1	mAP
PCB [46]	93.8	81.6
IANet [15]	94.9	83.1
AANet [47]	93.9	83.4
DSA-reID [61]	95.7	87.6
RGA-SC [62]	96.1	88.4
ISP [69]	95.3	88.6
Baseline (ResNet-50)	93.1	82.9
Ours	94.0	84.6

Table 4. Ablation study on the Celeb-reID dataset. "S→A" denotes the one-way cross-attention interaction from the body shape branch to the appearance branch, while "A → S" denotes the one-way cross-attention interaction from the appearance branch to the body shape branch.

Methods	SSE	Co-Attention	S → A	A → S	Rank-1	mAP
1 (Baseline)					52.86	9.92
2	✓				52.59	9.97
3	✓	✓			53.23	10.19
4	✓		✓		54.24	10.51
5	✓			✓	55.85	11.37
6	✓		✓	✓	55.92	11.27
7	✓	✓	✓		53.87	10.39
8	✓	✓		✓	57.17	12.09
9 (Ours)	✓	✓	✓	✓	**57.47**	**12.27**

4.3   Ablation Study

To verify the effectiveness of our method, detailed ablation experiments are carried out on each proposed module, on the large-scale long-term cloth-changing person Re-ID dataset Celeb-reID. Results are shown in Table 4.

The Effectiveness of SSE and Mutual Cross-Attention. Although body shape is more robust against clothing changes than color appearance, intuitively, we cannot distinguish a person only by his/her body shape. Experiments also demonstrate that the performance is quite low if we only use the body shape branch. The results of method 2 in Table 4 show that if we directly concatenate appearance features with body shape semantic features extracted by the SSE module, the performance is close to baseline. It shows the performance improvement is gained from our well-designed mutual cross-attention strategy, rather than just the extra introduction of the body shape branch. By comparing methods 2 and 6 in Table 4, we can see that our proposed mutual cross-attention strategy improves Rank-1 by 3.33%, and mAP by 1.30%. It also indicates that the SSE module has encoded useful body shape semantic information.

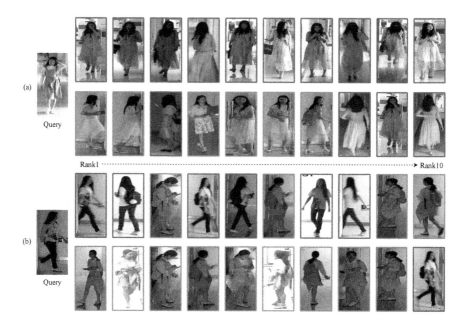

Fig. 3. Visualization of retrieval results. The left side of (a) and (b) is the input query image. For the right side, the first and the second row are the ordered matching results obtained by using the benchmark network ResNet-50 and our proposed network, respectively. Images with green borders and red borders indicate correct and error matching results, respectively. Best viewed in color and zoomed in. (Color figure online)

Design Effectiveness of Mutual Cross-Attention. Aiming for adequate interaction and information fusion between the two branches, we propose a symmetrical mutual cross-attention module. As shown in Table 4, compared with baseline, when we only apply the one-way cross-attention interaction either "S → A" or "A → S", the performance is improved. However, due to the unidirectional nature of information interaction, the network still cannot make full use of the information between the two modalities. When we apply the proposed mutual cross-attention interaction strategy, much greater improvement is achieved, which validates the effectiveness of our mutual cross-attention design.

It is worth noting that, as the results of methods 4 and 7 in Table 4 show, if only use the one-way cross-attention interaction, applying the co-attention alignment mechanism may not improve the performance effectively. Therefore, our proposed mutual cross-attention strategy is more stable and conducive to multi-modal feature fusion.

The Effectiveness of the Proposed Co-attention Alignment Module. As shown in Table 4, even without the mutual cross-attention module, compared with baseline, the co-attention alignment module can still improve the performance. Together with our proposed mutual cross-attention module,

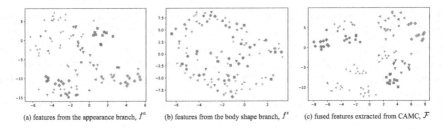

(a) features from the appearance branch, f^a (b) features from the body shape branch, f^s (c) fused features extracted from CAMC, $\mathcal{F}$

Fig. 4. t-SNE visualizations of features from the appearance branch and the body shape branch, as well as ones extracted from CAMC. Samples are randomly selected from the testing set of the LTCC dataset. Each color represents an identity, and different symbols indicate different clothes. Best viewed in color and zoomed in. (Color figure online)

the performance can be further improved. It is worth noting that, as the results of methods 4 $\sim$ 6 in Table 4 shown, when we discard our proposed co-attention alignment module, the mutual cross-attention strategy may not improve the performance effectively compared with the one-way cross-attention. The results confirm the necessity and effectiveness of our co-attention alignment operation. As the results of methods 7 $\sim$ 9 in Table 4 shown, when we apply our proposed co-attention aligned mutual cross-attention mechanism, the best results are achieved. It indicates features from various modalities, that are aligned with each other well, can interact and fuse more effectively, and better help the network to extract more discriminative and cloth-irrelevant identity features.

Visualization of Retrieval Results. With the introduction of the body shape semantic information and our proposed modality-aligned mutual fusion strategy, our method can help meet the challenge of changing clothes. To intuitively demonstrate this conclusion, we visualize the top 10 ranked retrieval results of the baseline network ResNet-50 and our proposed network under the cloth-changing setting on LTCC.

As shown in Fig. 3, our proposed network can better recognize the same person with different clothes. For example, in the second row of Fig. 3 (a), the top retrieval results have the same identity as the input query person, but with different clothes. However, for the matching results of ResNet-50, that is, the first row of Fig. 3 (a), the matching images have similar clothing textures to the input query image, but with different identities. For another example, we can see in Fig. 3 (b), the retrieved persons in the first row wear clothes with similar colors, resulting in some matching errors. However, benefiting from paying more attention to body shape information rather than volatile color appearance, our method can still correctly identify pedestrians even if they change clothes.

Visualization of Features. To verify our motivation and show the effectiveness of our proposed method, we use t-SNE [36] to visualize the learned features. As shown in Fig. 4, features from the appearance branch are relatively more chaotic than ones extracted from CAMC, indicating that different persons are misidentified under the influence of similar clothes. We can observe that although features from the body shape branch themselves are randomly distributed, our proposed CAMC framework can make use of them to obtain more discriminative fused features. More discussions and analyses are provided in the supplemental material.

5 Conclusion

In this paper, we study the more realistic and challenging long-term cloth-changing person Re-ID problem and propose a unified framework adopting Transformer to handle multiple modalities for the first time. Especially, with our proposed Shape Semantics Embedding (SSE) module, we can extract body shape semantic features, which are robust against clothing changes in the long term. To further integrate and make full use of the body shape semantic information, we propose a Co-attention Aligned Mutual Cross-attention (CAMC) framework and effectively fuse multiple modalities. As a result, features encoding useful appearance and body shape semantic information are distilled to an identity-related and discriminative feature, that is more robust to clothing changes. The effectiveness of our proposed method is validated through extensive experiments on several datasets.

Broader Impact. Our proposed CAMC framework can be easily used in existing person Re-ID methods to make long-term person Re-ID technology more practicable in intelligent video monitoring systems, and hopefully inspire more valuable and innovative studies. However, in reality, person Re-ID systems typically use unauthorized surveillance data, which may cause privacy breaches. As a result, governments and officials must take action to govern the use of person Re-ID data and technology, and researchers should avoid using datasets that may raise ethical concerns. For example, the dataset DukeMTMC [40] should no longer be used after it was shut down for violating data collection restrictions. It is worth noting that, all datasets used in our paper are publicly available and involve no ethical issues.

Acknowledgements. This work is supported by China Postdoctoral Science Foundation (2022M710746), the Science and Technology Major Project of Commission of Science and Technology of Shanghai (No. 21XD1402500), NSFC Project (62176061).

References

1. Ba, J.L., Kiros, J.R., Hinton, G.E.: Layer normalization. arXiv preprint arXiv:1607.06450 (2016)
2. Bansal, V., Foresti, G.L., Martinel, N.: Cloth-changing person re-identification with self-attention. In: Proceedings of the IEEE/CVF Winter Conference on Applications of Computer Vision, pp. 602–610 (2022)

3. Carion, N., Massa, F., Synnaeve, G., Usunier, N., Kirillov, A., Zagoruyko, S.: End-to-end object detection with transformers. In: Vedaldi, A., Bischof, H., Brox, T., Frahm, J.-M. (eds.) ECCV 2020. LNCS, vol. 12346, pp. 213–229. Springer, Cham (2020). https://doi.org/10.1007/978-3-030-58452-8_13

4. Chang, X., Hospedales, T.M., Xiang, T.: Multi-level factorisation net for person re-identification. In: CVPR, vol. 1, p. 2 (2018)

5. Chen, B., Deng, W., Hu, J.: Mixed high-order attention network for person re-identification. In: Proceedings of the IEEE/CVF International Conference on Computer Vision, pp. 371–381 (2019)

6. Chen, J., et al.: Learning 3D shape feature for texture-insensitive person re-identification. In: Proceedings of the IEEE/CVF Conference on Computer Vision and Pattern Recognition, pp. 8146–8155 (2021)

7. Chen, W., Chen, X., Zhang, J., Huang, K.: Beyond triplet loss: a deep quadruplet network for person re-identification. In: Proceedings of the CVPR, vol. 2 (2017)

8. Deng, J., Dong, W., Socher, R., Li, L.J., Li, K., Fei-Fei, L.: Imagenet: a large-scale hierarchical image database. In: IEEE Conference on Computer Vision and Pattern Recognition. CVPR 2009, pp. 248–255. IEEE (2009)

9. Dosovitskiy, A., et al.: An image is worth 16x16 words: transformers for image recognition at scale. arXiv preprint arXiv:2010.11929 (2020)

10. He, K., Zhang, X., Ren, S., Sun, J.: Deep residual learning for image recognition. In: CVPR (2015)

11. He, L., Liang, J., Li, H., Sun, Z.: Deep spatial feature reconstruction for partial person re-identification: Alignment-free approach. In: Proceedings of the IEEE Conference on Computer Vision and Pattern Recognition, pp. 7073–7082 (2018)

12. He, S., Luo, H., Wang, P., Wang, F., Li, H., Jiang, W.: Transreid: transformer-based object re-identification. In: Proceedings of the IEEE/CVF International Conference on Computer Vision, pp. 15013–15022 (2021)

13. Hermans, A., Beyer, L., Leibe, B.: In defense of the triplet loss for person re-identification. arXiv preprint arXiv:1703.07737 (2017)

14. Hong, P., Wu, T., Wu, A., Han, X., Zheng, W.S.: Fine-grained shape-appearance mutual learning for cloth-changing person re-identification. In: Proceedings of the IEEE/CVF Conference on Computer Vision and Pattern Recognition, pp. 10513–10522 (2021)

15. Hou, R., Ma, B., Chang, H., Gu, X., Shan, S., Chen, X.: Interaction-and-aggregation network for person re-identification. In: Proceedings of the IEEE/CVF Conference on Computer Vision and Pattern Recognition, pp. 9317–9326 (2019)

16. Hu, J., Shen, L., Sun, G.: Squeeze-and-excitation networks. In: Proceedings of the IEEE Conference on Computer Vision and Pattern Recognition, pp. 7132–7141 (2018)

17. Huang, G., Liu, Z., Weinberger, K.Q., van der Maaten, L.: Densely connected convolutional networks. In: Proceedings of the IEEE Conference on Computer Vision and Pattern Recognition, vol. 1, p. 3 (2017)

18. Huang, Y., Wu, Q., Xu, J., Zhong, Y.: SBSGAN: suppression of inter-domain background shift for person re-identification. In: Proceedings of the IEEE/CVF International Conference on Computer Vision, pp. 9527–9536 (2019)

19. Huang, Y., Wu, Q., Xu, J., Zhong, Y., Zhang, Z.: Clothing status awareness for long-term person re-identification. In: Proceedings of the IEEE/CVF International Conference on Computer Vision, pp. 11895–11904 (2021)

20. Huang, Y., Xu, J., Wu, Q., Zhong, Y., Zhang, P., Zhang, Z.: Beyond scalar neuron: adopting vector-neuron capsules for long-term person re-identification. IEEE Trans. Circuits Syst. Video Technol. 30, 3459–3471 (2019)

21. Isobe, T., Li, D., Tian, L., Chen, W., Shan, Y., Wang, S.: Towards discriminative representation learning for unsupervised person re-identification. In: Proceedings of the IEEE/CVF International Conference on Computer Vision, pp. 8526–8536 (2021)
22. Jin, X., et al.: Cloth-changing person re-identification from a single image with gait prediction and regularization. In: Proceedings of the IEEE/CVF Conference on Computer Vision and Pattern Recognition, pp. 14278–14287 (2022)
23. Jin, X., Lan, C., Zeng, W., Chen, Z., Zhang, L.: Style normalization and restitution for generalizable person re-identification. In: Proceedings of the IEEE/CVF Conference on Computer Vision and Pattern Recognition, pp. 3143–3152 (2020)
24. Kingma, D.P., Ba, J.: Adam: a method for stochastic optimization. arXiv preprint arXiv:1412.6980 (2014)
25. Koestinger, M., Hirzer, M., Wohlhart, P., Roth, P.M., Bischof, H.: Large scale metric learning from equivalence constraints. In: CVPR (2012)
26. Li, W., Zhu, X., Gong, S.: Harmonious attention network for person re-identification. In: CVPR, vol. 1, p. 2 (2018)
27. Li, Y.J., Chen, Y.C., Lin, Y.Y., Du, X., Wang, Y.C.F.: Recover and identify: a generative dual model for cross-resolution person re-identification. In: Proceedings of the IEEE/CVF International Conference on Computer Vision, pp. 8090–8099 (2019)
28. Li, Y.J., Lin, C.S., Lin, Y.B., Wang, Y.C.F.: Cross-dataset person re-identification via unsupervised pose disentanglement and adaptation. In: Proceedings of the IEEE/CVF International Conference on Computer Vision, pp. 7919–7929 (2019)
29. Li, Y.J., Luo, Z., Weng, X., Kitani, K.M.: Learning shape representations for clothing variations in person re-identification. arXiv preprint arXiv:2003.07340 (2020)
30. Li, Y., He, J., Zhang, T., Liu, X., Zhang, Y., Wu, F.: Diverse part discovery: occluded person re-identification with part-aware transformer. In: Proceedings of the IEEE/CVF Conference on Computer Vision and Pattern Recognition, pp. 2898–2907 (2021)
31. Liao, S., Hu, Y., Zhu, X., Li, S.Z.: Person re-identification by local maximal occurrence representation and metric learning. In: CVPR (2015)
32. Lin, T.-Y., et al.: Microsoft COCO: common objects in context. In: Fleet, D., Pajdla, T., Schiele, B., Tuytelaars, T. (eds.) ECCV 2014. LNCS, vol. 8693, pp. 740–755. Springer, Cham (2014). https://doi.org/10.1007/978-3-319-10602-1_48
33. Liu, F., Zhang, L.: View confusion feature learning for person re-identification. In: Proceedings of the IEEE/CVF International Conference on Computer Vision, pp. 6639–6648 (2019)
34. Luo, H., Gu, Y., Liao, X., Lai, S., Jiang, W.: Bag of tricks and a strong baseline for deep person re-identification. In: Proceedings of the IEEE/CVF Conference on Computer Vision and Pattern Recognition Workshops, pp. 0–0 (2019)
35. Luo, H., Jiang, W., Fan, X., Zhang, C.: Stnreid: deep convolutional networks with pairwise spatial transformer networks for partial person re-identification. IEEE Trans. Multimedia 22(11), 2905–2913 (2020)
36. Maaten, L.V.D., Hinton, G.: Visualizing data using t-SNE. J. Mach. Learn. Res. 9, 2579–2605 (2008)
37. Miao, J., Wu, Y., Liu, P., Ding, Y., Yang, Y.: Pose-guided feature alignment for occluded person re-identification. In: Proceedings of the IEEE International Conference on Computer Vision, pp. 542–551 (2019)

38. Qian, X., et al.: Pose-normalized image generation for person re-identification. In: Ferrari, V., Hebert, M., Sminchisescu, C., Weiss, Y. (eds.) ECCV 2018. LNCS, vol. 11213, pp. 661–678. Springer, Cham (2018). https://doi.org/10.1007/978-3-030-01240-3_40

39. Qian, X., et al.: Long-term cloth-changing person re-identification. In: Proceedings of the Asian Conference on Computer Vision (2020)

40. Ristani, E., Solera, F., Zou, R., Cucchiara, R., Tomasi, C.: Performance measures and a data set for multi-target, multi-camera tracking. In: European Conference on Computer Vision Workshop on Benchmarking Multi-Target Tracking (2016)

41. Shen, Y., Li, H., Yi, S., Chen, D., Wang, X.: Person re-identification with deep similarity-guided graph neural network. In: Ferrari, V., Hebert, M., Sminchisescu, C., Weiss, Y. (eds.) ECCV 2018. LNCS, vol. 11219, pp. 508–526. Springer, Cham (2018). https://doi.org/10.1007/978-3-030-01267-0_30

42. Suh, Y., Wang, J., Tang, S., Mei, T., Lee, K.M.: Part-aligned bilinear representations for person re-identification. In: Ferrari, V., Hebert, M., Sminchisescu, C., Weiss, Y. (eds.) Computer Vision – ECCV 2018. LNCS, vol. 11218, pp. 418–437. Springer, Cham (2018). https://doi.org/10.1007/978-3-030-01264-9_25

43. Sun, K., Xiao, B., Liu, D., Wang, J.: Deep high-resolution representation learning for human pose estimation. In: Proceedings of the IEEE/CVF Conference on Computer Vision and Pattern Recognition, pp. 5693–5703 (2019)

44. Sun, X., Zheng, L.: Dissecting person re-identification from the viewpoint of viewpoint. In: Proceedings of the IEEE Conference on Computer Vision and Pattern Recognition, pp. 608–617 (2019)

45. Sun, Y., et al.: Circle loss: a unified perspective of pair similarity optimization. In: Proceedings of the IEEE/CVF Conference on Computer Vision and Pattern Recognition, pp. 6398–6407 (2020)

46. Sun, Y., Zheng, L., Yang, Y., Tian, Q., Wang, S.: Beyond part models: person retrieval with refined part pooling (and a strong convolutional baseline). In: Ferrari, V., Hebert, M., Sminchisescu, C., Weiss, Y. (eds.) ECCV 2018. LNCS, vol. 11208, pp. 501–518. Springer, Cham (2018). https://doi.org/10.1007/978-3-030-01225-0_30

47. Tay, C.P., Roy, S., Yap, K.H.: Aanet: attribute attention network for person re-identifications. In: Proceedings of the IEEE Conference on Computer Vision and Pattern Recognition, pp. 7134–7143 (2019)

48. Vaswani, A., et al.: Attention is all you need. In: Advances in Neural Information Processing Systems, pp. 5998–6008 (2017)

49. Wan, F., Wu, Y., Qian, X., Chen, Y., Fu, Y.: When person re-identification meets changing clothes. In: Proceedings of the IEEE/CVF Conference on Computer Vision and Pattern Recognition Workshops, pp. 830–831 (2020)

50. Wang, C., Zhang, Q., Huang, C., Liu, W., Wang, X.: Mancs: a multi-task attentional network with curriculum sampling for person re-identification. In: Ferrari, V., Hebert, M., Sminchisescu, C., Weiss, Y. (eds.) ECCV 2018. LNCS, vol. 11208, pp. 384–400. Springer, Cham (2018). https://doi.org/10.1007/978-3-030-01225-0_23

51. Wang, D., Zhang, S.: Unsupervised person re-identification via multi-label classification. In: Proceedings of the IEEE/CVF Conference on Computer Vision and Pattern Recognition, pp. 10981–10990 (2020)

52. Wang, G., Yuan, Y., Chen, X., Li, J., Zhou, X.: Learning Discriminative Features with Multiple Granularities for Person Re-Identification. ArXiv e-prints, April 2018

53. Wang, Z., Wang, Z., Zheng, Y., Chuang, Y.Y., Satoh, S.: Learning to reduce dual-level discrepancy for infrared-visible person re-identification. In: Proceedings of the IEEE/CVF Conference on Computer Vision and Pattern Recognition, pp. 618–626 (2019)

54. Wu, Q., et al.: Discover cross-modality nuances for visible-infrared person re-identification. In: Proceedings of the IEEE/CVF Conference on Computer Vision and Pattern Recognition, pp. 4330–4339 (2021)

55. Yan, C., Pang, G., Jiao, J., Bai, X., Feng, X., Shen, C.: Occluded person re-identification with single-scale global representations. In: Proceedings of the IEEE/CVF International Conference on Computer Vision, pp. 11875–11884 (2021)

56. Yang, F., Yan, K., Lu, S., Jia, H., Xie, X., Gao, W.: Attention driven person re-identification. Pattern Recogn. **86**, 143–155 (2019)

57. Yang, Q., Wu, A., Zheng, W.S.: Person re-identification by contour sketch under moderate clothing change. IEEE Trans. Pattern Anal. Mach. Intell. **43**, 2029–2046 (2019)

58. Yu, H.X., Zheng, W.S., Wu, A., Guo, X., Gong, S., Lai, J.H.: Unsupervised person re-identification by soft multilabel learning. In: Proceedings of the IEEE/CVF Conference on Computer Vision and Pattern Recognition, pp. 2148–2157 (2019)

59. Yu, Q., Chang, X., Song, Y.Z., Xiang, T., Hospedales, T.M.: The devil is in the middle: exploiting mid-level representations for cross-domain instance matching. arXiv preprint arXiv:1711.08106 (2017)

60. Yu, S., Li, S., Chen, D., Zhao, R., Yan, J., Qiao, Y.: COCAS: a large-scale clothes changing person dataset for re-identification. In: Proceedings of the IEEE/CVF Conference on Computer Vision and Pattern Recognition, pp. 3400–3409 (2020)

61. Zhang, Z., Lan, C., Zeng, W., Chen, Z.: Densely semantically aligned person re-identification. In: Proceedings of the IEEE/CVF Conference on Computer Vision and Pattern Recognition, pp. 667–676 (2019)

62. Zhang, Z., Lan, C., Zeng, W., Jin, X., Chen, Z.: Relation-aware global attention for person re-identification. In: Proceedings of the IEEE/CVF Conference on Computer Vision and Pattern Recognition, pp. 3186–3195 (2020)

63. Zheng, L., Shen, L., Tian, L., Wang, S., Wang, J., Tian, Q.: Scalable person re-identification: a benchmark. In: ICCV (2015)

64. Zheng, L., Zhang, H., Sun, S., Chandraker, M., Tian, Q.: Person re-identification in the wild. arXiv preprint arXiv:1604.02531 (2016)

65. Zheng, W.S., Li, X., Xiang, T., Liao, S., Lai, J., Gong, S.: Partial person re-identification. In: Proceedings of the IEEE International Conference on Computer Vision, pp. 4678–4686 (2015)

66. Zheng, Z., Zheng, L., Yang, Y.: A discriminatively learned CNN embedding for person reidentification. ACM Trans. Multimedia Comput. Commun. Appl. (TOMM) **14**(1), 13 (2017)

67. Zhong, Z., Zheng, L., Kang, G., Li, S., Yang, Y.: Random erasing data augmentation. arXiv preprint arXiv:1708.04896 (2017)

68. Zhou, K., Yang, Y., Cavallaro, A., Xiang, T.: Omni-scale feature learning for person re-identification. In: Proceedings of the IEEE International Conference on Computer Vision, pp. 3702–3712 (2019)

69. Zhu, K., Guo, H., Liu, Z., Tang, M., Wang, J.: Identity-guided human semantic parsing for person re-identification. In: Vedaldi, A., Bischof, H., Brox, T., Frahm, J.-M. (eds.) ECCV 2020. LNCS, vol. 12348, pp. 346–363. Springer, Cham (2020). https://doi.org/10.1007/978-3-030-58580-8_21

70. Zhu, X., et al.: Deformable detr: deformable transformers for end-to-end object detection. arXiv preprint arXiv:2010.04159 (2020)

Affinity-Aware Relation Network for Oriented Object Detection in Aerial Images

Tingting Fang[1,2], Bin Liu[1,2(✉)], Zhiwei Zhao[1,2], Qi Chu[1,2], and Nenghai Yu[1,2]

[1] School of Cyber Science and Technology,
University of Science and Technology of China, Hefei, China
{fountain,zwzhao98}@mail.ustc.edu.cn, {flowice,qchu,ynh}@ustc.edu.cn
[2] Key Laboratory of Electromagnetic Space Information,
Chinese Academy of Science, Beijing, China

Abstract. Object detection in aerial images is a challenging task due to the oriented and densely packed objects. However, densely packed objects constitute a significant characteristic of aerial images: objects are not randomly scattered around in images but in groups sharing similar orientations. Such a recurring pattern of object arrangement could enhance the rotated features and improve the detection performance. This paper proposes a novel and flexible Affinity-Aware Relation Network based on two-stage detectors. Specifically, an affinity-graph construction module is adopted to measure the affinity among objects and to select bounding boxes sharing high similarity with the reference box. Furthermore, we design a dynamic enhancement module, which uses the attention to learn neighbourhood message and dynamically determines weights for feature enhancement. Finally, we conduct experiments on several public benchmarks and achieve notable AP improvements as well as state-of-the-art performances on DOTA, HRSC2016 and UCAS-AOD datasets.

1 Introduction

Oriented object detection of aerial images is a significant yet challenging task in computer vision. Unlike object detection in ordinary scenes, aerial images with high resolution often contain a larger number of densely packed objects. In this case, detection performance of horizontal object detection models [2, 3, 26, 29, 54] deteriorates considerably due to the intersection of axis-aligned receptive fields between objects. Existing methods mainly contribute to solving this challenge from two aspects: One is to optimize the extraction of rotated features [5, 10, 42, 43, 45], such as using rotation-equivariant backbones or enhancing the feature fusion. The other is to perform well-designed bounding box representations [7, 38, 44, 46, 52], such as using eight-parameter or convex-hull to represent boxes.

However, these densely packed objects form a pattern of object arrangements. For each object in each category, we count the average number of its similar objects with an angle difference less than 5 degrees inside a 1024×1024 image.

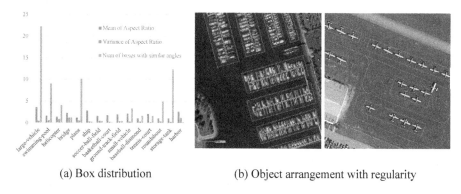

(a) Box distribution   (b) Object arrangement with regularity

Fig. 1. (a) For each object in each category, we count the average number (grey pillar) of its similar objects with an angle difference less than 5 degrees inside a 1024×1024 image. (b) Objects inside each red box share high similarity in categories and orientations. (Color figure online)

(a) Test image   (b) ReDet   (c) Our method

Fig. 2. (a) The edge of a tennis court is obscured by shadows. (b) and (c) are visualizations of ReDet [10] and our method on DOTA. Here red and green boxes represent predictions and ground truth. ReDet does not perform well on the obscured boundary, while our Affinity-Aware Relation Network perceives correct boundary information of the tennis court.

The result in Fig. 1(a) shows that each object can find 3–5 objects with similar categories and orientations on average, and some categories can even find more than 20 objects. Figure 1(b) displays the recurring pattern of object arrangement. Each object can be allocated to an imaginary red box, such that objects inside the same box share high similarity in categories and orientations. Therefore, the semantic information of one object can imply information of other objects in the same box, which can be utilized as an enhancement for the detection task.

This paper proposes an Affinity-Aware Relation Network(AARN) based on the two-stage detector, which aims to enhance the Rotated Region of Interest(RRoI) Align feature for classification and regression in the second stage. Specifically, the proposed AARN consists of two modules. One is a graph construction module, which measures the affinity among objects and dynamically selects bounding boxes sharing high similarity with the reference box. The other is a dynamic enhancement module, which use the attention module to learn neighbourhood message and dynamically determines weights for feature enhancement.

The effectiveness of the method can be simply illustrated by Fig. 2. The input image, detection results on ReDet [10] and results on our method are respectively shown in Fig. 2(a), (b) and (c). The edge of the tennis court on the right in Fig. 2(a) is obscured by the shadows. Figure 2(b) shows that ReDet cannot perceive the object shape correctly in this case. However, Fig. 2(c) shows our method performs well in understanding the accurate boundary of the tennis court, based on a semantic feature implying the height and width information from the other two tennis courts. Therefore, it is meaningful to construct a relation graph among objects and enhance the current object's feature using extra information aggregating from objects with high affinity. Our contributions can be summarized as follows:

- We propose an Affinity-Aware Relation Network, using the affinity among densely packed oriented objects to improve detection performance.
- A Graph Construction Module is proposed, designing KFIoU similarity to measure the affinity among objects and selecting high-quality neighbours for subsequent feature enhancement in a dynamic way.
- A Dynamic Enhancement Module is proposed, using the attention module to learn neighbourhood message and dynamically determining the weight for feature enhancement.
- Extensive experiments are conducted to show that the proposed two modules can notably improve detection performance based on two-stage methods.

2   Related Work

2.1   Oriented Object Detection

Existing oriented object detection methods mainly improve the detection accuracy from three aspects: enhancing rotated features, designing sampling assignment strategy and exploring the representation of bounding box.

Feature enhancement mainly aims at densely packed objects with arbitrary orientations. RoI Transformer [5] and ReDet [10] respectively design a detector with rotation-invariance and rotation-equivariance. R3Det [42] proposes a Feature Refinement Module (FRM), improving the single-stage method performance to a level comparable to two-stage ones. Mask OBB [32], CenterMap Net [33], SCRDet [45], SCRDet++ [43] introduce pixel-level semantic information and provides more granular feature fusion branch.

Well-designed assigner alleviates the inconsistency between classification and regression task. Both DAL [22] and CFC-Net [20] incorporate the Intersection of Union(IoU) [11] metric, which directly reflects the localization capability of predicted boxes, into the assignment strategy of positive samples. SASM [13] dynamically selects the IoU threshold for each object according to its shape.

Oriented RepPoints [17] selects sample points not only from the classification and localization but also from the orientation and point-wise feature correlation.

Studies on box representation and loss function mainly contribute to solving the boundary problem in regression-based methods. BBAVector [49] and PolarDet [51] represent the bounding box in coordinate systems. CFA [7] proposes a convex hull representation method. Gliding Vertex [38] predicts quadrilateral by learning the offset of the four corners of the horizontal bounding boxes. RIL [21] adopts the Hungarian loss. CSL [41] and DCL [40] transform the regression into a classification problem. GWD [44], KLD [46] and KFIoU [47] model the oriented object as a Gaussian distribution to construct a new loss function. P2PLoss [48] describes the spatial distance and morphological similarity of two convex polygons. Unlike our approach, none of these methods consider learning additional information from the affinity among objects for feature enhancement.

2.2   Graph Convolutional Neural Networks

The graph convolutional neural network extends the convolutional neural network to the non-Euclidean space. The graph convolutions fall into two categories: spectral [1, 4, 15, 16, 35] and spatial [6, 8, 23, 28] methods.

The spectral methods define the convolution in the spectral domain via the convolution theorem. The first graph convolutional neural network SCNN [1] defines its operator in the spectral domain. ChebNet [4] and GCN [15] parameterize the convolution kernel, significantly reducing the time and space complexity. The spatial methods define the node correlation in the spatial domain. GNN [12] selects a fixed number of neighbour nodes by a random walk algorithm. GraphSAGE [8] divides the convolution process into sampling and aggregation. GAT [31] uses the attention mechanism to differentiate the aggregation of neighbour nodes. PGC [39] defines convolution as the sum of a specific sampling function multiplied by a particular weight function. Our approach uses the idea of graph convolution for neighbour message learning and feature aggregation.

3   Methods

3.1   Overview

An overview of the proposed Affinity-Aware Relation Network is illustrated in Fig. 3. The model consists of a basic two-stage detector, a Graph Construction Module(GC-Module) and a Dynamic Enhancement Module(DE-Module).

An image is first fed into the pipeline of the basic detector. The GC-Module uses the proposal quintuples from RPN as well as RRoI features from RRoI Align to calculate the affinity matrix and dynamically determines the threshold to filter out low-quality neighbours. For each proposal, GC-Module selects proposals(neighbour) sharing high similarity with the current proposal(reference).

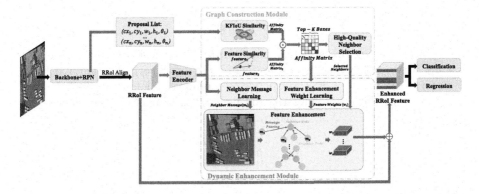

Fig. 3. An overview of the proposed AARN. Our approach is based on the basic two-stage detector ReDet.

The DE-Module consists of neighbour message learning and feature enhancement weight learning. Neighbour message learning performs an attention mechanism over the high-quality neighbours to obtain messages and weights for aggregation. Feature enhancement weight learning determines the feature enhancement factor in consideration of the proposal aspect ratios. Then the neighbour message is used for node aggregation to get the aggregation feature of each node, and the feature enhancement weight is used to dynamically add the aggregation feature to the original feature to obtain the final enhanced feature. Finally, the detection result is achieved after classification and box regression branches of the basic detector. Our proposed method is based on the two-stage model ReDet [10], which in fact can be easily applied to various modern two-stage detectors.

3.2   Graph Construction Module

This module aims to construct a graph to represent the relationship between proposal regions. Formally, given N_r proposal regions of the input image, the relationship among regions can be modeled as an undirected graph $\mathbf{G}(\mathbf{V}, \mathbf{E})$, where v_i in vertex set $\mathbf{V} = \{v_i\}_{i=1}^{N_r}$ corresponds to the i-th proposal and e_{ij} in $\mathbf{E} \in \mathbb{R}^{N_r \times N_r}$ quantifies the relationship between v_i and v_j. GC-Module calculates the affinity between proposal regions to filter out neighbours with low-similarity for each reference node, and then retains only the edges with high affinity in $\mathbf{G}$.

Affinity Matrix Calculation. Affinity matrix $\mathbf{M} \in \mathbb{R}^{N_r \times N_r}$ reflects the similarity between proposals. We should consider two aspects when calculating the affinity: the semantic similarity inside the proposal and the shape similarity of the bounding box.

Feature Similarity. RRoI features characterize object semantics. Given a visual feature $\mathbf{F} \in \mathbb{R}^{N_r \times D}$ extracted by RRoI Align, we first employ a nonlinear transformation $\psi(\cdot) : \mathbb{R}^{N_r \times D} \to \mathbb{R}^{N_r \times L}$, projecting $\mathbf{F}$ into the latent semantic space denoted by

$$\mathbf{F}' = \psi(\mathbf{F}) \tag{1}$$

where $\mathbf{F}' \in \mathbb{R}^{N_r \times L}$. We adopt a simple form of $\psi(\cdot)$ which is implemented by a stack of two fully-connected layers followed by layer normalization and ReLU in order. Each row $f_i' \in \mathbf{F}'$ corresponds to a proposal's latent semantic feature. Then we apply the cosine similarity to calculate the semantic affinity matrix $\mathbf{M}_1$ between $f_i'(i = 1, 2, .., N_r)$, as shown in Eq. (2).

$$\mathbf{M}_1[i][j] = \frac{f_i' f_j'}{\|f_i'\| \|f_j'\|} \tag{2}$$

where $\|\cdot\|$ is a modulus operation.

KFIoU Similarity. The calculation of shape similarity should involve the height, width and rotation angle of objects. As an evaluation metric, IoU well combines these factors. To overcome the high computational complexity of Skew-IoU, we approximate oriented boxes as Gaussian distributions and use the overlap of two Gaussian distributions to measure the shape similarity, as shown in Fig. 4. The conversion from a rotated box to a Gaussian distribution has been discussed in some previous works [14,44,47], described as follows.

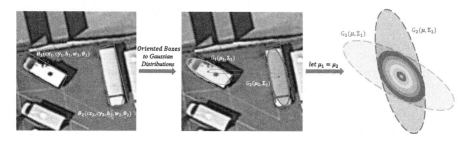

Fig. 4. First, we convert the oriented bounding boxes to Gaussian distributions. Then we make two Gaussian distributions be concentric and introduce Kalman Filter to simulate the distribution overlapping.

The oriented box can be represented by a quintuple $\mathbf{B}(cx, cy, h, w, \theta)$, where (cx, cy) are the center point coordinates. h, w and θ respectively refer to the height, width and rotation angle. The transformation from the proposal quintuple to the Gaussian distribution $\mathbf{N}(\mu, \Sigma)$ is shown in Eq. (3).

$$\Sigma^{\frac{1}{2}} = \begin{pmatrix} cos\theta & -sin\theta \\ sin\theta & cos\theta \end{pmatrix} \begin{pmatrix} \frac{w}{2} & 0 \\ 0 & \frac{h}{2} \end{pmatrix} \begin{pmatrix} cos\theta & sin\theta \\ -sin\theta & cos\theta \end{pmatrix}, \qquad \mu = (cx, cy) \tag{3}$$

After obtaining a 2D Gaussian distribution, we can easily calculate the box area from its covariance of the corresponding distribution.

$$\mathbf{S_B}(\Sigma) = 4\sqrt{\prod eig(\Sigma)} = 4 \cdot |\Sigma|^{\frac{1}{2}} \tag{4}$$

Then the overlapping distribution can be intuitively derived by multiplication of two Gaussian distributions. The probability density function of multiplying two Gaussian distributions $\mathbf{N_1}(\mu_1, \Sigma_1)$ and $\mathbf{N_2}(\mu_2, \Sigma_2)$ can be expressed as

$$\mathbf{f_1}(X)\mathbf{f_2}(X) = \mathbf{S_g} \cdot \frac{1}{\sqrt{2\pi\Sigma}} e^{-\frac{1}{2}(\mathbf{X}-\mu)^T \Sigma^{-1}(\mathbf{X}-\mu)} \tag{5}$$

$$\mathbf{S_g} = \frac{1}{\sqrt{2\pi(\Sigma_1 + \Sigma_2)}} e^{-\frac{1}{2}(\mu_1-\mu_2)^T(\Sigma_1+\Sigma_2)^{-1}(\mu_1-\mu_2)} \tag{6}$$

$$\mu = (\mu_2\Sigma_1 + \mu_1\Sigma_2)(\Sigma_1 + \Sigma_2)^{-1}, \qquad \Sigma = \Sigma_1\Sigma_2(\Sigma_1 + \Sigma_2)^{-1} \tag{7}$$

That is, the multiplication of two Gaussian distributions is equal to a compressed or enlarged Gaussian distribution. The constant $\mathbf{Sg}$ is a scaling factor.

Inspired by [47], we perform Kalman Filter to calculate the overlapping areas. Unlike the loss design in [47], the similarity should not be affected by the center distance. Therefore, we let $\mu_1 = \mu_2 = \mu$ to make two Gaussian distributions be concentric. In this case, $\mathbf{Sg}$ is decoupled from center points and IoU similarity can be calculated as Eq. (8).

$$\mathbf{IoU}(\mathbf{N_1}, \mathbf{N_2}) = \frac{\mathbf{S_B}(\Sigma)}{\mathbf{S_{B1}}(\Sigma_1) + \mathbf{S_{B2}}(\Sigma_2) - \mathbf{S_B}(\Sigma)} = \frac{\Sigma^{\frac{1}{2}}}{\Sigma_1^{\frac{1}{2}} + \Sigma_2^{\frac{1}{2}} - \Sigma^{\frac{1}{2}}} \tag{8}$$

The shape affinity matrix $\mathbf{M_2} \in \mathbb{R}^{N_r \times N_r}$ is obtained by

$$m_{ij} = \mathbf{IoU}(\varphi(cx_i, cy_i, h_i, w_i, \theta_i), \varphi(cx_j, cy_j, h_j, w_j, \theta_j)) \tag{9}$$

where $m_{ij} \in \mathbf{M_2}$ and φ represents the box to Gaussian distribution function.

We use min-max normalization to scale the value of $\mathbf{M_1}$ and $\mathbf{M_2}$ ranging from 0 to 1. The final affinity matrix $\mathbf{M}$ satisfies $\mathbf{M} = \mathbf{M_1} \odot \mathbf{M_2}$, where $\odot$ represents the point-wise multiplication.

High-Quality Neighbour Selection. Similar to ATSS [50], High-Quality Neighbour Selection is proposed to dynamically select high-quality neighbour nodes according to their statistical characteristics. We first keep the top-k largest values of each row in affinity matrix $\mathbf{M}$ for each proposal, and then use the mean and standard deviation of selected proposals' affinity values to determine the threshold $\mathbf{\Gamma_i}$ for i-th proposal.

$$\mathbf{\Gamma_i} = u_i + \sigma_i \tag{10}$$

$$u_i = \frac{1}{k}\sum_{j=idx_1}^{idx_k} m_{ij}, \qquad \sigma_i = \sqrt{\frac{1}{k}\sum_{j=idx_1}^{idx_k}(m_{ij} - u_i)^2} \tag{11}$$

where $m_{ij} \in \mathbf{M}$ and idx_i indicates the index of selected k boxes.

For each proposal, absorbing neighbours with inaccurate positions and shapes will degrade its detection performance due to the introduction of noise. Therefore, we perform a non-maximum suppression(NMS) on the neighbour nodes according to the score from the RPN, so that the aggregation nodes tend to be samples from different positions rather than overlapping proposals from the adjacent center points.

3.3   Dynamic Enhancement Module

After determining the reference nodes and neighbour nodes, we design a dynamic enhancement module consisting of neighbour message learning and feature enhancement weight learning. The former uses the attention to learn neighbour message for node aggregation, and the latter dynamically determines the weight for feature enhancement.

Neighbour Message Learning. We use an attention mechanism drawing global dependencies to learn the weighted messages between neighbour and reference nodes. As shown in Fig. 5, this module is implemented based on the Multi-Head Attention in [30]. Embedding Feature $\mathbf{F}' \in \mathbb{R}^{N_r \times L}$ is used as the query (Q), key (K) and value (V) of the Multi-Head Attention in [30].

$$Attention(\mathbf{Q}, \mathbf{K}, \mathbf{V}) = \mathbf{softmax}(\frac{\mathbf{Q}\mathbf{K}^T}{\sqrt{d_k}})\mathbf{V} \tag{12}$$

where the d_k is the channel dimension. $\mathbf{A} = \mathbf{softmax}(\frac{\mathbf{Q}\mathbf{K}^T}{\sqrt{d_k}}) \in \mathbb{R}^{N_r \times N_r}$ represents the attention weight matrix used for neighbourhood aggregation subsequently. It is worth noting that Eq. (12) only displays the structure of the single head. In practice, multiple heads are concatenated to get the *Multi-Head Attention*$(\mathbf{Q}, \mathbf{K}, \mathbf{V}) \in \mathbb{R}^{N_r \times L}$.

The final enhanced node features are obtained by residual connections, as shown in Eq. (13). Then the message m_{ij} delivered from the j-th neighbour to the i-th reference node can be expressed as Eq. (14).

$$Enhanced\_Node\_Feature = Attention(\mathbf{Q}, \mathbf{K}, \mathbf{V}) + Embedding\_Feature \tag{13}$$

$$m_{ij} = \mathbf{A}_{ij} \cdot Enhanced\_Node\_Feature_i \tag{14}$$

Feature Enhancement Weight Learning. Figure 6 displays two objects with the same angle offset ω. However, the box$_1$ with a lower aspect ratio outperforms box$_2$ on the IoU metric, indicating that objects with high aspect ratio are more sensitive to the angular deviation. Therefore, it is necessary to treat objects with high aspect ratios more cautiously in neighbourhood aggregation.

Intuitively, the message delivered to a reference node with high aspect ratio should be assigned a smaller weight before enhancement. Furthermore, for objects with drastic changes in aspect ratio, it tends to be difficult to learn a universal feature generalizing characteristic of all neighbour nodes. We should also tone down the enhancement of features from these objects.

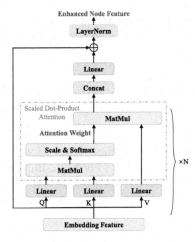

Fig. 5. Flowchart of the Multi-Head Attention Module. Enhanced Node Feature and Attention Weight Matrix respectively represent the neighbour message and the weight used for aggregation.

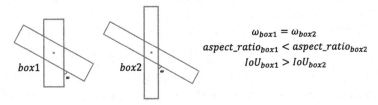

$$\omega_{box1} = \omega_{box2}$$
$$aspect\_ratio_{box1} < aspect\_ratio_{box2}$$
$$IoU_{box1} > IoU_{box2}$$

Fig. 6. The red and green box represent prediction and ground truth. The box$_1$ with a lower aspect ratio outperforms box$_2$ on the IoU metric, indicating that objects with high aspect ratio are more sensitive to the angular deviation. (Color figure online)

In response, we design the Feature Enhancement Weight Learning, which can dynamically adjust enhancement weight w_i according to the i-th object's aspect ratio, as shown in Eq. (15)-(16).

$$u_i^{ratio} = \frac{1}{N_r} \sum_{j=idx_{i1}}^{idx_{ik}} r_j, \qquad \sigma_i^{ratio} = \sqrt{\frac{1}{N_r} \sum_{j=idx_{i1}}^{idx_{ik}} (r_j - u_i)^2} \qquad (15)$$

$$w_i = (\alpha - e^{\frac{u_i^{ratio}}{\beta}}) \cdot e^{-\sigma_i^{ratio}} \qquad (16)$$

where $idx_i = \{idx_{ij}\}_{j=1}^k$ denotes indices of k boxes most relevant to the i-th reference box, selected in High-Quality Neighbour Selection. And r_j is the aspect ratio of the j-th proposal. Equation (15) computes the mean and standard deviation of the aspect ratios of the top-k boxes. The mean value reflects the estimated aspect ratio and the standard deviation implies the fluctuation of aspect ratio. Given $\alpha > 0$ and $\beta > 0$, w_i decreases as the mean or standard deviation increases.

Feature Enhancement and Final Prediction. The enhanced feature of the i-th proposal is obtained by combining the original feature with the aggregation feature, described as

$$Enhanced\_feature_i = feature_i + \varepsilon \cdot \frac{w_i}{N_i} \sum_{j=1}^{N_i} m_{ij} \tag{17}$$

where N_i is the number of selected neighbours of the i-th reference proposal. The meaning of m_{ij} and w_i are as same as mentioned above. ε is a learnable parameter with an initial value of 1.0, in order to implement a dynamic residual connection. Finally, the enhanced features are fed into the classification and regression branches of the basic detector to get prediction results.

4   Experiments

4.1   Experimental Setup

Datasets. *DOTA-v1.0* [36] is a large-scale dataset for oriented objects detection in aerial images, which contains 2806 images ranging from 800× 800 to 4k× 4k pixels, 188,282 instances and 15 categories: Plane (PL), Baseball diamond (BD), Bridge (BR), Ground track field (GTF), Small vehicle (SV), Large vehicle (LV), Ship (SH), Tennis court (TC), Basketball court(BC), Storage tank (ST), Soccer-ball field (SBF), Roundabout (RA), Harbor (HA), Swimming pool (SP), and Helicopter (HC). The proportions of the training, validation and testing set are 1/2, 1/6, and 1/3, respectively. All images of training and validation set are split into 1024× 1024 with an overlap of 200 pixels during training.

   HRSC2016 [18] is a high-resolution optical remote sensing dataset for ship recognition, which contains 1061 images (436 for training, 181 for validation and 444 for testing) ranging from 300×300 to 1,500×512 pixels. All images of training and validation set are resized to 800×512 pixels during training.

   UCAS-AOD [53] contains 1,510 images with approximately 659×1280 pixels, 14,596 instances and two categories: plane and car. Like other works [43,44,47], we randomly select 1100 images for training and 400 for testing.

Implementation Details. We use a two-stage detector ReDet [10] as our baseline and ReResNet-50 pretrained on ImageNet [27] following ReDet as our backbone. All modules before the RRoI Align follow the settings of ReDet.

   As for the implementation of AARN, we first use two linear layers of size 512 ($L = 512$) to learn the latent feature $\mathbf{F}'$ in Eq. (1). Then top $k = 9$ largest values of each row in the affinity matrix are kept to determine the threshold Γ_i for i-th proposal in Eq. (11). NMS with threshold $= 0.1$ is performed over selected neighbour nodes before DE-M to avoid the introduction of noise. For the neighbour message learning in DE-M, all linear layers of $\mathbf{Q}$, $\mathbf{K}$ and $\mathbf{V}$ produce outputs of dimension $d_{model} = L = 512$. And we employ $h = 8$ parallel attention heads so $d_k = d_{model}/h = 64$ in Eq. (12). For the feature enhancement weight learning, α and β in Eq. (16) are set to 2 and 3.5 respectively.

In the inference phase, RPN will generate 2000 proposals. If such a large number of proposals are input into AAFN, great noise will be introduced. Therefore, we set filter threshold as 0.9 in line with scores from RPN stage, so that only boxes with high confidence can participate in graph construction and feature enhancement. The weights of modules before RRoI Align are frozen during training. We adopt a stochastic gradient descent (SGD) optimizer with an initial learning rate of 0.0001, the momentum of 0.9 and weight decay of 0.0001. We train the model for 12, 40, 120 epochs on the DOTA, HRSC2016 and UCAS-AOD datasets. We use 2 TITAN RTX GPUs with a total batch size of 4 for training and one TITAN RTX GPU for inference.

4.2   Comparisons with the State-of-the-Art

Table 1 compares our method with the state-of-the-art detectors on DOTA-v1.0. Without random rotation and multi-scale data augmentation, we improve by **1.07%** AP over the baseline ReDet. Especially categories with low aspect ratios or less semantic information achieve more notable AP improvements: **4.7%** on roundabout (RA), **3.77%** on helicopter (HC), **2.21%** on soccer-ball-field (SF), and **2.04%** on ground-track-field (GTF). For multi-scale training with random rotation, our method achieves the state-of-the-art **80.79%** AP and the best performance on 6 categories. Figure 7 displays results of ReDet, results of our method, and visualization of Graph Construction Module results on DOTA. In Fig. 7(a) and (b), we mark some instances which are accurate under our method but inaccurate under ReDet with white circles. It shows a better performance of our method. In Fig. 7(c), the reference box (green) is connected to its selected neighbour boxes (red). It can be found that a reference box always share high similarity in category and orientation with its neighbour boxes.

Table 2 lists the performances of our method and state-of-the-art detectors on HRSC2016. Our method achieves the best performance of **90.57%** under the VOC2007 metric. Table 3 shows results on UCAS-AOD. Our method achieves the state-of-the-art **89.94%** and **97.45%** mAP under VOC2007 and VOC2012 metrics respectively, and the mAP of VOC2012 improves by **1.22%**.

4.3   Ablation Study

To prove the effectiveness of our proposed method, we choose ReDet as our baseline and perform a detailed ablative analysis on DOTA-v1.0 test set. Following previous works, random horizontal flipping without any other tricks is applied for data augmentation. Ablation study result is shown in Table 4, which demonstrates the effectiveness of each module.

Affinity Matrix Calculation. Affinity Matrix Calculation consists of feature similarity and shape similarity. As shown in Table 5, the absence of either component results in a lower performance than baseline. We also discuss the effectiveness of different ways to compute shape similarity. Theta similarity refers to

Table 1. AP for each class and AP$_{50}$ on DOTA-v1.0. R-50, RX-101 and H-104 respectively stand for ResNet-50, ResNeXt-101 and Hourglass-104. MS/RR denotes random rotation and multi-scale used for augmentation during training.

Method	Backbone	MS/RR	PL	BD	BR	GTF	SV	LV	SH	TC	BC	ST	SBF	RA	HA	SP	HC	AP$_{50}$
Single-stage/Anchor-free																		
Oriented RepPoints [17]	R-101		89.53	84.07	59.86	71.76	79.95	80.03	87.33	90.84	87.54	85.23	59.15	66.37	75.23	73.75	57.23	76.52
CFA [7]	R-152		89.08	83.20	54.37	66.87	81.23	80.96	87.17	90.21	84.32	86.09	52.34	69.94	75.52	80.76	67.96	76.67
O^2-DNet [34]	H-104	✓	89.31	82.14	47.33	61.21	71.32	74.03	78.62	90.76	82.23	81.36	60.93	60.17	58.21	66.98	61.03	71.04
DRN [24]	H-104	✓	89.71	82.34	47.22	64.10	76.22	74.43	85.84	90.57	86.18	84.89	57.65	61.93	69.30	69.63	58.48	73.23
BBAVectors [49]	R-101	✓	88.63	84.06	52.13	69.56	78.26	80.40	88.06	90.87	87.23	86.39	56.11	65.62	67.10	72.08	63.96	75.36
CSL [41]	R-152	✓	90.25	85.53	54.64	75.31	70.44	73.51	77.62	90.84	86.15	86.69	69.60	68.04	73.83	71.10	68.93	76.17
PolarDet [51]	R-101	✓	89.65	87.07	48.14	70.97	78.53	80.34	87.45	90.76	85.63	86.87	61.64	70.32	71.92	73.09	67.15	76.64
SASM [13]	R-101	✓	89.54	85.94	57.73	78.41	79.78	84.19	89.25	90.87	85.80	87.27	63.82	67.81	78.67	79.35	69.37	79.17
RetinaNet-P2P [48]	R-101	✓	89.22	86.12	55.23	81.39	80.34	83.45	88.25	90.87	86.63	87.08	71.74	69.87	77.34	76.01	59.59	79.15
Two/Refined-stage																		
Gliding Vertex [38]	R-101		89.64	85.00	52.26	77.34	73.01	73.14	86.82	90.74	79.02	86.81	59.55	70.91	72.94	70.86	57.32	75.02
SCRDet++ [43]	R-101		89.77	83.90	56.30	73.98	72.60	75.63	82.82	90.76	87.89	86.14	65.24	63.17	76.05	68.06	70.24	76.20
ReDet [10](baseline)	ReR-50		88.79	82.64	53.97	74.00	78.13	84.06	88.04	90.89	87.78	85.75	61.76	60.39	75.96	68.07	63.59	76.25
Oriented R-CNN [37]	R-101		88.86	83.48	55.27	76.92	74.27	82.10	87.52	90.90	85.56	85.33	65.51	65.62	66.82	74.36	57.28	76.28
KFIoU [47]	R-101		89.04	84.04	52.98	73.00	78.69	83.60	87.61	90.79	85.97	85.47	64.77	63.29	69.18	76.38	65.63	76.70
Mask OBB [11]	RX-101	✓	89.56	85.95	54.21	72.90	76.52	74.16	85.63	89.85	83.81	86.48	54.89	69.64	73.94	69.06	63.32	75.33
S^{2}A-Ne [9]	R-50	✓	89.07	82.22	53.63	69.88	80.94	82.12	88.72	90.73	83.77	86.92	63.78	67.86	76.51	73.03	56.60	76.38
RSDet-II [25]	R-152	✓	89.93	84.45	53.77	74.35	71.52	78.31	78.12	91.14	87.35	86.93	65.64	65.17	75.35	79.74	63.31	76.34
R^{3}Det [42]	R-152	✓	89.80	83.77	48.11	66.77	78.76	83.27	87.84	90.82	85.38	85.51	65.67	62.68	67.53	78.56	72.62	76.47
DAL [22]	R-50	✓	89.69	83.11	55.03	71.00	78.30	81.90	88.46	90.89	84.97	87.46	64.41	65.65	76.86	72.09	64.35	76.95
DCL [40]	R-152	✓	89.26	83.60	53.54	72.76	79.04	82.56	87.31	90.67	86.59	86.98	67.49	66.88	73.29	70.56	69.99	77.37
OSKDet [19]	R-101	✓	90.04	87.25	54.41	79.48	72.66	80.29	88.20	90.84	83.91	86.54	66.90	63.39	71.76	75.63	72.59	77.81
GWD [44]	R-152	✓	89.66	84.99	59.26	82.19	78.97	84.83	87.70	90.21	86.54	86.85	73.47	67.77	76.92	79.22	74.92	80.23
KLD [46]	R-152	✓	89.92	85.13	59.19	81.33	78.82	84.38	87.50	89.80	87.33	87.00	72.57	71.35	77.12	79.34	78.68	80.63
AARN(Ours)	ReR-50		89.18	84.31	52.65	76.04	78.22	84.29	88.29	91.87	86.82	86.85	63.97	65.09	74.64	70.33	67.36	77.32
AARN-MS(Ours)	ReR-50	✓	89.60	85.72	62.11	81.18	78.98	86.01	88.68	90.90	89.13	88.23	69.90	68.68	79.12	78.72	74.89	80.79

Table 2. Performances of AARN and state-of-the-art detectors on HRSC2016.

Method	RoI-Trans [5]	Gliding Vertex [38]	R^{3}Det [42]		CFC [20]	DAL [22]		GWD [44]
mAP(07)	86.20	88.20	89.26		89.70	89.77		89.85
Method	KLD [46]	S^{2}A-Net	Oriented RepPoints [17]	ReDet [10]	Oriented R-CNN [37]	AARN(Ours)
mAP(07)	89.97	90.17	90.38	90.46	90.50	**90.57**

angle cosine similarity. SkewIoU refers to the regular IoU calculation between skewed boxes. KFIoU refers to the Gaussian distribution overlapping method in this paper, which achieves the highest **77.32%** mAP. It shows that KFIoU similarity can better describe the affinity of objects.

Table 3. Performances of AARN and state-of-the-art detectors on UCAS-AOD.

Method	VOC2007			VOC2012		
	Car	Plane	mAP	Car	Plane	mAP
RIDet-O [21]	88.88	90.35	89.62	–	–	–
DAL [22]	89.25	90.49	89.87	–	–	–
R^{3}Det [42]	–	–	–	94.14	98.20	96.17
SCRDet++ [43]	–	–	–	94.97	98.93	96.95
OSKDet [19]	–	–	–	95.29	99.09	97.18
ReDet [10]	88.00	90.30	89.15	94.10	98.30	96.23
AARN(Ours)	**89.10**	**90.80**	**89.94**	96.30	98.60	**97.45**

Table 4. Ablation study for High-Quality Neighbour Selection (HQNS), Neighbour Message Learning (NML) and Feature Enhancement Weight Learning (FEWL) on DOTA-v1.0 test set.

Method	HQNS	NML	FEWL	AP$_{50}$	Individual improvement	Total improvement
ReDet	×	×	×	76.25	–	–
ReDet-AARN	✓	×	×	76.63	+0.38	+0.38
	✓	✓	×	77.04	+0.41	+0.79
	✓	×	✓	76.98	+0.35	+0.73
	✓	✓	✓	**77.32**	–	+1.07

High-Quality Neighbour Selection. We discuss the effects of different selection strategies and NMS thresholds on the performance of objects with different

(a)   (b)   (c)

Fig. 7. (a) ReDet (the basic detector) detection results. (b) ReDet with AARN detection results. (c) Visualization of graph construction module results.

Table 5. Performance of Affinity Matrix Calculation Module and comparisons of different methods to compute shape similarity.

Method	Feature Similarity	Shape Similarity			HQNS	NML	FEWL	AP_{50}
		Theta Similarity	SkewIoU Similarity	KFIoU Similarity				
ReDet	×	×			×	×	×	76.25
ReDet- AARN	✓	✓	×	×	✓	✓	✓	76.95
		×	✓	×				77.11
		×	×	✓				**77.32**
	✓	×						76.13
	×	✓						75.98

aspect ratios. We collect the aspect ratio distribution for each category in Fig. 1. As shown in Table 6, the dynamic selection strategy contributes to reducing the sensitivity of high aspect ratio objects, such as harbor (HA) and basketball-court (BC), to noisy neighbours. Furthermore, a lower neighbour NMS value, which corresponds to a more strict NMS strategy, aims to filter out more low-quality neighbour boxes with high aspect ratios.

Neighbour Message Learning. We compare the performance of three methods to learn message passing weight in Table 7. Gaussian refers to using a Gaussian distribution on similarity to model the edge weight as [23]. Affinity refers to aggregation with only affinity values. Attention refers to the aggregation with a multi-head attention, which achieves the highest **77.32%** mAP and **0.21%** mAP improvements than gaussian modeling. It shows that the attention module can better perceive neighbour features and represent neighbour messages.

Feature Enhancement Weight Learning. We compare the performance of different values of the two hyperparameters in Eq. (16). As shown in Table 8, the best performance is **77.32%** mAP when $\alpha = 2.0$ and $\beta = 3.5$. Especially for objects with low values and variances of aspect ratio such as roundabout (RA) and storage-tank (ST), the AP improvement is more obvious, with an increase of **1.12%** and **0.62%** AP compared to the situation without Feature Enhancement

Table 6. Comparisons of different selection strategies and effects of different NMS thresholds on detection performance.

Method	HQNS			NML	FE WL	$\frac{width}{height}\gg1$		$\frac{width}{height}\approx1$		AP_{50}
	Threshold		Neighbour NMS							
	Dynamic Threshold	Fixed Threshold				HA	BC	RA	BD	
ReDet	×	×	×	×	×	75.96	87.78	60.39	82.64	76.25
ReDet- AARN	×	0.5	×			71.45	83.72	62.89	83.22	75.97
	×	0.7	×			72.50	85.96	62.12	82.93	76.23
	×	0.9	×			73.69	86.24	61.37	82.76	76.31
	✓	×	×			**74.13**	**86.52**	**63.47**	**83.62**	**76.52**
	✓	×	0.5			74.40	86.60	63.63	83.78	76.56
	✓	×	0.3			74.36	86.69	63.65	83.82	76.58
	✓	×	0.1			**74.51**	**86.73**	**63.70**	**83.81**	**76.63**

Table 7. Comparisons of different methods learning neighbourhood message.

Method	HQNS	NML			FEWL	AP_{50}
		Gaussian	Affinity	Attention		
ReDet-AARN	✓	×	✓	×	✓	76.98
		✓	×	×		77.11
		×	×	✓		**77.32**

Table 8. The performances on different value of two hyperparameters in Eq. (16) and effects of Feature Enhancement Weight Learning (FEWL) on objects with low aspect ratios.

Method	HQNS	NML	FEWL		$\frac{width}{height}\approx1$ & std($\frac{width}{height}$) is low			AP_{50}
			Dynamic Feature Enhancement					
			α	β	RA	BD	ST	
ReDet-AARN	✓	✓	×		63.97	83.87	86.23	77.04
			2.0	4	64.93	84.35	86.73	77.27
				3.5	**65.09**	**84.31**	**86.85**	**77.32**
				3.0	65.00	84.04	86.94	77.23
			1.5	3.5	65.11	83.99	86.51	77.13
			2.5		65.07	84.29	86.32	77.09

Weight Learning. It shows this dynamic feature enhancement strategy is especially effective for objects with low aspect ratios and little semantic information.

5   Conclusions

In this paper, we propose an Affinity-Aware Relation Network, using the affinity among densely packed oriented objects, which consists of two parts: an affinity-graph construction module selecting bounding boxes sharing high similarity with the reference box, and a dynamic enhancement module using the attention module to learn neighbourhood message and dynamically determining the weight for feature enhancement. We conduct experiments on several public benchmarks and achieve the state-of-the-art performance.

Acknowledgements. This work is supported by the Natural Resources Science and Technology Project of Anhui Province (Grant No. 2021-K-14).

References

1. Bruna, J., Zaremba, W., Szlam, A., LeCun, Y.: Spectral networks and locally connected networks on graphs. arXiv preprint arXiv:1312.6203 (2013)
2. Chen, T., Saxena, S., Li, L., Fleet, D.J., Hinton, G.: Pix2seq: a language modeling framework for object detection. arXiv preprint arXiv:2109.10852 (2021)
3. Chen, X., Gupta, A.: An implementation of faster RCNN with study for region sampling. arXiv preprint arXiv:1702.02138 (2017)
4. Defferrard, M., Bresson, X., Vandergheynst, P.: Convolutional neural networks on graphs with fast localized spectral filtering. In: Advances in Neural information Processing Systems, vol. 29 (2016)
5. Ding, J., Xue, N., Long, Y., Xia, G.S., Lu, Q.: Learning ROI transformer for oriented object detection in aerial images. In: Proceedings of the IEEE/CVF Conference on Computer Vision and Pattern Recognition, pp. 2849–2858 (2019)
6. Gilmer, J., Schoenholz, S.S., Riley, P.F., Vinyals, O., Dahl, G.E.: Neural message passing for quantum chemistry. In: International Conference on Machine Learning, pp. 1263–1272. PMLR (2017)
7. Guo, Z., Liu, C., Zhang, X., Jiao, J., Ji, X., Ye, Q.: Beyond bounding-box: convex-hull feature adaptation for oriented and densely packed object detection. In: Proceedings of the IEEE/CVF Conference on Computer Vision and Pattern Recognition, pp. 8792–8801 (2021)
8. Hamilton, W., Ying, Z., Leskovec, J.: Inductive representation learning on large graphs. In: Advances in Neural Information Processing Systems, vol. 30 (2017)
9. Han, J., Ding, J., Li, J., Xia, G.S.: Align deep features for oriented object detection. IEEE Trans. Geosci. Remote Sens. **60**, 1–11 (2021)
10. Han, J., Ding, J., Xue, N., Xia, G.S.: Redet: a rotation-equivariant detector for aerial object detection. In: Proceedings of the IEEE/CVF Conference on Computer Vision and Pattern Recognition, pp. 2786–2795 (2021)
11. He, K., Gkioxari, G., Dollár, P., Girshick, R.: Mask R-CNN. In: Proceedings of the IEEE International Conference on Computer Vision, pp. 2961–2969 (2017)
12. Hechtlinger, Y., Chakravarti, P., Qin, J.: A generalization of convolutional neural networks to graph-structured data. arXiv preprint arXiv:1704.08165 (2017)
13. Hou, L., Lu, K., Xue, J., Li, Y.: Shape-adaptive selection and measurement for oriented object detection. In: Proceedings of the AAAI Conference on Artificial Intelligence (2022)
14. Huang, Z., Li, W., Xia, X.G., Tao, R.: A general gaussian heatmap label assignment for arbitrary-oriented object detection. IEEE Trans. Image Process. **31**, 1895–1910 (2022)
15. Kipf, T.N., Welling, M.: Semi-supervised classification with graph convolutional networks. arXiv preprint arXiv:1609.02907 (2016)
16. Klicpera, J., Bojchevski, A., Günnemann, S.: Predict then propagate: graph neural networks meet personalized pagerank. arXiv preprint arXiv:1810.05997 (2018)

17. Li, W., Chen, Y., Hu, K., Zhu, J.: Oriented reppoints for aerial object detection. In: Proceedings of the IEEE/CVF Conference on Computer Vision and Pattern Recognition, pp. 1829–1838 (2022)

18. Liu, Z., Yuan, L., Weng, L., Yang, Y.: A high resolution optical satellite image dataset for ship recognition and some new baselines. In: International Conference on Pattern Recognition Applications and Methods, vol. 2, pp. 324–331. SciTePress (2017)

19. Lu, D.: Oskdet: Towards orientation-sensitive keypoint localization for rotated object detection. arXiv preprint arXiv:2104.08697 (2021)

20. Ming, Q., Miao, L., Zhou, Z., Dong, Y.: CFC-net: a critical feature capturing network for arbitrary-oriented object detection in remote-sensing images. IEEE Trans. Geosci. Remote Sensing 60, 1–14 (2021)

21. Ming, Q., Miao, L., Zhou, Z., Yang, X., Dong, Y.: Optimization for arbitrary-oriented object detection via representation invariance loss. IEEE Geosci. Remote Sens. Lett. 19, 1–5 (2021)

22. Ming, Q., Zhou, Z., Miao, L., Zhang, H., Li, L.: Dynamic anchor learning for arbitrary-oriented object detection. In: Proceedings of the AAAI Conference on Artificial Intelligence, vol. 35, pp. 2355–2363 (2021)

23. Monti, F., Boscaini, D., Masci, J., Rodola, E., Svoboda, J., Bronstein, M.M.: Geometric deep learning on graphs and manifolds using mixture model CNNs. In: Proceedings of the IEEE Conference on Computer Vision and Pattern Recognition, pp. 5115–5124 (2017)

24. Pan, X., et al.: Dynamic refinement network for oriented and densely packed object detection. In: Proceedings of the IEEE/CVF Conference on Computer Vision and Pattern Recognition, pp. 11207–11216 (2020)

25. Qian, W., Yang, X., Peng, S., Yan, J., Guo, Y.: Learning modulated loss for rotated object detection. In: Proceedings of the AAAI Conference on Artificial Intelligence, vol. 35, pp. 2458–2466 (2021)

26. Redmon, J., Farhadi, A.: Yolov3: an incremental improvement. arXiv preprint arXiv:1804.02767 (2018)

27. Russakovsky, O., et al.: Imagenet large scale visual recognition challenge. Int. J. Comput. Vision 115(3), 211–252 (2015)

28. Scarselli, F., Gori, M., Tsoi, A.C., Hagenbuchner, M., Monfardini, G.: The graph neural network model. IEEE Trans. Neural Networks 20(1), 61–80 (2008)

29. Tian, Z., Shen, C., Chen, H., He, T.: FCOS: fully convolutional one-stage object detection. In: Proceedings of the IEEE/CVF International Conference on Computer Vision, pp. 9627–9636 (2019)

30. Vaswani, A., et al.: Attention is all you need. In: Advances in Neural Information Processing Systems, vol. 30 (2017)

31. Veličković, P., Cucurull, G., Casanova, A., Romero, A., Lio, P., Bengio, Y.: Graph attention networks. arXiv preprint arXiv:1710.10903 (2017)

32. Wang, J., Ding, J., Guo, H., Cheng, W., Pan, T., Yang, W.: Mask OBB: a semantic attention-based mask oriented bounding box representation for multi-category object detection in aerial images. Remote Sens. 11(24), 2930 (2019)

33. Wang, J., Yang, W., Li, H.C., Zhang, H., Xia, G.S.: Learning center probability map for detecting objects in aerial images. IEEE Trans. Geosci. Remote Sens. 59(5), 4307–4323 (2020)

34. Wei, H., Zhang, Y., Zhonghan, C., Li, H., Wang, H., Sun, X.: Oriented objects as pairs of middle lines. ISPRS J. Photogram. Remote Sens. 169, 268–279 (2020). https://doi.org/10.1016/j.isprsjprs.2020.09.022

35. Wu, F., Souza, A., Zhang, T., Fifty, C., Yu, T., Weinberger, K.: Simplifying graph convolutional networks. In: International Conference on Machine Learning, pp. 6861–6871. PMLR (2019)
36. Xia, G.S., et al.: DOTA: a large-scale dataset for object detection in aerial images. In: Proceedings of the IEEE Conference on Computer Vision and Pattern Recognition, pp. 3974–3983 (2018)
37. Xie, X., Cheng, G., Wang, J., Yao, X., Han, J.: Oriented R-CNN for object detection. In: Proceedings of the IEEE/CVF International Conference on Computer Vision, pp. 3520–3529 (2021)
38. Xu, Y., et al.: Gliding vertex on the horizontal bounding box for multi-oriented object detection. IEEE Trans. Pattern Anal. Mach. Intell. **43**(4), 1452–1459 (2020)
39. Yan, S., Xiong, Y., Lin, D.: Spatial temporal graph convolutional networks for skeleton-based action recognition. In: Thirty-Second AAAI Conference on Artificial Intelligence (2018)
40. Yang, X., Hou, L., Zhou, Y., Wang, W., Yan, J.: Dense label encoding for boundary discontinuity free rotation detection. In: Proceedings of the IEEE/CVF Conference on Computer Vision and Pattern Recognition, pp. 15819–15829 (2021)
41. Yang, X., Yan, J.: Arbitrary-oriented object detection with circular smooth label. In: Vedaldi, A., Bischof, H., Brox, T., Frahm, J.-M. (eds.) ECCV 2020. LNCS, vol. 12353, pp. 677–694. Springer, Cham (2020). https://doi.org/10.1007/978-3-030-58598-3_40
42. Yang, X., Yan, J., Feng, Z., He, T.: R3det: refined single-stage detector with feature refinement for rotating object. In: Proceedings of the AAAI Conference on Artificial Intelligence, vol. 35, pp. 3163–3171 (2021)
43. Yang, X., Yan, J., Liao, W., Yang, X., Tang, J., He, T.: Scrdet++: detecting small, cluttered and rotated objects via instance-level feature denoising and rotation loss smoothing. IEEE Trans. Pattern Anal. Mach. Intell. **45**, 2384–2399 (2022)
44. Yang, X., Yan, J., Ming, Q., Wang, W., Zhang, X., Tian, Q.: Rethinking rotated object detection with gaussian wasserstein distance loss. In: International Conference on Machine Learning, pp. 11830–11841. PMLR (2021)
45. Yang, X., et al.: SCRDET: towards more robust detection for small, cluttered and rotated objects. In: Proceedings of the IEEE/CVF International Conference on Computer Vision, pp. 8232–8241 (2019)
46. Yang, X., et al.: Learning high-precision bounding box for rotated object detection via kullback-leibler divergence. In: Advances in Neural Information Processing Systems, vol. 34 (2021)
47. Yang, X., et al.: The kfiou loss for rotated object detection. arXiv preprint arXiv:2201.12558 (2022)
48. Yang, Y., Chen, J., Zhong, X., Deng, Y.: Polygon-to-polygon distance loss for rotated object detection. In: AAAI Conference on Artificial Intelligence (2022)
49. Yi, J., Wu, P., Liu, B., Huang, Q., Qu, H., Metaxas, D.: Oriented object detection in aerial images with box boundary-aware vectors. In: Proceedings of the IEEE/CVF Winter Conference on Applications of Computer Vision, pp. 2150–2159 (2021)
50. Zhang, S., Chi, C., Yao, Y., Lei, Z., Li, S.Z.: Bridging the gap between anchor-based and anchor-free detection via adaptive training sample selection. In: Proceedings of the IEEE/CVF Conference on Computer Vision and Pattern Recognition, pp. 9759–9768 (2020)
51. Zhao, P., Qu, Z., Bu, Y., Tan, W., Guan, Q.: Polardet: a fast, more precise detector for rotated target in aerial images. Int. J. Remote Sens. **42**(15), 5831–5861 (2021)

52. Zhou, L., Wei, H., Li, H., Zhao, W., Zhang, Y., Zhang, Y.: Arbitrary-oriented object detection in remote sensing images based on polar coordinates. IEEE Access **8**, 223373–223384 (2020)
53. Zhu, H., Chen, X., Dai, W., Fu, K., Ye, Q., Jiao, J.: Orientation robust object detection in aerial images using deep convolutional neural network. In: 2015 IEEE International Conference on Image Processing (ICIP), pp. 3735–3739 (2015)
54. Zhu, X., Su, W., Lu, L., Li, B., Wang, X., Dai, J.: Deformable detr: deformable transformers for end-to-end object detection. arXiv preprint arXiv:2010.04159 (2020)

A Simple Strategy to Provable Invariance via Orbit Mapping

Kanchana Vaishnavi Gandikota[1]([envelope]), Jonas Geiping[2], Zorah Lähner[1],
Adam Czapliński[1], and Michael Möller[1]

[1] University of Siegen, Siegen, Germany
kanchana.Gandikota@uni-siegen.de
[2] University of Maryland, Maryland, USA

Abstract. Many applications require robustness, or ideally invariance, of neural networks to certain transformations of input data. Most commonly, this requirement is addressed by training data augmentation, using adversarial training, or defining network architectures that include the desired invariance by design. In this work, we propose a method to make network architectures provably invariant with respect to group actions by choosing one element from a (possibly continuous) orbit based on a fixed criterion. In a nutshell, we intend to 'undo' any possible transformation before feeding the data into the actual network. Further, we empirically analyze the properties of different approaches which incorporate invariance via training or architecture, and demonstrate the advantages of our method in terms of robustness and computational efficiency. In particular, we investigate the robustness with respect to rotations of images (which can hold up to discretization artifacts) as well as the provable orientation and scaling invariance of 3D point cloud classification.

1 Introduction

Deep neural networks have revolutionized the field of computer vision over the past decade. Yet, deep networks trained in a straight-forward way often lack desired robustness. In image classification, for instance, rotational, scale, and shift invariance are often highly desirable properties. While training deep networks with millions of realistic images in datasets like Imagenet [1] confers some degree of in/equi-variance [2–4], these properties however, cannot be guaranteed. On the contrary, networks are susceptible to adversarial attacks with respect to these transformations (see e.g. [5–8]), and small perturbations can significantly affect their predictions. To counteract this behavior, the two major directions of research are to either modify the training procedure or the network architecture. Modifications of the training procedure replace the common training of a network $\mathcal{G}$ with parameters θ on training examples (x^i, y^i) via a loss function $\mathcal{L}$,

$$\min_{\theta} \sum_{\text{examples } i} \mathcal{L}(\mathcal{G}(x^i; \theta); y^i), \tag{1}$$

Supplementary Information The online version contains supplementary material available at https://doi.org/10.1007/978-3-031-26348-4_23.

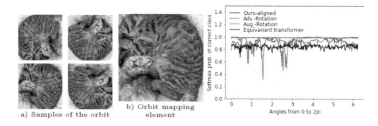

a) Samples of the orbit   b) Orbit mapping element

Fig. 1. (Left) Picture of a cat in 4 different rotation samples from the continuous orbit of rotations. Our orbit mapping selects the element with mean gradient direction (marked in red) along circle pointing upwards. (Right) Softmax probabilities of the true label when rotating an image by $0° − 360°$. Our method (in blue) is *robust for any angle*, which cannot be guaranteed through data augmentations (green) or adv. training (red). (Color Figure Online)

with a loss function that considers all perturbations in a given set S of transformations to be invariant towards. The most common choices are taking the mean loss of all predictions $\{\mathcal{G}(g(x^i); \theta) \mid g \in S\}$ (training with *data augmentation*), or the maximum loss among all predictions (*adversarial training*). However, such training schemes cannot guarantee provable invariance. In particular, training with data augmentation is far from being robust to transformations as illustrated in Fig. 1. The plot shows the softmax probabilities of the true label when feeding the exemplary image at rotations ranging from 0 to 2π into a network trained with rotational augmentation (green), adversarial training (red) and undoing rotations using a learned network (black). As we can see, rotational data augmentation is not sufficient to truly make a classification network robust towards rotations, and even the significantly more expensive adversarial training shows instabilities.

While modifications of the training scheme remain the best option for complex or hard-to-characterize transformations, more structured transformations, e.g., those arising from a group action, allow modifications to the network architecture to yield provable invariance. As opposed to previous works that largely rely on the ability to enlist all transformations of an input x (i.e., assume a finite *orbit*), we propose to make neural networks invariant by selecting a specific element from a (possibly infinite) orbit generated by a group action, through an application-specific *orbit mapping*. Simply put, we undo and fix the transformation or pose. Our proposed approach is significantly easier to train than adversarial training methods while being at least equally performant, robust, and computationally cheaper. We illustrate these findings on the rotation invariant classification of images (on which discretization artifacts from the interpolation after any rotation play a crucial role) as well as on the scale, rotation, and translation invariant classification of 3D point clouds. Our contributions can be summarized as follows:

- We present *orbit mapping*, a simple way to adapt neural networks to be in-(or equi)variant to transformations from sets S associated with a group action.
- We propose a gradient based orbit mapping strategy for image rotations, which can provably select unique orientation for continuous image models.
- Our proposed orbit mapping improves robustness of standard networks to transformations even *without* additional changes in training or architecture.

- Existing invariant approaches also demonstrate gain in robustness to discrete image rotations when combined with orbit mapping.
- We demonstrate orbit mappings to provable scale and orientation invariant 3D point cloud classification using well known scale normalization and PCA.

2 Related Work

Several approaches have been developed in the literature to encourage models to exhibit invariance or robustness to desired transformations of data. These include i) data augmentation using desired transformations, ii) regularization to encourage network output to be robust to transformations on the input [9], iii) adversarial training [10,11] and regularization [12], iv) unsupervised or self-supervised pretraining to learn transformation robust representations [13–17], v) parameterized learning of augmentations to learn invariances from training data[18,19], vi) use of hand-crafted invariant shallow [20–24] or deep [25–27] features for downstream classification tasks vii) incorporating desired invariance properties in to the network design [28–32], and viii) train time/test time data transformation. Recent works [33,34] have also explored certifying geometric robustness of networks. The approaches i)-v) can improve robustness but cannot yield provable invariance to transformations. Hand-crafting features can yield desired invariance, but is difficult and often sacrifices accuracy. Provable invariance to a finite number of transformations is achievable by applying all such transformations to the each input data point and pooling the corresponding features [35,36]. While this strategy can even be applied only during test time, it can not be extended to sets with infinitely many transformations. Recent approaches [28,30,37] incorporate in-/equivariances when the desired transformations of the data can be formulated as a group action, e.g. enforcing equivariance in each layer separately. Layer wise approaches for equivariance to finite groups such as [28] typically use all possible transformations at each layer.

Canonicalization. Closely related to our approach are methods which align input to a normalized or canonical pose. The use of PCA or scale renormalization are well known approaches to normalizing point clouds. However, PCA-based pose canonicalization is known to suffer from ambiguities, and learning based approaches [32,38,39] have been proposed for disambiguation. Several recent works directly leverage deep learning for 3d pose canonicalization, for example training with ground truth poses [40,41] or self-supervised learning [42–44]. For 2D images, PCA-based canonicalization is possible only with binary images [45]; the use of Radon transformations [46] requires an expensive, fine discretization of continuous rotations. The use of spatial transformer networks [47] is an alternate learning based approach to 2D/3D pose normalization which can be used along with an application-dependent coordinate transformation [48,49]. Such learning-based approaches, however, require additional training with data augmentation and cannot guarantee invariance. Since our orbit mappings essentially select a canonical group orbit element, our work can be interpreted as a formalization of canonicalization for group transformations. In contrast to learning based approaches, we select a canonical element from the orbit using simple analytical solutions, which can improve robustness even without data augmentations.

Provable Rotational In-/Equivariance in 2D. Several works [26–28,50–52] have considered layer wise equivariance to discrete rotations using multiple rotated versions of filters at each layer, which was formalized using group convolutions in [28]. While [28,50–52] learn these filters by training, [26,27] make use of rotated and scaled copies of fixed wavelet filters at each layer. For equivariance to continuous rotations, Worrall et al. [29] utilize circular harmonic filters at each layer. All these layer wise approaches for group equivariance in images were unified in a single framework in [30]. Instead of layer-wise approaches, [36,53,54] pool the features of multiple rotated copies of images input to the network.

Rotation Invariance in 3D. Due to the different representations of 3D data (e.g. voxels, point clouds, meshes), many strategies exist. Some techniques for image invariances can be adapted to voxel representations, e.g. probing several rotations at test time [55,56], use of rotationally equivariant convolution kernels [57–59]. Spatial transformers have also been used to learn 3D pose normalization, e.g. in the classical PointNet architecture [60], and its extension PointNet++ [61] which additionally considers hierarchical and neighborhood information. While point clouds do not suffer from discretization artifacts after rotations, they struggle with less clear neighborhood information due to unordered coordinate lists. [62] solve this by adding hierarchical graph connections to point clouds and using graph convolutions. However, the features learned using graph convolutions still depend on the rotation of the input data. [63,64] propose graph convolution networks equivariant to isometric transformations. [65,66] project point clouds onto 2D sphere and employ spherical convolutions to achieve rotational equivariance. [67] and [68] achieve rotation invariance on point clouds by considering pairs of features in the tangent plane of each point. While local operations and convolutions on the surface of triangular meshes are invariant to global rotations by definition [69], they however do not capture global information. MeshCNN [70] addresses this by adding pooling operations through edge collapse. [71] defines a representation independent network structure based on heat diffusion which can balance between local and global information.

3   Proposed Approach

Our idea is straightforward. We make neural networks invariant by consistently selecting a fixed element from the orbit of group transformations, i.e., we modify the input pose such that every element from the orbit of transformations maps to the same canonical element. For example, different rotated versions of an image are mapped to have the same orientation as visualized in Fig. 2. In conjunction with such *orbit mapping*, any standard network architecture can achieve provable invariance. In the following, we formalize our approach to achieve invariance.

3.1   Invariant Networks W.r.t. Group Actions

We consider a network $\mathcal{G}$ to be a function $\mathcal{G} : \mathcal{X} \times \mathbb{R}^p \to \mathcal{Y}$ that maps data $x \in \mathcal{X}$ from some suitable input space $\mathcal{X}$ to some prediction $\mathcal{G}(x; \theta) \in \mathcal{Y}$ in an output space $\mathcal{Y}$ where the way this mapping is performed depends on parameters

$\theta \in \mathbb{R}^p$. The question is how, for a given set $S \subset \{g : \mathcal{X} \to \mathcal{X}\}$ of transformations of the input data, we can achieve the *invariance* of $\mathcal{G}$ to S defined as

$$\mathcal{G}(g(x); \theta) = \mathcal{G}(x; \theta) \quad \forall x \in \mathcal{X}, \ g \in S, \ \theta \in \mathbb{R}^p. \tag{2}$$

The invariance of a network with respect to transformations in S is of particular interest when S induces a *group action*[1] on $\mathcal{X}$, which is what we will assume about S for the remainder of this paper. Of particular importance for the construction of invariant networks, is the set of all possible transformations of input data x,

$$S \cdot x = \{g(x) \mid g \in S\}, \tag{3}$$

which is called the *orbit of* x. A basic observation for constructing invariant networks is that any network acting on the orbit of the input is automatically invariant to transformations in S:

Fact 1. Characterization of Invariant Functions via the Orbit: *Let S define a group action on $\mathcal{X}$. A network $\mathcal{G} : \mathcal{X} \times \mathbb{R}^p \to \mathcal{Y}$ is invariant under the group action of S if and only if it can be written as $\mathcal{G}(x; \theta) = \mathcal{G}_1(S \cdot x; \theta)$ for some other network $\mathcal{G}_1 : 2^{\mathcal{X}} \times \mathbb{R}^p \to \mathcal{Y}$.*

The above observation is based on the fact that $S \cdot x = S \cdot g(x)$ holds for any $g \in S$, provided that S is a group. Although not taking the general perspective of Fact 1, approaches, like [36], which integrate (or sum over finite elements of) the mappings of $\mathcal{G}$ over a (discrete) group can be interpreted as instances of Fact 1 where $\mathcal{G}_1$ corresponds to the summation. Similar strategies of applying all transformations in S to the input x can be pursued for the design of equivariant networks, see supplementary material.

3.2   Orbit Mappings

While Fact 1 is stated for general (even infinite) groups, realizations of such constructions from the literature often assume a finite orbit. In this work we would like to include an efficient solution even for cases in which the orbit is not finite, and utilize Fact 1 in the most straight-forward way: We propose to construct provably invariant networks $\mathcal{G}(x; \theta) = \mathcal{G}_1(S \cdot x; \theta)$ by simply using an

$$\text{orbit mapping } h : \{S \cdot x \mid x \in \mathcal{X}\} \to \mathcal{X},$$

which uniquely selects a particular element from an orbit as a first layer in $\mathcal{G}_1$. Subsequently, we can proceed with any standard network architecture and Fact 1 still guarantees the desired invariance. A key in designing instances of orbit mappings is that they should not require enlisting all elements of $S \cdot x$ in order to evaluate $h(S \cdot x)$. Let us provide more concrete examples of orbit mappings.

Example 1 (Mean-subtraction). A common approach in data classification tasks is to first normalize the input by subtracting its mean. Considering $\mathcal{X} = \mathbb{R}^n$ and $S = \{g : \mathbb{R}^n \to \mathbb{R}^n \mid g(x) = x + a\mathbb{1}, \text{ for some } a \in \mathbb{R}\}$, with $\mathbb{1} \in \mathbb{R}^n$ being a vector of all ones, input-mean-subtraction is an orbit mapping that selects the unique element from any $S \cdot x$ which has zero mean.

---

[1] A (left) group action of a group S with the identity element e, on a set X is a map $\sigma : S \times X \to X$, that satisfies i) $\sigma(e, x) = x$ and ii) $\sigma(g, \sigma(h, x)) = \sigma(gh, x)$, $\forall g, h \in S$ and $\forall x \in X$. When the action being considered is clear from the context, we write $g(x)$ instead of $\sigma(g, x)$.

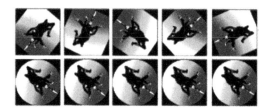

Fig. 2. Images of different orientations (top) are consistently aligned with the proposed gradient-based orbit mapping (bottom).

Example 2 (Permutation invariance via sorting). Consider $\mathcal{X} = \mathbb{R}^n$, and S to be all permutations of vectors in $\mathbb{R}^n$, i.e., $S = \{s \in \{0,1\}^{n \times n} \mid \sum_i s_{i,j} = 1 \, \forall j, \ \sum_j s_{i,j} = 1 \, \forall i\}$. We could define an orbit mapping that selects the element from an orbit whose entries are sorted by magnitude in an ascending order.

With the very natural condition that orbit mappings really select an element from the orbit, i.e., $h(S \cdot x) \in S \cdot x$, we can readily construct equivariant networks by applying the inverse mapping, see supplementary material. In our Example 2, undoing the sort operation at the end of the network allows to transfer from an invariant, to an equivariant network.

As a final note, our concept of orbit mappings can further be generalized by h not mapping to the input space $\mathcal{X}$, but to a different representation, which can be beneficial for particular, complex groups. In geometry processing, for instance, an important group action are isometric deformations of shapes. A common strategy to handle these (c.f. [72]) is to identify any shape with the eigenfunctions of its Laplace-Beltrami operator [73], which represents a natural (generalized) orbit mapping. We refer to [74–76] for exemplary deep learning applications.

4   Applications

We will now present two specific instances of orbit mappings for handling continuous rotations of images as well as for invariances in 3D point cloud classification.

4.1   Invariance to Continous Image Rotations

Images as Functions. Let us consider the important example of invariance to continuous rotations of images. To do so, consider $\mathcal{X} \subset \{u : \Omega \subset \mathbb{R}^2 \to \mathbb{R}\}$ to represent images as functions. For the sake of simplicity, we consider grayscale images only, but this extends to color images in a straight-forward way. In our notation $z \in \mathbb{R}^2$ represents spatial coordinates of an image (to avoid an overlap with our previous $x \in \mathcal{X}$, which we used for the input of a network). We set

$$S = \{g : \mathcal{X} \to \mathcal{X} \mid g \circ u(z) = u(r(\alpha)z), \text{ for } \alpha \in \mathbb{R}\},$$
$$\text{and } r(\alpha) = \begin{pmatrix} \cos(\alpha) & -\sin(\alpha) \\ \sin(\alpha) & \cos(\alpha) \end{pmatrix}. \tag{4}$$

As S has infinitely many elements, approaches that worked well for rotations by 90 degrees like [28] are not applicable anymore. We instead propose to uniquely select an element from the continuous orbit of rotation $g \in S$ by choosing a rotation that makes the average gradient of the image $\int_Z \nabla(g \circ u)(z) \, dz$ over a suitable set Z, e.g. a circle around the image center point upwards. It holds that

$$\nabla(g \circ u)(z) = r^T(\alpha)\nabla u\,(r(\alpha)z) \text{ such that}$$

$$\int_Z \nabla(g \circ u)(z)dz = \int_Z r^T(\alpha)\nabla u\,(r(\alpha)z) \, dz.$$

Substituting $\varphi = r(\alpha)z$, we obtain

$$\int_Z r^T(\alpha)\nabla u\,(r(\alpha)z) \, dz = \int_{r^T(\alpha)Z} r^T(\alpha)\nabla u\,(\varphi) \, d\varphi = r^T(\alpha) \int_Z \nabla u\,(\varphi) \, d\varphi \quad (5)$$

where we used that Z is rotationally invariant. Thus, choosing a rotation that makes $\int_Z \nabla(g \circ u)(z) \, dz$ point upwards is equivalent to solving

$$r(\hat{\alpha}) = \arg\max_{r(\alpha)} \left\langle \begin{pmatrix} 1 \\ 0 \end{pmatrix}, r^T(\alpha) \int_Z \nabla u(\varphi) \, d\varphi \right\rangle \quad (6)$$

whose solution is given by $\hat{\alpha}$ such that

$$\begin{pmatrix} \cos\hat{\alpha} \\ \sin\hat{\alpha} \end{pmatrix} = \begin{pmatrix} \dfrac{\int_Z \nabla u(z) \, dz}{\| \int_Z \nabla u(z) \, dz\|} \end{pmatrix}. \quad (7)$$

Note that (7) yields unique solution to the maximization problem. Since a consistent pose is always selected[2], it is an invariant mapping. When $\int_Z \nabla u(z) \, dz = 0$, any $g \in S$ maximizes (6). However, numerically $\int_Z \nabla u(z) \, dz$ rarely evaluates to exact zero and its magnitude of determines the stability of orbit mapping.

Discretization. For a discrete (grayscale) image given a matrix $\tilde{u} \in \mathbb{R}^{n_y \times n_x}$, we first apply Gaussian blur with a standard deviation of $\sigma = 1.5$ (to reduce the effect of noise and create a smooth image), and subsequently construct an underlying continuous function $u : \Omega \subset \mathbb{R}^2 \to \mathbb{R}$ by bilinear interpolation. For the set Z we choose two circles of radii 0.05 and 0.4 (for Ω being normalized to $[0,1]^2$). We approximate the integral by a sum over finite evaluations of the derivative along each circle, using exact differentiation of the continuous image model. This strategy can stabilize arbitrary rotations successfully as illustrated in Fig. 2. However, in practice, the magnitude of $\int_Z \nabla u(z) \, dz$ and interpolation artifacts affect the stability of the orbit mapping. We analyze the stability of the proposed gradient based orbit-mapping for discrete images in Sect. 3 of the supplementary, where we observe that use of forward or central differences to approximate gradients further deteriorates the stability of orbit mapping. Since the orbit mapping for discrete images has instabilities, exact invariance to rotations cannot be guaranteed. Even when the integral values are large leading to

---

[2] Note that $r^T(\alpha) = r(-\alpha)$, therefore if the predicted rotation for $u(z)$ is β, then for $u(r(\gamma)z)$, it is $\beta - \gamma$, i.e. the same element is consistently selected.

Table 1. Comparison of orbit mapping *(OM)* with training and architecture based methods. Robustness to rotations is compared using the average and worst case accuracies over 5 runs with test images rotated in steps of $1°$ using bilinear interpolation.

Method	OM(Ours)	CIFAR10			HAM10000			CUB200		
		Clean	Avg.	Worst	Clean	Avg.	Worst	Clean	Avg.	Worst
Std.	✗	**93.98**	40.06	1.31	93.82	91.73	82.52	**77.41**	53.45	8.07
	✓ Train+Test	87.99	84.12	68.60	93.31	91.38	87.96	71.19	71.56	58.80
RA	✗	85.54	75.99	44.71	93.30	90.81	82.30	69.89	70.12	41.01
	✓ Train+Test	85.40	81.82	71.09	93.41	92.13	88.55	70.35	70.72	57.54
STN	✗	83.74	78.86	54.03	–	–	–	–	–	–
ETN	✗	84.39	80.30	64.08	92.47	90.85	84.32	64.14	66.95	52.85
Adv.	✗	69.32	68.54	50.21	92.28	91.87	85.04	64.54	64.07	42.82
Mixed	✗	91.15	68.37	17.15	93.71	92.13	84.53	68.56	65.91	42.87
Adv.-KL	✗	72.28	70.29	51.05	92.54	91.79	85.42	64.47	64.65	43.04
Adv.-ALP	✗	71.25	70.30	52.29	92.89	91.84	85.98	64.63	64.34	43.63
TIpool	✗	93.56	66.46	20.22	93.19	91.87	88.16	76.80	74.90	59.04
	✓ Train+Test	91.94	**88.77**	76.26	**93.83**	92.05	**89.81**	76.82	**77.18**	**69.19**
TIpool-RA	✗	91.40	84.65	67.28	93.39	91.87	88.12	73.47	74.71	62.82
	✓ Train+Test	90.47	87.92	**80.07**	93.68	**92.78**	89.30	74.78	75.89	67.78

a stable orbit mapping, our approach does not need to give the same rotation angle for semantically similar content, for example, different cars are not necessarily rotated to have the same orientation. Due to these reasons, our approach can further benefit from augmentation.

Experiments. To evaluate our approach, we use orbit mapping in conjunction with image classification networks on three datasets: On CIFAR10, we train a Resnet-18 [77] from scratch. On the HAM10000 skin image dataset [78], we finetune an NFNet-F0 network [79], and on CUB-200 [80] we finetune a Resnet-50 [77], both pretrained on ImageNet. While the datasets CIFAR10 and CUB-200 have an inherent variance in orientation, for the HAM10000 skin lesion classification, exact rotation invariance is desirable. Finally, we also perform experiments with RotMNIST using state of the art E2CNN network [30]. The details of the protocol used for training all our networks as well as some additional experiments are provided in the supplementary material. We compare with following approaches on CIFAR10, HAM10000, and CUB-200: *i) adversarial training:* $\min_\theta \sum_{\text{examples } i} \mathcal{L}(\mathcal{G}(\hat{x}^i; \theta); y^i)$, for $\hat{x}^i = \arg\max_{z \in S \cdot x^i} \mathcal{L}(\mathcal{G}(z); y^i)$. This is approximated by selecting the worst out of 10 different random rotations for each image in every iteration, following [10]. It is referred to as Adv. in Table 1. *ii) mixed mode training:* $\min_\theta \sum_{\text{examples } i} \mathcal{L}(\mathcal{G}(\hat{x}^i; \theta); y^i) + \mathcal{L}(\mathcal{G}(x^i; \theta); y^i)$ which uses both natural and adversarial examples $\hat{x}^i$. *iii) adversarial training with regularization:* Use of adversarial logit pairing and KL-divergence regularizers [12] along with adversarial training (indicated as Adv.-ALP and Adv.-KL in Table 1):

a) *adversarial logit pairing (ALP):* $R_{ALP}(\mathcal{G}, x^i, y^i) = \|\mathcal{G}(x^i; \theta) - \mathcal{G}(\hat{x}^i; \theta)\|_2^2$,
b) *KL-divergence:*$R_{KL}(\mathcal{G}, x^i, y^i) = D_{KL}(\mathcal{G}(x^i; \theta)\|\mathcal{G}(\hat{x}^i; \theta))$.

iv) transformation invariant pooling (TIpool): which is a provably invariant approach for discrete rotations [36], where the features of multiple rotated copies of input image are pooled before the final classification. We use 4 rotated copies

Table 2. Effect of augmentation on robustness to rotations with different interpolations. Shown are clean accuracy on standard CIFAR10 test set, average and worst-case accuracies on rotated test set with mean and standard deviations over 5 runs.

Train	OM	Clean	Average			Worst-case		
			Nearest	Bilinear	Bicubic	Nearest	Bilinear	Bicubic
Std.	✗	**93.98 ± 0.32**	35.12 ± 0.81	40.06 ± 0.44	42.81 ± 0.50	0.79 ± 0.38	1.31 ± 0.13	2.22 ± 0.17
	✓ Train+Test	87.99 ± 0.43	72.40 ± 0.33	84.12 ± 0.55	86.61 ± 0.49	34.57 ± 0.94	68.60 ± 0.81	74.49 ± 0.84
RA	✗	85.54 ± 0.72	80.47 ± 0.74	75.99 ± 0.72	79.47 ± 0.65	45.50 ± 0.83	44.71 ± 0.74	50.50 ± 0.78
	✓ Test	79.26 ± 0.42	74.93 ± 0.51	69.31 ± 0.65	73.94 ± 0.63	48.93 ± 0.75	52.18 ± 0.91	58.69 ± 0.78
	✓ Train+Test	85.40 ± 0.57	84.37 ± 0.58	81.82 ± 0.59	84.82 ± 0.52	66.22 ± 0.75	71.09 ± 1.01	76.44 ± 0.89
RA-combined	✗	92.42 ± 0.21	80.90 ± 0.64	82.23 ± 0.74	82.71 ± 0.69	36.98 ± 1.27	48.07 ± 1.66	49.51 ± 1.47
	✓ Test	82.55 ± 0.86	76.33 ± 0.95	77.93 ± 0.68	78.42 ± 0.64	45.44 ± 1.32	60.23 ± 1.24	62.18 ± 1.33
	✓ Train+Test	86.69 ± 0.12	84.06 ± 0.21	85.27 ± 0.23	86.06 ± 0.20	61.75 ± 0.76	75.29 ± 0.42	77.25 ± 0.27
Adv.	✗	69.32 ± 1.61	61.73 ± 1.12	68.54 ± 0.68	68.00 ± 0.31	36.95 ± 0.97	50.21 ± 0.55	49.73 ± 0.98
Mixed	✗	91.15 ± 0.15	54.55 ± 0.40	68.37 ± 0.66	68.48 ± 0.37	3.86 ± 0.13	17.15 ± 1.25	16.85 ± 0.93
Adv.-KL	✗	72.28 ± 2.05	62.60 ± 1.72	70.29 ± 1.42	69.84 ± 1.29	32.60 ± 0.74	51.05 ± 2.47	51.11 ± 1.03
Adv.-ALP	✗	71.25 ± 0.97	62.36 ± 2.19	70.30 ± 1.50	69.71 ± 1.22	33.98 ± 1.44	52.29 ± 1.76	52.57 ± 1.57
STN	✗	83.74 ± 0.50	81.94 ± 0.51	78.86 ± 0.73	82.21 ± 0.55	48.93 ± 1.36	54.03 ± 1.36	59.65 ± 1.31
ETN	✗	84.39 ± 0.09	82.98 ± 0.28	80.30 ± 0.55	83.31 ± 0.31	59.40 ± 0.76	64.08 ± 0.78	68.75 ± 0.83
Augerino	✗	83.68 ± 0.76	80.17 ± 0.70	82.27 ± 0.69	81.69 ± 0.72	52.44 ± 0.66	60.36 ± 1.00	60.63 ± 0.94
TIpool	✗	93.56 ± 0.25	55.96 ± 0.39	66.46 ± 1.36	70.70 ± 0.77	3.14 ± 1.09	20.22 ± 1.51	27.88 ± 1.09
TIpool-RA	✗	91.40 ± 0.17	87.50 ± 0.24	84.65 ± 0.51	87.31 ± 0.29	66.52 ± 1.31	67.28 ± 1.03	72.35 ± 0.83
TIpool	✓Train+Test	91.94 ± 0.38	78.66 ± 0.83	88.77 ± 0.51	**90.76 ± 0.40**	42.01 ± 1.07	76.26 ± 1.12	81.46 ± 1.02
TIpool-RA	✓Train+Test	90.47 ± 0.36	**89.37 ± 0.36**	87.92 ± 0.36	89.91 ± 0.34	**74.51 ± 0.79**	80.07 ± 0.69	83.76 ± 0.60
TIpool-RA Combined	✓Train+Test	91.09 ± 0.40	89.02 ± 0.30	**90.13 ± 0.34**	90.64 ± 0.30	70.18 ± 1.12	**82.71 ± 0.62**	84.26 ± 0.41

of images rotated in multiples of 90°C. *v) Spatial transformer networks (STN):* which learns to undo the transformation by training using appropriate data augmentation [47]. *vi) Equivariant transformer networks (ETN):* which additionally uses appropriate coordinate transformation along with a learned spatial transformer to undo the transformation [48]. We also compare with the simple baseline of augmenting with random rotations, referred to as RA in Table 1. Additionally, we also compare with [19], an approach which learns distribution of augmentations on the task of rotated CIFAR10 classification, referred to as Augerino in Table 2. We use 4 samples from the learned distribution of augmentations during both training and test. We would also like to point out that adversarial training using the worst of 10 samples roughly increases the training effort of the underlying model by a factor of 5.

Results. We measure the accuracy on the original testset(*Clean*), as well as the average (*Avg.*) and worst-case (*Worst*) accuracies in the orbit of rotations discretized in steps of 1°C, where '*Worst*' counts an image as misclassified as soon as there exists a rotation at which the network makes a wrong prediction.

As we can see in Table 1, networks trained without rotation augmentation perform poorly in terms of both, the average and worst-case accuracy if the data set contains an inherent orientation. While augmenting with rotations during training results in improvements, there is still a huge gap ($\sim$ 30% for CIFAR10 and CUB200) between the average and worst-case accuracies. While adversarial training approaches [10,12] improve the performance in the worst case, there is a clear drop in the clean and average accuracies when compared to data augmentation.

Learned approaches to correct orientation i.e. STN [47], ETN [48] show an improvement over adversarial training schemes in terms of average and worst case accuracies, when training from scratch, with ETN demonstrating even higher robustness than plain STNs. While pooling over features of rotated versions of image provides provable invariance to discrete rotations, this approach is still susceptible to continuous image rotations. The robustness of this approach to continuous rotations is boosted by rotation augmentation, with improvements over even learned transformers. Note that using TI-pooling with 4 rotated copies increases the computation by 4 times. In contrast, our orbit mapping effortlessly leads to significant improvements in robustness even without augmenting with rotations, with performance better than adversarial training, learned transformers and discrete invariance based approaches. Since our orbit mapping for discrete images has some instabilities, our approach also benefits from augmentation with image rotations. Further, when combined with discrete invariant approach [36], we obtain the best accuracies for average and worst case rotations.

Even when finetuning networks, we observe that orbit mapping readily improves robustness to rotations over standard training, even without the use of augmentations. Furthermore, combination of orbit mapping with the discrete invariant approach of pooling over rotated features yields the best performance. For the birds dataset with inherent orientation, undoing rotations using ETN significantly improves robustness when compared to adversarial training schemes, which only marginally improve robustness over rotation augmentation. We found it difficult to train STN with higher accuracies (*Clean/Avg./Worst*) than plain augmentation with rotated images for CUB200 and HAM10000, despite extensive hyperparameter optimization, therefore we do not report the numbers here[3]. When the data itself does not contain a prominent orientation as in the HAM10000 data set, the general trend in accuracies still holds (*Clean>Avg.>Worst*), but the drops in accuracies are not drastic, and adversarial training schemes provide improvements over undoing transormations using ETN. Further, orbit mapping and pooling over rotated images provide comparable improvements in robustness, with their combination achieving the best results.

Discretization Artifacts: It is interesting to see that while consistently selecting a single element from the continuous orbit of rotations leads to provable rotational invariance when considering images as continuous functions, discretization artifacts and boundary effects still play a crucial role in practice, and rotations cannot be fully stabilized. As a result, there is still discrepancy between the average and worst case accuracies, and the performance is further improved when our approach also uses rotation augmentation. Motivated by the strong effect the discretization seems to have, we investigate different interpolation schemes used to rotate the image in more detail: Table 2 shows the results different training schemes with and without our orbit mapping (*OM*) obtained with a ResNet-18 architecture on CIFAR-10 when using different types of interpolation. Besides standard training (*Std.*), we use rotation augmentation (*RA*) using the Pytorch-default of nearest-neighbor interpolation, a combined augmentation scheme (*RA-combined*)

---

[3] We use a single spatial transformer as opposed to multiple STNs used in [47] and train on randomly rotated images.

Table 3. Effect of orbit mapping and rotation augmentation on RotMNIST classification using regular D4/C4 and D16/C16 E2CNN models. Shown are clean accuracy on standard test set and average and worst-case accuracies on test set rotated in steps of 1 °C, with mean and standard deviations over 5 runs.

Train.	OM	D4/C4			D16/C16		
		Clean	Avg	Worst	Clean	Avg	Worst
Std.	✗	98.73 ± 0.04	98.61 ± 0.04	96.84 ± 0.08	99.16 ± 0.03	99.02 ± 0.04	98.19 ± 0.08
Std.	✓(Train+Test)	98.86 ± 0.02	98.74 ± 0.03	98.31 ± 0.05	99.21 ± 0.01	99.11 ± 0.03	98.82 ± 0.06
RA.	✗	99.19 ± 0.02	99.11 ± 0.01	98.39 ± 0.05	99.31 ± 0.02	99.27 ± 0.02	98.89 ± 0.03
RA	✓(Train+Test)	98.99 ± 0.03	98.90 ± 0.01	98.60 ± 0.02	99.28 ± 0.02	99.23 ± 0.01	99.04 ± 0.02

that applies random rotation only to a fraction of images in a batch using at least one nearest neighbor, one bilinear and one bicubic interpolation. The adversarial training and regularization from [10,12] are trained using bilinear interpolation (following the authors' implementation).

Results show that interpolation used in image rotation impacts accuracies in all the baselines. Most notably, the worst-case accuracies between different types of interpolation may differ by more than 20%, indicating a huge influence of the interpolation scheme. Adversarial training with bi-linear interpolation still leaves a large vulnerability to image rotations with nearest neighbor interpolation. Further, applying an orbit mapping at test time to a network trained with rotated images readily improves its worst case accuracy, however, there is a clear drop in clean and average case accuracies, possibly due to the network not having seen doubly interpolated images during training. While our approach without rotation augmentation is also vulnerable to interpolation effects, it is ameliorated when using orbit mapping along with rotation augmentation. We observe that including different augmentations (RA-combined) improves the robustness significantly. Combining the orbit mapping with the discrete invariant approach [36] boosts the robustness, with different augmentations further reducing the gap between clean, average case and worst case performance.

Experiments with RotMNIST. We investigate the effect of orbit mapping on RotMNIST classification with the state of the art network from [30] employing regular steerable equivariant models[81]. This model uses 16 rotations and flips of the learned filters (with flips being restricted till layer3). We also compare with a variation of the same architecture with 4 rotations. We refer to these models as D16/C16 and D4/C4 respectively. We train and evaluate these models using their publicly available code[4]. Results in Tab. 3 indicate that even for these state of the art models, there is a discrepancy between the accuracy on the standard test set and the worst case accuracies, and their robustness can be further improved by orbit mapping. Notably, orbit mapping significantly improves worst case accuracy (by around 1.5%) for D4/C4 steerable model trained without augmenting using rotations, showing gains in robustness even over naively trained D16/C16 model of much higher complexity. Training with augmentation leads to improvement in robustness, with orbit mapping providing gains further in robustness. However, artifacts due to double interpolation affect performance of orbit mapping.

---

[4] code url https://github.com/QUVA-Lab/e2cnn_experiments.

Table 4. Scaling invariance in 3D pointcloud classification with PointNet trained on modelnet40, with and without data augmentation, with and without STNs or scale normalization. Mean and standard deviations over 10 runs are reported.

Augment.	Unscaling	With STN			Without STN		
		Clean	Avg	Worst	Clean	Avg	Worst
[0.8, 1.25]	✗	86.15 ± 0.52	24.40 ± 1.56	0.01 ± 0.02	85.31 ± 0.39	33.57 ± 2.00	2.37 ± 0.06
[0.8, 1.25]	✓(Train+Test)	**86.15 ± 0.28**	**86.15 ± 0.28**	**86.15 ± 0.28**	85.25 ± 0.43	85.25 ± 0.43	85.25 ± 0.43
[0.8, 1.25]	✓(Test)	86.15 ± 0.52	85.59 ± 0.79	85.59 ± 0.79	85.31 ± 0.39	83.76 ± 0.35	83.76 ± 0.35
[0.1, 10]	✗	85.40 ± 0.46	47.25 ± 1.36	0.04 ± 0.05	75.34 ± 0.84	47.58 ± 1.69	1.06 ± 0.87
[0.1, 10]	✓(Test)	85.40 ± 0.46	85.85 ± 0.73	85.85 ± 0.73	75.34 ± 0.84	81.45 ± 0.56	81.45 ± 0.56
[0.001, 1000]	✗	33.33 ± 7.58	42.38 ± 1.54	2.25 ± 0.22	5.07 ± 2.37	25.42 ± 0.73	2.24 ± 0.11
[0.001, 1000]	✓(Train+Test)	85.66 ± 0.39	85.66 ± 0.39	85.66 ± 0.39	85.05 ± 0.43	85.05 ± 0.43	85.05 ± 0.43

Table 5. Rotation and translation invariances in 3D pointcloud classification with PointNet trained on modelnet40, with and without rotation augmentation, with and without STNs or PCA. Mean and standard deviations over 10 runs are reported.

RA	STN	PCA	Clean	Rotation		Translation	
				Avg	Worst	Avg	Worst
✗	✓	✗	**86.15 ± 0.52**	10.37 ± 0.18	0.09 ± 0.07	10.96 ± 1.22	0.00 ± 0.00
✗	✗	✗	85.31 ± 0.39	10.59 ± 0.25	0.26 ± 0.10	6.53 ± 0.12	0.00 ± 0.00
✗	✓	✓(Train+Test)	74.12 ± 1.80	74.12 ± 1.80	74.12 ± 1.80	74.12 ± 1.80	74.12 ± 1.80
✗	✗	✓(Train+Test)	75.36 ± 0.70	**75.36 ± 0.70**	**75.36 ± 0.70**	**75.36 ± 0.70**	**75.36 ± 0.70**
✓	✓	✗	72.13 ± 5.84	72.39 ± 5.60	35.91 ± 4.87	5.35 ± 0.98	0.00 ± 0.00
✓	✗	✗	63.93 ± 0.65	64.75 ± 0.57	45.53 ± 0.29	3.90 ± 0.71	0.00 ± 0.00
✓	✓	✓(Test)	72.13 ± 5.84	72.96 ± 5.85	72.96 ± 5.85	72.96 ± 5.85	72.96 ± 5.85
✓	✗	✓(Test)	64.56 ± 0.91	64.56 ± 0.91	64.56 ± 0.91	64.56 ± 0.91	64.56 ± 0.91
✓	✓	✓(Train+Test)	72.84 ± 0.77	72.84 ± 0.77	72.84 ± 0.77	72.84 ± 0.77	72.84 ± 0.77
✓	✗	✓(Train+Test)	74.84 ± 0.86	74.84 ± 0.86	74.84 ± 0.86	74.84 ± 0.86	74.84 ± 0.86

4.2 Invariances in 3D Point Cloud Classification

Invariance to orientation and scale is often desired in networks classifying objects given as 3D point clouds. Popular architectures, such as PointNet [60] and its extensions [61], rely on the ability of spatial transformer networks to learn such invariances by training on large datasets and extensive data augmentations. We analyze the robustness of these networks to transformations with experiments using Pointnet on *modelnet40* dataset [55]. We compare the class accuracy of the final iterate for the clean validation set *(Clean)*, and transformed validation sets in the average *(Avg.)* and worst-case *(Worst)*. We show that PointNet performs better with our orbit mappings than with augmentation alone.

In this setting, $\mathcal{X} = \mathbb{R}^{d \times N}$ are N many d-dimensional coordinates (usually with $d = 3$). The desired group actions for invariance are left-multiplication with a rotation matrix, and multiplication with any number $c \in \mathbb{R}^+$ to account for different scaling. We also consider translation by adding a fixed coordinate $c_t \in \mathbb{R}^3$ to each entry in $\mathcal{X}$. Desired invariances in point cloud classification range from class-dependent variances to geometric properties. For example, the classification of airplanes should be invariant to the specific wing shape, as well as the scale or translation of the model. While networks can learn some invariance from training data, our experiments show that even simple transformations like scaling and translation are not learned robustly outside the scope of what was

Table 6. Combined Scale, rotation and translation invariances in 3D pointcloud classification with PointNet trained on modelnet40, with data augmentation and analytical inclusion of each invariance. Mean and standard deviations over 10 runs are reported.

Augmentation			STN	OM	Clean	Scaling		Rotation		Translation	
Scale	RA	Translation		All		Avg	Worst	Avg	Worst	Avg	Worst
[0.8, 1.25]	✓	[−0.1, 0.1]	✓	✗	72.13± 5.84	19.74± 4.01	0.16± 0.42	72.39± 5.60	35.91± 4.87	5.35±0.98	0.00±0.00
[0.8, 1.25]	✓	[−0.1, 0.1]	✓	✓ Test	67.38± 7.96	64.88± 12.16	64.88± 12.16	64.88± 12.16	64.88± 12.16	64.88± 12.16	64.88± 12.16
[0.8, 1.25]	✓	[−0.1, 0.1]	✓	✓ Train+Test	**77.52±1.03**	**77.52±1.03**	**77.52±1.03**	**77.52±1.03**	**77.52±1.03**	**77.52±1.03**	**77.52±1.03**
[0.8, 1.25]	✓	[−0.1, 0.1]	✗	✗	63.93±0.65	12.85±0.29	0.27±0.55	64.75±0.57	45.53±0.29	3.90±0.71	0.00±0.00
[0.8, 1.25]	✓	[−0.1, 0.1]	✗	✓ Test	64.71±0.92	57.10±1.14	57.10±1.14	57.10±1.14	57.10±1.14	57.10±1.14	57.10±1.14
[0.8, 1.25]	✓	[−0.1, 0.1]	✗	✓ Train+Test	74.41±0.58	74.41±0.58	74.41±0.58	74.41±0.58	74.41±0.58	74.41±0.58	74.41±0.58

provided in the training data, see Tables 4, 5, 6. This is surprising, considering that both can be undone by centering around the origin and re-scaling.

Scaling. Invariance to scaling can be achieved in the sense of Sect. 3 by scaling input point-clouds by the average distance of all points to the origin. Our experiments show that this leads to robustness against much more extreme transformation values without the need for expensive training, both for average as well as worst-case accuracy. We tested the worst-case accuracy on the following scales: $\{0.001, 0.01, 0.1, 0.5, 1.0, 5.0, 10, 100, 1000\}$. While our approach performs well on all cases, training PointNet on random data augmentation in the range of possible values actually reduces the accuracy on clean, not scaled test data. This indicates that the added complexity of the task cannot be well represented within the network although it includes spatial transformers. Even when restricting the training to a subset of the interval of scales, the spatial transformers cannot fully learn to undo the scaling, resulting in a significant drop in average and worst-case robustness, see Table 4. While training the original Pointnet including the desired invariance in the network achieves the best performance, dropping the spatial transformers from the architecture results in only a tiny drop in accuracy with significant gains in training and computation time[5]. This either indicates that in the absence of rigid deformation the spatial transformers do not add much knowledge and is strictly inferior to modeling invariance, at least on this dataset.

Rotation and Translation. In this section, we show that 3D rotations and translations exhibit a similar behavior and can be more robustly treated via orbit mapping than through data augmentation. This is even more meaningful than scaling as both have three degrees of freedom and sampling their respective spaces requires a lot more examples. For rotations, we choose the unique element of the orbit to be the rotation of $\mathcal{X}$ that aligns its principle components with the coordinate axes. The optimal transformation involves subtracting the center of mass from all coordinates and then applying the singular value decomposition $X = U\Sigma V$ of the point cloud X up to the arbitrary orientation of the principle axes, a process also known as PCA. Rotation and translation can be treated together, as undoing the translation is a substep of PCA. To remove the sign ambiguity in the principle axes, we choose signs of the first row of U and encode them into a diagonal matrix D, such that the final transform is

---

[5] Model size of PointNet with STNs is 41.8 MB, and without STNs 9.8 MB.

given by $\hat{X} = XV^\top D$. We apply this rotational alignment to PointNet with and without spatial transformers and evaluate its robustness to rotations in average-case and worst-case when rotating the validation dataset in 16×16 increments (i.e. with 16 discrete angles along each of the two angular degrees of freedom of a 3D rotation). We test robustness to translations in average-case and worst-case for the following shifts in each of x, y and z directions: $\{-10.0, -1.0, -0.5, -0.1, 0.1, 0.5, 1.0, 10.0\}$. Table 5 shows that PointNet trained without augmentation is susceptible in worst-case and average-case rotations and even translations. The vulnerability to rotations can be ameliorated in the average-case by training with random rotations, but the worst-case accuracy is still significantly lower, even when spatial transformers are employed. Also notable is the high variance in performance of Pointnets with STNs trained using augmentations. On the other hand, explicitly training and testing with stabilized rotations using PCA does provide effortless invariance to rotations and translations, even without augmentation. Interestingly, the best accuracy here is reached when training PointNet entirely without spatial transformers, which offer no additional benefits when the rotations are stabilized. The process for invariance against translation is well-known and well-used due to its simplicity and robustness. We show that this approach arises naturally from our framework, and that its extension to rotational invariance inherits the same numerical behavior, i.e., provable invariance outperforms learning to undo the transformation via data augmentation.

Combined invariance to Scaling, Rotation, Translation. Our approach can be extended to make a model simultaneously invariant to scaling, rotations and translations. In this setup, we apply a PCA alignment before normalizing the scale of input point cloud. Table 6 shows that PointNet trained with such combined orbit mapping does achieve the desired invariances.

5   Discussion and Conclusions

We proposed a simple and general way of incorporating invariances to group actions in neural networks by uniquely selecting a specific element from the orbit of group transformations. This guarantees provable invariance to group transformations for 3D point clouds, and demonstrates significant improvements in robustness to continuous rotations of images with a limited computational overhead. However, for images, a large discrepancy between the theoretical provable invariance (in the perspective of images as continuous functions) and the practical discrete setting remains. We conjecture that this is related to discretization artifacts when applying rotations that change the gradient directions, especially at low resolutions. Notably, such artifacts appear more frequently in artificial settings, e.g. during data augmentation or when testing for worst-case accuracy, than in photographs of rotating objects that only get discretized once. While we found a consistent advantage of enforcing the desired invariance via orbit mapping rather than training alone, combination of data augmentation and orbit mappings yields additional advantages (in cases where discretization artifacts prevent a provable invariance of the latter). Moreover, our orbit mapping can be combined with existing invariant approaches for improved robustness.

References

1. Russakovsky, O., et al.: ImageNet large scale visual recognition challenge. Int. J. Comput. Vision **115**(3), 211–252 (2015). https://doi.org/10.1007/s11263-015-0816-y
2. Tensmeyer, C., Martinez, T.: Improving invariance and equivariance properties of convolutional neural networks (2016)
3. Olah, C., Cammarata, N., Voss, C., Schubert, L., Goh, G.: Naturally occurring equivariance in neural networks. Distill **5**, e00024–004 (2020)
4. Lenc, K., Vedaldi, A.: Understanding image representations by measuring their equivariance and equivalence. Int. J. Comput. Vision **127**, 456–476 (2018)
5. Engstrom, L., Tsipras, D., Schmidt, L., Madry, A.: A rotation and a translation suffice: Fooling cnns with simple transformations. arXiv preprint arXiv:1712.02779 1 (2017)
6. Finlayson, S.G., Bowers, J.D., Ito, J., Zittrain, J.L., Beam, A.L., Kohane, I.S.: Adversarial attacks on medical machine learning. Science **363**, 1287–1289 (2019)
7. Zhao, Y., Wu, Y., Chen, C., Lim, A.: On isometry robustness of deep 3d point cloud models under adversarial attacks. In: IEEE/CVF Conference on Computer Vision and Pattern Recognition (2020)
8. Lang, I., Kotlicki, U., Avidan, S.: Geometric adversarial attacks and defenses on 3d point clouds. In: 2021 International Conference on 3D Vision (3DV) (2021)
9. Simard, P., Victorri, B., LeCun, Y., Denker, J.: Tangent prop-a formalism for specifying selected invariances in an adaptive network. In: Advances in Neural Information Processing Systems, vol. 4 (1991)
10. Engstrom, L., Tran, B., Tsipras, D., Schmidt, L., Madry, A.: Exploring the landscape of spatial robustness. In: International Conference on Machine Learning, pp. 1802–1811 (2019)
11. Wang, R., Yang, Y., Tao, D.: Art-point: Improving rotation robustness of point cloud classifiers via adversarial rotation. In: IEEE/CVF Conference on Computer Vision and Pattern Recognition (CVPR), pp. 14371–14380 (2022)
12. Yang, F., Wang, Z., Heinze-Deml, C.: Invariance-inducing regularization using worst-case transformations suffices to boost accuracy and spatial robustness. In: Advances in Neural Information Processing Systems, pp. 14757–14768 (2019)
13. Anselmi, F., Leibo, J.Z., Rosasco, L., Mutch, J., Tacchetti, A., Poggio, T.: Unsupervised learning of invariant representations. Theoret. Comput. Sci. **633**, 112–121 (2016)
14. Noroozi, M., Favaro, P.: Unsupervised learning of visual representations by solving jigsaw puzzles. In: Leibe, B., Matas, J., Sebe, N., Welling, M. (eds.) ECCV 2016. LNCS, vol. 9910, pp. 69–84. Springer, Cham (2016). https://doi.org/10.1007/978-3-319-46466-4_5
15. Komodakis, N., Gidaris, S.: Unsupervised representation learning by predicting image rotations. In: International Conference on Learning Representations (ICLR) (2018)
16. Zhang, L., Qi, G.J., Wang, L., Luo, J.: Aet vs. aed: Unsupervised representation learning by auto-encoding transformations rather than data. In: IEEE/CVF Conference on Computer Vision and Pattern Recognition, pp. 2547–2555 (2019)

17. Gu, J., Yeung, S.: Staying in shape: learning invariant shape representations using contrastive learning. In: de Campos, C., Maathuis, M.H., (eds.) Conference on Uncertainty in Artificial Intelligence, vol. 161, pp. 1852–1862. PMLR (2021)

18. Wilk, M.v.d., Bauer, M., John, S., Hensman, J.: Learning invariances using the marginal likelihood. In: Advances in Neural Information Processing Systems, pp. 9960–9970 (2018)

19. Benton, G.W., Finzi, M., Izmailov, P., Wilson, A.G.: Learning invariances in neural networks from training data. In: Advances in Neural Information Processing Systems (2020)

20. Sheng, Y., Shen, L.: Orthogonal fourier-mellin moments for invariant pattern recognition. J. Opt. Soc. Am. **11**, 1748–1757 (1994)

21. Yap, P.T., Jiang, X., Chichung Kot, A.: Two-dimensional polar harmonic transforms for invariant image representation. IEEE Trans. Pattern Anal. Mach. Intell. **32**, 1259–1270 (2010)

22. Tan, T.: Rotation invariant texture features and their use in automatic script identification. IEEE Trans. Pattern Anal. Mach. Intell. **20**, 751–756 (1998)

23. Lazebnik, S., Schmid, C., Ponce, J.: A sparse texture representation using local affine regions. IEEE Trans. Pattern Anal. Mach. Intell. **27**, 1265–1278 (2005)

24. Manthalkar, R., Biswas, P.K., Chatterji, B.N.: Rotation and scale invariant texture features using discrete wavelet packet transform. Pattern Recogn. Lett. **24**, 2455–2462 (2003)

25. Bruna, J., Mallat, S.: Invariant scattering convolution networks. IEEE Trans. Pattern Anal. Mach. Intell. **35**, 1872–1886 (2013)

26. Sifre, L., Mallat, S.: Rotation, scaling and deformation invariant scattering for texture discrimination. In: IEEE Conference on Computer Vision and Pattern Recognition (CVPR) (2013)

27. Oyallon, E., Mallat, S.: Deep roto-translation scattering for object classification. In: IEEE Conference on Computer Vision and Pattern Recognition (CVPR) (2015)

28. Cohen, T., Welling, M.: Group equivariant convolutional networks. In: International conference on machine learning, pp. 2990–2999 (2016)

29. Worrall, D.E., Garbin, S.J., Turmukhambetov, D., Brostow, G.J.: Harmonic networks: Deep translation and rotation equivariance. In: The IEEE Conference on Computer Vision and Pattern Recognition (CVPR) (2017)

30. Weiler, M., Cesa, G.: General E(2)-Equivariant Steerable CNNs. In: Advances in Neural Information Processing Systems (2019)

31. Zhang, J., Yu, M.Y., Vasudevan, R., Johnson-Roberson, M.: Learning rotation-invariant representations of point clouds using aligned edge convolutional neural networks. In: 2020 International Conference on 3D Vision (3DV), pp. 200–209. IEEE (2020)

32. Yu, R., Wei, X., Tombari, F., Sun, J.: Deep positional and relational feature learning for rotation-invariant point cloud analysis. In: Vedaldi, A., Bischof, H., Brox, T., Frahm, J.-M. (eds.) ECCV 2020. LNCS, vol. 12355, pp. 217–233. Springer, Cham (2020). https://doi.org/10.1007/978-3-030-58607-2_13

33. Balunovic, M., Baader, M., Singh, G., Gehr, T., Vechev, M.: Certifying geometric robustness of neural networks. In: Advances in Neural Information Processing Systems 32 (2019)

34. Fischer, M., Baader, M., Vechev, M.: Certified defense to image transformations via randomized smoothing. In: Advances in Neural information processing systems, vol. 33 (2020)

35. Manay, S., Cremers, D., Hong, B.W., Yezzi, A.J., Soatto, S.: Integral invariants for shape matching. IEEE Trans. Pattern Anal. Mach. Intell. **28**, 1602–1618 (2006)

36. Laptev, D., Savinov, N., Buhmann, J.M., Pollefeys, M.: Ti-pooling: transformation-invariant pooling for feature learning in convolutional neural networks. In: IEEE Conference On Computer Vision And Pattern Recognition, pp. 289–297 (2016)
37. Ravanbakhsh, S., Schneider, J., Poczos, B.: Equivariance through parameter-sharing. In: International Conference on Machine Learning, pp. 2892–2901. PMLR (2017)
38. Xiao, Z., Lin, H., Li, R., Geng, L., Chao, H., Ding, S.: Endowing deep 3d models with rotation invariance based on principal component analysis. In: IEEE International Conference on Multimedia and Expo (ICME). IEEE (2020)
39. Li, F., Fujiwara, K., Okura, F., Matsushita, Y.: A closer look at rotation-invariant deep point cloud analysis. In: International Conference on Computer Vision (ICCV), pp. 16218–16227 (2021)
40. Rempe, D., Birdal, T., Zhao, Y., Gojcic, Z., Sridhar, S., Guibas, L.J.: Caspr: Learning canonical spatiotemporal point cloud representations. Adv. Neural. Inf. Process. Syst. 33, 13688–13701 (2020)
41. Wang, H., Sridhar, S., Huang, J., Valentin, J., Song, S., Guibas, L.J.: Normalized object coordinate space for category-level 6d object pose and size estimation. In: IEEE/CVF Conference on Computer Vision and Pattern Recognition, pp. 2642–2651 (2019)
42. Sun, W., Tagliasacchi, A., Deng, B., Sabour, S., Yazdani, S., Hinton, G.E., Yi, K.M.: Canonical capsules: Self-supervised capsules in canonical pose. In: Advances in Neural Information Processing Systems 34 (2021)
43. Spezialetti, R., Stella, F., Marcon, M., Silva, L., Salti, S., Di Stefano, L.: Learning to orient surfaces by self-supervised spherical cnns. In: Advances in Neural Information Processing Systems 33 (2020)
44. Sajnani, R., Poulenard, A., Jain, J., Dua, R., Guibas, L.J., Sridhar, S.: Condor: Self-supervised canonicalization of 3d pose for partial shapes. In: IEEE/CVF Conference on Computer Vision and Pattern Recognition, pp. 16969–16979 (2022)
45. Rehman, H.Z.U., Lee, S.: Automatic image alignment using principal component analysis. IEEE Access 6, 72063–72072 (2018)
46. Jafari-Khouzani, K., Soltanian-Zadeh, H.: Radon transform orientation estimation for rotation invariant texture analysis. IEEE Trans. Pattern Anal. Mach. Intell. 27, 1004–1008 (2005)
47. Jaderberg, M., Simonyan, K., Zisserman, A., Kavukcuoglu, K.: Spatial transformer networks. In: Advances in Neural Information Processing Systems (2015)
48. Tai, K.S., Bailis, P., Valiant, G.: Equivariant transformer networks. In: International Conference on Machine Learning, pp. 6086–6095. PMLR (2019)
49. Esteves, C., Allen-Blanchette, C., Zhou, X., Daniilidis, K.: Polar transformer networks. In: International Conference on Learning Representations (2018)
50. Marcos, D., Volpi, M., Komodakis, N., Tuia, D.: Rotation equivariant vector field networks. In: IEEE International Conference on Computer Vision, pp. 5048–5057 (2017)
51. Veeling, B.S., Linmans, J., Winkens, J., Cohen, T., Welling, M.: Rotation equivariant CNNs for digital pathology. In: Frangi, A.F., Schnabel, J.A., Davatzikos, C., Alberola-López, C., Fichtinger, G. (eds.) MICCAI 2018. LNCS, vol. 11071, pp. 210–218. Springer, Cham (2018). https://doi.org/10.1007/978-3-030-00934-2_24
52. Marcos, D., Volpi, M., Tuia, D.: Learning rotation invariant convolutional filters for texture classification. In: International Conference on Pattern Recognition (ICPR), pp. 2012–2017. IEEE (2016)
53. Fasel, B., Gatica-Perez, D.: Rotation-invariant neoperceptron. In: International Conference on Pattern Recognition (ICPR), vol. 3, pp. 336–339. IEEE (2006)

54. Henriques, J.F., Vedaldi, A.: Warped convolutions: Efficient invariance to spatial transformations. In: International Conference on Machine Learning, pp. 1461–1469. PMLR (2017)
55. Wu, Z., et al.: 3d shapenets: A deep representation for volumetric shape modeling. In: IEEE Conference on Computer Vision and Pattern Recognition (CVPR) (2015)
56. Wang, P.S., Liu, Y., Guo, Y.X., Sun, C.Y., Tong, X.: O-CNN: Octree-based convolutional neural networks for 3D shape analysis. ACM Trans. Graph. (SIGGRAPH) **36** (2017)
57. Weiler, M., Geiger, M., Welling, M., Boomsma, W., Cohen, T.: 3d steerable cnns: learning rotationally equivariant features in volumetric data. In: Advances in Neural Information Processing Systems, pp. 10402–10413 (2018)
58. Thomas, N., et al.: Tensor field networks: Rotation-and translation-equivariant neural networks for 3d point clouds. arXiv preprint arXiv:1802.08219 (2018)
59. Fuchs, F., Worrall, D., Fischer, V., Welling, M.: Se (3)-transformers: 3d roto-translation equivariant attention networks. In: Advances in Neural Information Processing Systems 33 (2020)
60. Qi, C.R., Su, H., Mo, K., Guibas, L.J.: Pointnet: Deep learning on point sets for 3d classification and segmentation. In: Conference on Computer Vision and Pattern Recognition (CVPR) (2017)
61. Qi, C.R., Yi, L., Su, H., Guibas, L.J.: Pointnet++: Deep hierarchical feature learning on point sets in a metric space. In: Advances in Neural Information Processing Systems (2017)
62. Zhang, Y., Rabbat, M.: A graph-cnn for 3d point cloud classification. In: International Conference on Acoustics, Speech and Signal Processing (ICASSP) (2018)
63. Horie, M., Morita, N., Hishinuma, T., Ihara, Y., Mitsume, N.: Isometric transformation invariant and equivariant graph convolutional networks. In: International Conference on Learning Representations (2020)
64. Satorras, V.G., Hoogeboom, E., Welling, M.: E(n) equivariant graph neural networks. In Meila, M., Zhang, T., (eds.) International Conference on Machine Learning, vol. 139, pp. 9323–9332. PMLR (2021)
65. Esteves, C., Allen-Blanchette, C., Makadia, A., Daniilidis, K.: Learning SO(3) equivariant representations with spherical CNNs. In: Ferrari, V., Hebert, M., Sminchisescu, C., Weiss, Y. (eds.) ECCV 2018. LNCS, vol. 11217, pp. 54–70. Springer, Cham (2018). https://doi.org/10.1007/978-3-030-01261-8_4
66. Rao, Y., Lu, J., Zhou, J.: Spherical fractal convolutional neural networks for point cloud recognition. In: IEEE/CVF Conference on Computer Vision and Pattern Recognition, pp. 452–460 (2019)
67. Deng, H., Birdal, T., Ilic, S.: PPF-FoldNet: unsupervised learning of rotation invariant 3D local descriptors. In: Ferrari, V., Hebert, M., Sminchisescu, C., Weiss, Y. (eds.) ECCV 2018. LNCS, vol. 11209, pp. 620–638. Springer, Cham (2018). https://doi.org/10.1007/978-3-030-01228-1_37
68. Zhao, Y., Birdal, T., Deng, H., Tombari, F.: 3d point capsule networks. In: Conference on Computer Vision and Pattern Recognition (CVPR) (2019)
69. Monti, F., Boscaini, D., Masci, J., Rodolà, E., Svoboda, J., Bronstein, M.M.: Geometric deep learning on graphs and manifolds using mixture model cnns. In: IEEE Conference on Computer Vision and Pattern Recognition (CVPR) (2016)
70. Hanocka, R., Hertz, A., Fish, N., Giryes, R., Fleishman, S., Cohen-Or, D.: Meshcnn: A network with an edge. ACM Trans. Graph. (TOG) **38**, 90:1–90:12 (2019)
71. Sharp, N., Attaiki, S., Crane, K., Ovsjanikov, M.: Diffusion is all you need for learning on surfaces. CoRR abs/ arXiv: 2012.00888 (2020)

72. Ovsjanikov, M., Ben-Chen, M., Solomon, J., Butscher, A., Guibas, L.: Functional maps: a flexible representation of maps between shapes. ACM Trans. Graph **31** (2012)

73. Pinkall, U., Polthier, K.: Computing discrete minimal surfaces and their conjugates. Experimental Mathematics (1993)

74. Litany, O., Remez, T., Rodolà, E., Bronstein, A., Bronstein, M.: Deep functional maps: Structured prediction for dense shape correspondences. In: International Conference on Computer Vision (ICCV) (2017)

75. Eisenberger, M., Toker, A., Leal-Taixé, L., Cremers, D.: Deep shells: Unsupervised shape correspondence with optimal transport. In: Advances in Neural Information Processing Systems. (2020)

76. Huang, R., Rakotosaona, M.J., Achlioptas, P., Guibas, L., Ovsjanikov, M.: Operatornet: Recovering 3d shapes from difference operators. In: International Conference on Computer Vision (ICCV), (2019)

77. He, K., Zhang, X., Ren, S., Sun, J.: Deep residual learning for image recognition. In: IEEE Conference On Computer Vision And Pattern Recognition, pp. 770–778 (2016)

78. Tschandl, P., Rosendahl, C., Kittler, H.: The ham10000 dataset, a large collection of multi-source dermatoscopic images of common pigmented skin lesions. Sci. Data **5**, 1–9 (2018)

79. Brock, A., De, S., Smith, S.L., Simonyan, K.: High-performance large-scale image recognition without normalization. arXiv preprint arXiv:2102.06171 (2021)

80. Wah, C., Branson, S., Welinder, P., Perona, P., Belongie, S.: The caltech-ucsd birds-200-2011 dataset (2011)

81. Weiler, M., Geiger, M., Welling, M., Boomsma, W., Cohen, T.: 3d steerable cnns: Learning rotationally equivariant features in volumetric data. In: Advances in Neural Information Processing Systems (2018)

Datasets and Performance Analysis

AirBirds: A Large-scale Challenging Dataset for Bird Strike Prevention in Real-world Airports

Hongyu Sun, Yongcai Wang$^{(\boxtimes)}$, Xudong Cai, Peng Wang, Zhe Huang,
Deying Li, Yu Shao, and Shuo Wang

Renmin University of China, Beijing 100872, China
{sunhongyu,ycw,xudongcai,peng.wang,huangzhe21,
deyingli,sy492019,shuowang18}@ruc.edu.cn

Abstract. One fundamental limitation to the research of bird strike prevention is the lack of a large-scale dataset taken directly from real-world airports. Existing relevant datasets are either small in size or not dedicated for this purpose. To advance the research and practical solutions for bird strike prevention, in this paper, we present a large-scale challenging dataset AirBirds that consists of 118,312 time-series images, where a total of 409,967 bounding boxes of flying birds are manually, carefully annotated. The average size of all annotated instances is smaller than 10 pixels in 1920×1080 images. Images in the dataset are captured over 4 seasons of a whole year by a network of cameras deployed at a real-world airport, covering diverse bird species, lighting conditions and 13 meteorological scenarios. To the best of our knowledge, it is the first large-scale image dataset that directly collects flying birds in real-world airports for bird strike prevention. This dataset is publicly available at https://airbirdsdata.github.io/.

Keywords: Large-scale dataset · Bird detection in airport · Bird strike prevention

1 Introduction

Bird strike accidents cause not only financial debts but also human casualties. According to Federal Aviation Administration (FAA)[1], from 1990 to 2019, there have been more than 220 thousand wildlife strikes with civil aircraft in USA alone and 97% of all strikes involve birds. An estimated economic loss could be as high as $500 million per year. Furthermore, more than 200 human fatalities and 300 injuries attributed to bird strikes. Bird strikes happen most near or at airports during takeoff, landing and associated phrases. About 61% of bird strikes with civil aircraft occur during landing phases of flight (descent, approach and landing roll). 36% occur during takeoff run and climb[2]. It is the airspace that the airport

---

[1] https://www.faa.gov/airports/airport_safety/wildlife/faq/.
[2] https://en.wikipedia.org/wiki/Bird_strike.

Supplementary Information The online version contains supplementary material available at https://doi.org/10.1007/978-3-031-26348-4_24.

L. Wang et al. (Eds.): ACCV 2022, LNCS 13845, pp. 409–424, 2023.
https://doi.org/10.1007/978-3-031-26348-4_24

(a) camera deployment alongside a runway in a real-world airport

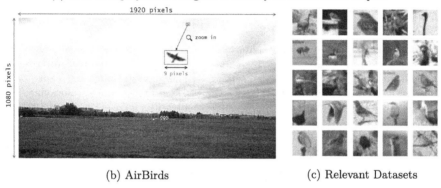

(b) AirBirds   (c) Relevant Datasets

Fig. 1. (a) The deployment of a network of cameras in a real-world airport (b) A bird example in AirBirds (c) Examples of birds in CUB [25, 27], Birdsnap [1], NABirds [24] and CIFAR10 [10].

should be responsible for so that the prevention of bird strikes is one of the most significant safety concerns. Although various systems are designed for preventing bird strikes, accidents keep occurring with increasing commercial activities and flights. Improving the performances of bird strike prevention systems remains a research challenge. One fundamental limitation to the performances is the lack of large-scale data collected at real-world airports. On the one hand, real-world airports have strict rules on security and privacy regarding camera system deployment. On the other hand, it is inevitably expensive to develop a large-scale dataset that involves a series of time-consuming and laborious tasks.

Existing relevant datasets are either small in size or not dedicated for bird strike prevention. The wildlife strike database created by FAA provides valuable information, while each record in this database only contains a few fields in text form, such as date and time, aircraft and airport information, environment conditions, lacking informative pictures and videos. The relevant dataset developed by Yoshihashi *et al* aims at preventing birds from hitting the blades of turbines in a wind farm [29], rather than in real-world airports, and its size is less than one seventh of ours. Well-known datasets like ImageNet [5], COCO [14], VOC [6], CIFAR [10] collects millions of common objects and animals, including birds, but they are developed for the research of general image recognition, object detection and segmentation. Another branch of datasets, such as CUB series [25, 27],

Birdsnap [1] and NABirds [24] containing hundreds of bird species, focus on fine-grained categorization and part localization. And the size of these datasets is less than 50% of ours. One of the most significant differences between the above-mentioned datasets and ours is that birds in previous datasets are carefully selected and tailored, which means they are often centered in the image, occupy the main part of an image and have clear outlines, referring to Fig. 1c.

However, it is unlikely that birds in the images captured in real-world airports have these idealized characteristics. The deployment of a network of cameras around a runway in a real-world airport is shown in Fig. 1a. Each camera is responsible for monitoring an area of hundreds of meters so that flying birds that appear are tiny in size even in a high-resolution image. For example, in our dataset, the average size of all annotated birds is smaller than 10 pixels in the 1920×1080 images, taking up only ∼0.5% of the image width, shown in Fig. 1b.

To advance the research and practical solutions for bird strike prevention, we collaborate with a real-world airport for two years and finally present AirBirds, a large-scale challenging dataset consisting of 118,312 time-series images with 1920×1080 resolution and 409,967 bounding box annotations of flying birds. The images are extracted from videos recorded by a network of cameras over one year, from September 2020 to August 2021, thus cover various bird species in different seasons. Diverse scenarios are also included in AirBirds, *e.g.*, changing lighting and 13 meteorological conditions. Planning, deployment and joint commissioning of the monitoring system last for one year. The data collection process takes another whole year and subsequent cleaning, labeling, sorting and experimental analysis consume parallel 12 months. To the best of our knowledge, AirBirds is the first large-scale challenging image dataset that collects flying birds in real airports for bird strike prevention. The core contributions of this paper are summarized as follows.

- A large-scale dataset, namely AirBirds, that consists of 118,312 time-series images with 1920×1080 resolution containing flying birds in real-world airports is publicly presented, where there exist 409,967 instances with carefully manual bounding box annotations. The dataset covers various kinds of birds in 4 different seasons and diverse scenarios that include day and night, 13 meteorological and lighting conditions, *e.g.*, overcast, sunny, cloudy, rainy, windy, haze, etc.
- To reflect significant differences with other relevant datasets, we make comprehensive statistics on AirBirds and compare it with relevant datasets. There are three appealing features. (i) The images in AirBirds are dedicatedly taken from a real-world airport, which provide rare first-hand sources for the research of bird strike prevention. (ii) Abundant bird instances in different seasons and changing scenarios are also covered by AirBirds as the data collection spans a full year. (iii) The distribution of AirBirds is distinctive with existing datasets since 88% of instances are smaller than 10 pixels, and the remaining 12% are more than 10 and less than 50 pixels in 1920×1080 images.
- To understand the difficulty of AirBirds, a wide range of strong baselines are evaluated on this dataset for bird discovering. Specifically, 16 detectors

Table 1. Comparisons of AirBirds and relevant datasets. Density is the average instances in each image. Duration refers to the period of data collection.

Dataset	Format	#Images	Resolution	#Instances	Density	Duration
FAA Database	text	-	–	227,005	–	30 years
CUB-200-2010 [27]	image	6,033	∼500×300	6,033	1.00	–
CUB-200-2011 [25]	image	11,788	∼500×300	11,788	1.00	–
Birdsnap [1]	image	49,829	various	∼49,829	1.00	–
NABirds [24]	image	48,562	various	∼48,562	1.00	–
Wind Farm [29]	image	16,200	∼5616×3744	32,000	1.97	3 days
VB100 [7]	video	-	∼848×464	1,416	–	–
AirBirds	image	118,312	1920×1080	409,967	3.47	1 year

are trained *from scratch* based on AirBirds with careful configurations and parameter optimization. The consistently unsatisfactory results reveal the non-trivial challenges of bird discovering and bird strike prevention in real-world airports, which deserve further investigation.

As far as we know, bird strike prevention remains a open research problem since it is not well solved by existing technologies. We believe AirBirds will benefit the researchers, facilitate the research field and push the boundary of practical solutions in real-world airports.

2   Related Work

In this section, we review the datasets that are either closely relevant to bird strike prevention or contain transferable information to this topic.

FAA Wildlife Strike Database. One of the most relevant datasets is the Wildlife Strike Database[3] maintained by FAA. This database contains more than 220K records of reported wildlife strikes since 1990 and 97% of strikes attribute to birds. The detailed descriptions for each incident can be divided into the following parts: bird species, date and time, airport information, aircraft information, environment conditions, etc. An obvious limitation is the contents in this database are mainly in text form, lacking informative pictures and videos.

Bird Dataset of a Wind Farm. Yoshihashi *et al.* develop this dataset for preventing birds striking the blades of the turbines in a wind farm [29]. 32,000 birds and 4,900 non-birds are annotated in total to conduct experiments of a two-class categorization. It is similar to us that the ratio of bird size and the image size is extremely small. However, compared to AirBirds' data collection process spanning a whole year, this dataset collects images only for 3 days so that the number of samples and scenarios are much less than those of AirBirds.

---

[3] https://wildlife.faa.gov.

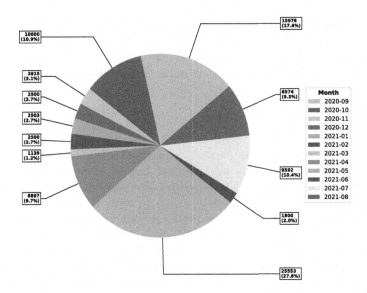

Fig. 2. The number of images per month in AirBirds.

Bird Datasets with Multiple Species. Bird species probably provide valuable information for bird strike prevention. Another branch of the relevant datasets, such as CUB series [25,27], Birdsnap [1], NABirds [24] and VB100 [7], focuses on fine-grained categorization of bird species. Images in these datasets are mainly collected from public sources, *e.g.*, Flickr[4], or by professionals. One of the most significant differences between the datasets in this branch and AirBirds is that birds in these datasets are carefully tailored, which means they are often centered in the image, occupy the main part and have clear outlines. However, it is unlikely for birds captured in real-world airports to have these wonderful characteristics. Moreover, bounding box annotations are absent in some of them, *e.g.*, VB100 [7], thus they are not suitable for the research of tiny bird detection.

Well-Known Datasets Containing Birds. Commonly used datasets in computer vision are also relevant as the `bird` belongs to one of the predefined categories in those datasets and there exist numbers of samples, such as ImageNet [5], COCO [14], VOC [6], CIFAR [10]. However, the above-mentioned datasets are dedicatedly designed for the research of general image classification, object detection and segmentation, not for bird strike prevention. And their data distributions differ from AirBirds, thus limited information can be transferred to this task.

The comparisons with related work are summarized in Table 1. AirBirds offers the most instances, the longest duration and the richest scenarios in image form.

---

[4] https://www.flickr.com/search/?text=bird.

3   AirBirds Construction

This section describes the process of constructing the AirBirds dataset, including raw data collection, subsequent cleaning, annotation, splits and sorting to complete it.

3.1   Collection

To cover diverse scenarios and prepare adequate raw data, we decide to record in a real-world airport (Shuangliu International Airport, Sichuan Province, China) over 4 seasons of a whole year. The process of data collection starts from September 2020 and ends in August 2021.

Considering frequent takeoffs and landings, airport runways and their surroundings are major monitoring areas. We deployed a network of high-resolution cameras along the runways, as Fig. 1a shows. All deployed cameras use identical configurations. The camera brand is AXIS Q1798-LE[5], recording 1920×1080 images at a frame rate of 25. Due to the vast volume of raw data but a limited number of disks, it is infeasible to save all videos. We split into two parallel groups, one group for data collection and the other for data processing, so disk spaces can be recycled once the second group finishes data processing.

3.2   Preprocessing

This step aims to process raw videos month by month and save 1920×1080 images in chronological order. 25 frames per second in raw videos lead to numerous redundant images. To avoid dense distribution of similar scenarios, a suitable sampling strategy is required. One crucial observation is that the video clips where flying birds appear are very sparse compared to other clips. Hence, at first, we manually locate all clips where there exist birds, then sample one every 5 continuous frames in previously selected clips instead of all of them, resulting in an average of 300+ images per day, ~10000 images per month, 118,312 in total. The number of images per month is shown in Fig. 2 and 13 meteorological conditions and the corresponding number of days are depicted in Fig. 3.

3.3   Annotation

To ensure quality and minimize costs, we divide the labeling process into three rounds. The first round that generates initial bounding box annotations for birds in the images is done by machines. The second round refines previous annotations manually by a team of employed workers. It should be noted that the team does not have to discover birds from scratch. In the third round, we are responsible for verifying those manual annotations and requiring further improvements of low-quality instances.

---

[5] https://www.axis.com/products/axis-q1798-le.

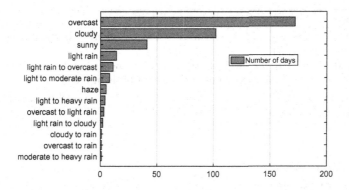

Fig. 3. The number of days of different weather in AirBirds.

It is not a simple task for humans to discover tiny birds from collected images with broad scenes. In the first round, we develop an algorithm for generating initial annotations and run on a computer. The idea of this algorithm is related to background subtraction in image processing. In our context, cameras are fixed in real-world airports, thus the background is static in the monitoring views. Since the images in each sequence are in chronological order, considering two consecutive frames, by computing the pixel differences between the first and second frame, the static part, namely the background, is removed while other moving targets, such as flying birds in the monitoring areas, are probably discovered. Algorithm 1 describes the detailed process. Initially, we treat the first frame as background, convert it to gray mode, apply Gaussian blur[6] to this gray image, and denote the output image as b, then remove the first image from the input sequence S. The set of initial bounding box annotations B is empty. Then we traverse the image I_i in S. In the loop, I_i is also converted to gray image g_i. After that, Gaussian blur is applied to g_i to generate a denoised image c_i. Then we compute differences between b and c_i, resulting in d. Fourth, regions in d whose pixel values are in the range of $[min, max]$ are considered as areas of interest, *e.g.*, if the pixel differences of the same area in those 2 consecutive frames are more than 30, there probably are moving targets in this area. The dilation operation is applied to those areas to expand contours for finding possible moving objects c_i, including flying birds. After that, heuristic rules are used to filter candidates according to the object size, *e.g.*, big targets like airplanes, working vehicles, workers, are removed, resulting in b_i. Then b_i is inserted into B. Finally, we set background b as c_i and move forward. The key steps of this algorithm are visualized in Fig. 4.

Refinement is required since previously discovered moving objects are not necessarily birds. In the second round, we cooperate with a team of workers to accomplish the task. According to the predefined instructions, every single image should be zoomed in to 250+% to check the initial annotations in detail

---

[6] https://en.wikipedia.org/wiki/Gaussian_blur.

Algorithm 1. The First Round of Annotations

Input: $\mathcal{S} = \{I_1, I_2, \ldots, I_n\}$, an image sequence, where n is the sequence length. Constants min and max

Output: $\mathcal{B} = \{\mathbf{b}_1, \mathbf{b}_2, \ldots, \mathbf{b}_n\}$, the bounding boxes set for birds in $\mathcal{S}$, where $\mathbf{b}_i = \{b_i^1, b_i^2, \ldots, b_i^m\}$, m is the number of detected birds in image I_i

1: $g \leftarrow \texttt{imRead}(I_1, 0)$ ▷ read image in gray mode
2: $b \leftarrow \texttt{gaussBlur}(g)$ ▷ denoise the image
3: $\mathcal{S} \leftarrow \mathcal{S} \setminus I_1$ ▷ remove I_1 from $\mathcal{S}$
4: $\mathcal{B} \leftarrow \emptyset$ ▷ initialize $\mathcal{B}$
5: **for** I_i in $\mathcal{S}$ **do**
6:      $g_i \leftarrow \texttt{imRead}(I_i, 0)$
7:      $c_i \leftarrow \texttt{gaussBlur}(g_i)$
8:      $d \leftarrow \texttt{Diff}(b, c_i)$ ▷ compute differences
9:      $d \leftarrow \texttt{Thresh}(d, min, max)$ ▷ apply threshold
10:      $d_i \leftarrow \texttt{Dilate}(d)$ ▷ dilate areas further
11:      $\mathbf{c}_i \leftarrow \texttt{findContours}(d_i)$ ▷ find candidates
12:      $\mathbf{b}_i \leftarrow \texttt{Filter}(\mathbf{c}_i)$ ▷ filter candidates
13:      $\mathcal{B} \leftarrow \mathcal{B} \cup \mathbf{b}_i$ ▷ insert annotations
14:      $b \leftarrow c_i$ ▷ move background next
15: **end for**

---

and the team mainly handles 3 types of issues that arose in the first round (i) add missed annotations, (ii) delete false-positive annotations, (iii) update inaccurate annotations. In the third round, we go through the annotations refined by the team, requiring further improvements where inappropriate.

3.4 Splits

To facilitate further explorations of bird strike prevention based on this dataset, it is necessary to split AirBirds into training and test set.

We need to pay attention to three key aspects when splitting the dataset. First, we should keep a proper ratio between the size of the training and the test set. Second, it is essential to ensure training and test sets have a similar distribution. Third, considering the characteristic of chronological order, we should put a complete sequence into either the training or the test set rather than split it into different sets.

At last, we divide 98,312 images into the training set and keep the remaining 20,000 images in the test set, a nearly 5:1 ratio. All images and labels are publicly available, but excluding the labels in the test set. The validation set is not explicitly distinguished as the primary evaluation should take place on the test set, and users can customize the ratio between training and validation set individually. We are actively building an evaluation server and the labels in the test will be kept there.

In addition, the images in AirBirds can also be divided into 13 groups according to 13 kinds of scenarios shown in Fig. 3. This division is easy to achieve since each image is recorded on a specific day and each day corresponds to one type of

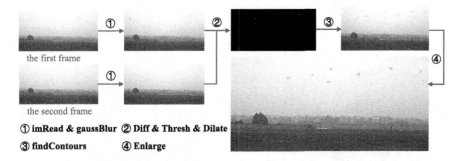

the first frame

the second frame

① imRead & gaussBlur ② Diff & Thresh & Dilate
③ findContours   ④ Enlarge

Fig. 4. Visualization of key steps in Algorithm 1.

meteorological condition, according to the official weather report. Based on the division, we can evaluate the difficulty of bird discovering in different scenarios in real-world airports.

4 Experiments

In this section, a series of comprehensive statistics and experiments based on AirBirds are presented. First, we investigate the data distribution in AirBirds and compare with relevant datasets to reflect their significant differences. Second, a wide range of SOTA detectors are evaluated on the developed dataset for bird discovering and the results are analyzed in detail to understand the non-trivial challenges of bird strike prevention. Third, the effectiveness of Algorithm 1 is evaluated since it plays an important role in the first round of annotations when constructing AirBirds.

4.1 Distribution

In this subsection, we investigate the distribution of AirBirds and compare with relevant datasets. Figure 5 shows the distribution of width and height of bounding box in different datasets. Obviously, objects in AirBirds have much smaller sizes. Further, Fig. 6 depicts the proportion of objects with various sizes in relevant datasets. 88% of all instances in AirBirds are smaller than 10 pixels and the rest 12% are mainly in the interval [10, 50). Therefore, data distribution in real-world airports is significantly different from that in web-crawled and tailor-made datasets.

4.2 Configurations

A wide range of detectors are tested on AirBirds for bird discovering. Before reporting their performances, it is necessary to elaborate on the specific modifications we made to accommodate the AirBirds dataset and the detectors. Concretely, we customize the following settings.

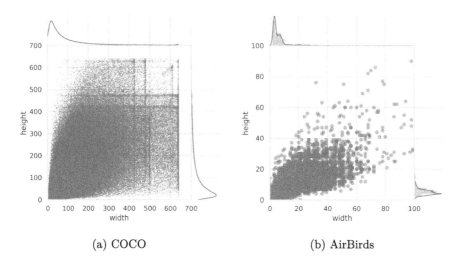

(a) COCO   (b) AirBirds

Fig. 5. Distributions of width and height of annotated instances in COCO and Air-Birds.

Models. To avoid AirBirds preferring a certain type of detectors, various kinds of strong baselines are picked for evaluation, including one-stage, multi-stage, transformer-based, anchor-free, and other types of models, referring to Table 2.

Devices. 6 NVIDIA RTX 2080Ti GPUs are used during training and a single GPU device is used during test for all models.

Data Format. The format of annotations in AirBirds is consistent with YOLO [16] style. Then we convert them to COCO format when training models other than YOLOv5.

Anchor Ratios and Scales. We need to adapt the ratios and scales for anchor-based detectors to succeed in custom training because objects in AirBirds have notable differences in size with that in the commonly used COCO dataset. The k-means clustering is applied to the labels of AirBirds, finally the ratios are set to $[\frac{8}{13}, \frac{9}{12}, \frac{11}{9}]$ and the scales are set to $[2^0, 2^{\frac{1}{3}}, 2^{\frac{2}{3}}]$.

Learning Schedules. All models are trained *from scratch* with optimized settings, *e.g.*, training epochs, learning rate, optimizer, batch size, etc. We summarize these settings in Sect. 2 in the supplementary material.

    Algorithm 1. The thresholds of pixel differences in Algorithm 1, *min* and *max*, are set to 25 and 255, respectively.

4.3   Results and Analysis

Both accuracy and efficiency are equivalently important for bird discovering in a real-world airport. The accuracy is measured by *average precision* (AP) and the efficiency is judged by *frames per second* (FPS). Results are recorded in Table 2.

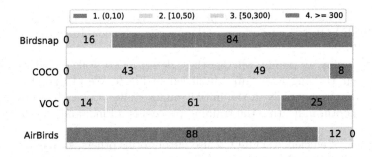

Fig. 6. Comparisons of the ratio of the number of objects with different sizes in the datasets Birdsnap, COCO, VOC and AirBirds. The numbers in each bar are in %. Here the object size in pixel level is divided into 4 intervals: (0,10), [10,50), [50, 300) and [300,+∞).

Table 2. Comparisons of various kinds of object detectors on AirBirds test set. The column AP_l has been removed as there are few large objects in AirBirds and the corresponding scores are all 0. EffiDet: EfficientDet-D2, Faster: Faster RCNN, Cascade: Cascade RCNN, Deform: Deformable DETR. These abbreviations have the same meaning in the following figure or table.

Method	Type	Backone	AP	AP@50	AP@75	AP_s	AP_m	FPS
FCOS [22]	one-stage	ResNet50 [9]	0.3	1.3	0.0	0.3	0.2	18.5
EffiDet [21]		EffiNet-B2 [20]	0.6	1.0	1.0	0.6	4.3	4.88
YOLOv3 [17]		DarkNet53 [15]	5.8	24.1	1.4	5.8	6.8	19.5
YOLOv5 [23]		CSPNet [26]	11.9	49.5	–	–	–	109.9
Faster [18]	multi-stage	ResNet50 [9]	7.1	26.9	1.3	7.1	0.2	16.0
Cascade [2]			6.8	24.0	1.8	6.8	1.8	13.4
DETR [3]	transformer	ResNet50 [9]	0.0	0.0	0.0	0.0	0.0	19.7
Deform [33]			0.4	2.2	0.0	0.4	1.0	11.6
FPN [12]	FPN	RetinaNet [13]	2.9	12.5	0.3	2.9	8.0	21.3
NASFPN [8]			3.0	12.5	0.4	2.9	15.8	25.0
RepPoints [28]	anchor-free	ResNet50 [31]	4.8	22.6	0.3	4.9	0.0	37.1
CornerNet [11]		HourglassNet	4.5	19.5	0.6	5.0	2.5	5.5
FreeAnchor [30]		ResNet101 [9]	6.6	26.5	1.1	6.7	9.0	53.5
HRNet [19]	high-resolution	HRNet [19]	8.9	33.0	1.3	9.0	0.3	21.7
DCN [4]	deformable	ResNet50 [9]	9.7	34.6	1.8	9.8	2.4	41.9
DCNv2 [32]		ResNet50 [9]	4.2	17.5	0.5	4.6	0.0	14.2

For accuracy, the primary metric AP seems unsatisfactory, *e.g.*, the highest score achieved by YOLOv5 is only 11.9, and the scores of all other models are less than 10. We also compare the performances of those detectors on COCO and AirBirds, shown in Fig. 7. Under the same detector, however, the performance gap is surprisingly large. For instance, the AP score of EfficientDet-D2 on COCO exceeds the one on AirBirds by 41.5(=42.1-0.6).

Besides, precision-recall relationship are also investigated, and results are shown in Fig. 8. The trend in all curves is that precision decreases with increased recall because more and more false-positive birds produce as more and more birds are recalled. YOLOv5 outperforms others while precision drops to 0 when recall reaches 0.7.

At this point, we wonder whether these detectors are well trained on AirBirds. Hence, their training losses are visualized in Fig. 9. We observe the losses of all detectors drop rapidly in the initial rounds of iterations, then progressively become smooth, indicating the training process is normal and converges to the target.

In terms of efficiency, YOLOv5 also outperforms others, surpassing 100 FPS on a 2080Ti GPU. However, most of detectors *fail to operate* in **real-time efficiency** even with GPU acceleration, which deviates a key principle of bird strike prevention.

We also wonder why a wide range of detectors work poorly on AirBirds. Reasons are detailed in Sect. 3 in the supplementary material due to space limitation.

In short, existing strong detectors show decent performances on commonly used datasets *e.g.* COCO, VOC etc. However, even with carefully customized configurations, they have room for significant improvements when validating on AirBirds. The results also imply the non-trivial challenges of the research of bird strike prevention in real-world airports, where AirBirds can serve as a valuable benchmark.

4.4   Effectiveness of the First Round of Annotations

As mentioned in Sect. 3, Algorithm 1 provides the first round of bounding box annotations for possible flying birds and the annotations are saved. Here we validate its effectiveness and compare it with the best performing YOLOv5. Different from *average precision* that sets strict IoU thresholds between detections and groundtruth, actually precision, recall and f1 score are more meaningful metrics for evaluating initial annotations.

Table 3 shows Algorithm 1 recalls more than 95% of birds in the initial round, which saves workers numerous efforts of discovering birds in subsequent rounds from scratch thus save costs. In addition, the results indicate that sequence information is helpful for tiny flying birds detection as the input images in Algorithm 1 are in chronological order. The star symbol in the second row in Table 3 means the results of Algorithm 1 are obtained on an ordinary computer(i5 CPU, 16GB memory), without GPU support.

Table 3. Comparisons of Algorithm 1 and YOLOv5 in terms of precision, recall and f1 score. Algorithm 1 runs on a common computer and YOLOv5 is tested with a 2080Ti GPU.

Method	Precision	Recall	F1	FPS
YOLOv5	**68.10%**	55.50%	61.16%	**109.89**
Algorithm 1	58.29%	**95.91%**	**72.51%**	67.44⋆

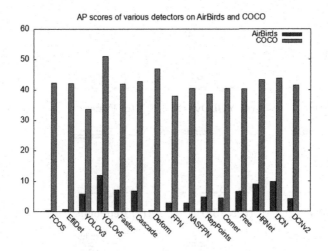

Fig. 7. Comparisons of the performances among representative detectors on AirBirds and COCO.

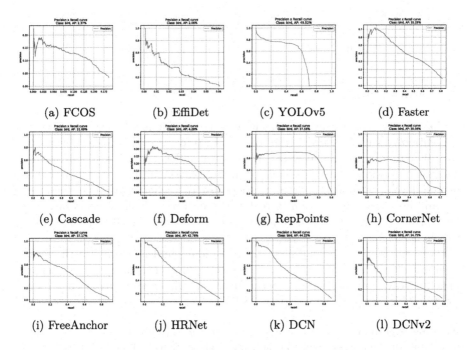

Fig. 8. Precision-Recall curves of different detectors in VOC [6] style.

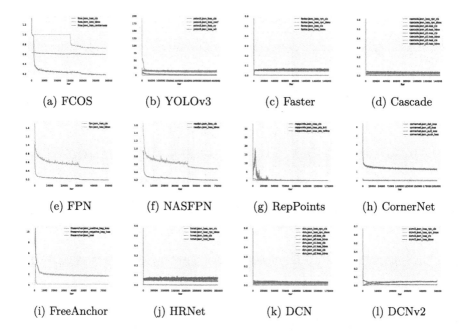

(a) FCOS   (b) YOLOv3   (c) Faster   (d) Cascade

(e) FPN   (f) NASFPN   (g) RepPoints   (h) CornerNet

(i) FreeAnchor   (j) HRNet   (k) DCN   (l) DCNv2

Fig. 9. Training losses of different types of detectors.

5   Conclusion

In this paper, we present AirBirds, a large-scale challenging dataset for bird strike prevention constructed directly from a real-world airport, to close the notable gap of data distribution between real world and other tailor-made datasets. Thorough statistical analysis and extensive experiments are conducted based on the developed dataset, revealing the non-trivial challenges of bird discovering and bird strike prevention in real-world airports, which deserves increasing and further investigation, where AirBirds can serve as a first-hand and valuable benchmark.

We believe AirBirds will alleviate the fundamental limitation of the lack of a large-scale dataset dedicated for bird strike prevention in real-world airports, benefit researchers and the field. In the future, we will develop advanced detectors for flying bird discovering based on AirBirds.

Acknowledgements. We thank all members who involved in the system deploying, data collecting, processing and labeling. This work was supported in part by the National Natural Science Foundation of China (Grant No. 61972404, 12071478).

References

1. Berg, T., Liu, J., Lee, S.W., Alexander, M.L., Jacobs, D.W., Belhumeur, P.N.: Birdsnap: large-scale fine-grained visual categorization of birds. In: Proceedings of the Conference on Computer Vision and Pattern Recognition (CVPR) (2014)
2. Cai, Z., Vasconcelos, N.: Cascade R-CNN: delving into high quality object detection. In: Proceedings of the IEEE Conference on Computer Vision and Pattern Recognition (CVPR) (2018)
3. Carion, N., Massa, F., Synnaeve, G., Usunier, N., Kirillov, A., Zagoruyko, S.: End-to-end object detection with transformers. In: Vedaldi, A., Bischof, H., Brox, T., Frahm, J.-M. (eds.) ECCV 2020. LNCS, vol. 12346, pp. 213–229. Springer, Cham (2020). https://doi.org/10.1007/978-3-030-58452-8_13
4. Dai, J., et al.: Deformable convolutional networks. In: Proceedings of the IEEE International Conference on Computer Vision (2017)
5. Deng, J., et al.: ImageNet: a large-scale hierarchical image database. In: 2009 IEEE Conference on Computer Vision and Pattern Recognition, pp. 248–255 (2009). https://doi.org/10.1109/CVPR.2009.5206848
6. Everingham, M., Van Gool, L., Williams, C.K.I., Winn, J., Zisserman, A.: The pascal visual object classes (voc) challenge. Int. J. Comput. Vision **88**(2), 303–338 (2010)
7. Ge, Z., et al.: Exploiting temporal information for DCNN-based fine-grained object classification. In: International Conference on Digital Image Computing: Techniques and Applications (2016)
8. Ghiasi, G., Lin, T.Y., Le, Q.V.: NAS-FPN: Learning scalable feature pyramid architecture for object detection. In: Proceedings of the IEEE/CVF Conference on Computer Vision and Pattern Recognition (CVPR) (2019)
9. He, K., Zhang, X., Ren, S., Sun, J.: Deep residual learning for image recognition. In: 2016 IEEE Conference on Computer Vision and Pattern Recognition (CVPR), pp. 770–778 (2016). https://doi.org/10.1109/CVPR.2016.90
10. Krizhevsky, A.: Learning multiple layers of features from tiny images. University of Toronto, Tech. Rep. (2009)
11. Law, H., Deng, J.: CornerNet: detecting objects as paired keypoints. In: Ferrari, V., Hebert, M., Sminchisescu, C., Weiss, Y. (eds.) Computer Vision – ECCV 2018. LNCS, vol. 11218, pp. 765–781. Springer, Cham (2018). https://doi.org/10.1007/978-3-030-01264-9_45
12. Lin, T.Y., Dollar, P., Girshick, R., He, K., Hariharan, B., Belongie, S.: Feature pyramid networks for object detection. In: Proceedings of the IEEE Conference on Computer Vision and Pattern Recognition (CVPR) (2017)
13. Lin, T.Y., Goyal, P., Girshick, R., He, K., Dollar, P.: Focal loss for dense object detection. In: Proceedings of the IEEE International Conference on Computer Vision (ICCV) (2017)
14. Lin, T.-Y., et al.: Microsoft COCO: common objects in context. In: Fleet, D., Pajdla, T., Schiele, B., Tuytelaars, T. (eds.) ECCV 2014. LNCS, vol. 8693, pp. 740–755. Springer, Cham (2014). https://doi.org/10.1007/978-3-319-10602-1_48
15. Redmon, J.: Darknet: Open source neural networks in c. http://pjreddie.com/darknet/ (2013-2016)
16. Redmon, J., Divvala, S., Girshick, R., Farhadi, A.: You only look once: unified, real-time object detection. In: 2016 IEEE Conference on Computer Vision and Pattern Recognition (CVPR), pp. 779–788 (2016). https://doi.org/10.1109/CVPR.2016.91
17. Redmon, J., Farhadi, A.: YOLOv3: An incremental improvement (2018)

18. Ren, S., He, K., Girshick, R., Sun, J.: Faster r-CNN: towards real-time object detection with region proposal networks. IEEE Trans. Pattern Anal. Mach. Intell. **39**(6), 1137–1149 (2017). https://doi.org/10.1109/TPAMI.2016.2577031

19. Sun, K., Xiao, B., Liu, D., Wang, J.: Deep high-resolution representation learning for human pose estimation. In: CVPR (2019)

20. Tan, M., Le, Q.: EfficientNet: rethinking model scaling for convolutional neural networks. In: Chaudhuri, K., Salakhutdinov, R. (eds.) Proceedings of the 36th International Conference on Machine Learning. Proceedings of Machine Learning Research, vol. 97, pp. 6105–6114. PMLR (2019). https://proceedings.mlr.press/v97/tan19a.html

21. Tan, M., Pang, R., Le, Q.V.: EfficientDet: scalable and efficient object detection. In: Proceedings of the IEEE/CVF Conference on Computer Vision and Pattern Recognition (CVPR) (2020)

22. Tian, Z., Shen, C., Chen, H., He, T.: Fcos: Fully convolutional one-stage object detection. arXiv preprint arXiv:1904.01355 (2019)

23. Ultralytics: YOLOv5. https://github.com/ultralytics/yolov5 (2021)

24. Van Horn, G.: Building a bird recognition app and large scale dataset with citizen scientists: The fine print in fine-grained dataset collection. In: Proceedings of the IEEE Conference on Computer Vision and Pattern Recognition (CVPR) (2015)

25. Wah, C., Branson, S., Welinder, P., Perona, P., Belongie, S.: The Caltech-UCSD Birds-200-2011 Dataset. Tech. Rep. CNS-TR-2011-001, California Institute of Technology (2011)

26. Wang, C.Y., Liao, H.Y.M., Wu, Y.H., Chen, P.Y., Hsieh, J.W., Yeh, I.H.: Cspnet: a new backbone that can enhance learning capability of CNN. In: Proceedings of the IEEE/CVF Conference on Computer Vision and Pattern Recognition (CVPR) Workshops (2020)

27. Welinder, P., et al.: Caltech-UCSD Birds 200. Tech. Rep. CNS-TR-2010-001, California Institute of Technology (2010)

28. Yang, Z., Liu, S., Hu, H., Wang, L., Lin, S.: Reppoints: Point set representation for object detection. In: The IEEE International Conference on Computer Vision (ICCV) (2019)

29. Yoshihashi, R., Kawakami, R., Iida, M., Naemura, T.: Construction of a bird image dataset for ecological investigations. In: 2015 IEEE International Conference on Image Processing (ICIP), pp. 4248–4252 (2015). https://doi.org/10.1109/ICIP.2015.7351607

30. Zhang, X., Wan, F., Liu, C., Ji, R., Ye, Q.: FreeAnchor: learning to match anchors for visual object detection. In: Neural Information Processing Systems (2019)

31. Zhou, X., Wang, D., Krähenbühl, P.: Objects as points (2019)

32. Zhu, X., Hu, H., Lin, S., Dai, J.: Deformable convnets v2: more deformable, better results. arXiv preprint arXiv:1811.11168 (2018)

33. Zhu, X., Su, W., Lu, L., Li, B., Wang, X., Dai, J.: Deformable DETR: deformable transformers for end-to-end object detection. In: International Conference on Learning Representations (2021). https://openreview.net/forum?id=gZ9hCDWe6ke

Rove-Tree-11: The Not-so-Wild Rover a Hierarchically Structured Image Dataset for Deep Metric Learning Research

Roberta Hunt[1]([✉])[ID] and Kim Steenstrup Pedersen[1,2][ID]

[1] Department of Computer Science, University of Copenhagen,
Universitetsparken 1, 2100 Copenhagen, Denmark
{r.hunt,kimstp}@di.ku.dk
[2] Natural History Museum of Denmark,
Øster Voldgade 5 - 7, 1350 Copenhagen, Denmark

Abstract. We present a new dataset of images of pinned insects from museum collections along with a ground truth phylogeny (a graph representing the relative evolutionary distance between species). The images include segmentations, and can be used for clustering and deep hierarchical metric learning. As far as we know, this is the first dataset released specifically for generating phylogenetic trees. We provide several benchmarks for deep metric learning using a selection of state-of-the-art methods.

Keywords: Phylogeny · Dataset · Tree · Hierarchy · Hierarchical dataset · Rove · Staphylinidae · Phylogenetic tree

1 Introduction

A phylogeny is a fundamental knowledge frame which hypothesizes how different species relate to each other [11]. A fully annotated phylogeny, i.e. a tree of life anchored in time scale, placed in the geographic context, and with a multitude of organismal traits mapped along the tree branches is an important tool in biology. It explains biodiversity changes over millennia or geological epochs, traces organismal movements in space and evolution of their properties, models populations response to climate change, navigates new species discovery and advises classification and taxonomy. An example phylogeny from our dataset is shown in Fig. 1 along with some example images from the most abundant species in the dataset.

Traditionally biologists generate phylogenies [9,10] using genetic data or morphological features (relating to the shape or development of the organism, for example the head shape, or the pattern of the veins on the wings). Despite genetic data dominating phylogenetic research in recent years, morphological features extracted by visual inspection of specimens are still of use. Fossils, for example,

---

Supplementary Information The online version contains supplementary material available at https://doi.org/10.1007/978-3-031-26348-4_25.

L. Wang et al. (Eds.): ACCV 2022, LNCS 13845, pp. 425–441, 2023.
https://doi.org/10.1007/978-3-031-26348-4_25

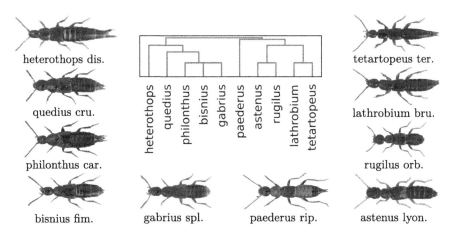

Fig. 1. Subset of phylogeny from the Rove-Tree-11 dataset, for the 10 genera with the most images in the dataset. Each leaf represents a genus. Genera which are closer together on the tree are more closely related, and nodes in the tree represent common ancestors. Nodes with more than two branches are considered not yet fully resolved. Many phylogenetic trees include estimations of time representing when the speciation event occurred (when the common ancestor split into two species). These dates are usually based on fossil evidence. This dated information is unfortunately not currently available for our ground truth tree. Example specimens from each genera are shown for reference.

contain no genetic data, but morphological features on the fossils can be used to relate them to existing biodiversity [26]. Occasionally morphological and genetic data are even combined to generate a so called 'total-evidence' phylogeny [34]. Morphological features are also of importance for species/specimens which lack good quality genetic data. Much of phylogenetic research on insects is done from museum specimens captured many years ago. Often the DNA of such specimens has degraded and is no longer of use. Genetic extraction is also expensive, time consuming, and a destructive process which can require completely destroying the specimen, particularly in the case of small insects.

However, the traditional process of generating morphological features is slow, meticulous and introduces some aspects of subjectivity by the researcher performing the analysis. Typically a phylogenetic researcher would generate a matrix of discrete traits (although the use of continuous traits has recently been explored [35]) which they hypothesize are of use in distinguishing the species and are evolutionary important. With thousands of new species of insects discovered each year [1], it is difficult for phylogeneticists to keep up.

Deep metric learning [22,38] is a proven technique to generate informative embedding matrices from images, and we posit that it can be used to generate morphological embeddings which more objectively represent the morphological features of a specimen. In this dataset we are unfortunately only looking at one view of the insect, in our case, the dorsal view (the back), whereas biologists would ideally examine and compare all external and internal features of the insect. However, we hypothesize that this can be offset by the model's ability to learn minute details. Our intention is that these methods could eventually be improved and used as a tool for biologists to inform their decision making process. Additionally, many natural history museums worldwide [8,20] are digitizing their collections, including in many cases, taking images of millions of museum insects. The Natural History Museum of Denmark (NHMD) alone estimates they have over 3.5 million pinned and dried insect specimens spanning 100,000 described species [32] and is in the process of digitizing their collections [31]. The importance of such digitization efforts have been studied from a biology research perspective [17,36]. Thus, given the increased data availability, we predict that phylogenetic generation from images will become a growing field of research within computer vision and related areas of artificial intelligence.

Despite the rapidly growing availability of images of pinned insects from natural history museums, the ongoing push from the biological community to generate phylogenies, and the increasing ability of deep learning to learn complex shapes and relationships, few publicly available datasets exist targeting the generation of phylogenies from images using deep learning techniques. There are several reasons for this, as we will explore in more detail in Sect. 2.1, when we compare with existing datasets. In brief, although the number of image analysis datasets is steadily growing, often the graphs which are included in the datasets are subjectively resolved (such as [5]) or the groupings they provide are too coarse-grained (such as [12]) or, particularly for biological datasets, the images are natural photos taken in the wild, meaning they are from various viewpoints and often obscured (such as [40,41]). This makes it difficult for the model to learn which distinct morphological features are more related to those from others species. Typical morphology based phylogenies are generated from careful inspection and comparison of features, meaning we expect direct comparison to be very important for this task.

In this paper we present 'Rove-Tree-11', a dataset of 13,887 segmented dorsal images of rove beetles along with a ground truth phylogeny down to genus level[1]. The species-level phylogeny is not included, because this level of information is not yet readily available. Our intention with releasing this data is that it can further research on deep hierarchical metric learning and computer vision solutions for building morphological phylogenies on interesting biological groups, leveraging the current digitization-wave that is gripping natural history museums worldwide.

The **contributions** of this paper are:

1. The release of a new hierarchically structured image dataset including segmentations and ground truth genus-level phylogeny
2. We provide baseline results on this dataset for the tasks of classification, clustering, and for predicting phylogenetic trees.

---

[1] To genus-level means that each species within a genus is considered unresolved, or equally likely to be related to any other species within that genus.

2   Related Work

2.1   Comparison with Existing Datasets

Hierarchically structured data is often found in computer vision related tasks. Examples include cognitive synonym relations between object categories such as clothing items [27] and is especially found in tasks concerning nature. However, current datasets which present a ground truth hierarchical grouping of the data are not intended for morphological phylogenetic research, and therefore poorly suited to the task.

There are several natural history related image datasets which do, or could easily be adapted to, include a taxonomy (i.e. IP102 [44], CUB-200-2011 [41], iNaturalist [40], Mammal dataset [12], PlantCLEF 2021 [13] and ImageNet [6]). With the exception of PlantCLEF, these are however all 'in the wild' images and identification has typically been done by non-experts with the naked eye. The phylogenies are also usually superficial - including only a few levels, and typically based only on the current taxonomy, which is not fine-grained and not necessarily representative of the state of the art phylogenetic tree, as taxonomies have a longer review process[2]. In the case of PlantCLEF the majority of the training images are of herbaria sheets, and therefore not 'in-the-wild', however only a shallow taxonomy is provided with the PlantCLEF dataset. In the case of IP102, the hierarchical tree is grouped by the plant the insect parasitizes, and is not related to ancestral traits at all. With the exception of CUB-200-2011, iNaturalist and PlantCLEF, the species are also easily identified by a layman/amateur by the images alone, which is not necessarily the case in our dataset, where many of the identifications traditionally require a microscope or dissection. It is also often the case that the taxonomy is not properly updated until years after the phylogeny has been altered, particularly in the case of entomology where new species are discovered regularly, so using the most recent taxonomy may not actually represent the state-of-the-art knowledge of the evolution of the species. In the case of iNaturalist, the dataset does include a tree with the same number of levels as Rove-Tree-11, however, this depth begins from kingdom-level, whereas ours begins from family level (four taxonomic ranks lower on the taxonomic hierarchy), and represents the most recent phylogeny.

Additionally there are non-biological hierarchical datasets, such as DeepFashion [27], for which others have created their own hierarchy [5]. This hierarchy is however based on loose groupings of clothing items which are highly subjective. For example, the top-level groupings are: top, bottom, onepiece, outer and special, where special includes fashion items such as kaftan, robe, and onesie, which might morphologically be more related to coats, which are in the 'outer' cate-

---

[2] The taxonomy represents how the organism is classified - ie which class, order, family the organism belongs to, and is a non binary tree. The phylogeny represents how related different species are together, and would ideally be a binary tree. In an ideal world the taxonomy would be a congruent to the phylogeny, but in reality they tend to diverge as taxonomic revisions take longer.

Table 1. Comparison of dataset properties. The table indicates number of images and categories, tree depth and whether or not the images are 'in the wild'. Tree depth is calculated as the maximum number of levels in the tree. For example, with iNaturalist this is 7 (corresponding to: kingdom, phylum, class, order, family, genus and species)

Dataset	No. Images	No. Cat	Tree Depth	Wild?	Year
Rove-Tree-11	13,887	215	11	No	2022
ImageNet [6]	14,197,122	21,841	2	Yes	2018
IP102 [44]	75,222	102	4	Yes	2019
CUB-200-2011 [41]	11,788	200	4 [2]	Yes	2011
Cars196 [24]	16,185	196	1	Yes	2013
INaturalist 2021 [40]	857,877	5,089	7	Yes	2021
PlantCLEF 2021 [13]	330,772	997	3	Mixed	2021
DeepFashion [27]	800,000	50	4[a]	Yes	2016

[a]hierarchy presented in [5]

gory. This kind of subjective hierarchy can be useful in other applications, but not particularly for research on generating relationships based on morphology.

The Rove-Tree-11 dataset on the other hand is a well-curated museum collection, where the identification has been done by experts, often using a microscope, and the ground truth phylogeny is as up to date as possible. Additionally, because the images are of museum collections and not 'in-the-wild', the specimen is always fully visible, and the dataset has been curated to include only whole dorsal images. Whether dorsal-view images are sufficient to generate a phylogeny remains to be seen. Typically biologists would use features from all over the body, including ventral and sometimes internal organs. We hypothesize that dorsal view may be sufficient given the ability of deep learning models to learn patterns which are difficult for the human eye to distinguish. Additionally results from our classification experiments shown in Table 3 suggest that distinguishing features can be learnt from the images, supporting our belief that phylogenies may be learnt from this dataset.

2.2   Related Methodologies

Classification. Classification is one of the most developed fields in computer vision and deep learning, with numerous new state of the art architectures and methods discovered each year. However, there are some architectures which have gained widespread usage in recent years, which we will use to give baselines for this dataset. In particular, we will compare classification results using ResNet [16] and EfficientNet B0 [39]. ResNet is a series of models, introduced in 2015, which uses residual convolution blocks. EfficientNet was introduced in 2019 and is known for achieving high accuracies with few parameters. Classification is not the main focus of this dataset, but we provide classification results for comparison with similar datasets.

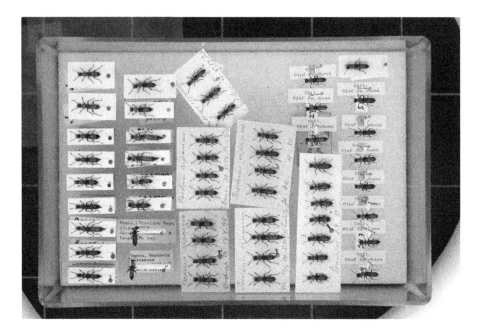

Fig. 2. Example image of museum unit tray from Stage 1 of image processing.

Deep Metric Learning. The goal of deep metric learning (DML) is to learn an embedding of the data which represents the dataset and distances between datapoints meaningfully. This could be through clustering related data together, or through creating independence and interpretability in the variables. Recent research into deep metric learning can be split into three groups [38]. **Ranking-based** methods attempt to pull instances from the same class (positive examples) closer together in the embedding space, and typically push examples from other classes further away (eg, [15] [43]). **Classification-based methods**, such as ArcFace [7], work by modifying the discriminative classification task. Finally **Proxy Based** methods, such as Proxy NCA [29] compare each sample with a learned distribution for each class.

In this paper we demonstrate results for this dataset using seven deep metric learning methods; Five ranking-based losses: margin loss [43], triplet loss [43], contrastive loss [15], multisimilarity loss [42], lifted loss [45], one classification-based loss: arcface loss [7] and one proxy-based loss: proxynca [29]. With many state of the art methods and variations on these, choosing which to use is difficult. We chose these firstly because they are all used in [38] as benchmarks, making our results directly comparable. Of the 23 described in [38], we focus on seven which represented some of the better results and show a variety of methods. For a detailed description of each loss we refer the reader to [38].

During training DML models are typically evaluated not just on the loss, but also on a number of clustering metrics. In our case, to do this the dataset is evaluated using nearest neighbors Recall@1 (R1) and Normalized Mutual Information

(NMI) after clustering using the k-means algorithm [28]. NMI is presented in our main results, and R1 in the supplemental material. NMI is a symmetric quantity measuring the overlap between clusters. A NMI of 1 indicates that the clusters are the same. Recall@1 is a measure of the % of results with a nearest neighbour in the same class. Both are described in further detail in [38].

Generating a Phylogeny from Embeddings. In order to use this dataset for deep phylogenetic generation, we need methods to generate binary graphs from embedding spaces. We could treat this as a classification problem, however, with only one graph to generate, this dataset is not large enough to perform direct graph generation. Instead, the graph can be generated indirectly from the embedding space and compared with the ground truth. This is analogous to how biologists would traditionally generate phylogenetic trees for small datasets using morphological matrices. Biologists use maximum parsimony or bayesian methods [10] to find the best-fitting tree based on discrete characters (either morphological or genetic). However, the use of continuous characters in improving phylogeny generation has been recently explored [35]. Therefore if we assume our embedding space represents morphological features and is a morphological space, this could similarly be used to generate a phylogeny using the same continuous trait bayesian phylogenetic inference methods. We use RevBayes [19], a popular bayesian inference package to complete the analysis. Similar methods have been used to generate phylogenetic trees [23].

Phylogenetic Comparison. The main purpose of this dataset is to allow exploration of methods for generating phylogenetic trees based on morphology. To do this, we need methods for comparing phylogenies. There are many standard methods of doing this in biology, a thorough comparison of them is provided in [25]. In brief, the metrics can be split into those which do and do not compare branch lengths. As branch lengths (i.e. evolutionary time) are not yet available in our ground truth phylogeny, we will focus on those which do not include branch lengths, called topology-only comparison methods. The most widely used of these is called the Robinson-Foulds (RF) metric, introduced in 1981 [37]. The RF metric defines the dissimilarity between two trees as the number of operations that would be required to turn one tree into another[3]. However, it has some notable disadvantages, including that apparently similar trees can have a disproportionately high RF score.

One of the more recently introduced metrics is called the Align Score [33]. The Align Score works in two stages. In the first stage, a 1:1 mapping of edges from each tree (T_1 and T_2) is assigned. This is done by calculating a similarity score $s(i, j)$ between the edges, i and j in T_1 and T_2 respectively, based on how similarly they partition the tree. More concretely, in tree T_1, edge i will partition the tree into two disjoint subsets P_{i0} and P_{i1}. The similarity scores can then by

---

[3] it is, however, different from the edit distance popular in computer science.

Fig. 3. Examples of specimen images before (above) and after (below) segmentation and rotation adjustment.

computed as:

$$s(i,j) = 1 - \max(\min(a_{00}, a_{11}), min(a_{01}, a_{10})) \tag{1}$$

where a_{rs} is the intersection over the union of the partitions:

$$a_{rs} = |\frac{P_{ir} \cap P_{js}}{P_{ir} \cup P_{js}}| \tag{2}$$

The munkres algorithm is then used to find the edge $j = f(i)$ that minimizes the assignment problem, and then the group with the minimum pairs are summed as follows to calculate the total align score for the two trees:

$$\sum_{i \in T_1} s(i, f(i)) \tag{3}$$

Unlike the RF score, for each set of partitions the align score calculates the similarity, $s(i,j)$, as a continuous variable instead of a binary value. That said, it has the disadvantage that the value is not normalized - a larger tree will likely have a larger align score, making the result difficult to interpret. Despite this, we choose to use it as it is a more accurate representation of the topological similarity between two trees[25].

3   An Overview of Rove-Tree-11

3.1   Image Collection

The images in the dataset were collected and prepared in 4 stages [14]:

Stage 1: Unit Tray Image Collection Rove-Tree-11 was collected by taking overview images of 619 unit trays from the entomology collection at Natural History Museum of Denmark, see Fig. 2. A Canon EOS R5 mounted on a camera stand with a macro lens was used to take images of 5760 × 3840 pixels (px) resolution. Since the camera height and focus were kept fixed, the images can be related to physical distance as approx. 400 px per cm. Artificial lighting was used to minimize lighting variance over the images.

Table 2. Species-level classification results on segmented and unsegmented images. We can see that using segmentations drastically reduces the accuracy, indicating that the model is learning from the background and not the morphology of the beetle, as desired. Top-1 and Top-5 represent accuracies. Uncertainties represent 95% confidence intervals.

Model	Dataset	Top-1	Top-5
ResNet-18 [16]	Segmented	90.9 ± 1.2	99.1 ± 1.2
ResNet-18 [16]	Unsegmented	99.1 ± 0.3	99.9 ± 0.3

Stage 2: Bounding Box Identification and Sorting. After image capture, bounding boxes for the individual specimens were then manually annotated using Inselect [18]. Images of 19,722 individual specimens were then sorted. Only dorsal views (views from the 'back' of the beetle) where the specimen was largely intact and limbs were mostly visible were included, resulting in images of 13,887 specimens in final dataset. See Fig. 3 for examples of bounding boxes around specimens. Estimates of body rotation were also annotated in 45°C increments which allows for coarse correction of the orientation of the crops.

Stage 3: Segmentation. Segmentations were then generated through an iterative process. First 200 images were manually segmented. Then U-Net was trained on these 200 images and was used to generate predictions for the rest of the images. 3000 of these segmentations were considered good enough. U-Net was then retrained with these images, then rerun and new segmentations produced. The final segmentations were then manually corrected. Examples of segmentation masks and final segmented specimens can be seen in Fig. 3 and Fig. 4. The dataset is released with both the original crops and the segmentation masks, however, as we show in Table 2, the segmentations are extremely important for phylogenetic analysis, as the background of the image is highly correlated with the species. This is because many of the same species were collected at the same time in the same place by the same person, meaning whether the specimen was glued to a card, the age and color of the card, could be correlated with the species, despite being unrelated to the phylogeny. The segmentations are not perfect. In particular they cut off some of the finer hairs on the body; It could therefore be the case that the segmentations are removing vital information which the model can use to complete classification. We consider this unlikely and suspect the model is instead learning from the backgrounds.

Stage 4: Rotation Adjustment. Rotations were corrected by finding the principal axis of inertia of the segmentation masks, (see [21] for details). Since all the beetles are more or less oval shaped, the minimal axis of rotation of their masks tends to line up well with their heads and tails. Using this we further standardized the rotations of the segmentations. This process is shown in Fig. 4.

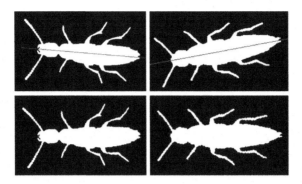

Fig. 4. Illustration of rotation adjustment algorithm. Example original masks (top) and rotated masks (bottom). The red line represents the principal axis of inertia found.

3.2   Preparation of Phylogeny

A current genus-level phylogeny of the closely related subfamilies Staphylininae, Xantholininae and Paederinae is provided for the sample of genera used in our analysis. The full phylogeny is visualized in Fig. 1 our supplementary material. A subset is shown in Fig. 1. This phylogeny represents the current state of knowledge as it was pieced together from the most relevant recently published phylogenetic analyses, such as [47] for sister-group relationships among all three subfamilies and the backbone topology of Xantholininae and Staphylininae, [3] for the subtribe Staphylinina, [4] for the subtribe Philonthina and [46] for the subfamily Paederinae. Below genus-level the phylogeny is considered unresolved as we were unable to find species-level phylogenies for the 215 species included in Rove-Tree-11. A newick file of the phylogeny is provided with the dataset.

3.3   Dataset Statistics

In total, 13,887 images of beetles from the family Staphylinidae, commonly known as rove beetles, are included from 215 species - spanning 44 genera, 9 tribes and 3 subfamilies. Example images are shown in Fig. 1.

The distribution of the dataset per genus is shown in Fig. 7. A species-level distribution is provided in the supplementary material. From this we can see that the dataset is not evenly distributed, with the species with the highest number of specimens having 261 examples and the lowest having 2 with the genus Philonthus accounting for 24.8% of the dataset. This is due to the number of specimens the museum had in the unit trays that were accessed and imaged at the time, although the curators also includes samples of species which were easily distinguishable from each other, and examples which were hard and can only usually be determined by genital extraction by experts (i.e. Lathrobium geminum and Lathrobium elongatum. Examples from these two species are shown in Fig. 5 to demonstate the difficulty of the task). The distribution of image sizes in the dataset is shown in Fig. 6. The majority of the images (82%) are under 500 × 250 pixels.

Fig. 5. Example images of Lathrobium geminum (top) and Lathrobium elongatum (bottom) from the dataset. Typically even experts need to dissect the specimen to complete the determination between these two species.

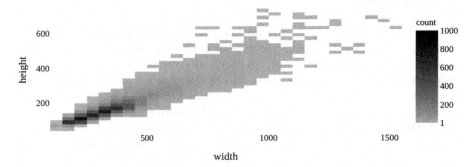

Fig. 6. Distribution of image sizes included in the dataset. The majority (82%) of images are under 500 × 250 pixels.

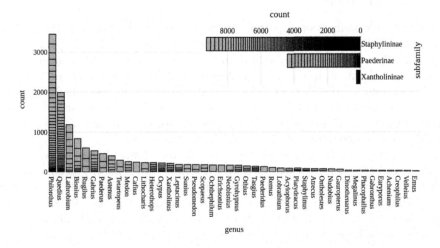

Fig. 7. Distribution of specimens per genus (bottom left) and per subfamily (top right). Each slice in the stacked bar chart represents a different species within that genus. Subfamily distribution is included as it is used to generate the validations and test sets for the clustering results in Sect. 4.2. A full species level distribution is shown in the supplemental material.

Table 3. Classification results using deep learning architectures. Top-1 and Top-5 represent accuracies. Uncertainties represent 95% confidence intervals based on 3 runs.

Model		Species		Genus	
	Params	Top-1	Top-5	Top-1	Top-5
ResNet-18 [16]	11.4 M	90.9 ± 1.2	99.2 ± 1.2	98.9 ± 0.3	100 ± 0.3
ResNet-50 [16]	23.9 M	89.4 ± 1.4	99.2 ± 1.4	98.2 ± 0.4	100 ± 0.4
EfficientNet B0 [39]	5.3 M	91.9 ± 1.8	99.3 ± 1.8	99.1 ± 0.2	100 ± 0.2

4   Evaluation

Here we evaluate the dataset by performing benchmark experiments. As stated previously, the main purpose of this dataset is for deep metric learning on hierarchical phylogenetic relationships, so this is also the focus of the benchmarks, although we also provide benchmarks for the classification and clustering tasks. The same augmentations were applied to the dataset as for CUB200 and Cars176 and as in [38], with the exception that the RandomHorizontalFlip was changed to a RandomVerticalFlip, as this makes more sense for the Rove-Tree-11 dataset. Gradient accumulation was also used in some cases due to memory constraints on the available clusters. The details of which experiments this was applied to are provided in the codebase.

4.1   Classification

Results from classification experiments are provided in Table 3. For these experiments the official pytorch implementations were used with default parameters: categorical cross entropy loss with an initial learning rate of 0.1, momentum of 0.9, weight decay of $1e - 4$ and SGD optimizer. Training details are released with the code for this dataset. The only alterations from the defaults were to reduce the batch-size to 32 due to memory constraints and to alter the data augmentations, detailed in the code. A species-stratified train/val/test split of 70/15/15 was used. The split is provided with the code.

As shown in Table 3, the models are able to achieve a top-1 species-level accuracy of 92% with no hyperparameter tuning, and a top-1 genus level of almost 100%. These results suggest that although this dataset could be used for classification tasks and might be useful as such for biologists, classification of this dataset is not particularly difficult, and this dataset is probably not ideal as a benchmark for classification in deep learning.

4.2   Clustering and Phylogenetic Results

In Table 4, we present benchmark results of applying state of the art methods for deep metric learning to the Rove-Tree-11 dataset and comparing phylogenies generated using phylogenetic bayesian methods on the embedding space to the ground truth phylogeny as described in Sect. 2. A more complete table showing R1 scores and Cars176 results, is provided for reference in the supplementary material (Table 1). The 'Random' row represents the align score of a randomly generated tree with the 9 genera leaves included in the test set, against the ground truth tree based on 5 random initializations. Since the align score is not normalized, this random baseline is useful to gauge our results and represents an upper bound our models should achieve. Following best practice, as described in [30], the dataset was split into three groups for training, validation and testing. To properly test the ability of the model to generalize, the groups were split at subfamily level, so the train, validation and test sets should be as phylogenetically distinct as possible, in the sense that they belong to different parts of the phylogenetic tree. This results in 8534 training images from the subfamily Staphylininae, 4399 validation images from the subfamily Paederinae and 954 test images from the subfamily Xantholininae.

All results on Rove-Tree-11 were generated using implementations used in [38], modified to calculate the align score. A forked codebase is provided as a submodule in the github repository.

Based on the clustering results in Table 4, we see that Rove-Tree-11 has similar NMI scores to CUB200, suggesting this dataset has a similar clustering difficulty to CUB200 and may be appropriate as a clustering benchmark. As with CUB200, the best models on Rove-Tree-11 are Triplet [43] and Multisimilarity [42]. We can also see that the align score results somewhat correspond with the NMI, with the best results being achieved with Triplet Loss. We can also see that the best test set align score of 4.0 is a marked improvement to the random align score baseline of 6.6, but still significantly far away from a perfect

Table 4. Benchmark clustering and Align-Score results on Rove-Tree-11 dataset. 'Random' represents the average align score of 5 randomly generated trees. This gives us a metric to compare our results with. A perfect align score would be 0. 95% confidence errors are provided based on 5 runs.

| | CUB200 | | Rove-Tree-11 | | | |
| | Test | | Validation | | Test | |
Loss	NMI	Align	NMI	Align	NMI	Align
Random	–	21.9 ± 0.2	–	15.8 ± 0.9	–	6.6 ± 0.5
Triplet	64.8 ± 0.5	9.9 ± 0.9	68.9 ± 0.4	$\mathbf{7.8 \pm 1.1}$	66.3 ± 0.3	4.1 ± 0.5
Margin	60.7 ± 0.3	10.6 ± 1.2	68.0 ± 0.7	8.2 ± 0.7	65.9 ± 0.5	4.2 ± 0.7
Lifted	34.8 ± 3.0	15.9 ± 2.0	55.0 ± 0.6	10.5 ± 0.7	56.0 ± 1.1	4.9 ± 0.8
Constrast.	59.0 ± 1.0	11.0 ± 1.2	66.7 ± 0.5	8.5 ± 1.0	65.4 ± 0.5	4.5 ± 0.6
Multisim.	$\mathbf{68.2 \pm 0.3}$	$\mathbf{8.6 \pm 0.8}$	$\mathbf{70.7 \pm 0.2}$	8.2 ± 0.4	$\mathbf{67.3 \pm 0.5}$	$\mathbf{4.0 \pm 0.5}$
ProxyNCA	66.8 ± 0.4	9.8 ± 0.8	67.5 ± 0.7	9.0 ± 0.8	65.5 ± 0.3	4.2 ± 0.4
Arcface	67.5 ± 0.4	9.8 ± 0.8	66.9 ± 0.9	8.5 ± 0.4	64.8 ± 0.5	4.1 ± 0.4

align score of 0, suggesting there is room for improvement. We find it surprising that the align score of the best model on the CUB200 dataset shows a 60% improvement to the random score, while on Rove-Tree-11 the improvement is only 40% on the test set and 51% on the validation set. This suggests that either CUB200 is an easier dataset to generate phylogenies from, or could be an artifact of the align score on trees of different depths (CUB200 has a depth of 4, while Rove-Tree-11 has a depth of 11). It is surprising that it could be an easier dataset, given that the images are in-the-wild, but this could also be due to phylogenetically close birds having similar backgrounds in the images (water-faring birds might typically have ocean backgrounds, for example, and be more closely phylogenetically related). The phylogenetic tree produced by the best model is provided in the supplementary material along with the ground truth tree for visual inspection.

5   Conclusions

In this paper we present Rove-Tree-11, a novel dataset of segmented images of and research-grade classifications of rove beetles for researching methods for generating phylogenies from images. We provide an eleven-level fine-grained ground truth phylogeny for the 44 (train, validation and test) genera included in this dataset.

We start by demonstrating the importance of the provided segmentations as the model can learn from the background. We show benchmark results on this dataset for classification, deep metric learning methods and tree alignment. We further demonstrate that this dataset shows similar clustering results to the CUB200 dataset suggesting it may be appropriate as an alternative clustering benchmark. Finally, we demonstrate how this dataset can be used to generate and compare phylogenies based on the align score, and show that while it is possible to generate such trees, there is plenty of room for improvement and we hope this will be a growing field of research. Code and data are available (code: https://github.com/robertahunt/Rove-Tree-11, data: http://doi.org/10.17894/ucph.39619bba-4569-4415-9f25-d6a0ff64f0e3).

Acknowledgements. Many people have been involved in this project. First we would like to thank David Gutschenreiter, Søren Bech and André Fastrup who took the photos of the unit trays and completed the initial segmentations of the images as part of their theses. Next, we would like to thank Alexey Solodovnikov of the Natural History Museum of Denmark for providing the specimens, the ground truth phylogeny and guidance for all things entomological. Also, thanks Francois Lauze and the entire Phylorama team for their input the project.

Ethical Concerns. Models similar to those described, if applied to images of faces, could be used to generate family trees for humans. This could result in public images being used to infer familial relationships which could have a negative societal impact. The authors strongly discourage this form of misuse of the proposed methods.

References

1. Bakalar, N.: Nicholas. The New York Times (2014). https://www.nytimes.com/2014/05/27/science/welcoming-the-newly-discovered.html
2. Bameri, F., Pourreza, H.R., Taherinia, A.H., Aliabadian, M., Mortezapour, H.R., Abdilzadeh, R.: TMTCPT: The tree method based on the taxonomic categorization and the phylogenetic tree for fine-grained categorization. Biosystems **195**, 104137 (2020). https://doi.org/10.1016/j.biosystems.2020.104137
3. Brunke, A., Smetana, A.: A new genus of staphylinina and a review of major lineages (staphylinidae: Staphylininae: Staphylinini). System. Biodiv. **17**, 745–758 (2019). https://doi.org/10.1080/14772000.2019.1691082
4. Chani-Posse, M.R., Brunke, A.J., Chatzimanolis, S., Schillhammer, H., Solodovnikov, A.: Phylogeny of the hyper-diverse rove beetle subtribe philonthina with implications for classification of the tribe staphylinini (coleoptera: Staphylinidae). Cladistics **34**(1), 1–40 (2018). https://doi.org/10.1111/cla.12188
5. Cho, H., Ahn, C., Min Yoo, K., Seol, J., Lee, S.g.: Leveraging class hierarchy in fashion classification. In: Proceedings of the IEEE/CVF International Conference on Computer Vision (ICCV) Workshops (October 2019)
6. Deng, J., Dong, W., Socher, R., Li, L.J., Li, K., Fei-Fei, L.: Imagenet: A large-scale hierarchical image database. In: 2009 IEEE Conference On Computer Vision And Pattern Recognitio, pp. 248–255. IEEE (2009)
7. Deng, J., Guo, J., Zafeiriou, S.: Arcface: Additive angular margin loss for deep face recognition. CoRR abs/ arXiv: 1801.07698 (2018)
8. DiSSCo: Distributed system of scientific collections (July 2022). https://www.dissco.eu/
9. Felsenstein, J.: Evolutionary trees from DNA sequences: A maximum likelihood approach. J. Mol. Evol. **17**, 368–376 (1981)
10. Felsenstein, J.: Statistical inference of phylogenies. J. Royal Statis. Soc. Ser. A (General) **146**(3), 246–262 (1983). https://doi.org/10.2307/2981654
11. Felsenstein, J.: Inferring phylogenies. Sinauer associates, Sunderland, MA (2003)
12. Fink, M., Ullman, S.: From aardvark to zorro: A benchmark for mammal image classification. Int. J. Comput. Vision **77**(1–3), 143–156 (2008). https://doi.org/10.1007/s11263-007-0066-8
13. Goëau, H., Bonnet, P., Joly, A.: Overview of plantclef 2021: cross-domain plant identification. In: Working Notes of CLEF 2021-Conference and Labs of the Evaluation Forum, vol. 2936, pp. 1422–1436 (2021)
14. Gutschenreiter, D., Bech, S.: Deep-learning methods on taxonomic beetle data Automated segmentation and classification of beetles on genus and species level. Master's thesis, University of Copenhagen (2021)
15. Hadsell, R., Chopra, S., LeCun, Y.: Dimensionality reduction by learning an invariant mapping. In: 2006 IEEE Computer Society Conference on Computer Vision and Pattern Recognition (CVPR'06). vol. 2, pp. 1735–1742 (2006). https://doi.org/10.1109/CVPR.2006.100
16. He, K., Zhang, X., Ren, S., Sun, J.: Deep residual learning for image recognition. arXiv preprint arXiv:1512.03385 (2015)
17. Hedrick, B.P., et al.: Digitization and the future of natural history collections. Bioscience **70**(3), 243–251 (2020). https://doi.org/10.1093/biosci/biz163
18. Hudson, L.N., et al.: Inselect: Automating the digitization of natural history collections. PLoS ONE **10**(11), 1–15 (2015). https://doi.org/10.1371/journal.pone.0143402

19. Höhna, L., Heath, B., Lartillot, M., Huelsenbeck, R.: Revbayes: Bayesian phylogenetic inference using graphical models and an interactive model-specification language. Syst. Biol. **65**, 726–736 (2016)
20. iDigBio: Integrated digitized biocollections (July 2022). https://www.idigbio.org/
21. J. Peraire, S.W.: Lecture notes from mit course 16.07 dynamics, fall 2008. l26–3d rigid body dynamics: The inertia tensor (2008). https://ocw.mit.edu/courses/16-07-dynamics-fall-2009/dd277ec654440f4c2b5b07d6c286c3fd_MIT16_07F09_Lec26.pdf
22. Kaya, M., Bilge, H.S.: Deep metric learning: A survey. Symmetry **11**(9) (2019). https://doi.org/10.3390/sym11091066, https://www.mdpi.com/2073-8994/11/9/1066
23. Kiel, S.: Assessing bivalve phylogeny using deep learning and computer vision approaches. bioRxiv (2021). https://doi.org/10.1101/2021.04.08.438943, https://www.biorxiv.org/content/early/2021/04/09/2021.04.08.438943
24. Krause, J., Stark, M., Deng, J., Fei-Fei, L.: 3d object representations for fine-grained categorization. In: 4th International IEEE Workshop on 3D Representation and Recognition (3dRR-13), Sydney, Australia (2013)
25. Kuhner, M.K., Yamato, J.: Practical performance of tree comparison metrics. System. Biol. **64**(2), 205–214 (2014). https://doi.org/10.1093/sysbio/syu085
26. Lee, M., Palci, A.: Morphological phylogenetics in the genomic age. Current Biol. **25**(19), R922–R929 (2015). https://doi.org/10.1016/j.cub.2015.07.009, https://www.sciencedirect.com/science/article/pii/S096098221500812X
27. Liu, Z., Luo, P., Qiu, S., Wang, X., Tang, X.: Deepfashion: Powering robust clothes recognition and retrieval with rich annotations. In: Proceedings of IEEE Conference on Computer Vision and Pattern Recognition (CVPR) (June 2016)
28. MacQueen, J.: Classification and analysis of multivariate observations. In: 5th Berkeley Symposium on Mathematical Statistics and Probability, pp. 281–297 (1967)
29. Movshovitz-Attias, Y., Toshev, A., Leung, T.K., Ioffe, S., Singh, S.: No fuss distance metric learning using proxies. In: 2017 IEEE International Conference on Computer Vision (ICCV), pp. 360–368 (2017). https://doi.org/10.1109/ICCV.2017.47
30. Musgrave, K., Belongie, S., Lim, S.-N.: A metric learning reality check. In: Vedaldi, A., Bischof, H., Brox, T., Frahm, J.-M. (eds.) ECCV 2020. LNCS, vol. 12370, pp. 681–699. Springer, Cham (2020). https://doi.org/10.1007/978-3-030-58595-2_41
31. Natural History Museum of Denmark: Digital nature: Giant grant makes the natural history collections of denmark accessible to everyone (2021). Newsletter
32. Natural History Museum of Denmark: (2022). Entomology - Dry and Wet Collections. Homepage
33. Nye, T., Lio, P., Gilks, W.: A novel algorithm and web-based tool for comparing two alternative phylogenetic trees. Bioinformatics (Oxford, England) **22**, 117–119 (2006). https://doi.org/10.1093/bioinformatics/bti720
34. Orlov, I., Leschen, R.A., Żyła, D., Solodovnikov, A.: Total-evidence backbone phylogeny of aleocharinae (coleoptera: Staphylinidae). Cladistics **37**(4), 343–374 (2021). https://doi.org/10.1111/cla.12444
35. Parins-Fukuchi, C.: Use of continuous traits can improve morphological phylogenetics. System. Biol. **67**(2), 328–339 (2017). https://doi.org/10.1093/sysbio/syx072
36. Popov, D., Roychoudhury, P., Hardy, H., Livermore, L., Norris, K.: The value of digitising natural history collections. Res. Ideas Outcomes **7**, e78844 (2021). https://doi.org/10.3897/rio.7.e78844

37. Robinson, D., Foulds, L.: Comparison of phylogenetic trees. Math. Biosci. **53**(1), 131–147 (1981). https://doi.org/10.1016/0025-5564(81)90043-2, https://www.sciencedirect.com/science/article/pii/0025556481900432
38. Roth, K., Milbich, T., Sinha, S., Gupta, P., Ommer, B., Cohen, J.P.: Revisiting training strategies and generalization performance in deep metric learning (2020)
39. Tan, M., Le, Q.: EfficientNet: Rethinking model scaling for convolutional neural networks. In: Chaudhuri, K., Salakhutdinov, R. (eds.) Proceedings of the 36th International Conference on Machine Learning. Proceedings of Machine Learning Research, vol. 97, pp. 6105–6114. PMLR (09–15 June 2019), https://proceedings.mlr.press/v97/tan19a.html
40. Van Horn, G., Cole, E., Beery, S., Wilber, K., Belongie, S., MacAodha, O.: Benchmarking representation learning for natural world image collections. In: Proceedings of the IEEE/CVF Conference on Computer Vision and Pattern Recognition, pp. 12884–12893 (2021)
41. Wah, C., Branson, S., Welinder, P., Perona, P., Belongie, S.: The caltech-ucsd birds-200-2011 dataset. Tech. Rep. CNS-TR-2011-001, California Institute of Technology (2011)
42. Wang, X., Han, X., Huang, W., Dong, D., Scott, M.R.: Multi-similarity loss with general pair weighting for deep metric learning. In: 2019 IEEE/CVF Conference on Computer Vision and Pattern Recognition (CVPR), pp. 5017–5025 (2019). https://doi.org/10.1109/CVPR.2019.00516
43. Wu, C.Y., Manmatha, R., Smola, A.J., Krähenbühl, P.: Sampling matters in deep embedding learning. In: 2017 IEEE International Conference on Computer Vision (ICCV), pp. 2859–2867 (2017). https://doi.org/10.1109/ICCV.2017.309
44. Wu, X., Zhan, C., Lai, Y., Cheng, M.M., Yang, J.: Ip102: A large-scale benchmark dataset for insect pest recognition. In: IEEE CVPR, pp. 8787–8796 (2019)
45. Yuan, Y., Chen, W., Yang, Y., Wang, Z.: In defense of the triplet loss again: Learning robust person re-identification with fast approximated triplet loss and label distillation (2019)
46. Żyła, D., Bogri, A., Hansen, A.K., Jenkins Shaw, J., Kypke, J., Solodovnikov, A.: A New Termitophilous Genus of Paederinae Rove Beetles (Coleoptera, Staphylinidae) from the Neotropics and Its Phylogenetic Position. Neotrop. Entomol. **51**(2), 282–291 (2022). https://doi.org/10.1007/s13744-022-00946-x
47. Żyła, D., Solodovnikov, A.: Multilocus phylogeny defines a new classification of staphylininae (coleoptera, staphylinidae), a rove beetle group with high lineage diversity. Syst. Entomol. **45**(1), 114–127 (2020). https://doi.org/10.1111/syen.12382

UTB180: A High-Quality Benchmark for Underwater Tracking

Basit Alawode[1], Yuhang Guo[1], Mehnaz Ummar[1], Naoufel Werghi[1], Jorge Dias[1], Ajmal Mian[2], and Sajid Javed[1(✉)]

[1] Department of Electrical Engineering and Computer Science, Khalifa University of Science and Technology, Abu Dhabi, UAE
{100060517,sajid.javed}@ku.ac.ae
[2] The University of Western Australia, Stirling Highway, Perth, Australia

Abstract. Deep learning methods have demonstrated encouraging performance on open-air visual object tracking (VOT) benchmarks, however, their strength remains unexplored on underwater video sequences due to the lack of challenging underwater VOT benchmarks. Apart from the open-air tracking challenges, videos captured in underwater environments pose additional challenges for tracking such as low visibility, poor video quality, distortions in sharpness and contrast, reflections from suspended particles, and non-uniform lighting. In the current work, we propose a new Underwater Tracking Benchmark (UTB180) dataset consisting of 180 sequences to facilitate the development of underwater deep trackers. The sequences in UTB180 are selected from both underwater natural and online sources with over 58,000 annotated frames. Video-level attributes are also provided to facilitate the development of robust trackers for specific challenges. We benchmark 15 existing pre-trained State-Of-The-Art (SOTA) trackers on UTB180 and compare their performance on another publicly available underwater benchmark. The trackers consistently perform worse on UTB180 showing that it poses more challenging scenarios. Moreover, we show that fine-tuning five high-quality SOTA trackers on UTB180 still does not sufficiently boost their tracking performance. Our experiments show that the UTB180 sequences pose a major burden on the SOTA trackers as compared to their open-air tracking performance. The performance gap reveals the need for a dedicated end-to-end underwater deep tracker that takes into account the inherent properties of underwater environments. We believe that our proposed dataset will be of great value to the tracking community in advancing the SOTA in underwater VOT. Our dataset is publicly available on Kaggle.

1 Introduction

Visual Object Tracking (VOT) is the task of estimating the trajectory and state of an arbitrary target object in a video sequence [20]. Given the location of the

Supplementary Information The online version contains supplementary material available at https://doi.org/10.1007/978-3-031-26348-4_26.

L. Wang et al. (Eds.): ACCV 2022, LNCS 13845, pp. 442–458, 2023.
https://doi.org/10.1007/978-3-031-26348-4_26

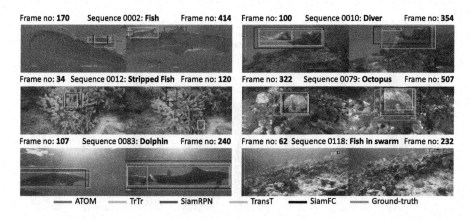

Fig. 1. Sample images of our proposed UTB180 benchmark dataset. The tracking results of some representative State-Of-The-Art (SOTA) trackers including ATOM [8], TrTr [41], SiamRPN [25,26], TransT [5], and SiamFC [1] are shown in terms of bounding boxes. The frame indexes and sequence names are also shown in each row.

target in the first frame, the main objective is to learn a robust appearance model to be used when searching for the target object in subsequent frames [1,18]. VOT has numerous open-air applications including autonomous driving, video surveillance, robotics, medical imaging, and sports video analysis [10,16]. In recent years, dominant deep learning trackers such as Siamese [1,25,26], correlation filters [2,8] and transformers [29] have advanced the SOTA performance in tracking. Despite the recent progress, VOT is still an open problem in computer vision because of its challenging nature [16].

Underwater video analysis is an emerging research area where VOT has significant importance in robotics applications including ocean exploration, homeland and maritime security, sea-life monitoring, search and rescue operations to name a few [4,14,16]. Over the years, considerable progress has been made by the tracking community in the development of SOTA end-to-end open-air trackers [16,17,19,19,21,22]. One of the main reasons behind this success is the availability of a variety of large-scale open-air tracking benchmarks such as LASOT [9], GOT-10K [15], and TrackingNet [31] to train, objectively evaluate and compare the different trackers. For instance, as shown in Fig. 2, these datasets exist in small and large scale from a few hundreds of video sequences such as the VOT dataset series [13,23,24], Object Tracking Benchmark (OTB100) [37], Unmanned Aerial Vehicle (UAV) [30], Temple Color (TC) [27] to several thousands of video sequences such as Large-Scale Single Object Tracking (LaSOT) [9], Generic Object Tracking (GOT-10K) [15], and TrackingNet [31]. These datasets provide high quality dense annotations (i.e. per frame) to ensure more accurate evaluations of open-air deep trackers [9,31]. As shown by the average sequence duration (see Fig. 2 and Table 1), they are available for both short-term (average sequence length less than 600 frames) [13,23,24,37] and long-term [9,30] tracking with video specific attributes to further enhance the tracking performance. Furthermore, the large number of video sequences and span variability

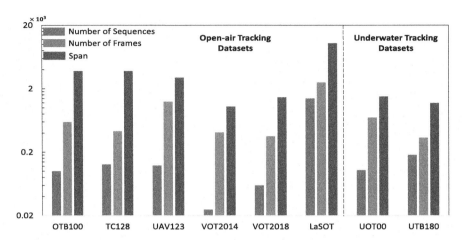

Fig. 2. Summary of open-air and underwater tracking benchmark datasets. Open-air datasets includes OTB100 [37], TC128 [27], UAV123 [30], VOT2014 [13], VOT2018 [24], and LaSOT [9]. Underwater datasets include UOT100 [32] and our proposed UTB180. Span means the difference between the minimum and the maximum number of frames per sequence.

have encouraged the direct training of deep open-air trackers for generic object tracking in these datasets [9,15,31].

All these aforementioned characteristics have immensely contributed towards open-air tracking. However, the same cannot be said for underwater VOT despite its importance. All solutions in this context have simply deployed open-air trackers directly on underwater visual data [32]. One of the major reasons for such stagnation is the unavailability of high-quality benchmarks for underwater tracking exhibiting the challenges of underwater scenes such as of poor visibility, non-uniform lighting conditions, scattering, absorption, blurring of light, flickering of caustic patterns, photometric artifacts, and color variations. To the best of our knowledge, Underwater Object Tracking (UOT100) benchmark is the only available dataset containing 100 underwater sequences covering diverse categories of objects [32]. Our frame-wise evaluation on UOT100 reveals that it belongs to the category of sparsely annotated benchmark datasets using a semi-automatic annotation approach i.e. manual annotations performed every 20 frames and the rest are generated by a tracker. While such an approach speeds up the annotation process, it often yields less accurate ground-truth bounding box predictions due to the propagation and accumulation of the tracker's prediction errors in subsequent frames. All the above motivate us to propose a novel high-quality benchmark dataset UTB180 for the tracking community.

1.1 Contributions

Following are the main contributions of this paper:

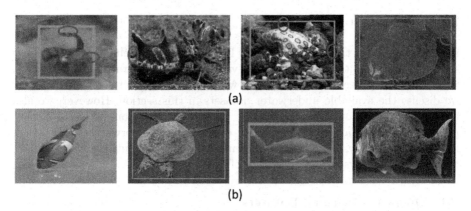

Fig. 3. Zoomed-in samples from the (a) UOT100 underwater benchmark dataset and (b) our proposed UTB180 dataset. The green rectangles show the ground-truth annotations. The red ovals highlight annotation errors in the UOT100 dataset. Our proposed dataset provides more accurate annotations of the target objects compared to UOT100. (Color figure online)

1. **Creation of a Dense and Diversified High-Quality Underwater Tracking Benchmark (UTB180) dataset**. Our dataset consists of 180 sequences with over 58,000 annotated frames. Each frame is carefully manually annotated and then visually verified to ensure its correctness. The dataset includes both natural and artificial videos under varying visibility levels sourced from our local marine facilities and several online underwater videos. Additionally, 10 underwater-adapted video-level attributes are also provided to benchmark the tracking performance under various challenges e.g. motion blur and occlusions etc. Although UOT100 has a larger average sequence length and span than UTB180 (Fig. 2), our proposed UTB180 provides more accurate, precise, and reliable annotations (Fig. 3) and a higher number of video sequences.
2. **Benchmarking.** We conducted an extensive benchmarking of 15 high-quality SOTA deep learning trackers on our UTB180 and the UOT100 datasets. Our experiments demonstrate that the majority of the SOTA trackers consistently show lower performance on several underwater challenging attributes revealing the more challenging nature of the proposed UTB180 dataset compared to existing ones (details in Sect. 4). Visual results comparison of some of the SOTA trackers is shown in Fig. 1 using six sequences captured from our proposed UTB180 dataset.
3. **Fine-Tuning Recent SOTA Trackers on UTB180 Benchmark.** We fine-tune five recent SOTA trackers on our dataset and show that performance improvements are small. This experiment demonstrates that there is still a significant performance gap of the existing trackers on underwater data compared to open-air data. This motivates the need to develop specialized end-to-end underwater trackers capable of handling the inherent challenges of underwater scenes.

2   Related Work

In recent years, the tracking community has put significant efforts towards open-air VOT, thanks to the availability of a variety of open-air tracking benchmarks. Since the main objective of the current work relates to underwater benchmarks, we discuss the available underwater datasets in this section. However, we also briefly explain the open-air datasets for comparison and completeness. Table 1 presents the summary of the available open-air and underwater VOT benchmarks. Surprisingly, 9 out of 11 presented datasets are utilized for open-air tracking. This shows the lagging state of underwater VOT.

2.1   Open-Air Tracking Datasets

Several open-air VOT datasets have been proposed in the past decade as shown in Table 1. For instance, Wu *et al.* proposed **OTB100** that consists of 100 videos with 11 tracking attributes with an average frame resolution of 356×530 pixels [37]. Liang *et al.* proposed **TC128** to evaluate the significance of color information for tracking [27]. This dataset consists of 128 video sequences with 11 distinct tracking challenges with an average resolution of 461×737 pixels. Muller *et al.* proposed **UAV123** dataset for short term tracking [30]. This dataset contains 123 short sequences of nine diverse object categories. The average resolution of each sequences is 1231×699 with 12 tracking attributes. The **VOT2014** [13], **VOT2016** [23], and **VOT2018** [24] are the datasets accompanying the VOT challenge competition to benchmark short-term and long-term tracking performance. As described in Table 1, the VOT2014-2018 series contain 25, 60, and 60 sequences and 12 tracking attributes with an average frame resolution of 757×480, 758×465, and 758×465 pixels, respectively. **LaSOT** [9], **GOT-10k** [15], and **TrackingNet** [31] are relatively larger open-air tracking benchmarks. LaSOT contains 1120 training sequences (2.8M frames) and 280 test sequences (685K frames). GOT-10k contains $10,000$ sequences in which $9,340$ are used for training and remaining 420 sequences used for testing purpose. Similarly, TrackingNet contains a total of $30,643$ sequences where $30,130$ sequences are used for training and remaining 511 sequences used for testing. These large-scale datasets also contain 14–16 distinct tracking attributes with average frame resolutions of 632×1089, 929×1638, and 591×1013 respectively. Due to the large diversity in these benchmarks, many SOTA open-air trackers have been entirely trained and tested on these datasets.

2.2   Underwater Tracking Datasets

Compared to open-air tracking benchmarks, underwater tracking datasets are scarcely available. To the best of our knowledge, the UOT100 is the only available underwater tracking benchmark [32]. This dataset comprises 104 underwater sequences selected from YouTube. It contains a total of $74,042$ annotated frames with 702 average number of frames per sequence. The dataset captures a wide variety of underwater distortions and non-uniform lighting conditions. However, this dataset is not sufficiently diverse for generic object tracking in underwater settings. Moreover, it is also sparsely annotated, containing annotation errors that lead to inaccuracies in tracking. In contrast, our proposed UTB180 dataset is more accurate and densely-annotated benchmark for underwater tracking.

Table 1. Summary of the existing open-air and underwater SOTA VOT benchmark datasets and our proposed UTB180 dataset.

Dataset/Publication	Video Sequences	Attributes	Min Frames	Average Frames	Max Frames	Open-Air	Under-water
OTB100 PAMI2015 [37]	100	11	71	598	3872	✓	
TC128 TIP2015 [27]	128	11	71	431	3872	✓	
UAV123 ECCV2016 [30]	123	12	109	1247	3085	✓	
VOT2014 ECCV-W2014 [13]	25	12	164	409	1210	✓	
VOT2016 ECCV-W2016 [23]	60	12	48		1507	✓	
VOT2018 ICCV-W2018 [24]	60	12	41	356	1500	✓	
LaSOT CVPR2019 [9]	1.4 k	14	1000	2506	11397	✓	
GOT-10k PAMI2019 [15]	10 k	6	51		920	✓	
TrackingNet ECCV2018 [31]	30.643 k	15	96	471	2368	✓	
UOT100 IEEE JOE 2022 [32]	104	3	264	702	1764		✓
Proposed UTB180	**180**	**10**	**40**	**338**	**1226**		✓

3   Proposed High-Quality UTB180 Benchmark

In this section, we explain our proposed Underwater Tracking Benchmark (UTB180) dataset in detail including data collection step, bounding box annotation process, and several video-level attributes included with the dataset.

3.1   Dataset

UTB180 consists of 180 videos selected from underwater environments offering dense (i.e. frame by frame), carefully, and manually annotated frames (58K bounding boxes). It spans sequences for both short-term and long-term underwater tracking. The minimum, average, and maximum number of frames per sequence are 40, 338, and 1226, respectively (shown in Table 1). Our dataset also contains a large variety of diverse underwater creature objects including diverse species of fishes (e.g., dwarf lantern shark, jelly fish, juvenile frog fish, cookie cutter shark, bristle mouths, angler fish, viper fish, grass carp, peruvian anchoveta, and silver carp etc.), crab, sea horse, turtle, squid, octopus, and seal. It aims to offer the tracking community a high-quality benchmark for underwater VOT.

3.2 Data Collection

UTB180 has been sourced from several publicly available online sources such as YouTube, pexel [33] and underwater change detection [34]. We also collected sequences from our marine observatory pond, adding thus more diversity to the dataset. The minimum, average, and maximum frame resolution of the sequences are 1520×2704, 1080×1920, and 959×1277 at 30 frames per second.

3.3 Annotation

To annotate target ground-truth bounding boxes in a sequence, each frame undergoes five sequential processes: 1) Rough estimate of the bounding box is done using a Computer Vision Annotation Tool (CVAT) [7], 2) Each bounding box is then manually and carefully examined, afterwards, to ensure accurate and precise bounding box values around each target object, 3) Each bounding box is then further inspected by a validator to ascertain the accurateness. If it fails at this validation step, it is returned to step 2. 4) For each video sequence, its attributes are labeled, and finally, 5) the sequence is validated with the attributes to ascertain the accurateness. Using these steps, we are able to create a high-quality error-free annotated sequences. It should be noted that each bounding box is a rectangle of four values using the format $[x, y, w, h]$, where x and y denotes the top and left coordinates, w and h denotes the width and height of the rectangle, respectively.

3.4 Underwater Tracking Attributes

Attributes, are video content's aspects that are used to better assess the trackers performance on specific challenges. In this work, we have carefully selected 10 underwater-adapted video-level attributes covering most of the essential variations expected in an underwater environment. These attributes are summarized as follows:

- **Unclear Water (UW):** It presents the low visibility tracking challenge indicating if the water is clear or not.
- **Target Scale Variation (SV):** It indicates whether or not the target varies in scale above a certain degree across the frames.

- **Out-of-View (OV):** It indicates that some portion of the target object leaves the scene.
- **Partial Occlusion (PO):** Accounts for partial occlusion of the target by other objects in the scene.
- **Full Occlusion (FO):** Indicates if the target is fully occluded by another object.
- **Deformation (DF):** It tells if the object is deformed probably due to camera angle or view.
- **Low Resolution (LR):** It indicates if a frame is of low resolution typically less than 300 dots per inch (dpi).
- **Fast Motion (FM):** It indicates if the target moves fast across the frames in the sequence.
- **Motion Blur (MB):** This indicates if the target is blurry.
- **Similar Object (SO):** This attribute indicates if there are object(s) similar to the target in the frames.

Note that all attributes assume binary values, i.e. 1 (presence) or 0 (absence). An attribute is considered present in a sequence if it is present in at least one frame. The sequence-level and frame-level distributions of the attributes in our proposed UTB180 dataset are shown in Fig. 4(a). Moreover, for each of the attributes, a sample image with red colored ground truth bounding box is also shown in Fig. 4(b) except for the UW attribute which shows three sample images illustrating the diverse and challenging nature of underwater visual data. In the next section, we benchmark and compare several SOTA trackers on the UTB180 dataset.

4   Experimental Evaluations

We evaluate and analyse the performance of existing trackers on our proposed UTB180 dataset and further compare with the publicly available underwater tracking benchmark UOT100 [32]. We also fine-tune 5 high-quality SOTA trackers on a training split of UTB180 dataset to improve their tracking performance. In addition, we analyse the attributes-wise tracking performance to further test the robustness of the trackers on specific challenges. All experiments are conducted on a workstation with a 128 GB of memory, CPU Intel Xeon E5-2698 V4 2.2 Gz (20-cores), and two Tesla V100 GPUs. All the trackers are implemented using the official source codes provided by the respective authors.

4.1   Evaluated Trackers

We evaluated the tracking performance of several popular SOTA deep tracking algorithms. These include end-to-end Discriminative Correlation Filters (DCFs)-based trackers such as ATOM [8], DiMP [2], and KYS [3], deep Siamese-based trackers such as SiamFC [1,38], SiamMask [36], SiamRPN [25,26], SiamCAR [12], DaSiamRPN [40], SiamBAN [6], and SiamGAT [11], and the recently proposed transformer-driven DCFs and Siamese-based trackers such as TrSiam [35], TrDimp [35], TrTr [41], TransT [5], and Stark [39].

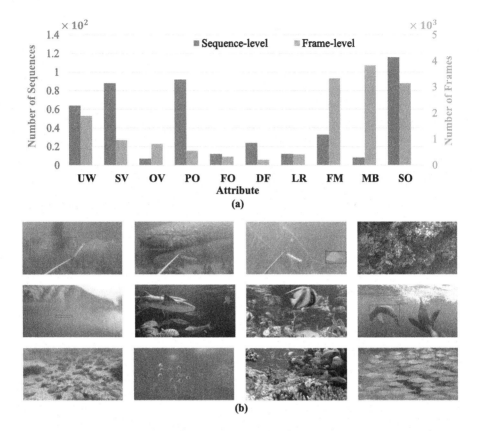

Fig. 4. Proposed UTB180 dataset statistics and sample images. (a) Statistics of sequence-level and frame-level attributes. (b) Sample images of distinct tracking attributes. From left to right, top row shows UW, UW, UW, and SV tracking challenges. Mid row represents OV, PO, FO, and DF attributes. Bottom row shows sample images involving LR, FM, MB, and SO attributes. A red bounding box shows the ground-truth target object. Three UW attribute samples are shown to illustrate the diverse and challenging nature of underwater visual data. (Color figure online)

4.2 Performance Metrics

Following popular tracking protocols developed in open-air tracking datasets e.g. OTB100 [37] and LaSOT [9], we performed the One-Pass Evaluation (OPE) on the benchmarks and measured the precision, normalized precision, and success of different tracking algorithms. The tracking performance metrics are defined as follows:

1. **Precision:** This metric is computed by estimating the distance between a predicted bounding box and a ground-truth bounding box in pixels. Similar to the protocols defined by Wu *et al.* [37], we ranked different trackers using this metric with a threshold of 20 pixels.
2. **Success:** Since the precision metric only measures the localization performance, it does not measure the scale of the predicted bounding boxes in

relation to the ground truth. The success metric takes this into account by employing the intersection over union (IOU) to evaluate the tracker [37]. The IOU is the ratio of the intersection and union of the predicted and the ground truth bounding box. The success plot is then generate by varying the IOU from 0 to 1. The trackers are ranked at a success rate of 0.5.

3. **Normalized Precision:** As the precision metric is sensitive to target size and image resolution, we also used the normalized precision as defined in [31]. With the normalized precision measure, we ranked the SOTA trackers using the area under the curve between 0 to 0.5. More details about this metric can be found in [31].

4.3   Evaluation Protocols

Similar to [9], we used two different protocols and evaluated the SOTA trackers on UTB180 dataset. In **Protocol I**, we used all 180 videos of UTB180 and evaluated the open-air pre-trained models of the SOTA tracking algorithms. This protocol aims to provide large-scale evaluations of the tracking algorithms. In **Protocol II**, we firstly divided the UTB180 into training and testing subsets and then fine-tuned recent SOTA trackers on the training split. Using a 70/30 split, we select 14 out of 20 videos in each category for training and the rest for testing. More specifically, the training subset contains 130 sequences with 41K frames, and the testing subset consists of 50 sequences with 17K frames. The evaluation of SOTA trackers is performed on the testing subset. This protocol aims to provide a large set of underwater videos for training and testing trackers.

4.4   Experiments on Protocol I: Pre-trained Trackers Evaluation

Overall Performance: In this experiment, we benchmark the pre-trained models of the SOTA trackers on the UTB180 and UOT100 [32] datasets. The overall performance in terms of success, normalized precision, and precision is shown in Table 2. Further, the success and precision plots are shown in Fig. 5(first row) and 5(second row) for the UOT100 and UTB180 respectively.

From the results, it can be observed that the Siamese and transformer-driven trackers achieved the best performance on UTB180. Among the compared SOTA trackers, TransT achieved the best results of 58.4% and 51.2% in terms of success and precision rates. In terms of normalized precision rate, SiamBAN achieved the best results of 67.9%. All compared trackers achieved consistently lower performance on all metrics on the UTB180 compared to the UOT100 despite the fact that UTB180 has fewer annotated frames compared to UOT100. The low performance obtained by the SOTA trackers evidenced the novel challenging scenarios in the UTB180 benchmark, and therefore, the need for the development of more powerful underwater trackers.

4.5   Experiments on Protocol II: Fine-Tuned Trackers Evaluation

Overall Performance: In this experiment, we investigated the ability of the open-air pre-trained trackers to generalize to underwater dataset. For this pur-

Table 2. Comparative results of pre-trained trackers on UTB180 and UOT100 benchmarks under protocol I. The best three trackers are shown in red, green, and blue colors, respectively.

Tracker	Sucess ↑		Norm Precision ↑		Precision ↑	
	UOT100	UTB180	UOT100	UTB180	UOT100	UTB180
SiamFC [1]	0.438	**0.350**	0.534	**0.412**	0.304	**0.228**
SiamRPN [25,26]	0.597	**0.534**	0.748	**0.635**	0.487	**0.419**
SiamBAN [6]	0.570	**0.562**	0.749	0.679	**0.522**	**0.462**
SiamMASK [36]	0.547	**0.523**	0.723	**0.640**	0.467	**0.418**
SiamCAR [12]	0.528	**0.461**	0.665	**0.549**	0.450	**0.389**
DaSiamRPN [40]	0.364	**0.355**	0.411	**0.370**	0.184	**0.180**
ATOM [8]	0.545	**0.477**	0.692	**0.555**	0.444	**0.348**
DiMP [2]	0.568	**0.467**	0.698	**0.529**	0.449	**0.332**
KYS [3]	0.585	**0.529**	0.729	**0.613**	0.480	**0.401**
KeepTrack [28]	**0.609**	**0.543**	0.779	**0.637**	0.515	**0.421**
Stark [39]	0.611	**0.482**	0.757	**0.542**	0.532	**0.400**
TrDiMP [35]	0.599	0.580	**0.759**	0.676	0.503	0.455
TrSiam [35]	0.598	**0.566**	0.752	**0.656**	0.492	**0.438**
TrTr [41]	0.535	**0.500**	0.713	**0.601**	0.486	**0.406**
TransT [5]	0.624	0.584	0.789	**0.672**	0.555	0.512

pose, we fine-tuned five SOTA trackers including SiamFC, SiamRPN, ATOM, TrTr, and TransT using the training split (130 videos) of UTB180 dataset. We froze the backbone of each pre-trained tracker for feature extraction and fine-tuned their prediction heads. For the most part during fine-tuning, the default training parameters were unchanged except for the learning rate which was reduced. The pre-trained and fine-tuned trackers performance evaluated on the testing split (50 videos) are presented in Table 3. The success and precision plots are also presented in the Fig. 6(first row) and 6(second row) respectively.

Table 3. Comparative results of the pre-trained and fine-tuned trackers on UTB180 benchmark under protocol II. The best two trackers are shown in red and green colors, respectively.

Tracker	Pretrained ↑			Finetuned ↑		
	Success	Norm	Precision	Success	Norm	Precision
SiamFC [1]	0.308	0.355	0.287	0.315	0.368	0.294
SiamRPN [25,26]	0.486	0.568	0.450	0.491	0.596	0.459
ATOM [8]	0.451	0.532	0.460	0.500	0.600	0.516
TrTr [41]	0.490	0.597	0.486	0.490	0.605	0.499
TransT [5]	0.492	0.562	0.508	0.494	0.570	0.510

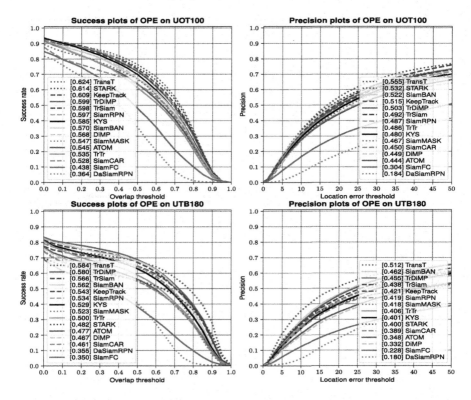

Fig. 5. Evaluation results on UOT100 and UTB180 benchmarks under protocol I using success and precision measures. The legend of precision plot contains threshold scores at 20 pixels, while the legend of success rate contains area-under-the-curve score for each tracker. Overall, the TransT tracker performs better against the SOTA trackers.

The following conclusions are drawn from this experiment:

1. While fine-tuning the trackers on underwater data slightly improved the tracking performance, it is still not comparable with the performance on open-air data. This suggests that specialized trackers are needed to be developed for underwater applications.
2. While the recent transformer-based trackers such as TrTr and TransT perform better, other trackers benefited more from the fine-tuning. This suggests that with the availability of enough data, trackers can be trained longer to achieve better performance.

4.6 Attribute-Wise Evaluation

We also investigated the attribute-wise performance on the UTB180 dataset. We selected a recently proposed TransT tracker since it achieved the best performance shown in Tables 3- 2. We benchmark the TransT on sequences belonging

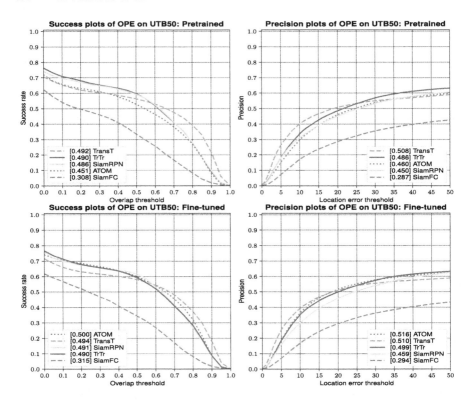

Fig. 6. Evaluation results on UTB50 dataset under protocol I (a) and II (b) using success and precision measure. The legend also contains area under the curve score for each tracker.

to each of the attributes discussed in Sect. 3.4. Table 4 shows the tracking performance in terms of success, normalized precision, and precision. The attribute-wise performance plots can be found in our supplementary material. It can be observed from the results that each of the attributes tend to degrade the performance when compared to the whole dataset. Overall, TransT achieved the best performance on UW attribute while it could hardly achieve 50% tracking performance on other attributes.

Conclusively, their is still a significant performance gap that needs to be filled for reliable and robust underwater tracking. The difficult target state estimation and several other environmental variations such as low visibility condition make the field of underwater VOT challenging.

Table 4. Attribute-wise performance of pre-trained TransT tracker on UTB180 dataset. The best three trackers are shown in red, green, and blue colors, respectively. More details can be found in the supplementary material.

Attribute	Acronym	Number of Videos	Success ↑	Norm Precision ↑	Precision ↑
UTB180		**180**	**0.584**	**0.672**	**0.512**
Unclear Water	UW	64	0.636	0.743	0.586
Scale Variation	SV	88	0.559	0.640	**0.478**
Out of View	OV	7	0.566	0.660	0.475
Partial Occlusion	PO	92	0.475	0.542	0.409
Full Occlusion	FO	12	0.342	0.375	0.330
Deformation	DF	24	**0.564**	**0.657**	0.402
Low Resolution	LR	12	0.489	0.583	0.390
Fast Motion	FM	33	0.515	0.593	0.486
Motion Blur	MB	8	0.485	0.540	0.417
Similar Objects	SO	116	0.513	0.583	0.472

5   Conclusion and Future Research Directions

5.1   Conclusion

In this work, we proposed a new VOT dataset dedicated to underwater scenes. It is a dense and diversified high-quality underwater VOT dataset with 180 video sequences and over 58,000 carefully and manually annotated frames. We benchmarked and fine-tuned existing SOTA Simaese and transformer trackers on the proposed dataset. Our results demonstrate that there is still a significant performance gap between open-air and underwater VOT. We showed that UTB180 presents more challenging sequences compared to the publicly available UOT100 dataset. It is expected that UTB180 will play an instrumental role in boosting the underwater VOT research.

5.2   Future Research Directions

When compared to open-air, the available underwater datasets are still insignificant. At the moment, the available underwater datasets can only be utilized for benchmarking and fine-tuning of the designed trackers. They are insufficient for the direct training of deep trackers. As such, we intend to extend this work to enable not only fine-tuning but also the training and testing of deep underwater trackers with underwater datasets.

From our experiments, we showed that recent transformer-based trackers consistently performed better than their DCFs and Siamese-based counterparts.

While this performance still lags compared to the open-air performance, it suggests that variants of transformer-based trackers could pave the way for the development of better underwater trackers. Improved backbone feature extraction, sophisticated target state estimation, and the role of implicit or explicit underwater video denoising approaches are required for robust end-to-end underwater VOT. Such extensions could lead to more generic algorithms suited for both open-air and underwater VOT.

Acknowledgments. This publication acknowledges the support provided by the Khalifa University of Science and Technology under Faculty Start Up grants FSU-2022–003 Award No. 84740 0 0401.

References

1. Bertinetto, L., Valmadre, J., Henriques, J.F., Vedaldi, A., Torr, P.H.S.: Fully-Convolutional Siamese Networks for Object Tracking. In: Hua, G., Jégou, H. (eds.) ECCV 2016. LNCS, vol. 9914, pp. 850–865. Springer, Cham (2016). https://doi.org/10.1007/978-3-319-48881-3_56
2. Bhat, G., Danelljan, M., Van Gool, L., Timofte, R.: Learning discriminative model prediction for tracking. Proc. In: IEEE Int. Conf. Comput. Vis. 2019-Oct (ICCV), pp. 6181–6190 (2019). https://doi.org/10.1109/ICCV.2019.00628
3. Bhat, G., Danelljan, M., Van Gool, L., Timofte, R.: Know Your Surroundings: Exploiting Scene Information for Object Tracking. In: Vedaldi, A., Bischof, H., Brox, T., Frahm, J.-M. (eds.) ECCV 2020. LNCS, vol. 12368, pp. 205–221. Springer, Cham (2020). https://doi.org/10.1007/978-3-030-58592-1_13
4. Boudiaf, A., et al.: Underwater image enhancement using pre-trained transformer. In: International Conference on Image Analysis and Processing, pp. 480–488. Springer (2022). https://doi.org/10.1007/978-3-031-06433-3_41
5. Chen, X., Yan, B., Zhu, J., Wang, D., Yang, X., Lu, H.: Transformer Tracking, pp. 8122–8131 (2021). https://doi.org/10.1109/cvpr46437.2021.00803
6. Chen, Z., Zhong, B., Li, G., Zhang, S., Ji, R.: Siamese Box Adaptive Network for Visual Tracking. Proc. In: IEEE Comput. Soc. Conf. Comput. Vis. Pattern Recognit, pp. 6667–6676 (2020). https://doi.org/10.1109/CVPR42600.2020.00670
7. CVAT: Computer Vision Annotation Tool. https://cvat.org
8. Danelljan, M., Van Gool, L., Timofte, R.: Probabilistic regression for visual tracking. Proc. In: IEEE Comput. Soc. Conf. Comput. Vis. Pattern Recognit, pp. 7181–7190 (2020). https://doi.org/10.1109/CVPR42600.2020.00721
9. Fan, H., et al.: LaSOT: a High-quality Large-scale Single Object Tracking Benchmark. Int. J. Comput. Vis. 129(2), 439–461 (2021). https://doi.org/10.1007/s11263-020-01387-y
10. Giraldo, J.H., Javed, S., Bouwmans, T.: Graph moving object segmentation. In: IEEE Trans. Pattern Anal. Mach. Intell. **44**(5), 2485-2503 (2020)
11. Guo, D., Shao, Y., Cui, Y., Wang, Z., Zhang, L., Shen, C.: Graph Attention Tracking. Proc. In: IEEE Comput. Soc. Conf. Comput. Vis. Pattern Recognit, pp. 9538–9547 (2021). https://doi.org/10.1109/CVPR46437.2021.00942
12. Guo, D., Wang, J., Cui, Y., Wang, Z., Chen, S.: SiamCAR: siamese Fully Convolutional Classification and Regression for Visual Tracking. Proc. In: IEEE Comput. Soc. Conf. Comput. Vis. Pattern Recognit, pp. 6268–6276 (2020). https://doi.org/10.1109/CVPR42600.2020.00630

13. Kristan, M., et al.: The Visual Object Tracking VOT2014 Challenge Results. In: Agapito, L., Bronstein, M.M., Rother, C. (eds.) ECCV 2014. LNCS, vol. 8926, pp. 191–217. Springer, Cham (2015). https://doi.org/10.1007/978-3-319-16181-5_14

14. Han, M., Lyu, Z., Qiu, T., Xu, M.: A Review on Intelligence Dehazing and Color Restoration for Underwater Images. In: IEEE Trans. Syst. Man, Cybern. Syst. 50(5), 1820–1832 (2020). https://doi.org/10.1109/TSMC.2017.2788902

15. Huang, L., Zhao, X., Huang, K.: Got-10k: a large high-diversity benchmark for generic object tracking in the wild. In: IEEE Trans. Pattern Anal. Mach. Intell. 43(5), 1562–1577 (2021). https://doi.org/10.1109/TPAMI.2019.2957464. https://www.scopus.com/inward/record.uri?eid=2-s2.0-85103976016&doi=10.1109%2FTPAMI.2019.2957464&partnerID=40&md5=3fd7d1e870e60df363a83a52a092c544

16. Javed, S., Danelljan, M., Khan, F.S., Khan, M.H., Felsberg, M., Matas, J.: Visual Object Tracking with Discriminative Filters and Siamese Networks: a Survey and Outlook 14(8), 1–20 (2021). http://arxiv.org/abs/2112.02838

17. Javed, S., Dias, J., Werghi, N.: Low-rank tensor tracking. In: 2019 IEEE/CVF International Conference on Computer Vision Workshop (ICCVW), pp. 605–614 (2019). https://doi.org/10.1109/ICCVW.2019.00074

18. Javed, S., Mahmood, A., Dias, J., Seneviratne, L., Werghi, N.: Hierarchical spatiotemporal graph regularized discriminative correlation filter for visual object tracking. In: IEEE Transactions on Cybernetics (2021)

19. Javed, S., Mahmood, A., Dias, J., Werghi, N.: Robust structural low-rank tracking. IEEE Trans. Image Process. 29, 4390–4405 (2020)

20. Javed, S., et al.: A novel algorithm based on a common subspace fusion for visual object tracking. IEEE Access 10, 24690–24703 (2022)

21. Javed, S., Zhang, X., Dias, J., Seneviratne, L., Werghi, N.: Spatial Graph Regularized Correlation Filters for Visual Object Tracking. In: Abraham, A., Ohsawa, Y., Gandhi, N., Jabbar, M.A., Haqiq, A., McLoone, S., Issac, B. (eds.) SoCPaR 2020. AISC, vol. 1383, pp. 186–195. Springer, Cham (2021). https://doi.org/10.1007/978-3-030-73689-7_19

22. Javed, S., Zhang, X., Dias, J., Seneviratne, L., Werghi, N.: Spatial Graph Regularized Correlation Filters for Visual Object Tracking. In: Abraham, A., Abraham, A., et al. (eds.) SoCPaR 2020. AISC, vol. 1383, pp. 186–195. Springer, Cham (2021). https://doi.org/10.1007/978-3-030-73689-7_19

23. Kristan, M., Leonardis, A., Matas, e.a.: The Visual Object Tracking VOT2016 Challenge Results, pp. 777–823. Springer International Publishing (2016). https://doi.org/10.1007/978-3-319-48881-3_54

24. Kristan, M., Leonardis, A., Matas, e.: The Sixth Visual Object Tracking VOT2018 Challenge Results. In: Leal-Taixé, L., Roth, S. (eds.) Comput. Vis. - ECCV 2018 Work, pp. 3–53. Springer International Publishing, Cham (2019) https://doi.org/10.1007/978-3-030-11009-3_1

25. Li, B., Wu, W., Wang, Q., Zhang, F., Xing, J., Yan, J.: SIAMRPN++: evolution of siamese visual tracking with very deep networks. Proc. IEEE Comput. Soc. Conf. Comput. Vis. Pattern Recognit. 2019-June, 4277–4286 (2019). https://doi.org/10.1109/CVPR.2019.00441

26. Li, B., Yan, J., Wu, W., Zhu, Z., Hu, X.: High Performance Visual Tracking with Siamese Region Proposal Network. In: IEEE Conf. Comput. Vis. Pattern Recognit, pp. 8971–8980 (2018)

27. Liang, P., Blasch, E., Ling, H.: Encoding color information for visual tracking: Algorithms and benchmark. IEEE Trans. Image Process. 24(12), 5630–5644 (2015)

28. Mayer, C., Danelljan, M., Pani Paudel, D., Van Gool, L.: Learning Target Candidate Association to Keep Track of What Not to Track (ICCV), 13424–13434 (2022). https://doi.org/10.1109/iccv48922.2021.01319

29. Meinhardt, T., Kirillov, A., Leal-Taixé, L., Feichtenhofer, C.: Trackformer: multiobject tracking with transformers. In: Proceedings of the IEEE/CVF Conference on Computer Vision and Pattern Recognition (CVPR), pp. 8844–8854 (2022)

30. Mueller, M., Smith, N., Ghanem, B.: A Benchmark and Simulator for UAV Tracking. In: Leibe, B., Matas, J., Sebe, N., Welling, M. (eds.) ECCV 2016. LNCS, vol. 9905, pp. 445–461. Springer, Cham (2016). https://doi.org/10.1007/978-3-319-46448-0_27

31. Müller, M., Bibi, A., Giancola, S., Alsubaihi, S., Ghanem, B.: TrackingNet: A Large-Scale Dataset and Benchmark for Object Tracking in the Wild. In: Ferrari, V., Hebert, M., Sminchisescu, C., Weiss, Y. (eds.) ECCV 2018. LNCS, vol. 11205, pp. 310–327. Springer, Cham (2018). https://doi.org/10.1007/978-3-030-01246-5_19

32. Panetta, K., Kezebou, L., Oludare, V., Agaian, S.: Comprehensive Underwater Object Tracking Benchmark Dataset and Underwater Image Enhancement with GAN. IEEE J. Ocean. Eng. 47(1), 59–75 (2022). https://doi.org/10.1109/JOE.2021.3086907

33. Pexel: 1,363+ Best Free Underwater 4K Stock Video Footage & Royalty-Free HD Video Clips. https://www.pexels.com/search/videos/underwater/

34. Underwaterchangedetection: Videos - Underwaterchangedetection. http://underwaterchangedetection.eu/Videos.html

35. Wang, N., Zhou, W., Wang, J., Li, H.: Transformer Meets Tracker: exploiting Temporal Context for Robust Visual Tracking. Proc. In: IEEE Comput. Soc. Conf. Comput. Vis. Pattern Recognit, pp. 1571–1580 (2021). https://doi.org/10.1109/CVPR46437.2021.00162

36. Wang, Q., Zhang, L., Bertinetto, L., Hu, W., Torr, P.H.: Fast online object tracking and segmentation: a unifying approach. Proc. In: IEEE Comput. Soc. Conf. Comput. Vis. Pattern Recognit. 2019-June, 1328–1338 (2019). https://doi.org/10.1109/CVPR.2019.00142

37. Wu, Y., Lim, J., Yang, M.H.: Object tracking benchmark. IEEE Trans. Pattern Anal. Mach. Intell. 37(9), 1834–1848 (2015). https://doi.org/10.1109/TPAMI.2014.2388226

38. Xu, Y., Wang, Z., Li, Z., Yuan, Y., Yu, G.: SiamFC++: Towards robust and accurate visual tracking with target estimation guidelines. AAAI 2020–34th AAAI Conf. Artif. Intell. 34(7), 12549–12556 (2020). https://doi.org/10.1609/aaai.v34i07.6944

39. Yan, B., Peng, H., Fu, J., Wang, D., Lu, H.: Learning Spatio-Temporal Transformer for Visual Tracking, pp. 10428–10437 (2022). https://doi.org/10.1109/iccv48922.2021.01028

40. Yu, Y., Xiong, Y., Huang, W., Scott, M.R.: Deformable Siamese Attention Networks for Visual Object Tracking. Proc. In: IEEE Comput. Soc. Conf. Comput. Vis. Pattern Recognit, pp. 6727–6736 (2020). https://doi.org/10.1109/CVPR42600.2020.00676

41. Zhao, M., Okada, K., Inaba, M.: TrTr: visual Tracking with Transformer (2021). http://arxiv.org/abs/2105.03817

The Eyecandies Dataset for Unsupervised Multimodal Anomaly Detection and Localization

Luca Bonfiglioli$^{(\boxtimes)}$, Marco Toschi , Davide Silvestri , Nicola Fioraio ,
and Daniele De Gregorio

Eyecan.ai, Bologna, Italy
{luca.bonfiglioli,marco.toschi,davide.silvestri,
nicola.fioraio,daniele.degregorio}@eyecan.ai
https://www.eyecan.ai

Abstract. We present Eyecandies, a novel synthetic dataset for unsupervised anomaly detection and localization. Photo-realistic images of procedurally generated candies are rendered in a controlled environment under multiple lightning conditions, also providing depth and normal maps in an industrial conveyor scenario. We make available anomaly-free samples for model training and validation, while anomalous instances with precise ground-truth annotations are provided only in the test set. The dataset comprises ten classes of candies, each showing different challenges, such as complex textures, self-occlusions and specularities. Furthermore, we achieve large intra-class variation by randomly drawing key parameters of a procedural rendering pipeline, which enables the creation of an arbitrary number of instances with photo-realistic appearance. Likewise, anomalies are injected into the rendering graph and pixel-wise annotations are automatically generated, overcoming human-biases and possible inconsistencies.

We believe this dataset may encourage the exploration of original approaches to solve the anomaly detection task, e.g. by combining color, depth and normal maps, as they are not provided by most of the existing datasets. Indeed, in order to demonstrate how exploiting additional information may actually lead to higher detection performance, we show the results obtained by training a deep convolutional autoencoder to reconstruct different combinations of inputs.

Keywords: Synthetic dataset · Anomaly detection · Deep learning

1 Introduction

Recent years have seen an increasing interest in visual unsupervised anomaly detection [34], the task of determining whether an example never seen before presents any aspects that deviate from a defect-free domain, which was learned during training. Similar to one-class classification [18,23,27], in unsupervised anomaly detection the model has absolutely no knowledge of the appearance of

---

L. Bonfiglioli and M. Toschi—Joint first authorship.

© The Author(s), under exclusive license to Springer Nature Switzerland AG 2023
L. Wang et al. (Eds.): ACCV 2022, LNCS 13845, pp. 459–475, 2023.
https://doi.org/10.1007/978-3-031-26348-4_27

anomalous structures and must learn to detect them solely by looking at *good* examples. There is a practical reason behind this apparent limitation: being anomalies rare by definition, collecting real-world data with enough examples of each possible deviation from a target domain may prove to be unreasonably expensive. Furthermore, the nature of all possible anomalies might even be unknown, so treating anomaly detection as a supervised classification task may hinder the ability of the model to generalize to new unseen types of defects.

Historically, a common evaluation practice for proposed AD methods was to exploit existing multi-class classification datasets, such as MNIST [21] and CIFAR [20], re-labeling a subset of related classes as inliers and the remaining as outliers [25]. The major drawback of this practice is that clean and anomalous domains are often completely unrelated, whereas in real-world scenarios, such as industrial quality assurance or autonomous driving, anomalies usually appear as subtle changes within a common scene, as for the anomalies presented in [6]. In the recent years this adaptation of classification datasets was discouraged in favor of using new datasets specifically designed for visual anomaly detection and localization, such as [9], which focuses on industrial inspection. However, most of the available datasets provide only color images with ground-truth annotations and very few add 3D information [8]. Moreover, all of them have to face the problem of manual labelling, which can be human-biased and error-prone, especially in the 3D domain.

The Eyecandies dataset is our main contribution to tackle these issues and provide a new and challenging benchmark for unsupervised anomaly detection, including a total of 90000 photo-realistic shots of procedurally generated synthetic objects, spanning across 10 classes of candies, cookies and sweets (cfr. Figure 1). Different classes present entirely different shapes, color patterns and materials, while intra-class variance is given by randomly altering parameters of the same model. The Eyecandies dataset comprises defect-free samples for training, as well as anomalous ones used for testing, each of them with automatically generated per-pixel ground-truth labels, thus removing the need for expensive (and often biased) manual annotation procedures. Of each sample, we also provide six renderings with different controlled lighting conditions, together with ground-truth depth and normal maps, encouraging the exploration and comparison of many alternative approaches.

We found that performance of existing methods on synthetic data are in line with the results obtained on *real* data, such as [9], though our dataset appears to be more challenging. Moreover, being the use of 3D data not common in the AD field, we deployed a deep convolutional autoencoder trained to reconstruct different combination of inputs, showing that the inclusion of 3D data results in better anomaly detection and localization performance.

To explore the data and evaluate a method, please go to https://eyecan-ai.github.io/eyecandies. Please refer to https://github.com/eyecan-ai/eyecandies for examples and tutorials on how to use the Eyecandies dataset.

Good samples   Anomalous samples   Good samples   Anomalous samples

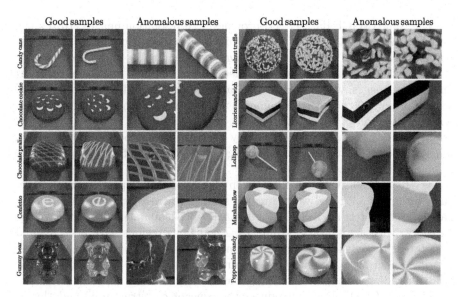

Fig. 1. Examples from the Eyecandies dataset. Each row shows good and bad samples from the same object category (best viewed in color). (Color figure online)

2   Related Work

Anomaly detection and localization on images (hereinafter AD) is an ubiquitous theme in many fields, from autonomous driving [12] to visual industrial inspection [9,30,33]. Likewise, the use of synthetic datasets to evaluate the performance of proposed methods has been already explored in many contexts [13,17,29,32]. However, very few works investigate how synthetic data can be effectively exploited to advance the AD field, which is indeed the focus of the dataset we are presenting. In the next sections we will first review the publicly available datasets for AD, then we will briefly analyze the most successful methods proposed to solve the AD task, showing how our work may help the research community.

2.1   Anomaly Detection Datasets

Different public AD datasets exists, some designed for industrial inspection of a very specialized type of objects, while other trying to be more generic. An example of the former group is the Magnetic Tile Dataset [33], composed by 952 anomaly-free images and 5 types of anomalies, for a total of 1501 manually annotated images of various resolutions. Despite being a reference in the field, this dataset comprises only a single texture category and it is limited to grayscale images. Though much larger, [30] is another similar dataset presented on Kaggle, focused on a single object class.

The NanoTWICE dataset [14] provides high resolution images (1024×696), although of little interest for deep learning approaches, since it is composed by only 5 anomaly-free images and 40 images with anomalies of different sizes.

In [22] the authors generate a synthetic dataset of 1000 good images and 150 anomalous images, with ground-truth labels approximated by ellipses. The test set comprises 2000 non defective images and 300 defective ones as 8-bit grayscale with a resolution of 512×512. Though larger than usual, it shows low texture variation and the ground-truth is very coarse. Instead, our synthetic pipeline aims at photo-realistic images with large intra-class variation and pixel-precise ground-truth masks.

MVTec AD [9], which is focused on the industrial inspection scenario, features a total of 5354 real-world images, spanning across 5 texture and 10 object categories. The test set includes 73 distinct types of anomalies (on average 5 per category) with a total of 1725 images. Anomalous regions have been manually annotated, though introducing small inconsistencies and unclear resolution of missing object parts. In our work we purposely avoid these undefined situations, while providing pixel-precise annotations in an automated way.

MVTec LOCO AD [6] introduces the concept of "structural" and "logical" anomalies: the former being local irregularities like scratches or dents, and the latter being violations of underlying logical constraints that require a deeper understanding of the scene. The dataset consists of 3644 images, distributed across 6 categories. Though interesting and challenging, the detection of logical anomalies is out of the scope of this work, where we focus on localized defects only. Moreover, such defects are usually specific for a particular object class, while we aim at automated and consistent defect generation. Finally, being the subject fairly new, there is no clear consensus on how to annotate the images and evaluate the performance of a method.

MVTec 3D-AD [8] has been the first 3D dataset for AD. Authors believe that the use of 3D data is not common in the AD field due to the lack of suitable datasets. They provide 4147 point clouds, acquired by an industrial 3D sensor, and a complementary RGB image for 10 object categories. The test set comprises 948 anomalous objects and 41 types of defects, all manually annotated. The objects are captured on a black background, useful for data augmentation, but not very common in real-world scenarios. Moreover, the use of a 3D device caused the presence of occlusions, reflections and inaccuracies, introducing a source of noise that may hinder a fair comparison of different AD proposals. Of course, our synthetic generation does not suffer from such nuisances.

Synthetic generation of defective samples is introduced in [13] to enhance the performance of an AD classifier. As in our work, they use Blender [11] to create the new data, though they focus on combining real and synthetic images, while we aim at providing a comprehensive dataset for evaluation and comparison. Also, the authors of [13] did not release their dataset.

In [29] another non-publicly available dataset is presented. They render 2D images from 3D models in a procedural way, where randomized parameters control defects, illumination, camera poses and texture. Their rendering pipeline is similar to ours, though, as in [13], their focus is on generating huge amounts of synthetic anomalous examples to train a model in a supervised fashion that could generalize to real data.

Finally, in [17] the authors propose to apply synthetic defects on the 3D reconstruction of a target object. The rendering pipeline shares some of our intuitions, such as parametric modeling of defects and rendering. However, the use of expensive hardware and the need of a physical object hinder general applicability to build up a comprehensive dataset. Moreover the approach followed in this and in the previously cited papers [13,17] about synthetic augmentation is different from ours since the model is trained on anomalous data.

2.2 Methods

In the last few years, many novel proposals emerged to tackle the AD task. Generally, methods can be categorized as discriminative or generative, where the former often model the distribution of features extracted, e.g., from a pre-trained neural network [7,15,26,35], while the latter prefer an end-to-end training [2,10, 28,33]. Due to the lack of diverse 3D dataset for AD, very few proposals are explicitly designed to exploit more than just a single 2D color image [5,31]. Therefore, we give the community a novel dataset to further investigate the use of 3D geometry, normal directions and lighting patterns in the context of AD.

3   The Eyecandies Dataset

The Eyecandies dataset comprises ten different categories of candies, chosen to provide a variety of shapes, textures and materials:

- Candy Cane
- Chocolate Cookie
- Chocolate Praline
- Confetto
- Gummy Bear
- Hazelnut Truffle
- Licorice Sandwich
- Lollipop
- Marshmallow
- Peppermint Candy.

Our pipeline generates a large number of unique instances of each object category, all of them differing in some controlled aspects. A subset of samples, labeled as defective, present one or more anomalies on their surface. An automatically annotated ground-truth segmentation mask provides pixel-precise classification labels.

In the next subsection we will generally describe the setup and the data produced by our pipeline, then in Sec. 3.2 we will present the available types of defects. More details on the data generation process are found in Sect. 4.

3.1   General Setup

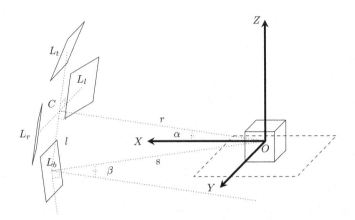

Fig. 2. An axonometric view of the camera and the surrounding lights. The object, represented as a cube, is placed at the center of the world (O). The camera (C) faces the object at a distance r making an angle α with the X axis. Four square-shaped lights (L_t, L_r, L_b and L_l) are placed on a circle of radius l, centered at C, at intervals of 90°. Every light is tilted of an angle β in order to face the center of the world. The value of β depends on the distances r and l.

We designed the Eyecandies dataset to be a full-fledged, multi-purpose source of data, fully exploiting the inherently controlled nature of a synthetic environment. First, we created a virtual scene resembling an industrial conveyor belt passing through a light box. Four light sources are placed on the four corners of the box, illuminating the whole scene. The camera is placed inside the light box, facing the conveyor at an angle, surrounded by other four square-shaped light sources, as depicted in Fig. 2). We will hereinafter use the terms "box lights" and "camera lights" meaning, respectively, the spotlights at the light box corners, and the light sources surrounding the camera. Among the many different conceivable lighting patterns, we chose to provide *six* shots for each sample object:

1. one with box lights only.
2. four with only one camera light (one per light).
3. one with all camera lights at the same time.

To the best of our knowledge, no other existing dataset for anomaly detection include multiple light patterns for each object, hence paving the way for novel and exciting approaches to solve the anomaly detection task. Indeed, as shown in Fig. 3, strong directional shadows enable the detection of surface irregularities otherwise hard to see: while the presence of a defect can appear unclear under some specific lighting conditions, when comparing multiple shots of the same object with different lighting, the detection becomes much easier.

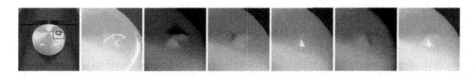

Fig. 3. An anomalous sample from *Peppermint Candy* category with a bump on its surface. The first figure shows the full picture with the anomalous area highlighted in red. The following figures all show the same crop around the bump, but each one with a different lighting condition.

Alongside RGB color images, depth and normal maps are rendered for each scene. Both are computed by ray-casting the mesh from the same camera point-of-view, i.e., the *depth* is the Z coordinate in camera reference frame, likewise normal directions are expressed as unit vectors in the camera reference frame as well (see an example in Fig. 4). The benefit of considering these additional data sources will be discussed in Sect. 5, where we will show that simply concate-nating color, depth and surface normals can boost the performance of a naive autoencoder. Interestingly, combining depth, normal and RGB images rendered in multiple lighting scenarios allows for addressing many more tasks than just anomaly detection and localization, such as photometric normal estimation and stereo reconstruction, depth from mono and scene relighting.

Since we do not aim at simulating the acquisition from any real devices, no noise is intentionally added to the data. Also, we do not claim that training mod-els on our synthetic dataset may somehow help solving real-world applications. Instead, we provide a clean, though challenging, benchmark to fairly compare existing and future proposals.

3.2   Synthetic Defects

Real-world defects come in various shapes and appearances, often tied to par-ticular object features and to the production process. However, we identified common properties and decided to focus on three general groups of anomalies that can occur on many different types of objects:

1. color alterations, such as stains and burns;
2. shape deformations, i.e., bumps or dents;
3. scratches and other small surface imperfections.

All these groups can be viewed as local anomalies applied on different input data, i.e., color alterations change the RGB image, shape deformations modify the 3D geometry and surface imperfections only alter the normal directions. We chose to include defects that modify the surface normals without affecting the

3D geometry to represent small imperfections hardly captured on a depth map, like, for example, a scratch on a metallic surface. Therefore, we actively change the mesh only in case of surface bumps or dents, while modifications to the normal map have a clear effect only on how the light is reflected or refracted. Finally, for each local defect we provide a corresponding pixel-wise binary mask, rendered directly from the 3D object model, to highlight the area a detector should identify as anomalous.

We purposely left for future investigations the inclusion of two class of anomalies. Firstly, we avoided class-specific defects, which would have been relevant for only one object class. An example might be altering the number of stripes of the *Marshmallow* or changing the text printed on the *Confetto*. However, this would introduce much more effort in designing how such anomalies should interact with the rendering pipeline, while current defects can be applied in an automated way with no prior information, as described in Sect. 4. Secondly, we did not include *logical* anomalies, as described in [6], because we believe there is no clear consensus on how to evaluate the localization performance of a detector on, e.g., finding missing object regions, and neither on how to annotate such regions in ground-truth anomaly masks.

Unlike many existing dataset, we do not require any human intervention, thus removing a possible source of biases and inconsistencies.

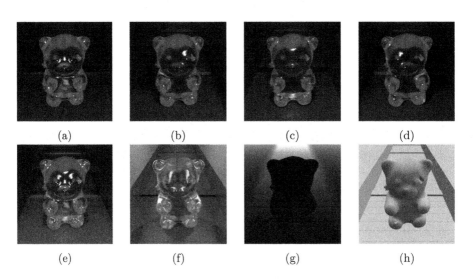

(a)	(b)	(c)	(d)
(e)	(f)	(g)	(h)

Fig. 4. Image data of a *Gummy Bear* sample from the test set. Renderings with a single light source are shown in Fig. a–d. In Fig. e all the camera lights are active. In Fig. f, the light comes instead from four spotlights at the corners of the surrounding lightbox, and the camera lights are switched off. Figure g and h show the rendered depth and normal maps.

4   Data Generation

We generated the Eyecandies dataset within the *Blender* framework [11], a popular 3D modeling software that provides a good level of interoperability with the Python programming language via the *BlenderProc* package [16], a useful tool for procedural synthetic data generation. Every candy class is modelled as a parametric *prototype*, i.e., a rendering pipeline where geometry, textures and materials are defined through a programming language. In this way, object features can be controlled by a set of scalar values, so that each combination of parameters leads to a unique output.

To generate different object instances, we treat all parameters as uniformly distributed random variables, where bounds are chosen to produce a reasonable variance among them, though preserving a realistic aspect for each candy. Furthermore, to achieve a high degree of photo-realism as well as intra-class variance, key to any procedural object generation pipeline is the use of *noise textures*, useful to produce slight random deformations (as in *Licorice Sandwich*) or fine-grained irregularities in a rough surface. Unlike the parameters described above, these textures are controlled by setting a generic *random seed*, hence letting blender generate them within a geometry modifier or shader. No other noise source is intentionally added, e.g., we do not model any real acquisition device. However, since Cycles, the chosen rendering engine, inevitably introduces random imperfections when computing surface colors, we counteract them by applying NVIDIA Optix^{TM} de-noising [24].

Anomalous samples differ in that they are given four more textures as parameters, one for each possible defect, i.e., following Sect. 3.2, *colors*, *bumps*, *dents* and *normals*. In order to get a realistic appearance, these textures are mapped onto the object mesh through a UV mapping and rendered as part of the object. However, non-trivial constraints must be applied, since simply generating a random blob on a black background would not suffice: it could end up outside of UV map islands, having no effect on the object, or worse, on the edge between two islands, resulting in multiple anomalies with spurious shape. Instead, we export the original UV map of the object from blender, compute a binary mask of the valid areas, then separate all connected components and for each of them compute the maximum inbound rectangle. A random blob is then generated and stitched into one of the inbound rectangles, chosen randomly. This ensures that the anomalies are always entirely visible and never lying on the border of a UV map island. The quality achieved can be appreciated in Fig. 1: modifying the 3D model produces more realistic images than artificially applying the defects on the 2D renderings, while still being able to auto-generate pixel-wise ground-truth masks with no human intervention.

Every object category contains a total of 1500 samples, split between training, validation and test sets consisting of, respectively, 1000, 100 and 400 samples. Training and validation sets provide good examples only, whereas half of the test samples are defective candies. Also, these 200 anomalous samples contain a balanced mixture of the four anomaly types: 40 examples for each one totalling to 160 examples, the remaining 40 containing all possible anomalies together.

5 Experiments

First, we evaluated existing methods for AD on Eyecandies and compared with the results obtained on MVTec AD [9]. In Table 1 the area under ROC curve (AUROC) is reported for Ganomaly (G) [2], Deep Feature Kernel Density Estimation (DFKDE) [4], Probabilistic Modeling of Deep Features (DFM) [1], Student-Teacher Feature Pyramid Matching (STFPM) [7] and PaDiM [15], all run within the Anomalib framework [3]. We notice a significant correlation between the performance on the *real* dataset MVTec AD and Eyecandies, thus suggesting that our proposal, though synthetic, is a valid approach to evaluate AD methods. Furthermore, all methods, with the exception of Ganomaly [2], show a large performance drop when trained and tested on Eyecandies, proving the increased complexity of the task w.r.t. popular AD datasets, such as MVTec AD.

Table 1. Image AUROC of existing AD methods on eyecandies dataset, considering only RGB images. We compared feature-based methods with both Resnet18 (r18) and Wide-Resnet50 (wr50) pre-trained backbones.

Category	G [2]	DFKDE [4]		DFM [1]		STFPM [7]		PaDiM [15]		Ours
		R18	Wr50	R18	Wr50	R18	Wr50	R18	Wr50	RGB
Candy Cane	0.485	0.537	0.539	0.529	0.532	0.527	0.551	0.537	0.531	0.527
Chocolate C.	0.512	0.589	0.577	0.759	0.776	0.628	0.654	0.765	0.816	0.848
Chocolate P.	0.532	0.517	0.482	0.587	0.624	0.766	0.576	0.754	0.821	0.772
Confetto	0.504	0.490	0.548	0.649	0.675	0.666	0.784	0.794	0.856	0.734
Gummy Bear	0.558	0.591	0.541	0.655	0.681	0.728	0.737	0.798	0.826	0.590
Hazelnut T.	0.486	0.490	0.492	0.611	0.596	0.727	0.790	0.645	0.727	0.508
Licorice S.	0.467	0.532	0.524	0.692	0.685	0.738	0.778	0.752	0.784	0.693
Lollipop	0.511	0.536	0.602	0.599	0.618	0.572	0.620	0.621	0.665	0.760
Marshmallow	0.481	0.646	0.658	0.942	0.964	0.893	0.840	0.978	0.987	0.851
Peppermint C.	0.528	0.518	0.591	0.736	0.770	0.631	0.749	0.894	0.924	0.730
Avg. Eyecandies	0.507	0.545	0.555	0.676	0.692	0.688	0.708	0.754	0.794	0.701
Avg. MVTecAD	0.421	0.762	0.774	0.894	0.891	0.893	0.876	0.891	0.950	

As for understanding the contribution of the 3D data to the AD task, we trained a different deep convolutional autoencoder on each of the 10 categories of the Eyecandies dataset. The model has been trained to reconstruct defect-free examples and evaluated on the test set containing a mixture of good and bad examples. Since the network sees only defect-free data during training, we expect anomalous areas to give large reconstruction errors. Thus, beside image-wise metrics, a per-pixel anomaly score can be computed as the L1 distance between the input image and its reconstruction, averaging over the image channels. Similarly, the image-wise anomaly score is computed as the maximum of the per-pixel score.

The model consists of two symmetrical parts, i.e., the encoder and the decoder, connected by a linear fully-connected bottleneck layer. Every encoder block increases the number of filters by a factor of 2 with respect to the previous one and halves the spatial resolution by means of strided convolutions. Each decoder block, on the other hand, halves the number of filters while doubling the spatial resolution. Both the encoder and the decoder comprise 4 blocks and the initial number of filters is set to 32, hence, starting from an input of size $3 \times 256 \times 256$, the bottleneck layer is fed with a $256 \times 16 \times 16$ tensor which is projected onto a latent space with 256 dimensions. Then, the decoder expands this feature vector back to $256 \times 16 \times 16$ and up to the same initial dimensions. The inner block layers are detailed in Fig. 5.

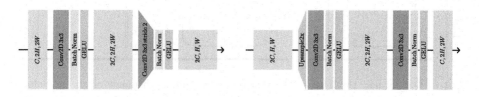

Fig. 5. The encoder (left) and decoder (right) blocks inner structure.

For every object category, we trained the autoencoder with different input combinations: RGB, RGB + Depth (RGBD), RGB + Depth + Normals (RGBDN), all downscaled to a fixed resolution of 256×256 pixels. To this end, being the color image as well as the depth and the normal maps of equivalent resolution, we simply concatenated them along the channel dimension, changing the number of input and output channels of the autoencoder accordingly. Therefore, the total number of channels is 3 for the RGB case, 4 when adding the depth and 7 when using all the available inputs, i.e., color, depth and normals. When using the depth maps, the values are re-scaled between 0 and 1 with a per-image min-max normalization. As for data augmentation, we used the following random transforms:

- random shift between -5% and 5% with probability 0.9;
- random scale between 0% and 5% with probability 0.9;
- random rotation between -2° and 2° with probability 0.9;
- HSV color jittering of 5° with probability 0.9 (RGB only).

Following [10], we defined the loss function as follows:

$$L\left(I,\hat{I}\right) = L_{L1}\left(I,\hat{I}\right) + L_{SSIM}\left(I,\hat{I}\right) \qquad (1)$$

where I is the autoencoder input, $\hat{I}$ its reconstruction, $L_{L1}\left(I,\hat{I}\right) = \left\|I - \hat{I}\right\|_1$ is the reconstruction error and L_{SSIM} the Multi-Scale Structural Similarity as defined in [10]. The SSIM window size is set to 7 pixels. When RGB, depth and normals maps are concatenated, the loss is computed as the sum of all the individual loss components:

$$L_{RGBDN}\left(I,\hat{I}\right) = L\left(RGB, R\hat{G}B\right) + L\left(D,\hat{D}\right) + L\left(N,\hat{N}\right) \qquad (2)$$

where RGB, D and N are, respectively, the color image, the depth map and the normals map, each reconstructed as $R\hat{G}B$, $\hat{D}$ and $\hat{N}$. All models are trained for 5000 epochs with Adam [19] optimizer with learning rate 0.001, β_1 set to 0.9 and β_2 to 0.999. The mini-batch size is set to 32, enforced by dropping the last batch of every epoch if the required batch size is not met.

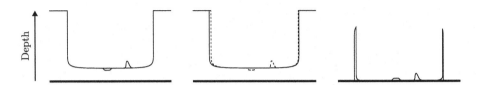

Fig. 6. On the left, the height map of a row from the depth image of an anomalous sample, containing a bump and a dent on its surface. On the center, its reconstruction with no surface anomalies, but with a slightly misaligned contour (the dashed line represents the original depth). On the right, the absolute difference of the two. The reconstruction error in the anomalous areas is negligible with respect to the error on the object contour.

The results, summarized in Table 2 and Table 3 as, respectively, image and pixel AUROC scores, suggest that a naive autoencoder trained on RGB data alone fails to effectively separate *good* and *bad* samples on most object categories. The worst performance is reached on *Candy Cane* and *Hazelnut Truffle*, where the results are comparable to a random classifier. In the former case, though the reconstruction quality is good, the anomalies may be too small to be effectively detected. In the latter, the detection possibly fails because of the low reconstruction quality, due to the roughness of the surface and known issues of convolutional autoencoders with high frequency spatial features [10]. Conversely, *Gummy Bear* exhibits a smooth and glossy surface, where edges are surrounded by bright glares, making the reconstruction from color images hardly conceivable. In fact, we achieve acceptable results only on objects with regular and opaque surfaces, thus easier to reconstruct, such as *Chocolate Cookie* and *Marshmallow*, where, moreover, anomalies appear on a fairly large scale.

Adding the depth map to the reconstruction task has little impact on the network performance. To understand the diverse reasons behind this, first consider that only two types of anomalies, namely bumps and dents, have an effect on the object depth, since color and normal alterations do not affect the geometry of the object. Furthermore, as depicted in Fig. 6, the average height of a bump or a dent is very small if compared to the size of the object, so that the reconstruction error of the anomalous regions is often negligible with respect to the error found on the silhouette of the object, where a naive autoencoder often produces slightly misaligned shapes. Also, extending the input and the output with depth information, the task becomes more complex and the reconstruction quality of the color image worsens when compared to the RGB-only scenario. This, eventually, leads to no improvement or even a drop in performance for most of the objects, being the color texture usually more informative.

Introducing the normals into the reconstruction task drastically improves the performance in almost every object category. Although normal maps do not provide any meaningful information to detect color alteration anomalies, they prove to be crucial to detect bumps, dents and surface normal alterations. Moreover, unlike the RGB image, normal maps do not suffer from glossy materials, reflections and sharp edged in the object texture, hence easing the reconstruction task.

We made an attempt to mitigate the problem with the low visibility of anomalies on the error maps obtained from depth reconstruction (the one described in Fig. 6), by clamping the depth maps between fixed lower and upper bounds. This helps reducing the contrast - and thus the reconstruction error - between pixels on the object silhouette and the background. Bounds were chosen so that the final depth range would be as small as possible, with the constraint to never affect pixels belonging to the object: the lower bound is always smaller than the closest point, the upper bound is always greater than the farthest visible point.

We repeated the RGB-D and RGB-D-N experiments described before, under the exact same settings, but clamping the depth maps. Results are summarized in the columns "RGB-cD" and "RGB-cD-N" of Table 2 and Table 3. As for RGB-D and RGB-cD, we can observe that the depth clamping results in slightly better performances across most object categories, both for image and pixel AUROC. When using also normal maps, the benefits of having a clamped depth range seem to have less relevance: only 6 categories out of 10 see an improvement in image AUROC, but pixel metrics improve only in 2 cases.

Table 2. Autoencoder image AUROC on the eyecandies dataset.

Category	RGB	RGB-D	RGB-D-N	RGB-cD	RGB-cD-N
Candy Cane	0.527	0.529	0.587	0.537	**0.596**
Chocolate Cookie	0.848	**0.861**	0.846	0.847	0.843
Chocolate Praline	0.772	0.739	0.807	0.748	**0.819**
Confetto	0.734	0.752	0.833	0.779	**0.846**
Gummy Bear	0.590	0.594	**0.833**	0.635	**0.833**
Hazelnut Truffle	0.508	0.498	0.543	0.511	**0.550**
Licorice Sandwich	0.693	0.679	0.744	0.691	**0.750**
Lollipop	0.760	0.651	**0.870**	0.699	0.846
Marshmallow	0.851	0.838	**0.946**	0.871	0.940
Peppermint Candy	0.730	0.750	0.835	0.740	**0.848**

Table 3. Autoencoder pixel AUROC on the eyecandies dataset.

Category	RGB	RGB-D	RGB-D-N	RGB-cD	RGB-cD-N
Candy Cane	0.972	0.973	**0.982**	0.975	0.980
Chocolate Cookie	0.933	0.927	**0.979**	0.939	**0.979**
Chocolate Praline	0.960	0.958	0.981	0.954	**0.982**
Confetto	0.945	0.945	**0.979**	0.957	0.978
Gummy Bear	0.929	0.929	**0.951**	0.933	**0.951**
Hazelnut Truffle	0.815	0.806	0.850	0.822	**0.853**
Licorice Sandwich	0.855	0.827	**0.972**	0.897	0.971
Lollipop	0.977	0.977	**0.981**	0.978	0.978
Marshmallow	0.931	0.931	**0.986**	0.940	0.985
Peppermint Candy	0.928	0.928	**0.967**	0.940	**0.967**

6   Conclusions and Future Works

We presented Eycandies, a novel synthetic dataset for anomaly detection and localization. Unlike existing datasets, for each unique object instance we provide RGB color images as well as depth and normals maps, captured under multiple light conditions. We have shown how unique candies are generated from a parametric reference model, with photo-realistic appearance and large intra-class variance. Likewise, random anomalies are carefully applied on color, depth and normals, then reprojected to 2D to get pixel-precise ground-truth data, avoiding any human intervention. Our experiments suggest that combining color and 3D data may open new possibilities to tackle the anomaly detection task and our dataset might be crucial to validate such new results.

As for future works, we see four main subjects to investigate. Firstly, we should expand the dataset by moving the camera around the target object, hence generating a full view. Secondly, we might add logical anomalies, such as missing parts, together with a sensible ground truth and a clear evaluation procedure. Thirdly, we could generate object-specific defects, such as wrong stripes on the *Candy Cane*, though the challenging part would be to leave the whole pipeline completely automated. Finally, we might model the noise, artifacts and deformations introduced by acquisition devices.

References

1. Ahuja, N.A., et al.: Probabilistic modeling of deep features for out-of-distribution and adversarial detection. https://arxiv.org/abs/1909.11786 (2019)
2. Akcay, S., Atapour-Abarghouei, A., Breckon, T.P.: GANomaly: semi-supervised anomaly detection via adversarial training. In: Jawahar, C.V., Li, H., Mori, G., Schindler, K. (eds.) ACCV 2018. LNCS, vol. 11363, pp. 622–637. Springer, Cham (2019). https://doi.org/10.1007/978-3-030-20893-6_39

3. Akcay, S., et al.: Anomalib: A deep learning library for anomaly detection. https://arxiv.org/abs/2202.08341 (2022)

4. Akcay, S., et al.: Deep feature kernel density estimation. http://github.com/openvinotoolkit/anomalib (2022)

5. Bengs, M., Behrendt, F., Krüger, J., Opfer, R., Schlaefer, A.: Three-dimensional deep learning with spatial erasing for unsupervised anomaly segmentation in brain mri. Int. J. Comput. Assist. Radiol. Surg. **16**(9), 1413–1423 (2021)

6. Bergmann, P., Batzner, K., Fauser, M., Sattlegger, D., Steger, C.: Beyond dents and scratches: Logical constraints in unsupervised anomaly detection and localization. Int. J. Comput. Vision **130**(1), 947–969 (2022)

7. Bergmann, P., Fauser, M., Sattlegger, D., Steger, C.: Uninformed students: Student-teacher anomaly detection with discriminative latent embeddings. In: 2020 IEEE/CVF Conference on Computer Vision and Pattern Recognition (CVPR), pp. 4182–4191 (2020). https://doi.org/10.1109/CVPR42600.2020.00424

8. Bergmann, P., Jin, X., Sattlegger, D., Steger, C.: The mvtec 3d-ad dataset for unsupervised 3d anomaly detection and localization. In: Proceedings of the 17th International Joint Conference on Computer Vision, Imaging and Computer Graphics Theory and Applications, vol. 5(1), pp. 202–213 (2022)

9. Bergmann, P., Kilian Batzner, M.F., Sattlegger, D., Steger, C.: The mvtec anomaly detection dataset: a comprehensive real-world dataset for unsupervised anomaly detection. Int. J. Comput. Vision **129**(4), 1038–1059 (2021)

10. Bergmann, P., Löwe, S., Fauser, M., Sattlegger, D., Steger, C.: Improving unsupervised defect segmentation by applying structural similarity to autoencoders. In: Proceedings of the 14th International Joint Conference on Computer Vision, Imaging and Computer Graphics Theory and Applications, vol. 130(1), pp. 372–380 (2019). https://doi.org/10.5220/0007364503720380

11. Blender Online Community: Blender - a 3D modelling and rendering package. Blender Foundation, Blender Institute, Amsterdam (2022). http://www.blender.org

12. Blum, H., Sarlin, P.E., Nieto, J.I., Siegwart, R.Y., Cadena, C.: Fishyscapes: A benchmark for safe semantic segmentation in autonomous driving. In: 2019 IEEE/CVF International Conference on Computer Vision Workshop (ICCVW), pp. 2403–2412 (2019)

13. Boikov, A., Payor, V., Savelev, R., Kolesnikov, A.: Synthetic data generation for steel defect detection and classification using deep learning. Symmetry 13(7) (2021). https://doi.org/10.3390/sym13071176, https://www.mdpi.com/2073-8994/13/7/1176

14. Carrera, D., Manganini, F., Boracchi, G., Lanzarone, E.: Defect detection in sem images of nanofibrous materials. IEEE Trans. Industr. Inf. **13**(2), 551–561 (2017). https://doi.org/10.1109/TII.2016.2641472

15. Defard, T., Setkov, A., Loesch, A., Audigier, R.: PaDiM: a patch distribution modeling framework for anomaly detection and localization. In: Del Bimbo, A., Cucchiara, R., Sclaroff, S., Farinella, G.M., Mei, T., Bertini, M., Escalante, H.J., Vezzani, R. (eds.) ICPR 2021. LNCS, vol. 12664, pp. 475–489. Springer, Cham (2021). https://doi.org/10.1007/978-3-030-68799-1_35

16. Denninger, M., et al.: Blenderproc. Arxiv (2019)

17. Gutierrez, P., Luschkova, M., Cordier, A., Shukor, M., Schappert, M., Dahmen, T.: Synthetic training data generation for deep learning based quality inspection (2021)

18. Kemmler, M., Rodner, E., Wacker, E.S., Denzler, J.: One-class classification with gaussian processes. Pattern Recogn. **46**(12), 3507–3518 (2013). https://doi.org/10.1016/j.patcog.2013.06.005, https://www.sciencedirect.com/science/article/pii/S0031320313002574

19. Kingma, D.P., Ba, J.: Adam: A method for stochastic optimization. arXiv preprint arXiv:1412.6980 (2014)

20. Krizhevsky, A., Nair, V., Hinton, G.: Learning Multiple Layers of Features From Tiny Images. University of Toronto, Tech. rep. (2009)

21. LeCun, Y., Cortes, C.: Mnist handwritten digit database (2010), http://yann.lecun.com/exdb/mnist/

22. M., W., T., H.: Weakly supervised learning for industrial optical inspection. In: 29th Annual Symposium of the German Association for Pattern Recognition (2007)

23. Moya, M.M., Hush, D.R.: Network constraints and multi-objective optimization for one-class classification. Neural Netw. **9**(3), 463–474 (1996)

24. Parker, S.G., et al.: Optix: A general purpose ray tracing engine. ACM Trans. Graph. **29**(4) (2010). https://doi.org/10.1145/1778765.1778803, https://doi.org/10.1145/1778765.1778803

25. Perera, P., Patel, V.M.: Deep transfer learning for multiple class novelty detection. In: Proceedings of the IEEE/CVF Conference on Computer Vision and Pattern Recognition (CVPR) (June 2019)

26. Roth, K., Pemula, L., Zepeda, J., Schölkopf, B., Brox, T., Gehler, P.: Towards total recall in industrial anomaly detection (2021). https://doi.org/10.48550/ARXIV.2106.08265, https://arxiv.org/abs/2106.08265

27. Sabokrou, M., Khalooei, M., Fathy, M., Adeli, E.: Adversarially learned one-class classifier for novelty detection. In: Proceedings of the IEEE Conference on Computer Vision and Pattern Recognition, pp. 3379–3388 (2018)

28. Schlegl, T., Seeböck, P., Waldstein, S.M., Langs, G., Schmidt-Erfurth, U.: f-anogan: Fast unsupervised anomaly detection with generative adversarial networks. Medical Image Anal. **54**, 30–44 (2019). https://doi.org/10.1016/j.media.2019.01.010, https://www.sciencedirect.com/science/article/pii/S1361841518302640

29. Schmedemann, O., Baaß, M., Schoepflin, D., Schüppstuhl, T.: Procedural synthetic training data generation for ai-based defect detection in industrial surface inspection. Procedia CIRP **107**, pp. 1101–1106 (2022). https://doi.org/10.1016/j.procir.2022.05.115, https://www.sciencedirect.com/science/article/pii/S2212827122003997, leading manufacturing systems transformation - Proceedings of the 55th CIRP Conference on Manufacturing Systems 2022

30. Severstal: Severstal: Steel defect detection (2019). https://www.kaggle.com/competitions/severstal-steel-defect-detection/overview

31. Simarro Viana, J., de la Rosa, E., Vande Vyvere, T., Robben, D., Sima, D.M., Investigators, C.E.N.T.E.R.-T.B.I.P.: Unsupervised 3D brain anomaly detection. In: Crimi, A., Bakas, S. (eds.) BrainLes 2020. LNCS, vol. 12658, pp. 133–142. Springer, Cham (2021). https://doi.org/10.1007/978-3-030-72084-1_13

32. Wang, W., et al.: Tartanair: A dataset to push the limits of visual slam. In: 2020 IEEE/RSJ International Conference on Intelligent Robots and Systems (IROS), pp. 4909–4916. IEEE (2020)

33. Huang, Y., Qiu, C., Yuan, K.: Surface defect saliency of magnetic tile. Vis Comput **36**, 85–96 (2020)

34. Yang, J., Xu, R., Qi, Z., Shi, Y.: Visual anomaly detection for images: a systematic survey. Procedia Comput. Sci. **199**, 471–478 (2022). https://doi.org/10.1016/j.procs.2022.01.057, https://www.sciencedirect.com/science/article/pii/S1877050922000576, the 8th International Conference on Information Technology and Quantitative Management (ITQM 2020 & 2021): Developing Global Digital Economy after COVID-19

35. Yu, J., et al.: Fastflow: Unsupervised anomaly detection and localization via 2d normalizing flows (2021). https://doi.org/10.48550/ARXIV.2111.07677, https://arxiv.org/abs/2111.07677

Two Video Data Sets for Tracking and Retrieval of Out of Distribution Objects

Kira Maag[1](✉), Robin Chan[2], Svenja Uhlemeyer[3], Kamil Kowol[3],
and Hanno Gottschalk[3]

[1] Ruhr University Bochum, Bochum, Germany
kira.maag@rub.de
[2] Bielefeld University, Bielefeld, Germany
rchan@techfak.uni-bielefeld.de
[3] University of Wuppertal, IZMD, Wuppertal, Germany
{suhlemeyer,kowol,hgottsch}@uni-wuppertal.de

Abstract. In this work we present two video test data sets for the novel computer vision (CV) task of out of distribution tracking (OOD tracking). Here, OOD objects are understood as objects with a semantic class outside the semantic space of an underlying image segmentation algorithm, or an instance within the semantic space which however looks decisively different from the instances contained in the training data. OOD objects occurring on video sequences should be detected on single frames as early as possible and tracked over their time of appearance as long as possible. During the time of appearance, they should be segmented as precisely as possible. We present the SOS data set containing 20 video sequences of street scenes and more than 1000 labeled frames with up to two OOD objects. We furthermore publish the synthetic CARLA-WildLife data set that consists of 26 video sequences containing up to four OOD objects on a single frame. We propose metrics to measure the success of OOD tracking and develop a baseline algorithm that efficiently tracks the OOD objects. As an application that benefits from OOD tracking, we retrieve OOD sequences from unlabeled videos of street scenes containing OOD objects.

Keywords: Computer vision · Video · Data sets · Out of distribution

1 Introduction

Semantic segmentation decomposes the pixels of an image into segments that adhere to a pre-defined set of semantic classes. In recent years, using fully convolutional deep neural networks [48] and training on publicly available data sets [14,18,28,29,68,82], this technology has undergone a remarkable learning curve. Recent networks interpret street scenes with a high degree of precision [17,78].

When semantic segmentation is used in open world scenarios, like in automated driving as area of application, objects could be present on images, which

Supplementary Information The online version contains supplementary material available at https://doi.org/10.1007/978-3-031-26348-4_28.

adhere to none of the semantic classes the network has been trained on and therefore force an error. Such objects from outside the network's semantic space form a specific class of out of distribution (OOD) objects. Naturally, it is desirable that the segmentation algorithm identifies such objects and abstains a decision on the semantic class for those pixels that are covered by the OOD object. At the same time, this additional requirement should not much deteriorate the performance on the primary segmentation task, if no OOD object is present. In other cases, an OOD object might be from a known class, however with an appearance that is very different from the objects of the same class in the training data, so that a stable prediction for this object is unrealistic. Also in this case, an indication as OOD object is preferable over the likely event of a misclassification. The computer vision (CV) task to mark the pixels of both kinds of objects can be subsumed under the notion of OOD segmentation. See [10,11,15,16,22,31,32,54,55] for recent contributions to this emerging field.

In many applications, images do not come as single frames, but are embedded in video sequences. If present, OOD objects occur persistently on subsequent frames. Tracking of OOD objects therefore is the logical next step past OOD segmentation. This ideally means identifying OOD objects in each frame on which they are present and give them a persistent identifier from the frame of first occurrence to the frame in which the OOD object leaves the image.

In this article we introduce the novel task of OOD tracking as a hybrid CV task inheriting from the established fields of OOD detection, OOD segmentation and object tracking. CV tasks often are dependent on suitable data sets, and OOD tracking is no exception in this regard. As our main contribution, we present two new labeled data sets of video sequences that will support the research effort in this field. The Street Obstacle Sequences (SOS) data set is a real world data set that contains more than 1,000 single frames in 20 video sequences containing one or two labeled OOD objects on streets along with further meta information, like distance or object ID. The SOS data set thus allows to evaluate the success of OOD tracking quantitatively for different kinds of OOD objects. As a second data set, we present CARLA-WildLife (CWL), a synthetic data set that consists of 26 fully annotated frames from the CARLA driving simulator in which a number of OOD objects from the Unreal Engine [27] collection of free 3D assets are introduced. Each frame in these video sequences contains in between 1 and 4 OOD instances. The meta data is consistent with SOS. In addition, the labeling policy is largely consistent with the single frame based road obstacle track in the SegmentMeIfYouCan benchmark [15]. Thereby, both data sets will also support standard OOD segmentation benchmarks. As a second contribution, we propose numerous metrics that can systematically measure the success of an OOD tracking algorithm. As a third contribution, we provide a first baseline that combines single frame OOD segmentation with tracking of segments. Using a single frame Nvidia DeepLabV3+ as a single frame segmentation network, we employ entropy maximization training for OOD segmentation with meta-classification to reduce the number of false positive OOD objects, following [16]. We then track the obtained OOD masks over the video sequences using an adjusted version of the light-weight tracking algorithm based on semantic masks introduced in [57–59]. We hope that this simple baseline will motivate researchers to develop their own OOD tracking algorithms and compare performance against our baseline.

It remains to show that OOD tracking is useful. Here we present an example from the context of automated driving and apply OOD tracking on the unsupervised retrieval of OOD objects. To this purpose, we combine our OOD tracking baseline with feature extractor based on DenseNet [42]. For each detected OOD object, we obtain a time series of feature vectors on which we employ a low dimensional embedding via the t-SNE algorithm [60]. Here the time series viewpoint makes it easy to clean the data and avoid false positives, e.g. by setting a filter to the minimum length. Clustering of similar objects, either on the basis of frames or on time series meta-clusters then enables the retrieval of previously unseen objects [69,77]. We apply this on the SOS and the CWL data sets as well as on self-recorded unlabeled data that contains OOD road obstacles. This provides a first method that enables the unsupervised detection of potentially critical situations or corner cases related to OOD objects from video data. The source code is publicly available at https://github.com/kmaag/OOD-Tracking and the datasets at https://zenodo.org/communities/buw-ood-tracking/.

This paper is organized as follows: Sect. 2 relates our work with existing OOD data sets as well as approaches in OOD segmentation, object tracking and object retrieval. The following Sect. 3 introduces our data sets for OOD tracking in street scenes and details on our labeling policy. In Sect. 4, we introduce a set of metrics to measure the success of OOD segmentation, tracking and clustering, respectively. The experiments are presented in Sect. 5 consisting of the method description, i.e., details of our OOD segmentation backbone, the tracking algorithm for OOD objects as well as OOD retrieval, and numerical results for the SOS as well as the CWL data set. Our findings are summarized in Sect. 6, where we also shortly comment on future research directions.

2   Related Work

OOD Data Sets. OOD detection in the field of CV is commonly tested by separating entire images that originate from different data sources. This includes e.g. separating MNIST [49] from FashionMNIST [80], NotMNIST [12], or Omniglot [46], and, as more complex task, separating CIFAR-10 [45] from SVHN [30] or LSUN [83]. Other data sets specifically designed to OOD detection in semantic segmentation are for instance Fishyscapes [10] and CAOS [38]. These two data sets either rely on synthetic data or generate OOD examples by excluding certain classes during model training. To overcome the latter limitations, data sets such as LostAndFound [71], RoadAnomaly [55], and also RoadObstacle21 [15] include images containing real OOD objects appearing in real world scenes. To this end, the established labeling policy of the semantic segmentation data set Cityscapes [18] serves as basis to decide whether an object is considered as OOD or not. However, all the outlined OOD data sets are based on single frames only. Although CAOS [38], LostAndFound [71], and RoadObstacle21 [15] include several images in the same scenes, they do not provide video sequences with (annotated) consecutive frames. In particular, mainly due to the labeling effort, none of the real world data sets provides a sufficient density of consecutive frames such that tracking of OOD objects could be applied and evaluated properly. One such but synthetic data set is StreetHazards [38]. This latter data set, however, mostly contains street scenes with OOD objects appearing in safety-irrelevant locations such as the background of the scene or in non-driveable areas.

In this work, we provide two novel video data sets with OOD objects on the road as region of interest. Therefore, our data sets can be understood to tackle the safety-relevant problem of obstacle segmentation [15]. While one of these two data sets consists of real-world images only, the other consists of synthetic ones. Both data sets include multiple sequences with pixel level annotations of consecutive frames, which for the first time enable tracking of OOD objects.

OOD Segmentation. OOD detection on image data was first tackled in the context of image classification. Methods such as [37,39,50,51,63] have proven to successfully identify entire OOD images by lowering model confidence scores. These methods can be easily extended to semantic segmentation by treating each pixel individually, forming common baselines for OOD detection in semantic segmentation [1,9], i.e., OOD segmentation. In particular, many of these OOD detection approaches are intuitively based on quantifying prediction uncertainty. This can also be accomplished e.g. via Monte-Carlo dropout [26] or an ensemble of neural networks [34,47], which has been extended to semantic segmentation in [3,44,65]. Another popular approach is training for OOD detection [21,40,63], which includes several current state-of-the-art works on OOD segmentation such as [7,8,16,22,32]. This type of approach relies on incorporating some kind of auxiliary training data, not necessarily real-world data, but disjoint from the original training data. In this regard, the most promising methods are based on OOD training samples generated by generative models as extensively examined in [19,54,55,66,79].

All existing methods are developed to operate on single frames. In this present work, we aim at investigating how such OOD segmentation methods could be extended to operate on video sequences with OOD objects appearing in multiple consecutive frames.

Object Tracking. In applications such as automated driving, tracking multiple objects in image sequences is an important computer vision task [64]. In instance segmentation, the detection, segmentation and tracking tasks are often performed simultaneously in terms of extending the Mask R-CNN network by an additional branch [6,81] or by building a variational autoencoder architecture on top [52]. In contrast, the tracking-by-detection methods first perform segmentation and then tracking using for example a temporal aggregation network [43] or the MOTSNet [72]. In addition, a more light-weight approach is presented in [13] based on the optical flow and the Hungarian algorithm. The tracking method introduced in [59] serves as a post-processing step, i.e., is independent of the instance segmentation network, and is light-weight solely based on the overlap of instances in consecutive frames. A modified version of this algorithm is used for semantic segmentation in [58].

Despite all the outlined works on object tracking, none of them were developed for OOD objects. In this present work, we therefore extend the post-processing method for tracking entire segments in semantic segmentation, that has originally been proposed in [58], to the unprecedented task of tracking OOD objects in image sequences.

Object Retrieval. Retrieval methods in general tackle the task of seeking related samples from a large database corresponding to a given query. Early works in this context aim to retrieve images that match best a query text or vice versa [2,33,41,62]. Another sub task deals with content-based image retrieval, which can be sub-categorized into instance- and category level retrieval. This is, given a query image depicting an object or scene, retrieving images representing the same object/scene or objects/scenes of the same category, respectively. To this end, these images must satisfy some similarity criteria based on some abstract description. In a first approach called QBIC [25], images are retrieved based on (global) low level features such as color, texture or shape. More advanced approaches utilize local level features [4,56], still they cannot fully address the problem of semantic gap [75], which describes the disparity between different representation systems [36]. Recent methods such as [61,67] apply machine/deep learning to learn visual features directly from the images instead of using hand-crafted features.

In this work, we do not directly retrieve images for some given query image, but instead we cluster all objects/images that are contained in our database based on their visual similarity, as it has been proposed in [69]. This particularly includes OOD objects. We extend this described single frame based approach to video sequences, i.e., we enhance the effectiveness by incorporating tracking information over multiple frames.

3   Data Sets

As already discussed in Sect. 2, in general there is a shortage of data sets that are dedicated to OOD detection in semantic segmentation. In particular, at the time of writing, there does not exist any OOD segmentation data set containing annotated video sequences. We therefore introduce the *Street Obstacle Sequences (SOS)*, *CARLA-WildLife (CWL)* and *Wuppertal Obstacle Sequences (WOS)* data sets. Example images and more details can be found in Appendix A.

3.1   Street Obstacle Sequences

The SOS data set contains 20 real-world video sequences in total. The given scenes are shown from a perspective of a vehicle that is approaching objects placed on the street, starting from a distance of 20 m to the street obstacle. The outlined street obstacles are chosen such that they could cause hazardous street scenarios. Moreover, each object corresponds to a class that is semantically OOD according to the Cityscapes labeling policy [18]. In SOS, there are 13 different

(a) Street Obstacle Sequences   (b) CARLA-WildLife

Fig. 1. Some exemplary OOD objects from our (a) SOS and (b) CWL data sets.

object types, which include e.g. bags, umbrellas, balls, toys, or scooters, cf. also Fig. 1(a). They represent potential causes of hazardous street scenarios, making their detection and localization particularly crucial in terms of safety.

Each sequence in SOS was recorded at a rate of 25 frames per second, of which every eighth frame is labeled. This yields a total number of 1,129 pixel-accurately labeled frames. As region of interest, we restrict the segmentation to the drivable area, i.e., the street. Consequently, SOS contains two classes, either

1) *street obstacle / OOD* ,   or   2) *street / not OOD* .

Note that image regions outside the drivable area are labeled as *void* and are ignored during evaluation.

Given the unique density of consecutive annotated frames, SOS allows for proper evaluation of tracking OOD objects besides their detection and pixel level localization. In this way, SOS facilitates the approach to the novel and practically relevant CV task of combining object tracking and OOD segmentation.

For a more in-depth evaluation, we further provide meta data to each obstacle in the SOS data set. This includes information such as the size of an object and their distance to the camera. In this regard, the size is approximated by the number of annotated pixels and the distance by markings on the street.

3.2   CARLA-WildLife

Since the generation of the SOS data set is time consuming and the selection of diverse real-world OOD objects is limited in practice, we additionally introduce a synthetic data set for OOD detection offering a large variety of OOD object types. The main advantage of synthetic data is that they can be produced inexpensively with accurate pixel-wise labels of full scenes, besides being able to manipulate the scenes as desired.

By adding freely available assets from Unreal Engine 4 [27] to the driving simulation software CARLA [23], we generate sequences in the same fashion as the SOS data set that we provide in the additional CWL data set. It contains 26 synthetic video sequences recorded at a rate of 10 frames per second with 18 different object types placed on the streets of CARLA. The objects include e.g. dogs, balls, canoes, pylons, or bags, cf. also Fig. 1(b). Again, the objects were chosen based on whether they could cause hazardous street scenarios. Since these objects are not included in the standard set of semantic labels provided by CARLA, each object type is added as extra class retroactively. In addition to the semantic segmentation based on the Cityscapes labeling policy (and including the OOD class), CWL further provides instance segmentation, i.e., individual OOD objects of the same class can be distinguished within each frame, and tracking information, i.e., the same object instance can be identified over the course of video frames. Moreover, we provide pixel-wise distance information for each frame of entire sequences as well as aggregated depth information per OOD object depicting the shortest distance to the ego-vehicle.

3.3   Wuppertal Obstacle Sequences

While the SOS data set considers video sequences where the camera moves towards the static OOD objects located on the street, we provide additional moving OOD objects in the *WOS* data set. It contains 44 real-world video sequences recorded from the viewpoint of a moving vehicle. The moving objects are mostly dogs, rolling or bouncing balls, skateboards or bags and were captured with either a static or a moving camera. This data set comes without labels and is used for test purposes for our OOD tracking and retrieval application.

4   Performance Metrics

In this section, we describe the performance metrics for the task of OOD tracking, i.e., OOD segmentation and object tracking, as well as clustering.

4.1   OOD Segmentation

Hereafter, we assume that the OOD segmentation model provides pixel-wise OOD scores s for a pixel discrimination between *OOD* and *not OOD*, see also Sect. 3. As proposed in [15], the separability of these pixel-wise scores is evaluated using the area under the precision recall curve (AuPRC) where precision and recall values are varied over some score thresholds $\tau \in \mathbb{R}$ applied to s. Furthermore, we consider the false positive rate at 95% true positive rate (FPR$_{95}$) as safety critical evaluation metric. This metric indicates how many false positive errors have to be made to achieve the desired rate of true positive predictions.

As already implied, the final OOD segmentation is obtained by thresholding on s. In practice, it is crucial to detect and localize each single OOD object. For this reason, we evaluate the OOD segmentation quality on segment level. To this end, we consider a connected component of pixels sharing the same class label in a segmentation mask as segment. From a practitioner's point of view, it is often sufficient to only recognize a fraction of OOD objects to detect and localize them. As quality measure to decide whether one segment is considered as detected, we stick to an adjusted version of the segment-wise intersection over union ($sIoU$) as introduced in [73]. Then, given some detection threshold $\kappa \in [0, 1)$, the number of true positive (TP), false negative (FN) and false positive (FP) segments can be computed. These quantities are summarized by the $F_1 = 2TP/(2TP + FN + FP)$ score, which represents a metric for the segmentation quality (for some fixed score threshold κ). As the numbers of TP, FN and FP depend on the detection threshold κ, we additionally average the F_1 score over different κ. This yields $\bar{F}_1$ as our main evaluation metric on segment level as it is less affected by the detection threshold.

For a more detailed description of the presented performance metrics for OOD segmentation, we refer to [15].

4.2 Tracking

To evaluate OOD object tracking, we use object tracking metrics such as multiple object tracking accuracy ($MOTA$) and precision ($MOTP$) as performance measures [5]. $MOTA$ is based on three error ratios: the ratio of false positives, false negatives and mismatches ($\overline{mme}$) over the total number of ground truth objects in all frames. A mismatch error is defined as the ID change of two predicted objects that are matched with the same ground truth object. $MOTP$ is the averaged distance between geometric centers of matched pairs of ground truth and predicted objects.

For the tracking measures introduced in [64], all ground truth objects of an image sequence are identified by different IDs and denoted by GT. These are divided into three cases: mostly tracked (MT) if it is tracked for at least 80% of frames (whether the object was detected or not), mostly lost (ML) if it is tracked for less than 20%, else partially tracked (PT). These common multiple object tracking metrics are created for the object detection task using bounding boxes and also applicable to instance segmentation. Thus, we can apply these measures to our detected OOD objects without any modification.

Moreover, we consider the tracking length metric l_t which counts the number of all frames where a ground truth object is tracked divided by the total number of frames where this ground truth object occurs. In comparison to the presented metrics which require ground truth information in each frame, the tracking length additionally uses non-annotated frames if present. Note that we find this case within the SOS data set where about every eighth frame is labeled. To this end, we consider frames $t, \ldots, t + i$, $i > 1$, with available labels for frames t and $t + i$. If the ground truth object in frame t has a match and the corresponding tracking ID of the predicted object occurs in consecutive frames $t + 1, \ldots, t + i - 1$, we increment the tracking length.

4.3 Clustering

The evaluation of OOD object clusters $C_i \in \{C_1, \ldots, C_n\}$, which contain the two-dimensional representatives of the segments k of OOD object predictions, depends on the differentiation level of these objects. We consider an instance level and a semantic level based on object classes. Let $\mathcal{Y} = \{1, \ldots, q\}$ and

$\mathcal{Y}^{\text{ID}} = \{1, \ldots, p\}$ denote the set of semantic class and instance IDs, respectively. For some given OOD segment k, y_k and y_k^{ID} correspond to the ground truth class and instance ID with which k has the highest overlap. On instance level, we aspire that OOD objects which belong to the same instance in an image sequence are contained in the same cluster. This is, we compute the relative amount of OOD objects per instance in the same cluster,

$$CS_{\text{inst}} = \frac{1}{p} \sum_{i=1}^{p} \frac{\max\limits_{C \in \{C_1, \ldots, C_n\}} |\{k \in C \mid y_k^{\text{ID}} = i\}|}{\sum\limits_{C \in \{C_1, \ldots, C_n\}} |\{k \in C \mid y_k^{\text{ID}} = i\}|} \in [0, 1] , \qquad (1)$$

averaged over all instances. On a semantic level, we pursue two objectives. The first concerns the semantic class impurity of the clusters,

$$CS_{\text{imp}} = \frac{1}{n} \sum_{i=1}^{n} |\{y_k | k \in C_i\}| \in [1, q] , \qquad (2)$$

averaged over all clusters $C_i \in \{C_1, \ldots, C_n\}$. Secondly, we aspire a low fragmentation of classes into different clusters

$$CS_{\text{frag}} = \frac{1}{q} \sum_{i=1}^{q} |\{C \in \{C_1, \ldots, C_n\} | \exists k \in C : y_k = i\}| , \qquad (3)$$

i.e., ideally, each class constitutes exactly one cluster. Here, we average over the semantic classes in $\mathcal{Y}$.

5   Experiments

In this section, we first introduce the methods which we use for OOD segmentation, tracking as well as retrieval and second, we show the numerical and qualitative results on our two main data sets, SOS and CWL. Qualitative results on OOD object retrieval from the WOS data set are given in the appendix.

5.1   Method

Our method consists of the CV tasks OOD segmentation and object tracking. For OOD segmentation, we consider the predicted region of interest and the entropy heatmap obtained by a semantic segmentation network. Via entropy thresholding, the OOD objects are created and the prediction quality is assessed by meta classification in order to discard false positive OOD predictions. In the next step, the OOD objects are tracked in an image sequence to generate tracking IDs. Furthermore, we study the retrieval of detected OOD objects using tracking information. An overview of our method is shown in Fig. 2.

OOD Object Segmentation. For the segmentation of OOD objects, we use the publicly available segmentation method that has been introduced in [16]. In the latter work, a DeepLabV3+ model [84], initially trained on Cityscapes [18], has been extended to OOD segmentation by including auxiliary OOD samples

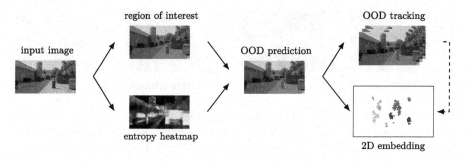

Fig. 2. Overview of our method. The input image is fed into a semantic segmentation network to extract the region of interest (here road) and the entropy heatmap. The resulting OOD prediction is used to produce the tracking IDs and a 2D embedding.

extracted from the COCO data set [53]. To this end, the model has been trained for high softmax entropy responses on the induced known unknowns (provided by COCO), which showed generalization capabilities with respect to truly unknown objects available in data sets such as LostAndFound [71] and RoadObstacle21 [15]. This outlined method is applied to single frames and utilizes the pixel-wise softmax entropy as OOD score.

Further, we apply meta classification [73,74] to OOD object predictions for the purpose of reducing false positive OOD indications. These false positives are identified by means of hand-crafted metrics, which are in turn based on dispersion measures like entropy as well as geometry and location information, see also [16]. These hand-crafted metrics form a structured data set where the rows correspond to predicted segments and the columns to features. Given this meta data set, we employ logistic regression with L^1-penalty on the weights (LASSO [76]) as post-processing (meta) model to remove false positive OOD object predictions, without requiring ground truth information at run time.

For more details on the construction of the structured data set, we refer the reader to [73,74]. An illustration of the single steps of the OOD object segmentation method can be found in Fig. 3.

OOD Object Tracking. In this section, we present the light-weight tracking approach that we use to track predicted OOD objects. This method has originally been introduced for semantic segmentation in [58] and does not require any training as it is an heuristic solely based on the overlap of OOD objects in consecutive frames. We assume that an OOD object segmentation is available for each frame x, as e.g. described in Sect. 5.1. The idea of employing this tracking method is to match segments based on their overlap (measured by the segment-wise intersection over union, shorthand IoU) and proximity of their geometric centers in consecutive frames.

ground truth   entropy heatmap   OOD segmentation   final prediction

Fig. 3. Segmentation of OOD objects (orange in ground truth) on the street via entropy thresholding & prediction quality rating via meta classification (green corresponds to a high confidence of being a correct OOD object prediction, red to a low one), resulting in final prediction mask. (Color figure online)

We apply the tracking approach sequentially to each frame $x \in \{x_t\}_{t=1}^{T}$ of an image sequence of length T. In more detail, the segments in the first frame, i.e., $t = 1$, are assigned with random IDs. Then, for each of the remaining frames $t, t > 1$, the segments are matched with the segment IDs of its respective previous frame $t - 1$. To this end, we use a tracking procedure consisting of five steps, which we will briefly describe in what follows. For a detailed description, we refer the reader to [58]. In step 1, OOD segments that are predicted in the same frame are aggregated by means of their distance. In steps 2 and 3, segments are matched if their geometric centers are close together or if their overlap is sufficiently large in consecutive frames, respectively. In step 4, linear regression is used to account for "flashing" segments (over a series of consecutive frames) or temporarily occluded as well as non-detected ones, i.e., false negatives. As final step 5, segments are assigned new IDs in case they have not received any in the steps 1–4 of the matching process.

OOD Object Retrieval. On top of the segmentation and tracking of OOD objects, we perform a method similar to content-based image retrieval in order to form clusters of the OOD objects that constitute novel semantic concepts. To this end we adapt an existing approach [69,77] to video sequences by incorporating the tracking information which we obtain e.g. as described in Sect. 5.1. This is, we require the tracking information to be available for each frame x and apply OOD object retrieval as a post-processing step which does not depend on the underlying semantic segmentation network nor on the OOD segmentation method but on given OOD segmentation masks.

For each frame x and OOD segment $k \in \hat{K}(x)$, let $\hat{y}_k^{ID}$ denote the predicted tracking ID. To diminish the number of the false positives, we only cluster predicted segments that are tracked over multiple frames of an image sequence $\{x_t\}_{t=1}^{T}$, based on some length parameter $\ell \in \mathbb{N}$ Further, each frame x is tailored to boxes around the remaining OOD segments k, which are vertically bounded by the pixel locations $\min_{(z_v,z_h) \in k} z_v$ and $\max_{(z_v,z_h) \in k} z_v$, horizontally by $\min_{(z_v,z_h) \in k} z_h$ and $\max_{(z_v,z_h) \in k} z_h$. Image clustering usually takes place in a lower dimensional latent space due to the curse of dimensionality. To this end, the image patches are fed into an image classification ResNet152 [35] (without its final classification layer) trained on ImageNet [20], which produces

Table 1. OOD object segmentation, tracking and clustering results for the SOS and the CWL data set.

Data set	AuPRC ↑	FPR$_{95}$ ↓	$\bar{F}_1$ ↑	$MOTA$ ↑	$\overline{mme}$ ↓	$MOTP$ ↓	GT	MT	PT	ML	l_t ↑
SOS	85.56	1.26	35.84	−0.0826	0.0632	12.3041	26	9	14	3	0.5510
CWL	79.54	1.38	45.46	0.4043	0.0282	16.4965	62	24	30	8	0.5389

	Without tracking ($\ell = 0$)			With tracking ($\ell = 10$)		
Data set	CS_{inst} ↑	CS_{imp} ↓	CS_{frag} ↓	CS_{inst} ↑	CS_{imp} ↓	CS_{frag} ↓
SOS	0.8652	2.5217	2.8182	0.8955	1.7917	1.9091
CWL	0.8637	2.8181	2.2500	0.8977	2.1739	1.8000

feature vectors of equal size regardless of the input dimension. These features are projected into a low-dimensional space by successively applying two dimensionality reduction techniques, namely principal component analysis (PCA [70]) and t-distributed stochastic neighbor embedding (t-SNE [60]). As final step, the retrieved OOD object predictions are clustered in the low-dimensional space, e.g., via the DBSCAN clustering algorithm [24].

5.2 Numerical Results

In this section, we present the numerical results on the novel task of OOD tracking. To this end, we apply simple baseline methods introduced in Sect. 5.1 on two labeled data sets of video sequences (SOS and CWL) and motivate the usefulness of OOD tracking using an unsupervised retrieval of OOD objects in the context of automated driving.

OOD Segmentation. For OOD segmentation, we apply the method described in Sect. 5.1, which provides pixel-wise softmax entropy heatmaps as OOD scores (see Fig. 3 (center left)). The pixel-wise evaluation results for the SOS and the CWL data sets are given in Table 1 considering AuPRC and FPR$_{95}$ as metrics (Sect. 4.1).

We achieve AuPRC scores of 85.56% and 79.54% as well as FPR$_{95}$ scores of 1.26% and 1.38% on SOS and CWL, respectively.

To obtain the OOD segmentation given some input image, thresholding is applied to the softmax entropy values. We choose the threshold τ by means of hyperparameter optimization, yielding $\tau = 0.72$ for SOS and $\tau = 0.81$ for CWL.

As next step, meta classification is used as post-processing to reduce the number of false positive OOD segments. We train the model on one data set and evaluate on the other one, e.g. for experiments on SOS the meta classification model is trained on CWL. The corresponding $\bar{F}_1$ scores on segment level are shown in Table 1. The higher $\bar{F}_1$ score of 45.46% is obtained for the CWL data set indicating that training the meta model on SOS and testing it on CWL is more effective than vice versa. In addition, we provide results for a different meta classification model which is trained and evaluated per leave-one-out cross

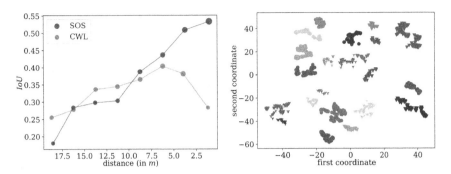

Fig. 4. *Left*: Discretized distance between ground truth objects and camera vs. mean *IoU* over all object types of the SOS and the CWL data set, respectively. The dot size is proportional to mean segment size. *Right*: Clustering of OOD segments predicted for the CWL data set with min. tracking frequency $\ell = 10$.

validation on the respective data set, see Appendix B. In Fig. 3, an example image of our OOD segmentation method is presented. The final prediction mask after entropy thresholding and meta classification contains only true OOD objects. In Appendix C and Appendix D, more numerical results evaluated for depth binnings and on individual OOD classes are presented, respectively.

OOD Tracking. Building upon the OOD segmentation masks obtained, in this subsection we report OOD tracking results. We consider several object tracking metrics (see Sect. 4.2) shown in Table 1 for the SOS and CWL data set. We observe a comparatively low *MOTA* performance for the SOS data set. The underlying reason is a high number of false positive segments that are accounted for in this metric, as also shown in the detection metric $\bar{F}_1$.

Furthermore, most of the ground truth objects are at least partially tracked, only 3 out of 26 and 8 out of 62 ground truth objects are largely lost out for SOS and CWL, respectively. Analogously, in Fig. 4, we observe that most ground truth objects are matched with predicted ones for the SOS data set. This plot shows the correlation between the *IoU* (of ground truth and predicted objects) and the distance of the ground truth objects to the camera as we provide meta data like depth for our data sets. We observe for both data sets that the *IoU* increases with decreasing distances, the only exception are very short distance objects to the ego-car for the CWL data set. Moreover, we provide video sequences[1] that visualize the final OOD segmentation and object tracking results. In Appendix D, more numerical results evaluated on individual OOD classes are presented.

Retrieval of OOD Objects. Finally, we evaluate the clustering of OOD segments obtained by the OOD object segmentation method introduced in Sect. 5.1. In Table 1, we report the clustering metrics CS_{inst}, CS_{imp} and CS_{frag}

---

[1] https://youtu.be/_DbV8XprDmc.

(see Sect. 4.3) with ($\ell = 10$) and without ($\ell = 0$) incorporating the OOD tracking information, respectively. For both, the CWL and the SOS data set, all clustering metrics improve when applying the OOD tracking as a pre-processing step. A reason for this is, that the tracking information "tidies up" the embedding space, e.g. by removing noise, which enhances the performance of the clustering algorithm. For CWL (with 18 object types), 1266/1026 OOD segments are clustered into 22/23 clusters without/with using tracking results, for SOS (with 13 object types), we obtain 23/24 clusters which contain 1437/888 OOD segments in total. For the clustering, we applied the DBSCAN algorithm with hyperparameters $\varepsilon = 4.0$ and minPts $= 15$. In Fig. 4, we exemplarily visualize the clustered embedding space for the CWL data set with $\ell = 10$. The remaining visualizations as well as additional results for the second meta classification model are provided in Appendix B. Furthermore, we visualize some clustering results for the WOS data set in Appendix E. As WOS comes without labels, we do not report any evaluation metrics, but provide some visualizations for the 5 largest clusters.

6   Conclusion and Outlook

We created a baseline for the CV task of tracking OOD objects by (a) publishing two data sets with 20 (SOS) and 26 (CWL) annotated video sequences containing OOD objects on street scenes and (b) presenting an OOD tracking algorithm that combines frame-wise OOD object segmentation on single frames with tracking algorithms. We also proposed a set of evaluation metrics that permit to measure the OOD tracking efficiency. As an application, we retrieved new, previously unlearned objects from video data of urban street scenes.

To go beyond this baseline, several directions of research seem to be promising. First, OOD segmentation on video data could benefit from 3D CNN acting on the spatial and temporal dimension, rather than combining 2D OOD segmentation with tracking. However, at least for those OOD segmentation algorithms that involve OOD training data, new and specific video data sets would be required. Similarly, genuine video sequence based retrieval algorithms should be developed to improve our revival baseline. Such algorithms could prove useful to enhance the coverage of urban street scenes in training data sets for AI-based perception in automated driving.

Acknowledgements. We thank Sidney Pacanowski for the labeling effort, Dariyoush Shiri for support in coding, Daniel Siemssen for support in the generation of CARLA data and Matthias Rottmann for interesting discussions. This work has been funded by the German Federal Ministry for Economic Affairs and Climate Action (BMWK) via the research consortia Safe AI for Automated Driving (grant no. 19A19005R), AI Delta Learning (grant no. 19A19013Q), AI Data Tooling (grant no. 19A20001O) and the Ministry of Culture and Science of the German state of North Rhine-Westphalia as part of the KI-Starter research funding program.

References

1. Angus, M., Czarnecki, K., Salay, R.: Efficacy of Pixel-Level OOD Detection for Semantic Segmentation. arXiv, pp. 1–13 (2019)
2. Arandjelović, R., Zisserman, A.: Multiple queries for large scale specific object retrieval. In: BMVC (2012)
3. Badrinarayanan, V., Kendall, A., Cipolla, R.: Bayesian SegNet: model uncertainty in deep convolutional encoder-decoder architectures for scene understanding. In: Proceedings of the British Machine Vision Conference (BMVC), pp. 1–12 (2017)
4. Bay, H., Tuytelaars, T., Van Gool, L.: SURF: speeded up robust features. In: Leonardis, A., Bischof, H., Pinz, A. (eds.) ECCV 2006. LNCS, vol. 3951, pp. 404–417. Springer, Heidelberg (2006). https://doi.org/10.1007/11744023_32
5. Bernardin, K., Stiefelhagen, R.: Evaluating multiple object tracking performance: the clear mot metrics. EURASIP J. Image Video Process. (2008)
6. Bertasius, G., Torresani, L.: Classifying, segmenting, and tracking object instances in video with mask propagation. arXiv abs/1912.04573 (2019)
7. Besnier, V., Bursuc, A., Picard, D., Briot, A.: Triggering failures: out-of-distribution detection by learning from local adversarial attacks in semantic segmentation. In: Proceedings of the IEEE/CVF International Conference on Computer Vision (ICCV), pp. 15701–15710 (2021)
8. Bevandić, P., Krešo, I., Oršić, M., Šegvić, S.: Simultaneous semantic segmentation and outlier detection in presence of domain shift. In: Proceedings of the German Conference on Pattern Recognition (GCPR), Dortmund, Germany, pp. 33–47 (2019)
9. Blum, H., Sarlin, P.E., Nieto, J., Siegwart, R., Cadena, C.: Fishyscapes: a benchmark for safe semantic segmentation in autonomous driving. In: Proceedings of the IEEE International Conference on Computer Vision (ICCV) Workshops, Seoul, Korea, pp. 2403–2412 (2019)
10. Blum, H., Sarlin, P.E., Nieto, J., Siegwart, R., Cadena, C.: The fishyscapes benchmark: measuring blind spots in semantic segmentation. Int. J. Comput. Vision **129**(11), 3119–3135 (2021)
11. Brüggemann, D., Chan, R., Rottmann, M., Gottschalk, H., Bracke, S.: Detecting out of distribution objects in semantic segmentation of street scenes. In: The 30th European Safety and Reliability Conference (ESREL), vol. 2 (2020)
12. Bulatov, Y.: Notmnist dataset. Google (Books/OCR), vol. 2, Technical report (2011). http://yaroslavvb.blogspot.it/2011/09/notmnist-dataset.html
13. Bullinger, S., Bodensteiner, C., Arens, M.: Instance flow based online multiple object tracking, pp. 785–789 (2017). https://doi.org/10.1109/ICIP.2017.8296388
14. Caesar, H., et al.: nuScenes: a multimodal dataset for autonomous driving. In: Proceedings of the IEEE/CVF Conference on Computer Vision and Pattern Recognition, pp. 11621–11631 (2020)
15. Chan, R., et al.: SegmentMeIfYouCan: a benchmark for anomaly segmentation. In: Thirty-Fifth Conference on Neural Information Processing Systems (NeurIPS) Datasets and Benchmarks Track (2021)
16. Chan, R., Rottmann, M., Gottschalk, H.: Entropy maximization and meta classification for out-of-distribution detection in semantic segmentation. In: Proceedings of the IEEE/CVF International Conference on Computer Vision (ICCV), pp. 5128–5137 (2021)

17. Chen, L.C., Papandreou, G., Kokkinos, I., Murphy, K.P., Yuille, A.L.: DeepLab: semantic image segmentation with deep convolutional nets, atrous convolution, and fully connected CRFs. IEEE Trans. Pattern Anal. Mach. Intell. **40**, 834–848 (2018)
18. Cordts, M., et al.: The cityscapes dataset for semantic urban scene understanding. In: 2016 IEEE Conference on Computer Vision and Pattern Recognition (CVPR), pp. 3213–3223 (2016)
19. Creusot, C., Munawar, A.: Real-time small obstacle detection on highways using compressive RBM road reconstruction. In: 2015 IEEE Intelligent Vehicles Symposium (IV), pp. 162–167 (2015)
20. Deng, J., Dong, W., Socher, R., Li, L.J., Li, K., Fei-Fei, L.: Imagenet: a large-scale hierarchical image database. In: CVPR (2009)
21. Devries, T., Taylor, G.W.: Learning confidence for out-of-distribution detection in neural networks. arXiv abs/1802.04865 (2018)
22. Di Biase, G., Blum, H., Siegwart, R., Cadena, C.: Pixel-wise anomaly detection in complex driving scenes. In: Proceedings of the IEEE/CVF Conference on Computer Vision and Pattern Recognition (CVPR), pp. 16918–16927 (2021)
23. Dosovitskiy, A., et al.: CARLA: an open urban driving simulator. In: Proceedings of CoRL, Mountain View, USA, pp. 1–16 (2017)
24. Ester, M., Kriegel, H.P., Sander, J., Xu, X.: A density-based algorithm for discovering clusters in large spatial databases with noise. In: KDD (1996)
25. Flickner, M., et al.: Query by image and video content: the QBIC system. Computer **28**(9), 23–32 (1995). https://doi.org/10.1109/2.410146
26. Gal, Y., Ghahramani, Z.: Dropout as a Bayesian approximation: representing model uncertainty in deep learning. In: Proceedings of the 33rd International Conference on Machine Learning. Proceedings of Machine Learning Research, New York, USA, vol. 48, pp. 1050–1059. PMLR (2016)
27. Games, E.: Unreal engine (2004–2022). https://www.unrealengine.com
28. Geiger, A., Lenz, P., Stiller, C., Urtasun, R.: Vision meets robotics: the KITTI dataset. Int. J. Robot. Res. **32**(11), 1231–1237 (2013)
29. Geyer, J., et al.: A2D2: Audi autonomous driving dataset. arXiv abs/2004.06320 (2020)
30. Goodfellow, I.J., Bulatov, Y., Ibarz, J., Arnoud, S., Shet, V.: Multi-digit number recognition from street view imagery using deep convolutional neural networks. arXiv preprint arXiv:1312.6082 (2013)
31. Grcić, M., Bevandić, P., Šegvić, S.: Dense open-set recognition with synthetic outliers generated by real nvp. arXiv preprint arXiv:2011.11094 (2020)
32. Grcić, M., Bevandić, P., Šegvić, S.: Dense anomaly detection by robust learning on synthetic negative data. arXiv preprint arXiv:2112.12833 (2021)
33. Guadarrama, S., et al.: Open-vocabulary object retrieval. In: Robotics: Science and Systems (2014)
34. Gustafsson, F.K., Danelljan, M., Schön, T.B.: Evaluating scalable Bayesian deep learning methods for robust computer vision. In: Proceedings of the IEEE/CVF Conference on Computer Vision and Pattern Recognition (CVPR) Workshops, pp. 1289–1298. Virtual Conference (2020)
35. He, K., Zhang, X., Ren, S., Sun, J.: Deep residual learning for image recognition. CoRR abs/1512.03385 (2015). http://arxiv.org/abs/1512.03385
36. Hein, A.M.: Identification and bridging of semantic gaps in the context of multi-domain engineering. In: Proceedings 2010 Forum on Philosophy, Engineering & Technology (2010)

37. Hein, M., Andriushchenko, M., Bitterwolf, J.: Why ReLU networks yield high-confidence predictions far away from the training data and how to mitigate the problem. In: Proceedings of the IEEE/CVF Conference on Computer Vision and Pattern Recognition (CVPR), Long Beach, CA, USA, pp. 41–50 (2019)

38. Hendrycks, D., Basart, S., Mazeika, M., Mostajabi, M., Steinhardt, J., Song, D.: Scaling out-of-distribution detection for real-world settings (2020)

39. Hendrycks, D., Gimpel, K.: A baseline for detecting misclassified and out-of-distribution examples in neural networks. In: 5th International Conference on Learning Representations, ICLR 2017, Toulon, France, 24–26 April 2017, Conference Track Proceedings (2017)

40. Hendrycks, D., Mazeika, M., Dietterich, T.: Deep anomaly detection with outlier exposure. In: Proceedings of the International Conference on Learning Representations (ICLR), New Orleans, LA, USA, pp. 1–18 (2019)

41. Hu, R., Xu, H., Rohrbach, M., Feng, J., Saenko, K., Darrell, T.: Natural language object retrieval. In: 2016 IEEE Conference on Computer Vision and Pattern Recognition (CVPR), pp. 4555–4564 (2016)

42. Huang, G., Liu, Z., Weinberger, K.Q.: Densely connected convolutional networks. In: 2017 IEEE Conference on Computer Vision and Pattern Recognition (CVPR), pp. 2261–2269 (2017)

43. Huang, X., Xu, J., Tai, Y.W., Tang, C.K.: Fast video object segmentation with temporal aggregation network and dynamic template matching. In: Proceedings of the IEEE/CVF Conference on Computer Vision and Pattern Recognition (CVPR) (2020)

44. Kendall, A., Gal, Y.: What uncertainties do we need in Bayesian deep learning for computer vision? In: NIPS (2017)

45. Krizhevsky, A., Nair, V., Hinton, G.: The CIFAR-10 dataset, vol. 55, no. 5 (2014). http://www.cs.toronto.edu/kriz/cifar.html

46. Lake, B.M., Salakhutdinov, R., Tenenbaum, J.B.: Human-level concept learning through probabilistic program induction. Science **350**, 1332–1338 (2015)

47. Lakshminarayanan, B., Pritzel, A., Blundell, C.: Simple and scalable predictive uncertainty estimation using deep ensembles. In: NIPS (2017)

48. Lateef, F., Ruichek, Y.: Survey on semantic segmentation using deep learning techniques. Neurocomputing **338**, 321–348 (2019)

49. Lecun, Y.: The MNIST database of handwritten digits (2010). http://yann.lecun.com/exdb/mnist/. https://ci.nii.ac.jp/naid/10027939599/en/

50. Lee, K., Lee, K., Lee, H., Shin, J.: A simple unified framework for detecting out-of-distribution samples and adversarial attacks. In: Proceedings of the Conference on Neural Information Processing Systems (NIPS/NeurIPS), Montréal, QC, Canada, pp. 7167–7177 (2018)

51. Liang, S., Li, Y., Srikant, R.: Enhancing the reliability of out-of-distribution image detection in neural networks. In: International Conference on Learning Representations (2018)

52. Lin, C.C., Hung, Y., Feris, R., He, L.: Video instance segmentation tracking with a modified VAE architecture. In: Proceedings of the IEEE/CVF Conference on Computer Vision and Pattern Recognition (CVPR) (2020)

53. Lin, T.-Y., et al.: Microsoft COCO: common objects in context. In: Fleet, D., Pajdla, T., Schiele, B., Tuytelaars, T. (eds.) ECCV 2014. LNCS, vol. 8693, pp. 740–755. Springer, Cham (2014). https://doi.org/10.1007/978-3-319-10602-1_48

54. Lis, K., Honari, S., Fua, P., Salzmann, M.: Detecting road obstacles by erasing them. arXiv preprint arXiv:2012.13633 (2020)

55. Lis, K., Nakka, K., Fua, P., Salzmann, M.: Detecting the unexpected via image resynthesis. In: Proceedings of the IEEE/CVF International Conference on Computer Vision, pp. 2152–2161 (2019)
56. LoweDavid, G.: Distinctive image features from scale-invariant keypoints. Int. J. Comput. Vision **60**, 91–110 (2004)
57. Maag, K.: False negative reduction in video instance segmentation using uncertainty estimates. In: 2021 IEEE 33rd International Conference on Tools with Artificial Intelligence (ICTAI), pp. 1279–1286. IEEE (2021)
58. Maag, K., Rottmann, M., Gottschalk, H.: Time-dynamic estimates of the reliability of deep semantic segmentation networks. In: 2020 IEEE 32nd International Conference on Tools with Artificial Intelligence (ICTAI), pp. 502–509 (2020)
59. Maag, K., Rottmann, M., Varghese, S., Hueger, F., Schlicht, P., Gottschalk, H.: Improving video instance segmentation by light-weight temporal uncertainty estimates. arXiv preprint arXiv:2012.07504 (2020)
60. Van der Maaten, L., Hinton, G.: Visualizing data using t-SNE. J. Mach. Learn. Res. **9**(11) (2008)
61. Maji, S., Bose, S.: CBIR using features derived by deep learning. ACM/IMS Trans. Data Sci. (TDS) **2**, 1–24 (2021)
62. Mao, J., Huang, J., Toshev, A., Camburu, O.M., Yuille, A.L., Murphy, K.P.: Generation and comprehension of unambiguous object descriptions. In: 2016 IEEE Conference on Computer Vision and Pattern Recognition (CVPR), pp. 11–20 (2016)
63. Meinke, A., Hein, M.: Towards neural networks that provably know when they don't know. In: Proceedings of the International Conference on Learning Representations (ICLR), pp. 1–18. Virtual Conference (2020)
64. Milan, A., Leal-Taixé, L., Reid, I.D., Roth, S., Schindler, K.: MOT16: a benchmark for multi-object tracking. arXiv abs/1603.00831 (2016)
65. Mukhoti, J., Gal, Y.: Evaluating Bayesian deep learning methods for semantic segmentation. arXiv abs/1811.12709 (2018)
66. Munawar, A., Vinayavekhin, P., De Magistris, G.: Limiting the reconstruction capability of generative neural network using negative learning. In: Proceedings of the IEEE International Workshop on Machine Learning for Signal Processing (MLSP), Tokyo, Japan, pp. 1–6 (2017)
67. Naaz, E., Kumar, T.: Enhanced content based image retrieval using machine learning techniques. In: 2017 International Conference on Innovations in Information, Embedded and Communication Systems (ICIIECS), pp. 1–12 (2017)
68. Neuhold, G., Ollmann, T., Bulò, S.R., Kontschieder, P.: The mapillary vistas dataset for semantic understanding of street scenes, pp. 5000–5009 (2017)
69. Oberdiek, P., Rottmann, M., Fink, G.A.: Detection and retrieval of out-of-distribution objects in semantic segmentation. In: 2020 IEEE/CVF Conference on Computer Vision and Pattern Recognition Workshops (CVPRW), pp. 1331–1340 (2020)
70. Pearson F.R.S., K.: LIII. On lines and planes of closest fit to systems of points in space. Philos. Mag. Lett. **2**(11), 559–572 (1901)
71. Pinggera, P., Ramos, S., Gehrig, S., Franke, U., Rother, C., Mester, R.: Lost and found: detecting small road hazards for self-driving vehicles. In: International Conference on Intelligent Robots and Systems (IROS), Daejeon, Korea, pp. 1099–1106 (2016)

72. Porzi, L., Hofinger, M., Ruiz, I., Serrat, J., Bulo, S.R., Kontschieder, P.: Learning multi-object tracking and segmentation from automatic annotations. In: Proceedings of the IEEE/CVF Conference on Computer Vision and Pattern Recognition (CVPR) (2020)

73. Rottmann, M., Colling, P., Hack, T.P., Hüger, F., Schlicht, P., Gottschalk, H.: Prediction error meta classification in semantic segmentation: detection via aggregated dispersion measures of softmax probabilities. In: 2020 International Joint Conference on Neural Networks (IJCNN), pp. 1–9 (2020)

74. Rottmann, M., Schubert, M.: Uncertainty measures and prediction quality rating for the semantic segmentation of nested multi resolution street scene images. In: 2019 IEEE/CVF Conference on Computer Vision and Pattern Recognition Workshops (CVPRW), pp. 1361–1369 (2019)

75. Smeulders, A.W.M., Worring, M., Santini, S., Gupta, A., Jain, R.C.: Content-based image retrieval at the end of the early years. IEEE Trans. Pattern Anal. Mach. Intell. **22**, 1349–1380 (2000)

76. Tibshirani, R.: Regression shrinkage and selection via the lasso. J. Roy. Stat. Soc. B **58**, 267–288 (1996)

77. Uhlemeyer, S., Rottmann, M., Gottschalk, H.: Towards unsupervised open world semantic segmentation (2022)

78. Wang, J., et al.: Deep high-resolution representation learning for visual recognition. IEEE Trans. Pattern Anal. Mach. Intell. **43**, 3349–3364 (2021)

79. Xia, Y., Zhang, Y., Liu, F., Shen, W., Yuille, A.L.: Synthesize then compare: detecting failures and anomalies for semantic segmentation. In: Vedaldi, A., Bischof, H., Brox, T., Frahm, J.-M. (eds.) ECCV 2020. LNCS, vol. 12346, pp. 145–161. Springer, Cham (2020). https://doi.org/10.1007/978-3-030-58452-8_9

80. Xiao, H., Rasul, K., Vollgraf, R.: Fashion-MNIST: a novel image dataset for benchmarking machine learning algorithms. arXiv preprint arXiv:1708.07747 (2017)

81. Yang, L., Fan, Y., Xu, N.: Video instance segmentation, pp. 5187–5196 (2019)

82. Yu, F., et al.: BDD100K: a diverse driving dataset for heterogeneous multitask learning. In: 2020 IEEE/CVF Conference on Computer Vision and Pattern Recognition (CVPR), pp. 2633–2642 (2020)

83. Yu, F., Seff, A., Zhang, Y., Song, S., Funkhouser, T., Xiao, J.: LSUN: construction of a large-scale image dataset using deep learning with humans in the loop. arXiv preprint arXiv:1506.03365 (2015)

84. Zhu, Y., et al.: Improving semantic segmentation via video propagation and label relaxation. In: 2019 IEEE/CVF Conference on Computer Vision and Pattern Recognition (CVPR), pp. 8848–8857 (2019)

DOLPHINS: Dataset for Collaborative Perception Enabled Harmonious and Interconnected Self-driving

Ruiqing Mao[1], Jingyu Guo[1], Yukuan Jia[1], Yuxuan Sun[1,2], Sheng Zhou[1]([✉]), and Zhisheng Niu[1]

[1] Department of Electronic Engineering, Tsinghua University, Beijing 100084, China
{mrq20,guojy18,jyk20}@mails.tsinghua.edu.cn,
{sheng.zhou,niuzhs}@tsinghua.edu.cn
[2] School of Electronic and Information Engineering, Beijing Jiaotong University, Beijing 100044, China
yxsun@bjtu.edu.cn

Abstract. Vehicle-to-Everything (V2X) network has enabled collaborative perception in autonomous driving, which is a promising solution to the fundamental defect of stand-alone intelligence including blind zones and long-range perception. However, the lack of datasets has severely blocked the development of collaborative perception algorithms. In this work, we release DOLPHINS: Dataset for cOLlaborative Perception enabled Harmonious and INterconnected Self-driving, as a new simulated *large-scale various-scenario multi-view multi-modality* autonomous driving dataset, which provides a ground-breaking benchmark platform for interconnected autonomous driving. DOLPHINS outperforms current datasets in six dimensions: temporally-aligned images and point clouds from both vehicles and Road Side Units (RSUs) enabling both *Vehicle-to-Vehicle (V2V)* and *Vehicle-to-Infrastructure (V2I)* based collaborative perception; 6 typical scenarios with dynamic weather conditions make the most *various* interconnected autonomous driving dataset; meticulously selected viewpoints providing *full coverage* of the key areas and every object; 42376 frames and 292549 objects, as well as the corresponding 3D annotations, geo-positions, and calibrations, compose the *largest* dataset for collaborative perception; Full-HD images and 64-line LiDARs construct *high-resolution* data with sufficient details; well-organized APIs and open-source codes ensure the *extensibility* of DOLPHINS. We also construct a benchmark of 2D detection, 3D detection, and multi-view collaborative perception tasks on DOLPHINS. The experiment results show that the raw-level fusion scheme through V2X communication can help to improve the precision as well as to reduce the necessity of expensive LiDAR equipment on vehicles when RSUs exist, which may accelerate the popularity of

This work is sponsored in part by the Nature Science Foundation of China (No. 62022049, No. 61871254), and by the project of Tsinghua University-Toyota Joint Research Center for AI Technology of Automated Vehicle.

Supplementary Information The online version contains supplementary material available at https://doi.org/10.1007/978-3-031-26348-4_29.

interconnected self-driving vehicles. DOLPHINS dataset and related codes are now available on www.dolphins-dataset.net.

Keywords: Collaborative perception · Interconnected self-driving · Dataset

1   Introduction

One major bottleneck of achieving ultra-reliability in autonomous driving is the fundamental defect of stand-alone intelligence due to the single perception viewpoint. As illustrated in Fig. 1(a), the autonomous vehicle could not detect the pedestrians in its blind zone caused by the truck, which may lead to a severe accident. Great efforts have been put into single-vehicle multi-view object detection with multiple heterogeneous sensors [6,8] or homogeneous sensors [20, 35], but the intrinsic limitation of stand-alone intelligence still exists.

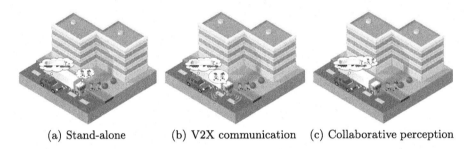

(a) Stand-alone   (b) V2X communication   (c) Collaborative perception

Fig. 1. An illustration of the advantages of collaborative perception over stand-alone intelligence.

Thanks to the Vehicle-to-Everything (V2X) network [42], *interconnected autonomous driving* is a highly-anticipated solution to occlusions, and thus enables advanced autonomous driving capabilities in complex scenarios such as intersections and overtaking. A vehicle can exchange the local sensor data with other terminals (as shown in Fig. 1(b)), including other vehicles and Road Side Units (RSUs), and then perform the object detection by fusing data from multiple viewpoints. The shared sensor data might contain information about the object in the blind zones of the ego vehicle, potentially enhancing the perception reliability [40] as in Fig. 1(c). This procedure is named as *collaborative perception*, which can be categorized into three levels: raw-level (early fusion, e.g. [5]), feature-level (middle fusion, e.g. [4,34]), and object-level (late fusion, e.g. [17]).

However, the lack of large-scale datasets for collaborative autonomous driving has been seriously restricting the research of collaborative perception algorithms. Traditional datasets focus on a single viewpoint, i.e., the ego vehicle. In the past decade, KITTI [13], nuScenes [1], and Waymo Open [33] have successfully accelerated the development of stand-alone self-driving algorithms with a huge amount of multi-modality data. But all of the information is collected from the ego vehicle view. Unfortunately, the most challenging but the greatest beneficial issue is the large parallax due to strong perspective changes between different terminals, i.e., aux vehicles and RSUs, as illustrated in Fig. 2. The large parallax leads to various occlusion relationships between objects, which may help the terminals to fulfill the blind zones, but also put forward the matching of the same object from different perspectives. Recently, some pioneer works have concentrated on datasets with multiple viewpoints, such as OPV2V [38], V2X-Sim [21], and DAIR-V2X [41]. Nevertheless, either data from aux vehicles (Vehicle-to-Vehicle, V2V) and RSUs (Vehicle-to-Infrastructure, V2I) are not provided simultaneously, or only an intersection scenario is considered. A more comprehensive dataset is required to fully support the development of V2X-based collaborative autonomous driving algorithms.

Fig. 2. An example of multi-view object detection in DOLPHINS dataset. There is a right merging lane in front of the ego vehicle. Because of the occlusion, the ego vehicle can hardly detect the purple vehicle (*red box*) on the branch and the police car (*blue box*). The auxiliary vehicle is in front of the ego vehicle, which can see both object vehicles distinctly. Additionally, the RSU can detect another two vehicles (*purple box*) on the branch. (Color figure online)

To meet the demands, we present **DOLPHINS**, a new **D**ataset for c**OL**labor-ative **P**erception enabled **H**armonious and **I**nterconnected **S**elf-driving. We use the CARLA simulator [9] to complete this work, which can provide us with realistic environment modeling and real-time simulations of the dynamics and sensors of various vehicles. Figure 3 briefly demonstrates the advantages of DOLPHINS in six dimensions.

V2X DOLPHINS contains temporally-aligned images and point clouds from both aux vehicles and RSUs simultaneously, which provides a universal out-of-the-box benchmark platform for the development and verification of

V2V and V2I enabled collaborative perception without extra generation of data.

Variety DOLPHINS includes 6 typical autonomous driving scenarios, which is second only to real-world single-vehicle datasets [1,13]. Our dataset includes urban intersections, T-junctions, steep ramps, highways on-ramps, and mountain roads, as well as dynamic weather conditions. Different scenarios raise different challenges to autonomous driving, such as dense traffic, ramp occlusions, and lane merging. More detailed information on traffic scenarios is presented in Sect. 3.1.

Viewpoints Considering the actual driving situation, 3 different viewpoints are meticulously set for each scenario, including both RSUs and vehicles.   The data collected from viewpoints can achieve full coverage of key areas in each scenario as illustrated in Fig. 4. More specific locations of each viewpoint are illustrated in Fig. 5.

Scale In total, temporally-aligned images and point clouds are recorded over 42376 frames from each viewpoint, which is much larger than any other dataset for collaborative perception. 3D information of 292549 objects is annotated in KITTI format for ease of use, along with the geo-positions and calibrations. Statistical analysis of objects is provided in Sect. 3.4.

Resolution DOLPHINS furnishes high-resolution images and point clouds to maintain sufficient details. Full − HD(1920 × 1080)cameras and 64 − line LiDARs equipped on both vehicles and RSUs, which are both among the highest quality in all datasets. Detailed descriptions of sensors are stated in Sect. 3.2.

Extensibility We also release the related codes of DOLPHINS, which contains the well-organized API to help researchers to generate additional data on demand, which makes DOLPHINS easily extensible and highly flexible.

We also conduct a comprehensive benchmark of state-of-the-art algorithms on DOLPHINS. Three typical tasks are considered: 2D object detection, 3D

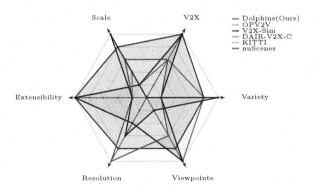

Fig. 3. A comparison with 3 brand new collaborative perception datasets: OPV2V [38], V2X-Sim [21], and DAIR-V2X-C [41], as well as 2 well-known single-vehicle autonomous driving datasets: KITTI [13] and nuScenes [1]. A detailed comparison is provided in Sect. 2.

(a) Ego vehicle   (b) RSU   (c) Aux vehicle

Fig. 4. An illustration of temporary-aligned images and point clouds from three viewpoints. The position of each viewpoint is demonstrated in Fig. 2.

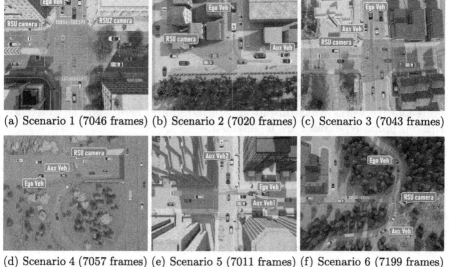

(a) Scenario 1 (7046 frames) (b) Scenario 2 (7020 frames) (c) Scenario 3 (7043 frames)

(d) Scenario 4 (7057 frames) (e) Scenario 5 (7011 frames) (f) Scenario 6 (7199 frames)

Fig. 5. All ego vehicles are driving along a pre-defined route (*green arrows*), while each RSU camera is settled with a fixed direction and range (*blue or brown sector mark*). We also mark positions where the ego vehicle or possible auxiliary vehicles are initialized. Among all scenarios, (a) and (e) are two intersection scenarios; (b) is the scenario of a T-junction with moderate rain; (c) is also a crossroads while the ego vehicle is on a steep ramp; (d) is a scenario existing a right merging lane on the expressway, and the weather is foggy; (f) is the scenario of a mountain road. All scenarios have plenty of occlusion situations. (Color figure online)

object detection, and multi-view collaborative perception. Other tasks, such as tracking, are also supported in DOLPHINS but not exhibited here. Besides, we construct two raw-level fusion schemes: the point clouds from the ego vehicle and the other viewpoint, and the image from the ego vehicle and point clouds from the RSU. The results of the raw-level fusion algorithms reveal the dual character of interconnected self-driving: enhancing the precision with more information or reducing the cost of sensors on the self-driving vehicles within the same precision.

As a new *large-scale various-scenario multi-view multi-modality* dataset, we hope this work brings a new platform to discover the potential benefits of connected intelligence. Our main contributions are summarized as:

i. release DOLPHINS dataset with different scenarios, multiple viewpoints, and multi-modal sensors, aiming to inspire the research of collaborative autonomous driving;
ii. provide open source codes for on-demand generation of data;
iii. benchmark several state-of-the-art methods in 2D object detection, 3D object detection, and multi-view collaborative perception, illustrating the possibility of solving blind zones caused by occlusions as well as cutting the cost of self-driving vehicles by V2V and V2I communication.

2   Related Works

There are many relative research areas, such as object detection, collaborative perception, and autonomous driving dataset. Due to the space limitation, some representative works which inspire us are introduced here, and the differences with our proposed dataset are highlighted.

Object detection is one of the most important tasks in autonomous driving. Typically, there are two kinds of object detectors, distinguished by whether to generate region proposals before the object detection and bounding box regression. R-CNN family [14–16,30] is the representative of two-stage detectors, which exhibits epoch-making performance. On the other hand, the single-stage detectors, such as SSD [23] and YOLO [27–29], focus on the inference time and perform significantly faster than the two-stage competitors. Recently, CenterNet [10] and CornerNet [19] propose a new detection method without anchor generation. They directly predict the key points per-pixel, which makes the detection pipeline much simpler. DETR [2] firstly brings transformer architecture into object detection tasks.

Collaborative perception is a growing topic in the intelligent transportation society. Due to the 3D information provided by point clouds, LiDAR-based data fusion and object detection have been widely discussed. [5] proposes a raw-level fusion on point clouds with a deep network for detection. [26,36] aim to use deep neural networks to enhance the perception outputs for sharing. V2VNet [34] considers the feature-level fusion and uses a compressor to reduce the size of original point clouds, which is vital in bandwidth-limited V2X communications. V2V-ViT [37] introduces vision transformer to conquer the noises from communication. [25] and [24] consider the pure image fusion on feature-level and conduct a re-identification task during the detection. The latter uses Graph Neural Network (GNN) for perception and clustering.

Autonomous driving datasets are the key to evaluating the performance of detection methods. The commonly used KITTI [13] and nuScenes [1] only contain data from the ego vehicle. Pasadena Multi-view ReID dataset is proposed in [25], which contains data from different viewpoints of a single object. However, the objects are only street trees, which is not enough for autonomous driving. OPV2V [38] uses CARLA simulator [9] to produce multi-view autonomous driving data, but it only considers V2V communication. V2X-Sim [21] is also a CARLA-based simulated dataset. The first version of V2X-Sim only contains point clouds from different vehicles, which can only be applied for V2V communication. The second version of V2X-Sim contains both RGB images and the infrastructure viewpoints. Nevertheless, it still only considers the intersections scenario, and the BEV Lidar on the infrastructure is not realistic. By late February 2022, a new real-world connected autonomous driving dataset DAIR-V2X [41] is released. It consists of images and point clouds from one vehicle and one RSU, and contains both high-ways and intersections. However, DAIR-V2X is not capable of V2V data fusion or any other scenarios with more than two terminals. Our proposed dataset is generated by the CARLA simulator with six different scenarios and reasonable settings of RSUs and aux vehicles. Besides, with the related codes (which will also be released with the dataset), researchers can add any type and any number of sensors at any location as needed. A comparison to the above datasets is provided in Table 1.

Table 1. A detailed comparison between datasets. For DAIR-V2X, we choose DAIR-V2X-C since only this part is captured synchronously by both vehicles and infrastructure sensors.

Dataset	Year	V2X	Scenarios	Viewpoints	Frames	Extensibility	Resolution
KITTI	2012	none	–	1	15 k	×	1382 × 512 64 lines
nuScenes	2019	none	–	1	1.4 M	×	1600 × 1200 32 lines
OPV2V	2021	V2V	6	2-7 (avg. 3)	11.5 k	✓	800 × 600 64 lines
V2X-Sim	2022	V2V+V2I	1	2-5	10 k	✓	1600 × 900 32 lines
DAIR-V2X-C	2022	V2I	1	2	39 k	×	1920 × 1080 I: 300 lines; V: 40 lines
DOLPHINS (Ours)	2022	V2V+V2I	6	3	42 k	✓	1920 × 1080 64 lines

3   DOLPHINS Dataset

3.1   Settings of Traffic Scenarios

We select six typical autonomous driving scenarios and several common types of weather from the preset scenarios of the CARLA simulator (as shown in Fig. 5). In each scenario, we set three units (RSU or vehicles) to collect both images and point cloud information. The first unit is attached to the vehicle we drive, namely, the ego vehicle, which provides us with the main viewpoint. In each simulation round, we initialized it at a specific location. The other two units will also be set up at appropriate positions. They are set on the RSUs or the auxiliary vehicles selected from the scenario and initialized at a specially designated point with a stochastic vehicle model. We initially set 20–30 vehicles as well as 10–15 pedestrians within a specific area around the ego vehicle, which is 100–150 meters in length and 100 m in width. The initial locations are randomly selected from the preset locations provided by the CARLA simulator, which guarantees that no collisions will happen.

In each scenario, our ego vehicle chooses a specific route. At the same time, we collect the information of all sensors synchronously every 0.5 s in the simulation environment, i.e., at the rate of 2 fps. After the vehicle passes through the specific scenario, we wind up the current simulation round, reinitialize the scenario and start a new one. During each round, except for our ego vehicle and the possible auxiliary vehicle, all other traffic participants appear in a reasonable position randomly at the beginning and choose their route by themselves freely.

3.2   Settings of Sensors

We equip each unit with a LiDAR and an RGB camera, whose parameters are listed in Table 2. For the convenience of calibration between different sensors, we install both camera and LiDAR on the same point. The position of sensors on the vehicle is illustrated in the supplementary material.

Table 2. Parameters of sensors on different units

Sensor type	Parameter attributes	RSU	Vehicle
RGB Camera	Horizontal field of view in degrees	90	90
	Resolution	1920×1080	1920×1080
	Height in meters	4	$0.3 + h_{veh}$ [a]
LiDAR	Number of lasers	64	64
	Maximum distance to measure in meters	200	200
	Points generated by all lasers per second	2.56×10^6	2.56×10^6
	LiDAR rotation frequency	20	20
	Angle in degrees of the highest laser	0	2
	Angle in degrees of the lowest laser	−40	−24.8
	General proportion of points that are randomly dropped	0.1	0.1

[a] h_{veh} denotes the height of the ego vehicle

3.3   Extra Data and Calibrations

For each scenario, We divide our data into the training set and the test set at the ratio of 8:2. Each set contains the original pictures taken by the camera, the point cloud information generated by LiDAR, and the ground truth labels, and the calibration files. The labels include the following information: (i) 2D bounding box of the object in the image, (ii) 3D object dimensions and location, (iii) the value of *alpha* and *rotation_y* which are defined in the KITTI Vision Benchmark [13]. Except for the above data, we further introduce two extra pieces of information in DOLPHINS: the locations of key vehicles and the context-aware labels. These two kinds of data are essential for collaborative autonomous driving. The geo-positions of vehicles can greatly help to align the perceptual information from different perspectives through coordinate transformations. Actually, to the best of our knowledge, all the published multi-view collaborative perception algorithms are based on the locations of each vehicle, no matter image-based [24,25] or LiDAR-based [4,5,34]. Besides, the interconnected autonomous vehicles can have wider perception fields with the help of other transportation participants and the RSUs, which means they can detect invisible objects. Most of the datasets only provide the labels of those who are in the view angle of sensors, which is not enough for the vehicles to make safe and timely decisions. We provide the labels of all traffic participants within 100 m in front of or behind the ego unit, as well as 40 m in the left and right side directions.

3.4   Data Analysis

To further analyze the data components of the dataset, we calculate the number of cars and pedestrians in each scenario both in the training dataset and the test dataset (as illustrated in Table 3). What's more, we categorize each object into three detection difficulty levels based on the number of laser points reflected by it in the point clouds. Easy objects reflect more than 16 points, as well as hard objects have no visible point, and the remaining objects are defined as moderate ones. In other words, the difficulty level actually indicates the occlusion level of each object. Since it is unlikely for us to manually annotate the occlusion level, such kind of definition is a suitable and convenient approximation. From the statistical analysis, it turns out that there is no pedestrian in scenarios 4 and 6, i.e., on high-way and mountain roads, which is self-consistent. Scenario 1 contains the most cars and pedestrians, as it is a crowded intersection. Scenario 2 is a T-junction, which has fewer directions for vehicles to travel. Scenario 3 is a steep ramp, which will be the hardest scenario along with Scenario 6, because of the severe occlusions caused by height difference.

Table 3. Statistical analysis of objects in DOLPHINS training and test dataset

Scenario	Training dataset						Test dataset					
	Car			Pedestrians			Car			Pedestrians		
	Easy	Moderate	Hard	Easy	Moderate	Hard	Easy	Moderate	Hard	Easy	Moderate	Hard
1	27548	4423	1090	12370	2117	349	7048	1096	296	3079	579	96
2	15428	1290	567	5641	2281	314	3895	330	155	1481	579	79
3	14365	4029	4291	3003	3462	584	3631	1068	1049	789	889	150
4	34012	11771	4089	0	0	0	8497	2937	1053	0	0	0
5	31648	6201	1993	4797	9734	1476	7918	1578	440	1161	2446	394
6	14035	2203	8531	0	0	0	3531	513	2150	0	0	0
Total	137036	29917	20561	25811	17594	2723	34520	7522	5143	6510	4493	719

4   Benchmarks

In this section, we provide benchmarks of three typical tasks on our proposed DOLPHINS dataset: 2D object detection, 3D object detection, and multi-view collaborative perception. For each task, we implement several classical algorithms.

4.1   Metrics

We first aggregate the training datasets of six scenarios altogether. The composed dataset will contain various background characteristics and occlusion relationships, which help the model to have better generalization ability. The training dataset is split for training and validation at the ratio of 5:3, then the performance of each detector is examined on the test dataset. Similar to KITTI [13], we use Average Precision (AP) at Intersection-over-Union (IoU) threshold of 0.7 and 0.5 to illustrate the goodness of detectors on cars, as well as IoU of 0.5 and 0.25 for the pedestrians since the pedestrians are much smaller than cars. The degree of difficulty is cumulative in the test, that is, the ground truths of easy objects are also considered in moderate and hard tests.

4.2   Experiment Details

We use MMDetection [3] and MMDetection3D [7] to construct the training and test pipeline. As for 2D object detection tasks, we finetune the COCO [22] pretrained models on our dataset. We also provide the GPU memory consumption and the inference speed to illustrate the differences between different methods, where the experiment is set with a batch size equal to 1. All the experiments are performed on 8 RTX 3090 GPUs.

4.3   2D Object Detection

As mentioned in Sect. 2, there are four typical detection paradigms: anchor-based two-stage detectors, anchor-based one-stage detectors, anchor-free one-stage detectors, and the vision transformer. In this part, we select Faster R-CNN [30] as the representative of anchor-based two-stage detectors, YOLOv3 [29] for the anchor-based one-stage detectors, YOLOX [12] and TOOD [11] for the anchor-free one-stage detectors, and DETR [2] for the vision transformer. Specifically, we set the backbone network of Faster-RCNN and YOLOv3 to be Resnet-50, so that the size of these networks is close to each other. The experiment results are illustrated in Table 4 (left) and Table 5. It shows that all the detectors can have good knowledge of different scenarios. However, the modern anchor-free detectors can significantly speed up the entire inference procedure without loss of precision. One abnormal result is the surprising rise of AP in hard and moderate tasks compared with easy tasks, especially in the pedestrian detection. A reasonable explanation is the large proportion of moderate and hard objects due to the characteristics of different scenarios. For example, in Scenario 6, which is a mountain road, nearly half of the objects are severely occluded due to the undulating planes. Thus, the detectors tend to propose much more candidate objects to match those hard objects, which leads to low AP in easy tasks because of the false positives. It is proved by the high recall scores in easy tasks. The same thing happens in pedestrian detection, where the pedestrians are smaller and thus more likely to be hard ones. However, low AP is not equal to poor performance. On the contrary, meeting the ultra-reliability demands of self-driving, a higher recall rate is much more meaningful than the AP, which can alert the vehicles to the potential dangers in blind zones. More detailed analysis can be found in the supplementary material.

Table 4. 2D and 3D object detection analysis on speed and cost

2D Method	Inference speed (fps)	Memory usage (MB)
Faster-RCNN	35.6	2513
YOLOv3	50.7	2285
YOLOX-S	58.1	2001
YOLOX-L	36.5	2233
TOOD	26.8	2247
DETR	26.3	2419

3D Method	Inference speed (fps)	Memory usage (MB)
SECOND	45.7	2433
PointPillars	36.8	3483
PV-RCNN	13.1	2899
MVX-Net	11.0	11321

4.4   3D Object Detection

As for the 3D object detection tasks, different modals of sensors lead to different detector architectures. We choose SECOND [39], PointPillars [18], and PV-RCNN [31] as SOTA LiDAR-based methods in this part. What's more, the multi-modal detectors can combine the segmentation information from images and the depth information from LiDARs, which is an advantage to the detection of small objects which reflect few points, e.g. pedestrians. We also test MVX-Net [32] on our multi-modality dataset. The experiment results are illustrated in Table 4 (right) and Table 6. The results show that Scenarios 3 and 6 are the corner cases where the AP is significantly lower than in other scenarios. Due to the steep ramp, the LiDAR on the ego vehicle is hard to detect the opposite vehicles and pedestrians, which is the fundamental defect of stand-alone intelligence. What's more, PV-RCNN [31] gains significantly better performance at the cost of taking nearly four times as long as SECOND [39]. MVX-Net [32] is inferior to those pure LiDAR-based methods, but it achieves surprising performance in pedestrians, which means the rich segmentation of information from images is profitable for the detection of small objects.

4.5   Multi-view Collaborative Perception

Based on the information to exchange, collaborative perception can be categorized into three levels: raw-level (early fusion), feature-level (middle fusion), and

Table 5. 2D object detection results on DOLPHINS

Scenario	Method	Car AP@IoU=0.7			Car AP@IoU=0.5			Pedestrian AP@IoU=0.5			Pedestrian AP@IoU=0.25		
		Easy	Moderate	Hard	Easy	Moderate	Hard	Easy	Moderate	Hard	Easy	Moderate	Hard
1	Faster R-CNN	89.18	80.73	80.62	90.07	89.72	88.76	87.12	90.77	90.82	87.12	90.79	90.88
	YOLOv3	84.76	79.36	79.19	86.93	89.90	89.79	85.22	90.41	90.54	85.40	90.66	90.79
	YOLOX-S	76.21	69.18	68.31	87.57	84.79	82.44	87.19	90.21	89.54	87.46	90.63	90.40
	YOLOX-L	84.72	79.58	78.79	88.95	88.90	87.42	87.76	90.66	90.70	87.82	90.73	90.81
	TOOD	88.30	80.04	79.97	89.76	90.07	89.71	89.02	90.83	90.87	89.03	90.83	90.88
	DETR	82.29	77.76	75.61	89.40	88.43	86.79	87.16	89.44	89.19	87.77	90.17	90.14
2	Faster R-CNN	90.50	90.31	90.19	90.82	90.81	90.80	87.36	90.49	90.18	87.36	90.66	90.60
	YOLOv3	89.33	89.46	89.48	90.17	90.27	90.47	75.96	90.05	90.11	75.96	90.47	90.59
	YOLOX-S	86.46	79.98	79.53	90.20	89.85	89.54	86.99	86.70	80.76	87.24	89.66	87.92
	YOLOX-L	89.48	86.87	80.85	90.61	90.47	90.25	86.29	89.51	87.65	86.29	89.98	89.72
	TOOD	90.39	90.20	90.10	90.85	90.84	90.82	88.12	90.75	90.53	93.67	90.80	90.76
	DETR	89.90	88.52	87.30	90.60	90.53	90.47	86.80	89.63	87.12	89.83	90.51	90.12
3	Faster R-CNN	85.23	88.55	80.25	86.54	90.18	81.51	75.89	89.77	90.67	75.89	89.80	90.76
	YOLOv3	84.64	78.65	71.03	87.46	88.97	80.94	44.31	88.77	89.23	44.63	89.80	90.74
	YOLOX-S	85.95	76.67	66.72	89.42	87.02	77.75	82.54	87.99	81.18	86.54	89.91	85.83
	YOLOX-L	88.55	80.41	71.25	89.64	89.70	80.39	86.55	90.39	89.65	89.02	90.43	90.37
	TOOD	88.25	80.47	71.50	89.62	90.30	81.26	85.60	90.15	90.69	85.60	90.18	90.83
	DETR	87.94	85.00	79.18	89.17	88.71	86.74	85.73	87.49	85.87	85.73	88.69	87.56
4	Faster R-CNN	89.33	81.21	81.09	89.40	89.68	88.70	N/A	N/A	N/A	N/A	N/A	N/A
	YOLOv3	82.43	78.22	77.66	83.95	89.21	89.50	N/A	N/A	N/A	N/A	N/A	N/A
	YOLOX-S	88.97	76.37	70.29	89.65	86.24	84.43	N/A	N/A	N/A	N/A	N/A	N/A
	YOLOX-L	89.96	80.81	79.58	90.07	89.20	88.25	N/A	N/A	N/A	N/A	N/A	N/A
	TOOD	90.08	81.19	80.22	90.22	90.04	81.63	N/A	N/A	N/A	N/A	N/A	N/A
	DETR	88.17	76.00	74.13	89.44	86.50	86.00	N/A	N/A	N/A	N/A	N/A	N/A
5	Faster R-CNN	89.70	80.86	81.04	90.25	89.34	88.82	68.07	80.10	81.06	71.49	80.17	81.41
	YOLOv3	83.57	78.58	78.80	85.45	89.64	89.77	38.03	87.20	86.48	40.79	89.25	90.52
	YOLOX-S	79.04	75.71	70.72	88.53	86.92	85.07	65.48	79.95	77.82	65.54	85.47	79.99
	YOLOX-L	87.88	80.14	80.18	89.20	89.53	88.75	64.05	87.73	81.18	64.05	88.61	87.24
	TOOD	89.12	85.12	80.73	89.87	89.92	89.96	77.25	88.18	81.41	78.71	89.16	88.43
	DETR	84.26	78.03	75.95	89.58	88.35	87.00	74.59	81.21	77.55	77.04	86.34	83.82
6	Faster R-CNN	77.97	79.60	90.10	78.09	79.74	90.55	N/A	N/A	N/A	N/A	N/A	N/A
	YOLOv3	79.67	79.68	88.85	80.00	80.12	90.35	N/A	N/A	N/A	N/A	N/A	N/A
	YOLOX-S	81.63	79.53	79.73	82.34	81.28	88.61	N/A	N/A	N/A	N/A	N/A	N/A
	YOLOX-L	76.59	76.87	87.35	76.85	77.32	89.07	N/A	N/A	N/A	N/A	N/A	N/A
	TOOD	82.54	81.62	90.63	82.65	81.77	89.93	N/A	N/A	N/A	N/A	N/A	N/A
	DETR	85.09	83.27	81.81	85.47	84.03	89.89	N/A	N/A	N/A	N/A	N/A	N/A

object-level (late fusion). Due to the 3D information provided by point clouds, LiDAR-based data fusion and object detection have been widely discussed. We realize a raw-level fusion algorithm based on DOLPHINS LiDAR data through the superposition of point clouds from different perspectives. However, not all the LiDAR-based 3D detection algorithms can be adapted to raw-level fusion schemes. Since many detectors use voxels to represent the point clouds of a district, the height of voxels is limited to reduce the computation complexity. The limitation will not be violated when the cooperators are on the same horizontal plane, as in [5] and [38]. However, when the data are from RSUs or from vehicles on a mountain road (as in Scenario 3 and 6 in Fig. 5), the height of the aggregated point clouds will be too large to tackle through traditional voxel processing. In our experiment settings, PointPillars [18] is the only algorithm to be compatible with the raw-level fusion scheme.

Table 6. 3D object detection results on DOLPHINS

Scenario	Method	Car AP@IoU=0.7			Car AP@IoU=0.5			Pedestrian AP@IoU=0.5			Pedestrian AP@IoU=0.25		
		Easy	Moderate	Hard	Easy	Moderate	Hard	Easy	Moderate	Hard	Easy	Moderate	Hard
1	SECOND	95.65	90.37	87.36	98.79	96.05	92.97	74.17	70.08	68.19	96.78	95.86	93.47
	PointPillar	96.63	92.09	88.83	98.55	96.19	93.13	70.12	65.41	63.56	95.57	94.38	91.98
	PV-RCNN	98.14	93.87	90.69	98.90	96.18	92.98	83.67	80.36	78.37	97.31	96.50	94.09
	MVX-Net	89.25	84.3	83.99	89.62	89.51	87.02	91.36	88.93	86.49	99.61	99.55	97.06
2	SECOND	96.33	91.32	88.77	98.48	97.50	95.56	66.71	59.35	57.27	95.07	91.00	87.94
	PointPillar	97.39	93.06	91.70	98.23	97.33	95.48	59.65	53.33	51.35	92.52	87.58	84.49
	PV-RCNN	98.57	94.76	91.15	99.12	97.44	94.33	78.74	71.59	68.94	97.37	93.91	90.60
	MVX-Net	93.77	91.04	88.63	96.33	95.91	93.72	89.12	82.22	79.78	99.37	98.72	94.19
3	SECOND	80.30	67.16	54.96	85.51	75.08	64.29	49.87	29.31	27.05	92.73	60.26	55.88
	PointPillar	78.94	68.29	56.78	85.67	75.75	66.75	37.49	22.04	20.00	82.64	52.55	48.25
	PV-RCNN	85.93	73.85	60.59	87.54	77.18	64.30	63.95	37.46	34.53	90.93	58.96	54.32
	MVX-Net	68.96	58.47	48.73	71.80	61.67	56.25	71.96	41.36	37.10	93.99	56.62	53.85
4	SECOND	97.81	92.11	84.47	99.33	97.35	90.50	N/A	N/A	N/A	N/A	N/A	N/A
	PointPillar	98.07	94.00	86.52	98.79	97.57	91.01	N/A	N/A	N/A	N/A	N/A	N/A
	PV-RCNN	99.37	95.54	87.70	99.50	97.78	89.98	N/A	N/A	N/A	N/A	N/A	N/A
	MVX-Net	91.76	86.39	83.73	91.96	89.19	86.55	N/A	N/A	N/A	N/A	N/A	N/A
5	SECOND	96.49	91.41	87.44	98.68	96.36	92.71	75.33	65.03	58.90	97.37	94.29	87.52
	PointPillar	97.45	92.92	89.30	98.81	96.52	92.94	71.36	62.07	56.27	97.32	93.59	85.70
	PV-RCNN	98.57	94.39	90.49	99.23	97.21	93.15	90.47	79.07	71.64	98.94	95.45	87.29
	MVX-Net	91.69	86.67	84.17	94.45	91.94	89.39	84.87	75.83	68.91	99.52	99.36	91.81
6	SECOND	90.53	82.60	56.05	97.54	82.11	68.15	N/A	N/A	N/A	N/A	N/A	N/A
	PointPillar	89.31	82.30	57.32	97.44	92.66	70.62	N/A	N/A	N/A	N/A	N/A	N/A
	PV-RCNN	95.95	89.29	62.05	98.23	93.95	69.28	N/A	N/A	N/A	N/A	N/A	N/A
	MVX-Net	87.53	75.29	52.60	90.76	80.55	57.97	N/A	N/A	N/A	N/A	N/A	N/A

What's more, we also extend the MVX-Net to the collaborative autonomous driving scenarios. With the help of the point clouds from the LiDARs on the RSUs, whose locations are usually much higher, the ego vehicle can have a wider view with fewer occlusions. In addition, a single LiDAR on the RSU could free all the nearby autonomous vehicles from the necessity of equipping expensive LiDARs by sharing its point clouds through the V2I network, which brings great benefits to the realization of Level-5 autonomous driving. In this work, we use the point clouds from the RSU (or the aux vehicle 1 in Scenario 5) instead of the ego vehicle by transforming the coordinates.

Table 7 illustrates the multi-view collaborative perception on PointPillars [18] and MVX-Net [32], and the corresponding AP difference compared with stand-alone detection. It turns out that as for the superposition of raw point clouds, the ego vehicle can gain plentiful benefits from the richer information directly from another perspective. Under those circumstances with severe occlusions such as Scenario 3 and 6 and for those hard objects, the cooperative perception-based PointPillars [18] achieves up to 38.42% increment in AP. However, the extra noise also infects the detection of small objects, which is also discussed in the supplementary material. On the other hand, as for the MVX-Net with the local camera and RSU LiDAR, the performance is nearly the same as the one with stand-alone sensors. It shows the opportunity to enable high-level autonomous driving on cheap, LiDAR-free vehicles through the sensors on infrastructures.

Table 7. Multi-view collaborative perception results on DOLPHINS

Scenario	Method	Car AP@IoU=0.7			Car AP@IoU=0.5			Pedestrian AP@IoU=0.5			Pedestrian AP@IoU=0.25		
		Easy	Moderate	Hard	Easy	Moderate	Hard	Easy	Moderate	Hard	Easy	Moderate	Hard
1	PointPillar CP	97.13	95.19	94.57	97.63	96.40	95.89	72.73	70.60	70.23	94.01	93.69	93.31
	Difference	0.52%	3.37%	6.46%	−0.93%	0.22%	2.96%	3.72%	7.93%	10.49%	−1.63%	−0.73%	1.45%
	MVX-Net CP	89.34	84.30	84.02	89.64	89.49	87.00	90.77	86.08	85.92	99.61	99.53	97.04
	Difference	0.10%	0.00%	0.04%	0.02%	−0.02%	−0.02%	−0.65%	−3.20%	−0.66%	0.00%	−0.02%	−0.02%
2	PointPillar CP	97.97	97.03	96.42	98.52	97.89	97.55	57.51	51.89	50.92	91.94	88.52	86.54
	Difference	0.60%	4.27%	5.15%	0.30%	0.58%	2.17%	−3.59%	−2.70%	−0.84%	−0.63%	1.07%	2.43%
	MVX-Net CP	93.75	90.99	88.47	96.41	96.19	93.81	88.52	81.72	79.25	99.29	98.76	94.13
	Difference	−0.02%	−0.05%	−0.18%	0.08%	0.29%	0.10%	−0.67%	−0.61%	−0.66%	−0.08%	0.04%	−0.06%
3	PointPillar CP	81.48	72.73	66.30	86.88	77.69	72.33	32.62	19.89	18.47	73.24	48.12	44.80
	Difference	3.22%	6.50%	16.77%	1.41%	2.56%	8.36%	−12.99%	−9.75%	−7.65%	−11.37%	−8.43%	−7.15%
	MVX-Net CP	69.04	58.48	48.77	71.61	61.60	54.18	70.82	39.76	37.43	91.87	56.60	51.76
	Difference	0.12%	0.02%	0.08%	−0.26%	−0.11%	−3.68%	−1.58%	−3.87%	0.89%	−2.26%	−0.04%	−3.88%
4	PointPillar CP	97.60	96.22	94.40	97.93	97.00	95.74	N/A	N/A	N/A	N/A	N/A	N/A
	Difference	−0.48%	2.36%	9.11%	−0.87%	−0.58%	5.20%	N/A	N/A	N/A	N/A	N/A	N/A
	MVX-Net CP	91.70	86.38	83.69	91.92	89.22	86.58	N/A	N/A	N/A	N/A	N/A	N/A
	Difference	−0.07%	−0.01%	−0.05%	−0.04%	0.03%	0.03%	N/A	N/A	N/A	N/A	N/A	N/A
5	PointPillar CP	96.38	94.24	92.77	96.87	95.82	94.38	65.50	61.53	58.69	93.00	91.62	87.82
	Difference	−1.10%	1.42%	3.89%	−1.96%	−0.73%	1.55%	−8.21%	−0.87%	4.30%	−4.44%	−2.10%	2.47%
	MVX-Net CP	91.63	86.57	84.06	94.49	91.94	89.40	82.56	72.02	67.19	99.78	96.93	89.44
	Difference	−0.07%	−0.12%	−0.13%	0.04%	0.00%	0.01%	−2.72%	−5.02%	−2.50%	0.26%	−2.45%	−2.58%
6	PointPillar CP	94.46	91.43	79.34	97.77	96.63	87.17	N/A	N/A	N/A	N/A	N/A	N/A
	Difference	5.77%	11.09%	38.42%	0.34%	4.28%	23.44%	N/A	N/A	N/A	N/A	N/A	N/A
	MVX-Net CP	87.10	74.85	52.34	90.45	80.43	59.52	N/A	N/A	N/A	N/A	N/A	N/A
	Difference	−0.49%	−0.58%	−0.49%	−0.34%	−0.15%	2.67%	N/A	N/A	N/A	N/A	N/A	N/A

5   Conclusions

In this paper, we present a new large-scale various-scenario multi-view multi-modality autonomous driving dataset, DOLPHINS, to facilitate the research on collaborative perception-enabled connected autonomous driving. All the data are temporally-aligned and generated from three viewpoints, including both vehicles and RSUs, in six typical driving scenarios, along with the annotations, calibrations, and the geo-positions. What's more, we benchmark several SOTA algorithms on traditional 2D/3D object detection and brand-new collaborative perception tasks. The experiment results suggest that not only the extra data from V2X communication can eliminate the occlusions, but also the RSUs at appropriate locations can provide equivalent point clouds to the nearby vehicles, which can greatly reduce the prime cost of self-driving cars. In the future, we are going to further extend the number of infrastructures and aux vehicles, and construct more realistic maps of the downtown.

References

1. Caesar, H., et al.: nuScenes: a multimodal dataset for autonomous driving. In: Proceedings of the IEEE/CVF Conference on Computer Vision and Pattern Recognition, pp. 11621–11631 (2020)
2. Carion, N., Massa, F., Synnaeve, G., Usunier, N., Kirillov, A., Zagoruyko, S.: End-to-end object detection with transformers. In: Vedaldi, A., Bischof, H., Brox, T., Frahm, J.-M. (eds.) ECCV 2020. LNCS, vol. 12346, pp. 213–229. Springer, Cham (2020). https://doi.org/10.1007/978-3-030-58452-8_13

3. Chen, K., et al.: MMDetection: open MMLab detection toolbox and benchmark. arXiv preprint arXiv:1906.07155 (2019)

4. Chen, Q., Ma, X., Tang, S., Guo, J., Yang, Q., Fu, S.: F-cooper: feature based cooperative perception for autonomous vehicle edge computing system using 3D point clouds. In: Proceedings of the 4th ACM/IEEE Symposium on Edge Computing, SEC 2019, pp. 88–100. Association for Computing Machinery, New York (2019). https://doi.org/10.1145/3318216.3363300

5. Chen, Q., Tang, S., Yang, Q., Fu, S.: Cooper: cooperative perception for connected autonomous vehicles based on 3D point clouds. In: 2019 IEEE 39th International Conference on Distributed Computing Systems (ICDCS), pp. 514–524. IEEE (2019)

6. Chen, X., Ma, H., Wan, J., Li, B., Xia, T.: Multi-view 3D object detection network for autonomous driving. In: Proceedings of the IEEE Conference on Computer Vision and Pattern Recognition, pp. 1907–1915 (2017)

7. Contributors, M.: MMDetection3D: OpenMMLab next-generation platform for general 3D object detection (2020). https://github.com/open-mmlab/mmdetection3d

8. Deng, J., Czarnecki, K.: MLOD: a multi-view 3D object detection based on robust feature fusion method. In: 2019 IEEE Intelligent Transportation Systems Conference (ITSC), pp. 279–284. IEEE (2019)

9. Dosovitskiy, A., Ros, G., Codevilla, F., Lopez, A., Koltun, V.: CARLA: an open urban driving simulator. In: Proceedings of the 1st Annual Conference on Robot Learning, pp. 1–16 (2017)

10. Duan, K., Bai, S., Xie, L., Qi, H., Huang, Q., Tian, Q.: Centernet: keypoint triplets for object detection. In: Proceedings of the IEEE/CVF International Conference on Computer Vision, pp. 6569–6578 (2019)

11. Feng, C., Zhong, Y., Gao, Y., Scott, M.R., Huang, W.: TOOD: task-aligned one-stage object detection. In: ICCV (2021)

12. Ge, Z., Liu, S., Wang, F., Li, Z., Sun, J.: YOLOX: exceeding yolo series in 2021. arXiv preprint arXiv:2107.08430 (2021)

13. Geiger, A., Lenz, P., Urtasun, R.: Are we ready for autonomous driving? The KITTI vision benchmark suite. In: Conference on Computer Vision and Pattern Recognition (CVPR) (2012)

14. Girshick, R.: Fast R-CNN. In: Proceedings of the IEEE International Conference on Computer Vision (ICCV) (2015)

15. Girshick, R., Donahue, J., Darrell, T., Malik, J.: Rich feature hierarchies for accurate object detection and semantic segmentation. In: Proceedings of the IEEE Conference on Computer Vision and Pattern Recognition, pp. 580–587 (2014)

16. He, K., Gkioxari, G., Dollár, P., Girshick, R.: Mask R-CNN. In: Proceedings of the IEEE International Conference on Computer Vision, pp. 2961–2969 (2017)

17. Kim, S.W., Liu, W., Ang, M.H., Frazzoli, E., Rus, D.: The impact of cooperative perception on decision making and planning of autonomous vehicles. IEEE Intell. Transp. Syst. Mag. 7(3), 39–50 (2015)

18. Lang, A.H., Vora, S., Caesar, H., Zhou, L., Yang, J., Beijbom, O.: Pointpillars: fast encoders for object detection from point clouds. In: Proceedings of the IEEE Conference on Computer Vision and Pattern Recognition, pp. 12697–12705 (2019)

19. Law, H., Deng, J.: Cornernet: detecting objects as paired keypoints. In: Proceedings of the European Conference on Computer Vision (ECCV), pp. 734–750 (2018)

20. Li, P., Chen, X., Shen, S.: Stereo R-CNN based 3D object detection for autonomous driving. In: Proceedings of the IEEE/CVF Conference on Computer Vision and Pattern Recognition, pp. 7644–7652 (2019)

21. Li, Y., An, Z., Wang, Z., Zhong, Y., Chen, S., Feng, C.: V2X-sim: a virtual collaborative perception dataset for autonomous driving (2022). https://doi.org/10.48550/ARXIV.2202.08449. https://arxiv.org/abs/2202.08449

22. Lin, T.Y., et al.: Microsoft coco: common objects in context (2014). https://doi.org/10.48550/ARXIV.1405.0312. https://arxiv.org/abs/1405.0312

23. Liu, W., et al.: SSD: single shot MultiBox detector. In: Leibe, B., Matas, J., Sebe, N., Welling, M. (eds.) ECCV 2016. LNCS, vol. 9905, pp. 21–37. Springer, Cham (2016). https://doi.org/10.1007/978-3-319-46448-0_2

24. Nassar, A.S., D'Aronco, S., Lefèvre, S., Wegner, J.D.: GeoGraph: graph-based multi-view object detection with geometric cues end-to-end. In: Vedaldi, A., Bischof, H., Brox, T., Frahm, J.-M. (eds.) ECCV 2020. LNCS, vol. 12352, pp. 488–504. Springer, Cham (2020). https://doi.org/10.1007/978-3-030-58571-6_29

25. Nassar, A.S., Lefèvre, S., Wegner, J.D.: Simultaneous multi-view instance detection with learned geometric soft-constraints. In: Proceedings of the IEEE/CVF International Conference on Computer Vision, pp. 6559–6568 (2019)

26. Rawashdeh, Z.Y., Wang, Z.: Collaborative automated driving: a machine learning-based method to enhance the accuracy of shared information. In: 2018 21st International Conference on Intelligent Transportation Systems (ITSC), pp. 3961–3966. IEEE (2018)

27. Redmon, J., Divvala, S., Girshick, R., Farhadi, A.: You only look once: unified, real-time object detection. In: Proceedings of the IEEE Conference on Computer Vision and Pattern Recognition, pp. 779–788 (2016)

28. Redmon, J., Farhadi, A.: Yolo9000: better, faster, stronger. In: Proceedings of the IEEE Conference on Computer Vision and Pattern Recognition, pp. 7263–7271 (2017)

29. Redmon, J., Farhadi, A.: Yolov3: an incremental improvement. arXiv preprint arXiv:1804.02767 (2018)

30. Ren, S., He, K., Girshick, R., Sun, J.: Faster R-CNN: towards real-time object detection with region proposal networks. In: Advances in Neural Information Processing Systems, vol. 28 (2015)

31. Shi, S., et al.: PV-RCNN: point-voxel feature set abstraction for 3D object detection. In: Proceedings of the IEEE/CVF Conference on Computer Vision and Pattern Recognition, pp. 10529–10538 (2020)

32. Sindagi, V.A., Zhou, Y., Tuzel, O.: MVX-Net: multimodal VoxelNet for 3D object detection. In: 2019 International Conference on Robotics and Automation (ICRA), pp. 7276–7282. IEEE (2019)

33. Sun, P., et al.: Scalability in perception for autonomous driving: Waymo open dataset. In: Proceedings of the IEEE/CVF Conference on Computer Vision and Pattern Recognition, pp. 2446–2454 (2020)

34. Wang, T.-H., Manivasagam, S., Liang, M., Yang, B., Zeng, W., Urtasun, R.: V2VNet: vehicle-to-vehicle communication for joint perception and prediction. In: Vedaldi, A., Bischof, H., Brox, T., Frahm, J.-M. (eds.) ECCV 2020. LNCS, vol. 12347, pp. 605–621. Springer, Cham (2020). https://doi.org/10.1007/978-3-030-58536-5_36

35. Wang, Y., Guizilini, V.C., Zhang, T., Wang, Y., Zhao, H., Solomon, J.: DETR3D: 3D object detection from multi-view images via 3D-to-2D queries. In: Conference on Robot Learning, pp. 180–191. PMLR (2022)

36. Xiao, Z., Mo, Z., Jiang, K., Yang, D.: Multimedia fusion at semantic level in vehicle cooperactive perception. In: 2018 IEEE International Conference on Multimedia & Expo Workshops (ICMEW), pp. 1–6. IEEE (2018)

37. Xu, R., Xiang, H., Tu, Z., Xia, X., Yang, M.H., Ma, J.: V2X-VIT: vehicle-to-everything cooperative perception with vision transformer (2022). https://doi.org/10.48550/ARXIV.2203.10638. https://arxiv.org/abs/2203.10638
38. Xu, R., Xiang, H., Xia, X., Han, X., Li, J., Ma, J.: OPV2V: an open benchmark dataset and fusion pipeline for perception with vehicle-to-vehicle communication (2021). https://doi.org/10.48550/ARXIV.2109.07644. https://arxiv.org/abs/2109.07644
39. Yan, Y., Mao, Y., Li, B.: Second: sparsely embedded convolutional detection. Sensors **18**, 3337 (2018)
40. Yang, Q., Fu, S., Wang, H., Fang, H.: Machine-learning-enabled cooperative perception for connected autonomous vehicles: challenges and opportunities. IEEE Network **35**(3), 96–101 (2021)
41. Yu, H., et al.: DAIR-V2X: a large-scale dataset for vehicle-infrastructure cooperative 3D object detection. In: Proceedings of the IEEE/CVF Conference on Computer Vision and Pattern Recognition, pp. 21361–21370 (2022)
42. Zeadally, S., Javed, M.A., Hamida, E.B.: Vehicular communications for its: standardization and challenges. IEEE Communi. Stand. Mag. **4**(1), 11–17 (2020)

Energy-Efficient Image Processing Using Binary Neural Networks with Hadamard Transform

Jaeyoon Park[ID] and Sunggu Lee[(✉)][ID]

Pohang University of Science and Technology (POSTECH), Pohang, South Korea
{jaeyoonpark,slee}@postech.ac.kr

Abstract. Binary neural networks have recently begun to be used as a highly energy- and computation-efficient image processing technique for computer vision tasks. This paper proposes a novel extension of existing binary neural network technology based on the use of a *Hadamard transform* in the input layer of a binary neural network. Previous state-of-the-art binary neural networks require floating-point arithmetic at several parts of the neural network model computation in order to maintain a sufficient level of accuracy. The Hadamard transform is similar to a Discrete Cosine Transform (used in the popular JPEG image compression method) except that it does not include expensive multiplication operations. In this paper, it is shown that the Hadamard transform can be used to replace the most expensive floating-point arithmetic portion of a binary neural network. In order to test the efficacy of this proposed method, three types of experiments were conducted: application of the proposed method to several state-of-the-art neural network models, verification of its effectiveness in a large image dataset (ImageNet), and experiments to verify the effectiveness of the Hadamard transform by comparing the performance of binary neural networks with and without the Hadamard transform. The results show that the Hadamard transform can be used to implement a highly energy-efficient binary neural network with only a miniscule loss of accuracy.

Keywords: Binary neural network · Hadamard transformation · DCT

1 Introduction

Although deep neural networks have resulted in highly accurate image classification, object recognition, and other computer vision tasks, such networks typically involve excessive amounts of numerical computation with excessive memory storage requirements, making them difficult to use in energy or computation capability constrained environments. A popular neural network compression method that can be used in such cases is binarization, in which 32-bit floating point parameters are approximated using single-bit numbers.

Binary Neural Networks (BNNs) are neural networks that use extensive levels of binarization throughout the network to achieve extreme network compression with a concomitant but relatively small loss of accuracy. In order to maintain

L. Wang et al. (Eds.): ACCV 2022, LNCS 13845, pp. 512–526, 2023.
https://doi.org/10.1007/978-3-031-26348-4_30

acceptable accuracy levels, such BNNs typically use a mixture of highly accurate numbers (such as 32-bit floating point) and highly inaccurate binary numbers for different types of parameters and/or different layers of the neural network. For example, binarization of AlexNet [14] through the method proposed by Hubara et al. [10], which is one of the early BNNs, can reduce the model size by 32 times at the cost of a 28.7% reduction in accuracy [25] on the ImageNet dataset [2]. Later research works on BNNs attempted to reduce this extremely high accuracy gap. The current state-of-the-art (SOTA) BNN [16] has approximately the same model size as [10] with only a 1.9% reduction in accuracy on the ImageNet dataset when compared to the equivalent non-binarized neural network model. A standard method for measuring the inference cost of a neural network has been proposed by Zhang et al. [31]. Referred to as *arithmetic computation effort (ACE)*, it counts the number of multiply-accumulate (MAC) operations, which are the most computationally expensive operations used in a neural network, weighted by the bit-widths of the operands used in those MAC operations.

In almost all previous state-of-the-art (SOTA) BNN models, the input layer uses floating-point arithmetic. This is because binarization of the input layer severely degrades the accuracy of a BNN [16–19,25]. However, due to its use of floating-point arithmetic, the input layer has been found to be *the major* contributor to the computation cost of a SOTA BNN. For example, when using the popular SOTA BNN referred to as ReActNet [16], the input layer contributes to approximately 65% out of the entire network ACE.

Previous studies on CNN have found that input layer extracts abstract features such as colors and various edge directions in images [7,29]. The filters of input layer resemble the Gabor filter [20] which analyzes specific frequency components in each local area, so that can detect edge of image. The discrete cosine transform (DCT), which is the encoding method used in the popular JPEG compression format, also computes in a similar manner. Using this fact, Guegeun et al. proposed to feed discrete cosine transformed data directly into a CNN, without first decoding that JPEG compressed image into a raw image [7]. This enabled the first few layers of the neural network to be pruned without any accuracy loss.

The Hadamard transform, which is also known as the Walsh-Hadamard transform, is similar transformation to the DCT. The main difference is that the Hadamard transform is multiplication-free, and it only requires add/subtract operations [23].

In this paper, we propose a new input layer using the Hadamard transform for an energy-efficient BNN. The proposed layer is fed with raw images, and it can replace conventional expensive floating-point MAC operations with light 8-bit add/subtract and logical operations. The input layer is expected to reduce energy consumption of BNN, which can be measured by ACE metric, and to achieve acceptable level of accuracy degradation. Experiments were conducted to reveal a possibility of proposed input layer for BNN. First of all, a generality of the layer was tested by applying it on binarized versions of two widely used CNN architectures, MobileNetV1 [9] and ResNet-18 [8], on a small image dataset, *i.e.*, CIFAR-10. Secondly, accuracy drop evaluation was performed on ReActNet [16], which is the SOTA BNN in terms of the accuracy gap from its real-valued counterpart. The test was conducted using the ImageNet dataset, which consists of large-scale real images, so that the practicality of new layer for real-world problem can be demonstrated. Lastly, the proposed input layer structure with trainable weight filters, instead of the Hadamard Transformation, was investigated on ReActNet to validate the efficacy of the Hadamard transform for BNNs.

2   Related Work

2.1   Binary Neural Networks

Parameter quantization is one of the methods of compressing convolutional neural networks. It is a method representing weight parameters and activations of neural network, which are normally 32-bit floating point numbers, with fewer number of N-bit width, such as 8, 4, and 2-bit. A size of the compressed network can be reduced by $32/N$ times, and an improvement of inference speed can be obtained [24].

A binary neural network is a special case of quantization with a single-bit precision, which is the smallest bit width in computer system. The process of quantization is binarization and it can be simply implemented using signum function, which outputs the sign bit of input values. An exceptional advantage of BNN over other quantized neural networks is in a convolution operation. The convolution operation with the 1-bit operands requires bit-wise logical operator, which is fast and energy-efficient, instead of expensive and relatively slow floating-point MAC unit [25]. Binary convolution refers to the convolution with operands of single-bit precision. Therefore, BNN can save massive energy consumption and reduce the size of deep neural network. Although early studies on BNN achieved comparable level of accuracy on tiny image dataset [10], such as MNIST [3] and SVHN [21], training results of BNNs on large-scale image showed poor image classification accuracy [25]. So until recently, most BNN studies have tried to mitigate the accuracy degradation.

Authors in [25], proposed that binarization error from real-valued operands to its binarized version can be reduced by introducing scaling factor. [15] designed a convolution layer with multiple binary convolution bases. The multiple outputs from the multiple bases are accumulated to enhance representability of BNN. [18] suggested adding high-precision values before binarization to the output of binary convolution via short-cut and it improved model capacity. [17] proposed that activations with high-precision should be binarized not simply by their signs but by a threshold, which determines a value to be +1 or -1. The author implemented signum function with trainable parameters to learn the appropriate threshold during training time. The network with the method achieved the smallest accuracy loss, which is caused by binarization of original full-precision network.

In addition to increasing the accuracy of BNNs, there is a study to reduce inference cost of BNN. [31] presented an energy-efficient convolution block for BNN. They also proposed arithmetical computing efficiency(ACE), which is a metric to measure efficiency of neural network. It calculates energy consumption on neural network inference by counting the number of MAC operations and weighting the bit width of operands. Table 4 shows a summary of the accuracy of this method as well as the previous methods described in this section.

2.2   Input Layer of Convolutional Neural Networks

CNN consists of convolutional layers which extract features of spatial data. Each layer receives an data in the form of feature maps, extracts specific features, and passes them to subsequent layer. What features to be extracted are determined through training process. More specifically, filters of convolutional layer are shaped differently by training dataset. Interestingly, input layer of the neural network captures general features, such as color and texture. It is relatively independent of the dataset used for training [29]. Subsequent layers are learned to extract more detailed features based on the general features.

It is known that the general filters in the input layer resemble the Gabor filter in image processing [7,29]. The Gabor filter is mainly used to extract edge and texture of image [20]. Parameters in the filter, such as angle, width, and repetition period of edge to be extracted from an image, can be selected by engineer. The general filters of input layer, which are acquired from network training, are similar to a set of several the Gabor filters with various combinations of parameters.

Based on this fact, there are studies that apply the image processing technique to CNN's input layer. For example, discrete cosine transform(DCT) is proved its usefulness as an input layer by [7]. The paper showed that using DCT as input layer, instead of conventional trainable input layer, can achieve better accuracy for ResNet-50 architecture. In addition, the authors attempted to train input layer with a regularizer, whose role is guiding the filters of input layer to resemble DCT. But they concluded that training DCT-like filter is hard and inefficient.

In case of BNN, all of the aforementioned methods for increasing the accuracy of the BNN are not applied to input layer. This is because binarization of input layer directly can degrades model's performance severely [30], while the improvement of execution time is small [25]. Recently, [30] suggested to transform input image with thermometer encoding, which contains division, ceiling and rounding operations, so it can avoid direct binarization of input layer. Alternatively, [31] proposed 8-bit quantization for input layer rather than binarization.

2.3   Hadamard Transform

The Hadamard transform is used as a feature extractor in the field of image and video processing. In [5], specific basis vectors of the Hadamard transform is selected to detect shot boundary of videos. The transform can be used for image compression [4,6,28]. [4] suggested image compression method by taking advantage of simple and efficient property of the Hadamard transform. A recent study on CNN proposed a layer with Hadamard transform, which is designed to replace 1×1 convolution layer of neural networks, to achieve faster network inference [22].

3   Hadamard Transform as an Input Layer

3.1   Hadamard Matrix

Hadamard transform is one of the linear image transforms [26]. The transform is an operator capable of processing 2D images and has averaging property [23]. And the transformed image can be inversely transformed into the original spatial domain. Hadamard transform can perform the same function as DCT more efficiently. This is because they are both orthogonal transforms [26] but Hadamard transform uses Hadamard matrix, whose entries are $+1$ and -1, making the operation simpler.

Hadamard matrix(H) is in square array form, and the matrix of $N=2^n$ order can be obtained using the Kronecker product(1). When $N=1$, the entry is one with 1, and for $N=2^n \geq 2$, the Hadamard matrix can be derived by recursively utilizing the matrix of $N=2^{(n-1)}$ order. For example, when $N=4$, the Hadamard matrix H_4 consists of four H_2, and the H_2 holds four H_1 which is one 1 with appropriate sign of entries according to equation(1).

$$H_{2^n} = H_2 \otimes H_{2^{n-1}} = \begin{bmatrix} H_{2^n-1} & H_{2^n-1} \\ H_{2^n-1} & -H_{2^n-1} \end{bmatrix} \tag{1}$$

$$H_1 = \begin{bmatrix} 1 \end{bmatrix} \quad H_2 = \begin{bmatrix} 1 & 1 \\ 1 & -1 \end{bmatrix} \quad H_4 = \begin{bmatrix} 1 & 1 & 1 & 1 \\ 1 & -1 & 1 & -1 \\ 1 & 1 & -1 & -1 \\ 1 & -1 & -1 & 1 \end{bmatrix} \xrightarrow[\text{ordered}]{} \begin{bmatrix} 1 & 1 & 1 & 1 \\ 1 & 1 & -1 & -1 \\ 1 & -1 & -1 & 1 \\ 1 & -1 & 1 & -1 \end{bmatrix} \tag{2}$$

The matrix has several properties. The first property is that the matrix is symmetric. Second, each row is orthogonal to each other. Third, how many times the sign of the entries of the row changes is called sequency, and the Hadamard matrix of N order consists of rows with sequency from 0 to $N-1$. If the rows are ordered in ascending, it is exactly same as the Walsh matrix [23]. In this paper, we refers to Hadamard matrix as the matrix with the rows of the ascending ordered.

3.2   Hadamard Transform

Two-dimensional image can be processed with Hadamard transform using Eq. (4). In the equation, the original image in spatial domain is denoted by $s(x,y)$ and transformed image is represented by $G(u,v)$. The size of processed

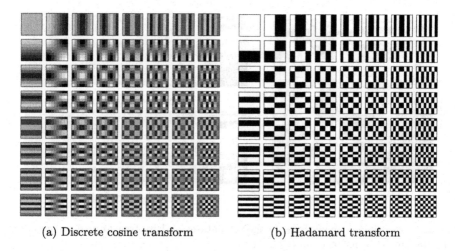

(a) Discrete cosine transform   (b) Hadamard transform

Fig. 1. Visualized 2D kernels of (a) DCT and (b) Hadamard transform for block size of 8. Ordered Hadamard matrix is used to obtain (b).

image is the number of 2D kernels of Hadamard transform. The 2D kernels can be obtained by outer product of rows and columns of Hadamard matrix. For example, from the ordered H_4, the first two kernels are 4×4 as they are in (5). In a same way, DCT on 2D image can be done using Eq. (3) and the kernels from the DCT can be obtained. Figure 1 shows that the two kernel sets from DCT and Hadamard transform are similar to each other. Additionally, transformed images using the transforms are illustrated in Fig. 2 to provide qualitative comparison.

$$F(u, v) = \sum_{x=0}^{N-1} \sum_{y=0}^{N-1} s(x, y) \exp\left(-\frac{2\pi i}{N}(ux + vy)\right) \qquad (3)$$

$$G(u, v) = \sum_{x=0}^{N-1} \sum_{y=0}^{N-1} s(x, y) g_{(u,v)}(x, y) \qquad (4)$$

$$g(1,1) = \begin{bmatrix} 1 & 1 & 1 & 1 \\ 1 & 1 & 1 & 1 \\ 1 & 1 & 1 & 1 \\ 1 & 1 & 1 & 1 \end{bmatrix} \quad g(1,2) = \begin{bmatrix} 1 & 1 & 1 & 1 \\ 1 & 1 & 1 & 1 \\ -1 & -1 & -1 & -1 \\ -1 & -1 & -1 & -1 \end{bmatrix} \qquad (5)$$

The characteristic of Hadamard transform is that no multiplication is required, which results in efficient computation. Secondly, the energy before and after transformation is preserved (6). And the computational result using the zero sequency kernel($g(1,1)$) means the average brightness of the spatial domain image (7). It is the same operation with average pooling layer in neural networks. Moreover, the energy of most images is concentrated in this area, and for the

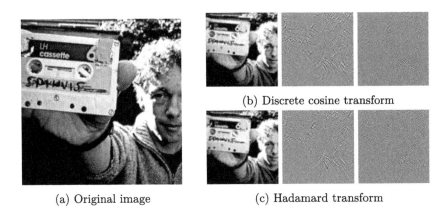

(a) Original image   (c) Hadamard transform

Fig. 2. (a) Original image from ImageNet dataset and its transformed images using (b) Hadamard transform and (c) DCT.

higher sequency kernels, relatively small amount of energy is held [23], which enables image compression [4,6,28].

$$\sum_{x=0}^{N-1}\sum_{y=0}^{N-1}|s(x,y)|^2 = \frac{1}{N^2}\sum_{u=0}^{N-1}\sum_{v=0}^{N-1}|G(u,v)|^2 \tag{6}$$

$$G(0,0) = \frac{1}{N^2}\sum_{x=0}^{N-1}\sum_{y=0}^{N-1}s(x,y) \tag{7}$$

3.3 Proposed Input Layer

As DCT can be used as input layer of CNN [7], and the DCT and Hadamard transform are functionally same, it is possible to use kernels of Hadamard transform input layer of the BNN. However, there are several considerations to materialize it.

Assume that there is a single-channel 2D image. Hadamard transform processes on N by N size blocks of image in the spatial domain, where the blocks are non-overlapped. In terms of convolution operation, this is same as windowing weight filters with stride step of N and producing N^2 output channels. However, to utilize Hadamard transform in input layer, the transform must be implemented to overlap the N by N block size. This is to provide feature maps of particular dimension, which can be different from structures of existing BNNs but can not be covered with stride N, for subsequent layer. At the same time, the overlapping should properly extracts features without hurting network's performance. It has been proved that overlapping 2D kernels on the spatial domain image, which is called modified DCT(MDCT) [27], can extract features well in CNN [11,27]. Considering the same functionality of DCT and Hadamard, it is possible to adopting the MDCT manner on Hadamard transform.

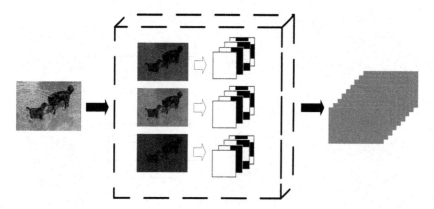

Fig. 3. Hadamard transform in proposed input layer. The operation can be done for each channels with same N^2 kernels. The number of kernels can be vary depending on order of Hadamard matrix. The transformed images are concatenated in channel-wise.

In addition, an input image with 3-channels is normally fed into neural network. Therefore, the kernels should be applied to each channel, which can be regarded as grouped convolution [14], and the transform will eventually output $3 \times N^2$ channels. The process is depicted in Fig. 3. However, the aforementioned particular dimension for subsequent layer also includes the number of channels(or depth). Therefore, the channel of $3 \times N^2$ size need to flexibly modified depending on possible BNN structures. To address this issue, pointwise binary convolution with shortcut [17], which operates in bit-wise operators, is followed by the Hadamard transform. When the dimension of shortcut and output of pointwise convolution is not matched, channel-wise zero padding can used for the shortcut. This proposed input layer is illustrated in Fig. 4.

Moreover, it is not necessary to have N^2 kernels for transformation in the input layer. Hadamard transform preserves the energy of the pre-transformed spatial domain, while the high-sequency kernels could result fewer energy portions. Even if these high-sequency kernels are discarded, the energy of the spatial domain does not change significantly. This concept is used in one of the lossy compression method, JPEG. Thus, the number of operations can be reduced by ignoring insignificant energy loss. For example, when $N = 4$, the number of 2D kernels to preserve entire energy is 16. However, our experiments witnessed that no accuracy drop occurred using 10 kernels of low frequencies.

4   Evaluation

Experiments on proposed input layer have three parts. First of all, a generality of proposed layer is validated. We took two representative used BNNs, whose full-precision networks are based on ResNet and MobileNetV1 respectively, and tested the proposed input layer on them. Also, the experiment includes comparison between Hadamard transform to DCT as input layer. Next, in order to test

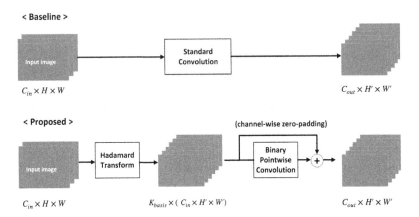

Fig. 4. Structures of conventional and proposed input layer. The Hadamard transform in proposed structure is performed by grouped convolution with 8-bit integer. Output dimension of the operation depends on $K_{basis} \leq N^2$, which is the number of 2-d kernels of Hadamard transform. In this paper, we used $N=4$ and $K_{basis}=10$. The binary pointwise convolution can be operated by bitwise XNOR and bit count. And shortcuts may requires channel-wise zero padding to match the dimension of the binary pointwise convolution.

practicality of proposed layer, we replaced input layer of ReActNet [16], which showed the best performance regarding accuracy degradation in this field, with the proposed layer. The network with proposed layer was tested on real-world large scale images, the ImageNet [2]. Lastly, we replaced the kernels of Hadamard from the proposed input layer with binarized filters through training. Datasets used in the experiments are CIFAR-10 [13] and ImageNet [2]. CIFAR-10 is a representative small image(32×32) dataset and has 10 categories in it. The dataset consists 50K training images and 10K test images. ImageNet contains 1.2M training images and 50K validation images each of which can be categorized in 1K classes. Unlike CIFAR-10, the image sizes are different from one another, so they are normally resized to fit a particular size(e.g. 224×224) for training and validation. Image classification accuracy and energy consumption of the MAC operation are considered to compare the efficacy of proposed input layer. To measure the energy consumption, we used ACE which is proposed by [31]. The metric counts the number of MAC operations and each operation are weighted by bit width of the operands. ACE for different precision is summarized in the Table 1.

4.1   Generality of the Hadamard Transform as an Input Layer

Implementation Details. We implemented two BNNs based on ResNet and MobileNet, which are widely used in BNN studies so far [15,17,19,30,31], and trained them with CIFAR10 [13]. The binarization techniques, which are used in this experiment, follow ReActNet [17]. On top of them, minor modifications

Table 1. ACE metric [31]

Precision	float			int				
	32	**16**	**bfloat 16**	**32**	**8**	**4**	**4**	**1**
ACE	1024	256	256	1024	64	16	4	1

of BNN models were processed. Specifically, when experimenting with a ResNet model, the ResNet-18 structure was used instead of ResNet-20, which has 3x3 kernel size at input layer. The same structure was used in [1]. Afterward we refer to this network as ReActNet-18. And when testing MobileNetV1 based BNN, we took the structure proposed by ReActNet [16] and reduced the stride step of input layer from 2 to 1. The MobileNet-based BNN will be referred to ReActNet-A, regardless of stride size at input layer.

There are three differences between baseline input layer and proposed input layer. For the baseline, kernel size is 3×3 and standard 2d convolution is used. In other words, group size is 1. And operands are high-precision with 32-bit floating point. On the other hand, proposed input layer has a kernel size of 4×4, and the grouped convolution with group size of the input channel(RGB channels for conventional input image). Each groups take 10 2D kernels of Hadamard transform. And binary point wise convolution is followed by the transform, to flexibly control the number of output channels.

Two-stage training strategy, which is widely used in BNN training [17–19,31], is adopted for training the BNNs. In the first step, only activations are binarized, and weights remain 32-bit floating point number. In the second step, the previously trained model becomes the initial state, and then additional binarization function for weights are added in the network. Thus in this step, both activations and weights are binarized. Adam [12] optimizer was used, and hyperparameters were set as follows. Training 100K steps for each stage with 256 epochs, batch size of 128 and learning rate of 5e-4. Weight decay is used in the first stage of learning, but not in the second stage [16].

Results. ReActNet-18 with conventional input layer, shows accuracy of 93.49% on CIFAR10. ACE for the network is $2.36G$ and the input layer accounts for 76.63% out of the entire network ACE. On the other hand, in the case of using the proposed layer, the accuracy dropped by 1.17% resulting in accuracy level of 92.51%. The network saved 75.13% of ACE compared to baseline. In ReActNet-A, the trend were same as the ResNet-18. The accuracy of the baseline is 90.74%, and the ACE is $1.31G$. The baseline input layer occupies 69.18% of entire ACE. However, when the input layer is replaced with the proposed layer, accuracy level is 89.60% which is loss of 1.39% point, and the ACE is decreased by 66.70% compared to the baseline. Additionally to compare Hadamard transform with DCT, we implemented proposed input layer with DCT instead of Hadamard transform. As the two transformations are same in terms of functionality, accuracy levels achieved with DCT are similar to with Hadamard transform. However, DCT consumes more energy than Hadamard transform because the latter is multiplication-free. The results are summarized in Table 2.

Table 2. Results on CIFAR10

Network	Input layer	Accuracy	ΔAcc.(%p)	ACE(1e9)	Δ ACE
ReActNet-18	Baseline	93.94%	–	2.36	–
	DCT	92.97%	−0.97	1.06	−55.18%
	Proposed	92.51%	−1.43	0.59	−75.13%
ReActNet-A	Baseline	90.74%	–	1.31	–
	DCT	89.41%	−1.33	0.91	−30.67%
	Proposed	89.35%	−1.39	0.44	−66.70%

4.2 BNN with Proposed Input Layer on ImageNet

Implementation Details. The baseline for this experiment is exactly same as proposed in [16]. Stride step is 2 for both baseline and proposed input layer. The two-stage strategy was applied on this experiment. The proposed network was trained with 256 epochs, batch size of 256 and learning rate of 5e-6 for each stage as the authors in [16] suggested. Weight decay was set 5e-6 and used only for the first step.

Results. The baseline has a validation accuracy of 70.5% and ACE of 16.96G, where input layer alone accounts for 65.42%. However, with the proposed input layer, the ACE decrease by 63.08% at the cost of 1.38% of accuracy loss which is summarized in Table 3. Compared to ReActNet's real-valued counter part, it was finally reduced by 3.28% point. The gap is superior to FracBNN [30], which showed the second best result in the accuracy gap. As a result, the SOTA BNN with the proposed input layer still showed the smallest accuracy gap from real-valued counterpart and achieved better energy-efficiency. This result is summarized in Table 4.

4.3 Hadamard Transform vs. Trained Binary Weights

As mentioned in Sect. 3, using the Hadamard transform as an input layer means using the transform's 2D kernels as weight filters. Since the filters consist of only +1 and −1, the convolution operation consists of only add/sub without multiplication. Thanks to this, we were able to implement energy-efficient BNNs. However, this result can be attributed not to the kernels of Hadamard, but to

Table 3. Results on ImageNet

Network	Input layer	Accuracy	ΔAcc.(%p)	ACE(1e9)	Δ ACE
ReActNet-A	Baseline	70.5%	–	16.9	–
	Proposed	69.12%	−1.39	6.26	−63.08%

Table 4. Top-1 accuracy of BNNs on ImageNet.

Network	Method	Top-1 accuracy(%)	Gap(%)
AlexNet	Full-precision	56.6	–
	BinaryNet [10]	27.9	−28.7
	XNOR-Net [25]	44.2	−12.4
ResNet-18	Full-precision	69.3	–
	ABC-Net (5 bases) [15]	65.0	−4.3
	ABC-Net (1 base)	42.7	−28.6
	Bi-RealNet [18]	56.4	−12.9
	ReActNet [17]	65.5	−3.8
PokeBNN	Full-precision	79.2	–
	PokeBNN-1.0 [31]	73.4	−5.8
FracBNN	Full-precision	75.6	–
	FracBNN [30]	71.8	−3.8
ReActNet-A	Full-precision	72.4	–
	ReActNet(Adam) [16]	70.5	−1.9
	Ours	**69.12**	**−3.28**

filters composed of +1 and -1. To make it clear, we created the same structure as the proposed input layer and trained binarized weights from the scratch.

The experiment was conducted on ReActNet-A with CIFAR10 dataset. The result showed that training filter was less accurate than proposed input layer and the training curve was largely fluctuated as illustrated in Fig. 5. If the learning process is unstable, the final accuracy can be deteriorated [16]. In general, it is a phenomenon that occurs when the sign of binarized weights changes only at a few steps, because the real-valued weights before binarization are close to zero during the training time [16]. Empirically, this case can be solved by lowering the learning rate, but BNN already uses a much lower learning rate than full-precision network learning, so overall learning time can be increased to achieve the same result.

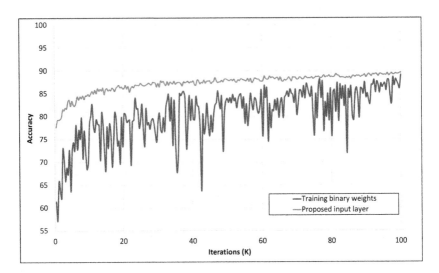

Fig. 5. Training curves of ReActNet-A on CIFAR10. The blue line is the network with training input layer with binary weights, which is unstable. The orange line is the network with the proposed input layer. Unlike the blue line, the accuracy of proposed network increases without fluctuation even the filters are binary values.

5 Conclusion

The input layer of in state-of-the-art binary neural networks (BNNs) typically use floating-point arithmetic because of the resulting steep drop in accuracy when quantized and its negligible effect on inference speed. However, from an energy consumption perspective, the layer consumes an abnormal amount of energy. To address this issue, we proposed an energy-efficient input layer for binary neural networks using a Hadamard transform. The proposed input layer has been tested on ReActNet-A and ReActNet-18, which are MobileNetV1 and ResNet-18 based BNN respectively. The energy consumption of BNN was measured by ACE, and with the proposed input layer, the networks' ACE value was reduced by up to 75%. In addition, the accuracy degradation caused by this input layer was less than 1.5%.

Acknowledgements. This work was supported by Institute of Information & communications Technology Planning & Evaluation (IITP) grant funded by the Korea government(MSIT) (No.2019-0-01906, Artificial Intelligence Graduate School Program(POSTECH))

References

1. Chen, T., Zhang, Z., Ouyang, X., Liu, Z., Shen, Z., Wang, Z.: "BNN-BN=?": Training binary neural networks without batch normalization. In: Proceedings of the IEEE/CVF conference on computer vision and pattern recognition, pp. 4619–4629 (2021)
2. Deng, J., Dong, W., Socher, R., Li, L.J., Li, K., Fei-Fei, L.: ImageNet: a large-scale hierarchical image database. In: 2009 IEEE Conference on Computer Vision and Pattern Recognition, pp. 248–255. IEEE (2009)
3. Deng, L.: The MNIST database of handwritten digit images for machine learning research. IEEE Signal Process. Mag. **29**(6), 141–142 (2012)
4. Diana Andrushia, A., Thangarjan, R.: Saliency-based image compression using Walsh–Hadamard transform (WHT). In: Hemanth, J., Balas, V.E. (eds.) Biologically Rationalized Computing Techniques For Image Processing Applications. LNCVB, vol. 25, pp. 21–42. Springer, Cham (2018). https://doi.org/10.1007/978-3-319-61316-1_2
5. GG, L.P., Domnic, S.: Walsh-Hadamard transform kernel-based feature vector for shot boundary detection. IEEE Trans. Image Process. **23**(12), 5187–5197 (2014)
6. Ghrare, S.E., Khobaiz, A.R.: Digital image compression using block truncation coding and Walsh Hadamard transform hybrid technique. In: 2014 International Conference on Computer, Communications, and Control Technology (I4CT), pp. 477–480. IEEE (2014)
7. Gueguen, L., Sergeev, A., Kadlec, B., Liu, R., Yosinski, J.: Faster neural networks straight from JPEG. In: Advances in Neural Information Processing Systems 31 (2018)
8. He, K., Zhang, X., Ren, S., Sun, J.: Deep residual learning for image recognition. In: Proceedings of the IEEE Conference on Computer Vision and Pattern Recognition, pp. 770–778 (2016)
9. Howard, A.G., et al.: MobileNets: efficient convolutional neural networks for mobile vision applications. arXiv preprint arXiv:1704.04861 (2017)
10. Hubara, I., Courbariaux, M., Soudry, D., El-Yaniv, R., Bengio, Y.: Binarized neural networks. In: Advances in Neural Information Processing Systems 29 (2016)
11. Ju, S., Lee, Y., Lee, S.: Convolutional neural networks with discrete cosine transform features. IEEE Trans. Comput. **71**, 3389–3395 (2022)
12. Kingma, D.P., Ba, J.: Adam: a method for stochastic optimization. arXiv preprint arXiv:1412.6980 (2014)
13. Krizhevsky, A., Hinton, G., et al.: Learning multiple layers of features from tiny images. Technical Report (2009)
14. Krizhevsky, A., Sutskever, I., Hinton, G.E.: ImageNet classification with deep convolutional neural networks. In: Advances in Neural Information Processing Systems 25 (2012)
15. Lin, X., Zhao, C., Pan, W.: Towards accurate binary convolutional neural network. In: Advances in Neural Information Processing Systems 30 (2017)
16. Liu, Z., Shen, Z., Li, S., Helwegen, K., Huang, D., Cheng, K.T.: How do Adam and training strategies help BNNs optimization. In: International Conference on Machine Learning, pp. 6936–6946. PMLR (2021)
17. Liu, Z., Shen, Z., Savvides, M., Cheng, K.-T.: ReActNet: towards precise binary neural network with generalized activation functions. In: Vedaldi, A., Bischof, H., Brox, T., Frahm, J.-M. (eds.) ECCV 2020. LNCS, vol. 12359, pp. 143–159. Springer, Cham (2020). https://doi.org/10.1007/978-3-030-58568-6_9

18. Liu, Z., Wu, B., Luo, W., Yang, X., Liu, W., Cheng, K.-T.: Bi-Real Net: enhancing the performance of 1-bit CNNs with improved representational capability and advanced training algorithm. In: Ferrari, V., Hebert, M., Sminchisescu, C., Weiss, Y. (eds.) ECCV 2018. LNCS, vol. 11219, pp. 747–763. Springer, Cham (2018). https://doi.org/10.1007/978-3-030-01267-0_44

19. Martinez, B., Yang, J., Bulat, A., Tzimiropoulos, G.: Training binary neural networks with real-to-binary convolutions. arXiv preprint arXiv:2003.11535 (2020)

20. Mehrotra, R., Namuduri, K.R., Ranganathan, N.: Gabor filter-based edge detection. Pattern Recogn. **25**(12), 1479–1494 (1992)

21. Netzer, Y., Wang, T., Coates, A., Bissacco, A., Wu, B., Ng, A.Y.: Reading digits in natural images with unsupervised feature learning. In: NIPS Workshop on Deep Learning and Unsupervised Feature Learning (2011)

22. Pan, H., Badawi, D., Cetin, A.E.: Fast Walsh-Hadamard transform and smooththresholding based binary layers in deep neural networks. In: Proceedings of the IEEE/CVF Conference on Computer Vision and Pattern Recognition, pp. 4650–4659 (2021)

23. Pratt, W.K., Kane, J., Andrews, H.C.: Hadamard transform image coding. Proc. IEEE **57**(1), 58–68 (1969)

24. Qin, H., Gong, R., Liu, X., Bai, X., Song, J., Sebe, N.: Binary neural networks: a survey. Pattern Recogn. **105**, 107281 (2020)

25. Rastegari, M., Ordonez, V., Redmon, J., Farhadi, A.: XNOR-Net: ImageNet classification using binary convolutional neural networks. In: Leibe, B., Matas, J., Sebe, N., Welling, M. (eds.) ECCV 2016. LNCS, vol. 9908, pp. 525–542. Springer, Cham (2016). https://doi.org/10.1007/978-3-319-46493-0_32

26. Salomon, D.: Data compression: the complete reference. Springer Science & Business Media (2004). https://doi.org/10.1007/978-1-84628-603-2

27. Ulicny, M., Krylov, V.A., Dahyot, R.: Harmonic convolutional networks based on discrete cosine transform. Pattern Recogn. **129**, 108707 (2022)

28. Valova, I., Kosugi, Y.: Hadamard-based image decomposition and compression. IEEE Trans. Inf Technol. Biomed. **4**(4), 306–319 (2000)

29. Yosinski, J., Clune, J., Bengio, Y., Lipson, H.: How transferable are features in deep neural networks? In: Advances in Neural Information Processing Systems 27 (2014)

30. Zhang, Y., Pan, J., Liu, X., Chen, H., Chen, D., Zhang, Z.: FracBNN: accurate and fpga-efficient binary neural networks with fractional activations. In: The 2021 ACM/SIGDA International Symposium on Field-Programmable Gate Arrays, pp. 171–182 (2021)

31. Zhang, Y., Zhang, Z., Lew, L.: PokeBNN: a binary pursuit of lightweight accuracy. In: Proceedings of the IEEE/CVF Conference on Computer Vision and Pattern Recognition, pp. 12475–12485 (2022)

Author Index

L. Wang et al. (Eds.): ACCV 2022, LNCS 13845, pp. 527–528, 2023.
https://doi.org/10.1007/978-3-031-26348-4

Printed in the United States
by Baker & Taylor Publisher Services

Printed in the United States
by Baker & Taylor Publisher Services